D1331466

MATHS IN ACTION

Mathematics in Action Group

Members of the Mathematics in Action Group associated with this book:
D. Brown, J. L. Hodge, R. D. Howat, J. Hunter, E. C. K. Mullan, K. Nisbet, A. G. Robertson

WITHDRAWN

STUDENTS' BOOK

093971
Aberdeen College Library

Thomas Nelson and Sons Ltd
Nelson House Mayfield Road
Walton-on-Thames Surrey
KT12 5PL UK

© Mathematics in Action Group 1995

First published by Blackie and Son Ltd 1987
New edition published by Thomas Nelson and Sons Ltd 1995

ISBN 0-17-431434-5
NPN 9 8 7 6

All rights reserved. No paragraph of this publication may be reproduced, copied or transmitted save with written permission or in accordance with the provisions of the Copyright, Design and Patents Act 1988, or under the terms of any licence permitting limited copying issued by the Copyright Licensing Agency, 90 Tottenham Court Road, London, W1P 9HE.

Any person who does any unauthorised act in relation to this publication may be liable to criminal prosecution and civil claims for damages.

Printed in China.

CONTENTS

WITHDRAWN

INTRODUCTION

Maths in Action—New Edition provides a course in mathematics that covers the Mathematics 5-14 National Guidelines, Standard Grade and Higher Grade in Scotland, the Northern Ireland Curriculum and the National Curriculum in England and Wales.

The new edition builds on experience gained in the classroom with the original series, and particular attention has been paid to providing a differentiated course at every stage. Book 3B provides a course for Standard Grade at Credit/General levels and for NIC/NC at levels 5–8, while Book 3A aims mainly at General/Foundation and levels 4–6. Each chapter starts with a Looking Back exercise, which can be used for revision and to assess readiness for the topic, and ends with a Check-up exercise giving a further element of revision and assessment. Investigative work features prominently in each chapter in the many puzzles, projects, challenges, brainstormers and investigations. Answers to every question (except puzzles, challenges, brainstormers, investigations, and 'Topics to Explore') are to be found at the end of this book.

Each *Students' Book* is supported by a *Teacher's Resource Book* and, in the first two years, by revised books of *Extra Questions* and *Further Questions*.

The *Teacher's Resource Book* contains Standard Grade, Northern Ireland Curriculum and National Curriculum references for every chapter, photocopiable worksheets, notes and suggestions for further activities, and the answers to the puzzles, challenges, brainstormers, investigations and 'Topics to Explore' in the *Students' Book*. In addition, there are grids which may be photocopied and used to record and assess students' progress.

1 CALCULATIONS AND CALCULATORS

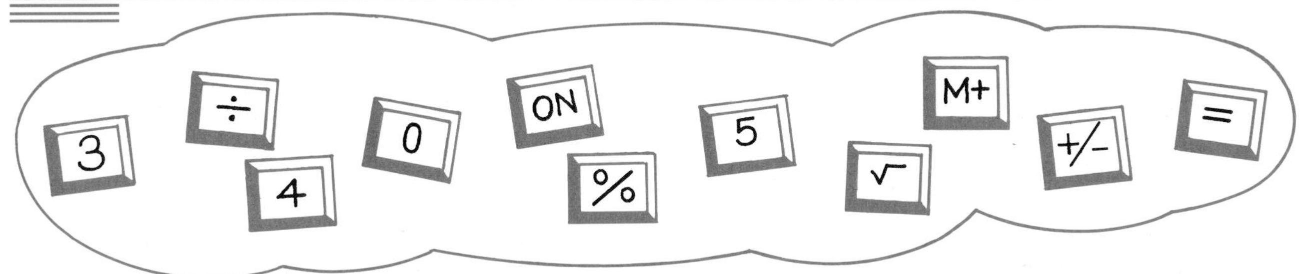

LOOKING BACK

1 a In 1989, the population of the United Kingdom was fifty-seven million, two hundred and eighteen thousand, to the nearest thousand. Write this number in figures.

b The area of the UK is 244 755 km^2. Calculate the average number of people per square kilometre, correct to 3, 2 and 1 significant figures.

2

Exchange rate: £1 = \$1.4972

a Write the exchange rate correct to 3, 2 and 1 decimal places.

b Change to dollars, to the nearest cent (2 decimal places):
(i) £10 (ii) £100 (iii) £1000.

3 Ian measured the radius of his Frisbee and found it was 11.6 cm. On his calculator he worked out that the area of the Frisbee was 422.732 707 5 cm^2. What number of cm^2 should he use as a reasonable value for the area?

4 a *Estimate* the perimeter and area of this square metal plate.

b *Calculate* its perimeter and area to 2 significant figures.

43 cm

5 Calculate the value of:
a 10^2 **b** 10^3 **c** 2.5×10^2 **d** 0.5×10
e $36 \div 10$ **f** $0.5 \div 100$

6 Find the value of n, given:
a $2^n = 8$ **b** $5^n = 25$ **c** $10^n = 1\,000\,000$

7 Find pairs of equal fractions in this collection.

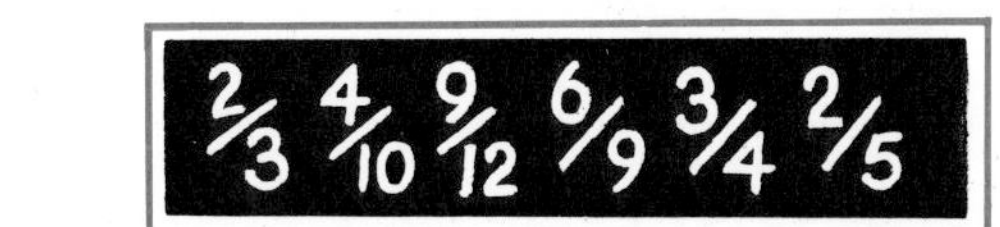

8 Write each fraction below in its simplest form:
a $\frac{3}{9}$ **b** $\frac{8}{10}$ **c** $\frac{15}{20}$ **d** $\frac{7}{14}$ **e** $\frac{16}{20}$

9 Find the least common multiple (lcm) of:
a 4 and 5 **b** 6 and 9 **c** 8 and 12

10 Estimate, then calculate correct to 2 decimal places:
a $\dfrac{220 \times 73}{195}$ **b** $3.65^2 + 6.35^2$ **c** $\dfrac{22.53}{1.07 + 8.94}$

INVESTIGATIONS

1 *Key the numbers 1 to 9 into your calculator in order, one at a time.*

Use either the [+] *key or the* [×] *key between the numbers, but make sure that each time your answer is an odd number.*

For example, ***1*** [×] ***2*** [=] *2 won't do.*
1 [+] ***2*** [=] *3* [+] ***3*** [=] *6 won't do.*

Remember to press [=] *after each calculation.*

a *Explain your method.*
b *What number is displayed after you key in 9?*

2 *Some patterns to explore, using your calculator:*

a
1
1 + 2 + 1
1 + 2 + 3 + 2 + 1
. . .

From the pattern of answers, can you work out:
1 + 2 + 3 + . . . + 99 + 100 + 99 + . . . + 3 + 2 + 1?

b
1
1 + 3
1 + 3 + 5
. . .
. . .

What is the sum of the first 500 odd numbers?
What, then, is the sum of the first 500 even numbers, starting with 2?

c *In the same way, investigate these patterns.*
(i) 1 − 3 + 5 − . . . (ii) 1 − 2 + 3 − . . .

CALCULATOR PRACTICE

There are ten different types of calculation in Exercise 1.
One calculation in each has been done for you, using values (i) of a, b and c in this table.

	a	b	c
(i)	2	3	4
(ii)	1.4	2.9	5.5
(iii)	121	212	139

1 Check the answer for Type 1, correct to 3 significant figures.

2 Then do the calculation with values (ii) and (iii) of a, b and c in the table.

3 Repeat **1** and **2** for Types 2, 3, 4, . . . , 10.

EXERCISE 1

1 **2**

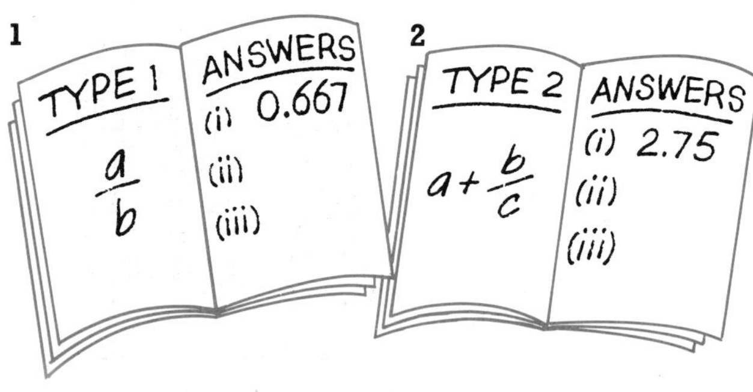

3 **4**

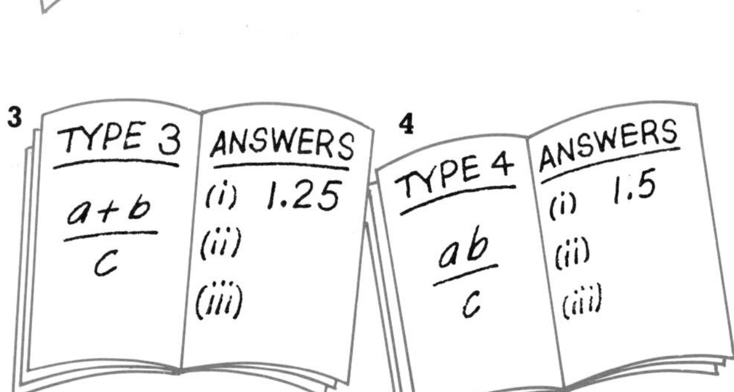

5 **6**

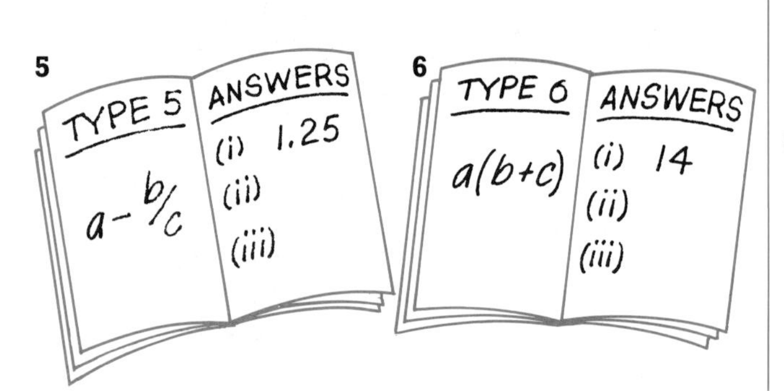

7

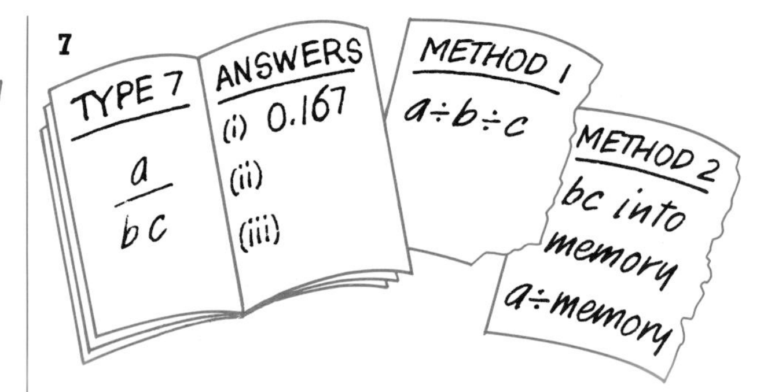

8

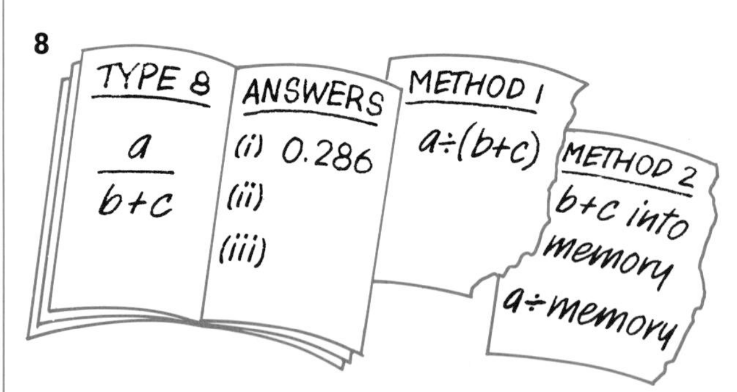

9 **10**

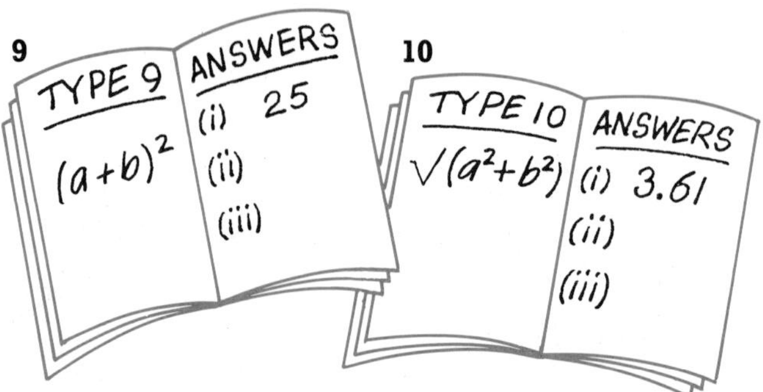

REPEATED CALCULATIONS

You can make repeated calculations using either the memory or the constant facility.
Example The currency exchange rate for Spain is £1 = 198.75 pesetas.
Convert to pesetas: **a** £12 **b** £140 **c** £1048
Key 198.75 [x→M], then: **a** 12 [×] [RM] [=] or [RCL] **b** 140 [×] [RM] [=] **c** 1048 [×] [RM] [=]

giving 2385 pesetas 27 825 pesetas 208 290 pesetas

EXERCISE 2

1 Use your calculator's memory, or constant, to check the answers to calculations **a–c** in the box above.

2 Key 28.75 into the memory, or use the constant, to calculate, correct to 2 decimal places:
a 28.75×19 **b** 28.75×187 **c** $350 \div 28.75$
d $1000 \div 28.75$

Use your calculator's memory, or constant, for the calculations in questions **3–7**.

3 The only ruler George could find had 1.8 cm broken off the end. So he has to add 1.8 cm to all the measurements he makes, in centimetres.

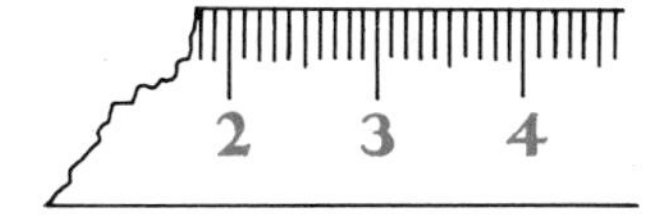

Correct his list: 12.9, 23.2, 16.8, 22.8, 29.

4 Jen's watch was 2.8 minutes fast. She had been keeping a note of the times when cars passed her census point, so she had to subtract 2.8 from all her observed times, in minutes. Correct these times: 8.7, 9.4, 9.9, 10.6, 11.2.

5 Kath did her shopping at the Cash'n'Carry. The prices quoted did not include VAT, so she had to multiply each price by 1.175 to find how much she had to pay. Find the cost, to the nearest penny, of items marked: £70, £30.50, £120, £780, £1005.

6 A recent exchange rate for American dollars was £1 = $1.4625.
a Convert to dollars, correct to 2 decimal places: £44, £583, £1000 000.
b Convert to £s, correct to the nearest penny: $10, $230, $9852.

7 Calculate the value of each of these in three different ways:
(i) by multiplication (ii) using the memory or constant facility (iii) using the [y^x] key.

a 3^5 **b** 12^4 **c** 156^3 **d** 0.5^{10} **e** π^5

CALCULATING AND ESTIMATING

$22 \times 366 = 802$, **or** 8050, **or** 8052, **or** 80 052. Which is correct?

a Estimate the answer, using 1 significant figure.
$20 \times 400 = 8000$.
So the answer could be 8050 or 8052.

b In $2\mathbf{2} \times 36\mathbf{6}$, $\mathbf{2} \times \mathbf{6} = \mathbf{12}$,
so the answer ends in 2.
It must be 8052.
So $22 \times 366 = 8052$.

EXERCISE 3

1 Calculate mentally, then check with your calculator:
a 9×1000 **b** 0.6×100 **c** 0.01×1000
d $700 \div 10$ **e** $3000 \div 100$ **f** $8 \div 10$

2 As in question **1**:
a 20×0.4 **b** 500×0.2 **c** 1000×0.1
d $50 \div 0.1$ **e** $100 \div 0.5$ **f** $60 \div 0.02$

3 a Which number below is the best estimate for 18×81?
(i) 160 (ii) 1600 (iii) 16 000
b One of the following is the correct answer. Which one? (Don't use your calculator yet!)
(i) 188 (ii) 1590 (iii) 1458 (iv) 16 008
c Now check with your calculator.

4 a Which number is the best estimate for 37×23?
(i) 800 (ii) 8000 (iii) 80 000
b Which of these could be correct?
(i) 8010 (ii) 7091 (iii) 851 (iv) 80 051
c Check with your calculator.

5 Katherine measures the dimensions of the rectangle, to the nearest cm.

a Estimate the area of the rectangle.
b Calculate its area. How many significant figures should you have in your answer?

6 Estimate, then calculate:
a the number of minutes in 18 hours
b the number of days in 29 weeks
c the cost of one book, if 52 books cost £429
d the average of 18.4 kg and 28.6 kg.

7 The distance d metres which a skydiver falls in t seconds is given by the formula $d = \frac{1}{2}gt^2$, where g, the gravity acceleration, is $9.8\,m/s^2$.

Estimate, then calculate, d to an appropriate number of significant figures, when:
a $t = 3.0$ **b** $t = 6.7$

LARGE AND SMALL NUMBERS—STANDARD FORM

The distance to the sun is 93 million miles.
Key this number into your calculator.
Calculate the distance to a star which is:
(i) 10 times as far away (ii) 10 times further away than this
(iii) 10 times further away again (iv) 10 times further away still.

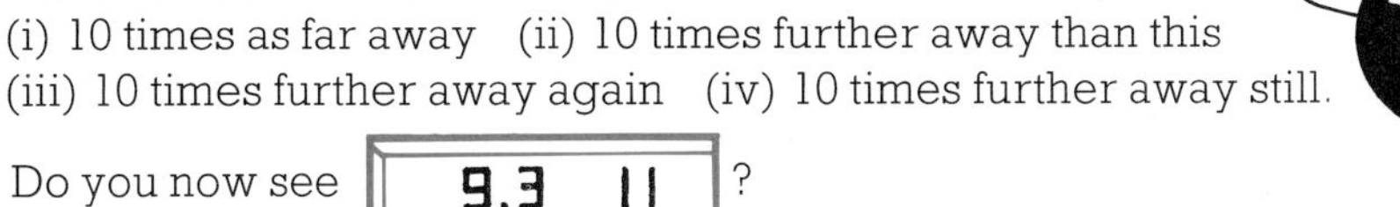

Do you now see **9.3 11** ?

This means that the number is too big for the screen, so 930 000 000 000 is shown as 9.3×10^{11}.
9.3×10^{11} is **the standard form**, or **scientific notation**, for 930 000 000 000.

> Every number can be written in the form $a \times 10^n$, where a is a number between 1 and 10 (including 1), and n is zero or a positive or negative whole number.

$12.5 = 1.25 \times 10$ $\quad 0.25 = 2.5 \times \frac{1}{10} = 2.5 \times 10^{-1}$

$125 = 1.25 \times 10^2$ $\quad 0.025 = 2.5 \times \frac{1}{100} = 2.5 \times 10^{-2}$

$1250 = 1.25 \times 10^3$ $\quad 0.0025 = 2.5 \times \frac{1}{1000} = 2.5 \times 10^{-3}$

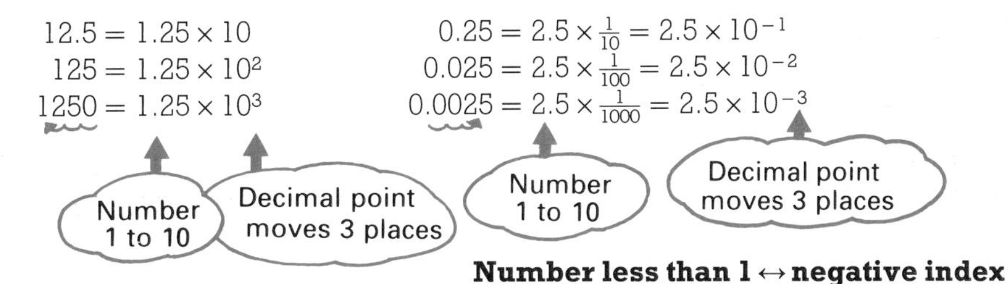

Number less than 1 ↔ negative index

Examples

1 $120\,000\,000 = 1.2 \times 10^8$ **2** $0.000\,073 = 7.3 \times 10^{-5}$ **3** $10\,000 = 1 \times 10^4$

EXERCISE 4

Write the numbers in questions **1–3** in standard form, $a \times 10^n$.

1 **a** 40 **b** 400 **c** 4000 **d** 230
e 1230 **f** 204 **g** 67 000 **h** 110 000

2 **a** 0.4 **b** 0.04 **c** 0.004 **d** 0.65
e 0.103 **f** 0.0005 **g** 0.1 **h** 0.000 002

3 **a** 97 **b** 0.42 **c** 1080 **d** 0.04
e 63 000 **f** 0.006 **g** 1230 000 **h** 0.000 011

4 Rewrite the numbers in these sentences in standard form.
a The speed of light is 300 000 000 metres per second.
b The diameter of the Earth is about 12 680 km.
c The thickness of a sheet of paper is 0.113 mm.
d A Building Society's funds totalled £2 150 000 000.
e The radius of the orbit of an electron is 0.000 000 05 mm.
f A space probe reached a speed of 149 125 mph in 1976.
g A film of oil is 0.000 000 8 mm thick.
h The weight of the Earth is about 6 600 000 000 000 000 000 000 tonnes.
i The average human life-span is 1 000 000 000 seconds.
j Sound travels across a room in about 0.01 second.

5 **SUPERNOVA MAY SOLVE SPACE RIDDLE**
British scientists have discovered a new exploding star—a supernova—in a galaxy which is five billion light years from the earth.

Express this distance in scientific notation (a billion is a thousand million).

> $4.2 \times 10^4 = 42\,000$
> $8.5 \times 10^{-4} = 0.000\,85$
> These are given in **normal**, or **floating point, form**.

6 Write these numbers in normal, or floating point, form:
a 4×10^2 **b** 6×10^3 **c** 7×10^{-2}
d 9×10^{-3} **e** 9.8×10^{-1} **f** 5×10^0

7 Write the numbers in the sentences below in normal, or floating point, form:

a The national expenditure is £2.4×10^{11}.
b The area of the Sahara desert is 8.6×10^{6} km^2.
c Pluto's mass is 2.5×10^{-3} times the Earth's mass.
d There are about 1.65×10^{5} species of butterfly.
e The wavelength of radiation used in a microwave is 2.45×10^{3} MHz.
f The mass of a neutron is 1.675×10^{-27} kg.

8 a A centillion is 10^{600}. How many zeros does it have?
b The mass of a photon is 3×10^{-53} g. How many zeros are there after the decimal point?

CALCULATORS CAN COPE WITH VERY LARGE AND VERY SMALL NUMBERS

EXERCISE 5A

1 a Fill your calculator display with 9s.

b What do you expect to see if you key [+] [1] [=]?
c Do you see [1.10] or [1.08] meaning 1×10^{10} or 1×10^{8}?
d Write the number out in full.

2 a Use the constant facility to key $1.2 \times 10 \times 10 \times 10 \times \ldots$ until the display is full.
b Now key $\times 10$ once more. Write down the number on the display, in standard form. Next write it out in full.

3 Write these numbers in full:

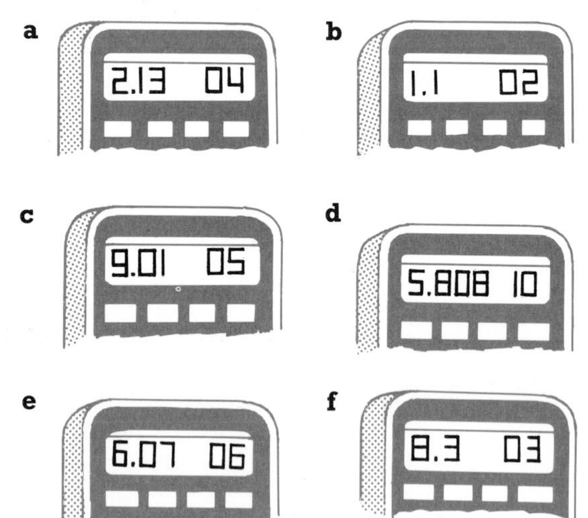

4 a Keep keying $5.3 \div 10 \div 10 \div 10 \ldots$ until the display is full.
b Then key it in once more. Write out the number in standard form, and in full.

5 Write these numbers in full:

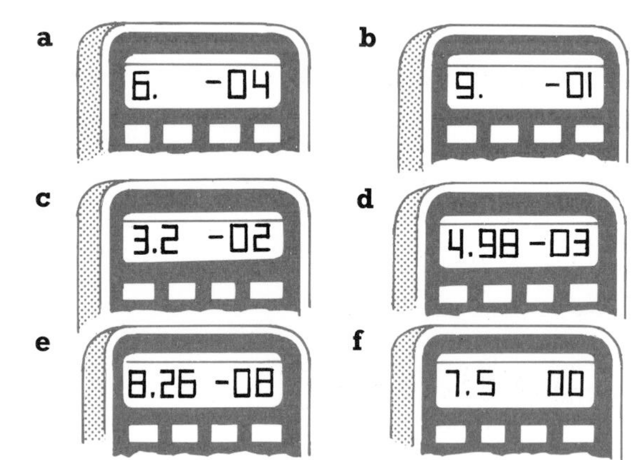

6 Write these numbers in full:

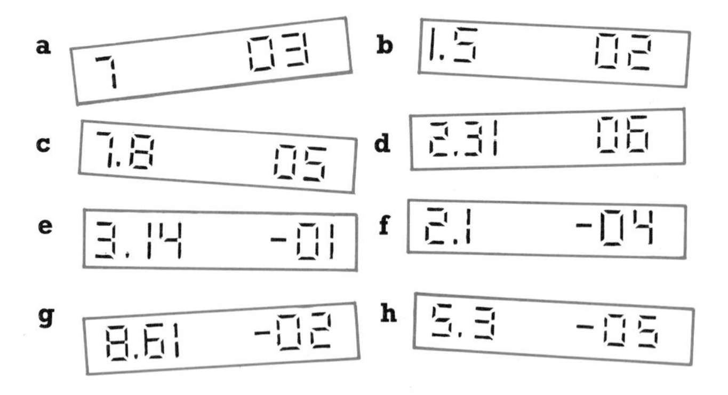

7 Write down the greatest and least numbers in this collection:

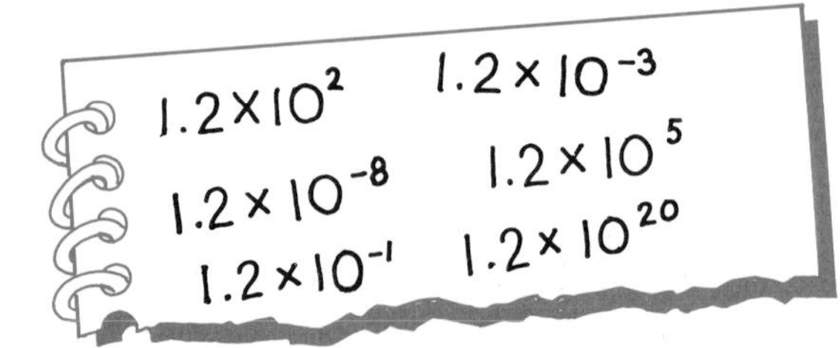

Express your answers to questions **8–10**:
(i) in full (ii) in standard form.

8 The most popular daily newspaper sells 3 600 000 copies on average each day. How many copies are sold in 310 days?

9 The population of the United Kingdom is approximately 57 million. On average each person uses 140 litres of water a day. Calculate the total volume of water used in the UK:
a each day **b** in a year.

10 The diameter of the Sun is 1380 000 km. Calculate its circumference.

INVESTIGATION

*Scientific calculators can display a number in either standard form or floating point form. Find out how your calculator does this, and try questions **4** and **7** of Exercise 4 again to test your skill (and your calculator).*

EXERCISE 5B

Calculating in standard form

Example
Light travels at a speed of 3×10^8 m/s. There are 3.15×10^7 seconds in a year. How far does light travel in a year?

Distance = speed × time
$= 3 \times 10^8 \times 3.15 \times 10^7$

Key 3 EXP 8 × 3.15 EXP 7

to get 9.45 15

The distance is 9.45×10^{15} m.

1 $A = 6.5 \times 10^5$, $B = 2.6 \times 10^3$ and $C = 5.0 \times 10^{-4}$.
Calculate in standard form:
a AB **b** BC **c** $\frac{B}{A}$ **d** $\frac{A}{C}$

2 An aircraft flying at Mach 1 has a speed of 3.315×10^2 m/s. Calculate its speed in standard form at Mach 2.9.

3 There are 5×10^9 red blood cells in 1 ml of blood. Calculate, in scientific notation, the number of cells in: **a** 1 litre of blood **b** 7.5 litres.

4 Light travels 9.46×10^{12} km in a year. Calculate the distance it travels in 1.5×10^3 years.

5 The mass of the Earth can be calculated in kilograms by using the formula $m = \frac{gR^2}{G}$, where $g = 9.8$, $R = 6.37 \times 10^6$ and $G = 6.67 \times 10^{-11}$.
Show that the Earth's mass is 5.96×10^{24} kg, to 3 significant figures.

6 The mass of the Moon is 7.3×10^{22} kg. Calculate, in standard form:
a the difference between the masses of the Earth (question **5**) and the Moon
b the ratio of their masses, to the nearest whole number.

7 The galaxy we live in is roughly lens shaped, and spins as shown. Its diameter is given in light years (1 light year $= 9.5 \times 10^{12}$ km).

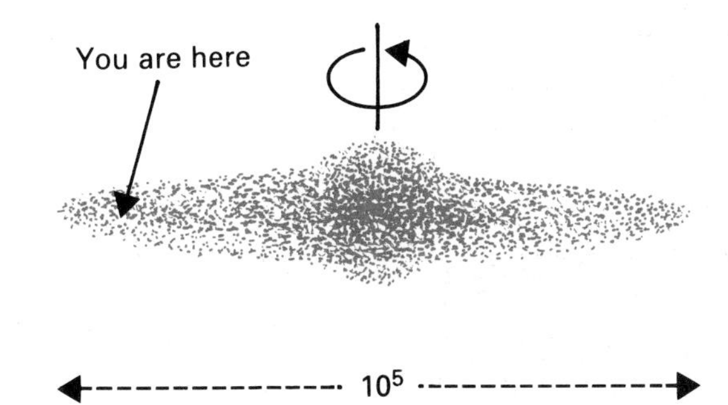

10^5

a Calculate the circumference of our galaxy in kilometres.
b The Earth is at a point 3×10^4 light years from the centre. How far does the Earth travel in one turn of the galaxy?

INVESTIGATION

a *Look for a pattern as you calculate:*
(i) 9999 × 2 (ii) 9999 × 3 (iii) 9999 × 4
b *Now write down the answers for multiplying 9999 × 5 . . . to . . . 9999 × 9.*

CHALLENGE

1 Using only the $\boxed{\times}$ *key on your calculator, can you find the value of x, correct to 1 decimal place, if* $x^3 = 20$*? Copy and complete:*

$2^3 = \ldots$ } so *x* is between
$3^3 = \ldots$ } 2 and 3, nearer . . .
$2.7^3 = \ldots$ } so *x* is between
$2.8^3 = \ldots$ } . . . and . . . , nearer . . .
$x = \ldots$, correct to 1 decimal place.

2 In the same way, find a value for x, correct to 1 decimal place, if:

a $7x = 30$ **b** $6x = 47$

c *x cm is the length of the side of a square of area* (i) 11 cm^2 (ii) 53 cm^2

d *x cm is the length of the edge of a cube of volume* (i) 12 cm^3 (ii) 50 cm^3

e (i) $x^2 + x = 7$ (ii) $x^2 - x = 8$ (iii) $x^3 + x = 90$

COPING WITH FRACTIONS

If your calculator has a fraction key you can use it, if you wish, in the rest of this chapter in questions marked **C**

(i) Mixed numbers

Examples

1 Change $\frac{5}{3}$ to a mixed number.

$\frac{5}{3} = \frac{3}{3} + \frac{2}{3} = 1\frac{2}{3}$, *or* $\frac{5}{3} = 5 \div 3 = 1\frac{2}{3}$

C Key 5  3  

2 Change $5\frac{3}{4}$ to a fraction.

$5\frac{3}{4} = \frac{20}{4} + \frac{3}{4} = \frac{23}{4}$, *or* $5\frac{3}{4} = \frac{4 \times 5 + 3}{4} = \frac{23}{4}$

C Key 5 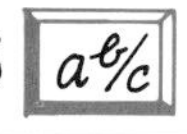 3 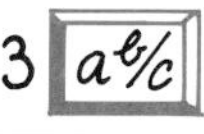 4  $a^b/_c$ 23r4

EXERCISE 6

1 Write as mixed numbers:
a $\frac{3}{2}$ **b** $\frac{9}{4}$ **c** $\frac{4}{3}$ **d** $\frac{7}{2}$ **e** $\frac{12}{5}$

2 Write as fractions:
a $3\frac{1}{2}$ **b** $1\frac{3}{4}$ **c** $3\frac{1}{5}$ **d** $5\frac{1}{3}$ **e** $1\frac{1}{10}$

3C Write as mixed numbers:
a $\frac{11}{8}$ **b** $\frac{16}{3}$ **c** $\frac{24}{5}$ **d** $\frac{33}{10}$ **e** $\frac{100}{9}$

4C Write as fractions:
a $4\frac{2}{3}$ **b** $5\frac{3}{8}$ **c** $2\frac{7}{10}$ **d** $1\frac{1}{8}$ **e** $9\frac{7}{8}$

5C A record spins at $33\frac{1}{3}$ rpm. Write $33\frac{1}{3}$ as a proper fraction.

6C Copy and complete this table. The lengths are in inches.

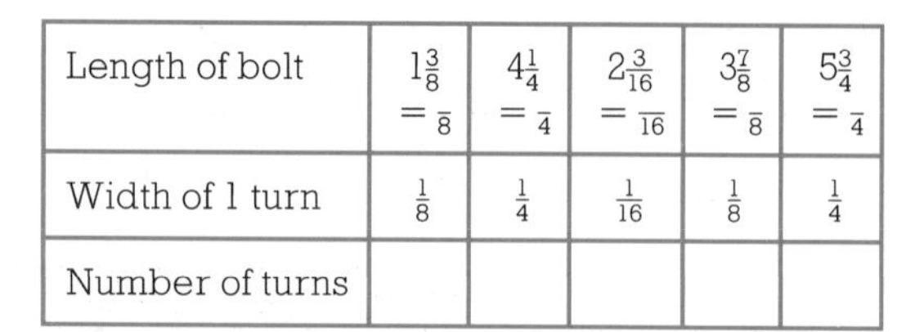

Length of bolt	$1\frac{3}{8} = \frac{}{8}$	$4\frac{1}{4} = \frac{}{4}$	$2\frac{3}{16} = \frac{}{16}$	$3\frac{7}{8} = \frac{}{8}$	$5\frac{3}{4} = \frac{}{4}$
Width of 1 turn	$\frac{1}{8}$	$\frac{1}{4}$	$\frac{1}{16}$	$\frac{1}{8}$	$\frac{1}{4}$
Number of turns					

7C In each pair, find which fraction is greater by changing them to mixed numbers:
a $\frac{3}{2}$ or $\frac{5}{4}$ **b** $\frac{4}{3}$ or $\frac{6}{5}$ **c** $\frac{13}{10}$ or $\frac{7}{5}$ **d** $\frac{13}{9}$ or $\frac{15}{11}$

8C Check your answers to question **7C** by changing each fraction to decimal form. Which method is better?

9C Flags are placed every $\frac{1}{5}$ km along the route of a 'fun run'. The course is $4\frac{3}{5}$ km long. How many flags are there, including the ones at the start and finish of the race?

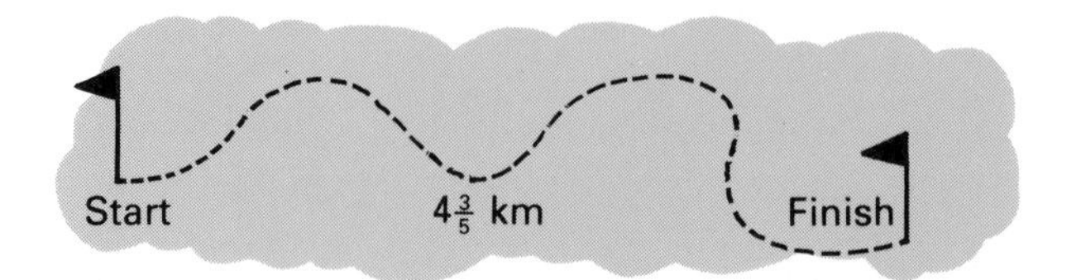

A PUZZLE

These three wheels have 20, 5 and 10 teeth.

What happens to:

a *B, if A is given a clockwise quarter turn?*
b *C, if A is given a clockwise quarter turn?*
c *B, if C is given* $1\frac{1}{2}$ *turns anti-clockwise?*
d *A, if C is given* $2\frac{1}{2}$ *turns anti-clockwise?*

(ii) Adding and subtracting fractions

Examples

1

$\frac{1}{4}+\frac{2}{3}$

$=\frac{3}{12}+\frac{8}{12}$ (lcm of 4 and 3 is 12)

$=\frac{11}{12}$

or

$\frac{1}{4}+\frac{2}{3}$

$=\dfrac{3\times 1+4\times 2}{4\times 3}$

$=\frac{11}{12}$

C Key 1 [a b/c] 4 [+] 2 [a b/c] 3 [=] [11r12]

2

$5\frac{1}{2}-2\frac{3}{5}$

$=\frac{11}{2}-\frac{13}{5}$ (change to proper fractions)

$=\frac{55}{10}-\frac{26}{10}$ (lcm of 2 and 5 is 10)

$=\frac{29}{10}$

$=2\frac{9}{10}$

C Key 5 [a b/c] 1 [a b/c] 2 [−] 2 [a b/c] 3 [a b/c] 5 [=] [2r9r10]

EXERCISE 7A

Calculate **1–4C** in their simplest form:

1 **a** $\frac{1}{5}+\frac{2}{5}$ **b** $\frac{1}{4}+\frac{1}{4}$ **c** $\frac{3}{4}-\frac{1}{4}$ **d** $\frac{7}{8}-\frac{1}{8}$
e $\frac{1}{2}+\frac{1}{3}$ **f** $\frac{2}{3}+\frac{1}{4}$ **g** $\frac{3}{5}-\frac{1}{2}$ **h** $\frac{5}{6}-\frac{1}{2}$

2 **a** $1\frac{3}{5}+1\frac{1}{5}$ **b** $4-\frac{1}{2}$ **c** $2\frac{1}{4}-\frac{3}{4}$ **d** $3\frac{2}{3}-2\frac{1}{3}$

3C **a** $\frac{3}{4}+\frac{3}{5}$ **b** $\frac{1}{3}+\frac{3}{4}$ **c** $\frac{2}{3}-\frac{1}{4}$ **d** $\frac{4}{5}-\frac{3}{4}$
e $\frac{2}{5}+\frac{1}{3}$ **f** $\frac{5}{6}+\frac{1}{3}$ **g** $\frac{5}{7}-\frac{1}{2}$ **h** $\frac{7}{8}-\frac{3}{4}$

4C **a** $2\frac{1}{2}+1\frac{1}{4}$ **b** $3\frac{7}{8}-1\frac{3}{4}$ **c** $4\frac{3}{4}+2\frac{1}{2}$ **d** $6\frac{1}{2}-5\frac{2}{3}$

5C A car's petrol tank holds 30 litres. $18\frac{1}{2}$ litres are used. How much is left?

6C

How far is it from:
a the village to the coast
b the village to the lighthouse
c the coast to the lighthouse?

7C A video cassette lasts 3 hours.

a A programme lasting $\frac{3}{4}$ hour is recorded. How much time is left?
b How much time is left after another $\frac{3}{4}$ hour programme is recorded?

8C One morning half the students at Jane's school arrived before 8.55, and three eighths of the students arrived from 8.55 am to just before 9 am. What fraction of the students arrived:
a before 9 am **b** at 9 am or later?

9C In a year group election, $\frac{1}{2}$ the students voted for Janice, $\frac{1}{4}$ for Lana and $\frac{1}{5}$ for Sean. The rest voted for Andrew.
a What fraction of the votes did Andrew get?
b 100 students voted. How many votes did each person get?

EXERCISE 7B

1C A video tape lasts for 4 hours. Programmes lasting $1\frac{1}{2}$ hours, $\frac{3}{4}$ hour, $\frac{1}{2}$ hour and $\frac{3}{4}$ hour are recorded. How much time is left?

2C Nails $1\frac{7}{10}$ cm and $2\frac{3}{10}$ cm long are driven into a piece of wood $3\frac{1}{10}$ cm thick. How far is the point of each from the other side of the wood?

3C Calculate the new rates of interest in Britbank's savings account.

Old rate (%)	$7\frac{1}{2}$	$4\frac{3}{10}$	$6\frac{2}{3}$	$5\frac{1}{4}$	$3\frac{9}{10}$
Change (%)	$+\frac{3}{4}$	$-\frac{7}{10}$	$+1\frac{2}{3}$	$-2\frac{3}{4}$	$+1\frac{3}{5}$

4C How far is it from:
a the church to the stables
b the town to the farm
c the farm to the stables
d the town to the church?

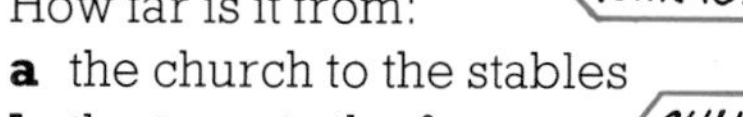

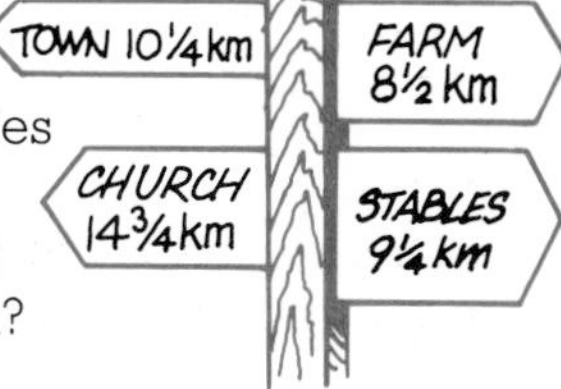

5C The smallest of a set of spanners is $\frac{1}{4}$ inch, and each spanner is $\frac{1}{16}$ inch larger than the previous one.

a If the $\frac{1}{2}$ inch one is just too small, which size should you try next?

b How much bigger is a $\frac{5}{8}$ inch spanner than a $\frac{1}{4}$ inch one?

6C (i) Calculate the perimeter of each photo frame.
(ii) What is the difference between the length and breadth of each?

a

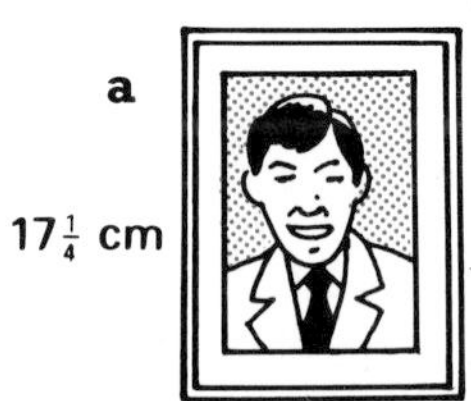

b

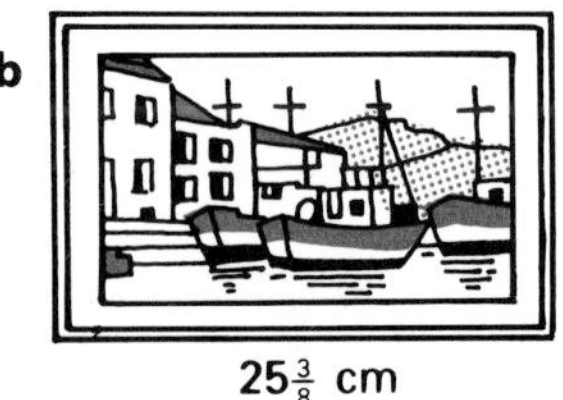

BRAINSTORMER

A wealthy merchant died, leaving seventeen valuable paintings to his three sons. The eldest son was to have one half of the paintings, the middle son one third, and the youngest son one ninth. The three sons were in despair. How could they divide out the seventeen paintings?
They explained their problem to a rich uncle, who said that there was an easy answer to the problem. He added a valuable painting of his own to the seventeen others. The eldest son took half of the eighteen paintings, the middle son took his third, and the youngest his ninth. The uncle then took back his own painting, and everyone was happy—or were they?

(iii) Multiplying by mixed numbers and fractions

Examples

1 $6 \times \frac{3}{4}$
$= \frac{6}{1} \times \frac{3}{4}$
$= \frac{18}{4}$
$= \frac{9}{2}$
$= 4\frac{1}{2}$

2 $\frac{3}{4} \times \frac{1}{2}$
$= \frac{3 \times 1}{4 \times 2}$
$= \frac{3}{8}$

3 $2\frac{1}{2} \times 1\frac{1}{3}$
$= \frac{5}{2} \times \frac{4}{3}$ (Change to fractions)
$= \frac{20}{6}$
$= \frac{10}{3}$
$= 3\frac{1}{3}$

C Key 2 [a b/c] 1 [a b/c] 2 [×] 1 [a b/c] 1 [a b/c] 3 [=]

EXERCISE 8A

Calculate **1**–**5C** in their simplest form:

1 **a** $8 \times \frac{1}{2}$ **b** $6 \times \frac{2}{3}$ **c** $5 \times \frac{1}{3}$ **d** $\frac{5}{8} \times 4$

2 **a** $\frac{1}{2} \times \frac{1}{4}$ **b** $\frac{1}{2} \times \frac{3}{4}$ **c** $\frac{3}{4} \times \frac{2}{3}$ **d** $\frac{1}{8} \times \frac{4}{5}$
e $\frac{1}{4}$ of $\frac{4}{5}$ **f** $\frac{1}{3}$ of $\frac{3}{8}$ **g** $\frac{2}{5}$ of $\frac{5}{6}$ **h** $\frac{2}{3}$ of $\frac{3}{8}$

3 **a** $2\frac{1}{2} \times 1\frac{1}{2}$ **b** $1\frac{1}{4} \times 1\frac{1}{2}$ **c** $1\frac{1}{2} \times 1\frac{2}{3}$ **d** $1\frac{3}{5} \times 2\frac{1}{2}$

4C **a** $\frac{2}{3} \times \frac{3}{4}$ **b** $\frac{4}{5} \times \frac{5}{4}$ **c** $\frac{1}{6} \times \frac{2}{3}$ **d** $\frac{5}{8} \times \frac{2}{15}$

5C **a** $4\frac{1}{2} \times 1\frac{1}{3}$ **b** $3\frac{1}{3} \times 2\frac{1}{4}$ **c** $1\frac{1}{8} \times \frac{5}{6}$ **d** $2\frac{2}{3} \times 1\frac{1}{8}$

6C A bottle of orange juice contains $\frac{3}{4}$ litre. Calculate the volume of juice in:
a (i) 4 bottles (ii) 3 bottles (iii) 12 bottles
b (i) $\frac{1}{3}$ bottle (ii) $\frac{1}{6}$ bottle (iii) $2\frac{1}{2}$ bottles

7C A record makes $33\frac{1}{3}$ revolutions per minute. Calculate the number of revolutions it makes in:
a 3 minutes **b** 5 minutes.

8C Deepa's tape recorder has three speeds: $7\frac{1}{2}$, $3\frac{3}{4}$ and $1\frac{7}{8}$ inches per second.
a How many inches of tape will be used at each speed in 8 seconds?
b What is the connection between the three speeds?

9C A map has a scale of 1 inch to $3\frac{1}{4}$ miles. Calculate the actual distances between places which are 4 inches, 10 inches and 36 inches apart on the map.

10C Calculate the area of the square and the rectangle.

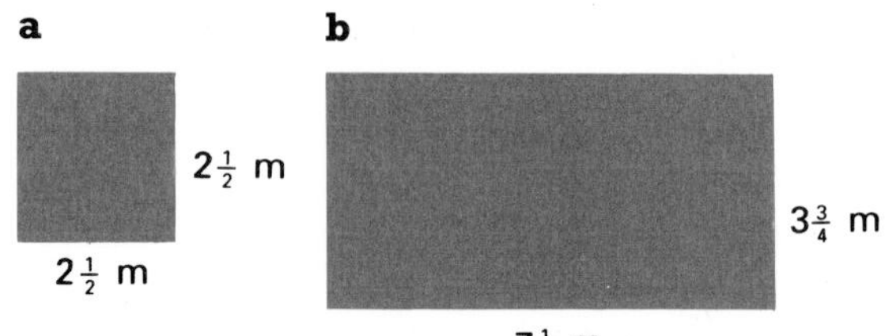

EXERCISE 8B

1C Lucy puts $3\frac{1}{2}$ gallons of petrol in her car. 1 gallon is about $4\frac{1}{2}$ litres. How many litres does she put in?

2C 30 metres of carpet are needed for a stair $\frac{3}{4}$m wide.
a Calculate the area of carpet needed.
b The carpet comes in a 1 metre width. Calculate the area of carpet which would have to be trimmed off.

3C Mrs Samson chose a patterned material to make a dress for herself and a skirt for each of her twin daughters. She bought $6\frac{3}{4}$m at £9.60 a metre, and used $\frac{2}{3}$ of it for her dress and the rest for her daughters' skirts.

How much did the dress and each skirt cost?

4C This table gives the circumferences of some model car wheels.

Type	Circumference (cm)
A	$3\frac{3}{5}$
B	$4\frac{1}{2}$
C	$5\frac{3}{10}$

Calculate the distance travelled by:
a type A in 10 turns **b** type B in $3\frac{1}{3}$ turns
c type C in $2\frac{1}{2}$ turns.

5C Use $3\frac{1}{7}$ for π to calculate the area of a circle with radius $3\frac{1}{2}$ cm.

6C Use the formula $C = \frac{5}{9}(F - 32)$ to calculate C when:
a $F = 36\frac{1}{2}$ **b** $F = 31\frac{1}{4}$.
Give answers as fractions.

CHALLENGE

Rule 1 *Everyone at a table shares the cakes on that table equally.*
Rule 2 *Each person arriving has to go to the table where he or she will get the most cake, after sharing.*
Rule 3 *If there is a choice of 'equal' tables, go to the one with fewest people.*

Alice arrives first and goes to Table 1. She would have four cakes if no one else arrives.
Bernie is next. If he goes to Table 1 he would get two cakes, so he goes to Table 2.
Investigate the choice of table and share of cakes as Carl, Dave, Eileen, Finlay, Grace, Hannah, Imogen and Josh arrive one by one, in that order.
When the nineteenth person arrives they start eating. What fraction of a cake does the nineteenth get?

(iv) Dividing by fractions

Examples

1 $6 \div \frac{3}{4} = \dfrac{\frac{6}{1}}{\frac{3}{4}} = \dfrac{\frac{6}{1}\times 4}{\frac{3}{4}\times 4} = \frac{24}{3} = 8$, **or** $6 \div \frac{3}{4} = \frac{6}{1} \times \frac{4}{3} = \frac{24}{3} = 8$ (inverting the fraction and *multiplying*)

2 $5\frac{1}{2} \div 2\frac{2}{3} = \frac{11}{2} \div \frac{8}{3} = \frac{11}{2} \times \frac{3}{8} = \frac{33}{16} = 2\frac{1}{16}$

C Key 5 [a b/c] 1 [a b/c] 2 [÷] 2 [a b/c] 2 [a b/c] 3 [=] [2⌟1⌟16]

EXERCISE 9B

1C Calculate:

a $\frac{1}{2} \div \frac{1}{4}$ **b** $\frac{1}{2} \div \frac{1}{8}$ **c** $\frac{3}{4} \div \frac{1}{2}$ **d** $\frac{1}{2} \div \frac{1}{2}$

e $2\frac{1}{8} \div \frac{1}{8}$ **f** $2 \div \frac{1}{4}$ **g** $3 \div \frac{3}{4}$ **h** $1\frac{3}{4} \div \frac{1}{4}$

i $4\frac{1}{2} \div 1\frac{1}{2}$ **j** $15 \div 2\frac{1}{2}$ **k** $10 \div \frac{1}{4}$ **l** $3\frac{3}{4} \div 2\frac{1}{2}$

2C Henry, the cat, eats a $\frac{1}{2}$ tin of Catto each day. How many days would a pack of 24 tins last him?

3C How many $\frac{1}{4}$ litre cartons of juice can be filled from a 48 litre tank?

4C How many $\frac{3}{4}$ hour programmes could be recorded on:

a a 3 hour tape **b** a $1\frac{1}{2}$ hour tape

c a 4 hour tape?

5C Fully loaded, a tip-up truck can take $4\frac{7}{8}$ tonnes.

a Calculate the least number of loads needed to remove 25 tonnes of earth.

b What is the weight of the final load?

6C A photograph is enlarged $2\frac{1}{2}$ times.

a The original is $2\frac{1}{2}$ cm wide. What is the width of the enlargement?

b The enlargement is $11\frac{1}{4}$ cm high. What is the height of the original?

7C Isobel has a weekly piano lesson of $1\frac{1}{4}$ hours. She sat her first examination after a total of 25 hours of piano lessons. How many lessons did she need?

8C In music the lengths of sounds are shown by notes of different shapes.
These notes, with their shapes and lengths, are shown compared to the length of a semibreve.

Note	Shape	Value in terms of a semibreve
Semibreve	𝅝	1
Minim	𝅗𝅥	$\frac{1}{2}$
Crochet	𝅘𝅥	$\frac{1}{4}$
Quaver	𝅘𝅥𝅮	$\frac{1}{8}$
Semi-quaver	𝅘𝅥𝅯	$\frac{1}{16}$
Demi-semi-quaver	𝅘𝅥𝅰	$\frac{1}{32}$

How many semiquavers are there in each of the notes shown below?

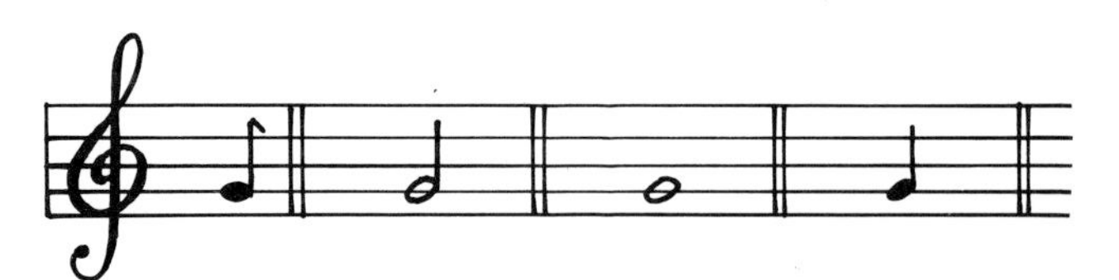

9C Calculate the approximate number of:

a pounds in 5 kg **b** kg in 12 pounds

c litres in $2\frac{1}{2}$ gallons **d** miles in 20 km

e inches in 1 metre **f** inches in 12 cm.

1 inch $\doteqdot 2\frac{1}{2}$ centimetres
1 gallon $\doteqdot 4\frac{1}{2}$ litres
1 kilogram $\doteqdot 2\frac{1}{4}$ pounds
1 kilometre $\doteqdot \frac{5}{8}$ mile

10C A square has an area of $11\frac{1}{9}$ cm². Calculate the length of its side.

CHECK-UP ON CALCULATIONS AND CALCULATORS

1C $u = 126$, $v = 88$ and $w = 53$. Calculate, correct to 3 significant figures, the values of:

a $\dfrac{u+v}{w}$ **b** $u+\dfrac{v}{w}$ **c** $u(v-w)$

d $\sqrt{(v^2+w^2)}$ **e** $(u-v)^3$ **f** $\dfrac{1}{u+v+w}$

2C (i) *Estimate* the area of the rectangle and the volume of the cuboid.
(ii) Then *calculate* the area and volume, to a suitable number of significant figures.

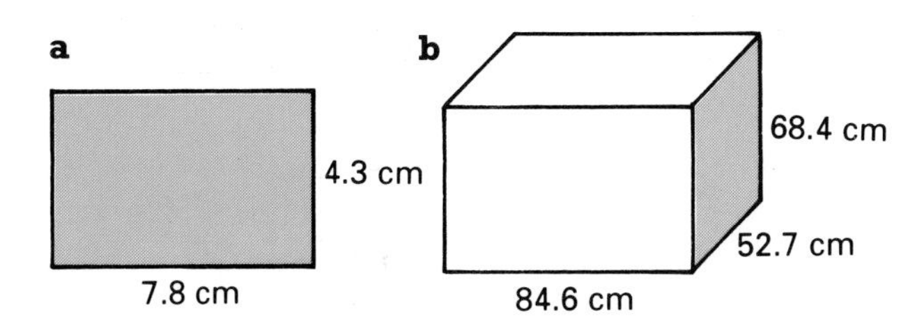

3C **a** Which of these numbers do you think could equal 125×78?
(i) 850 (ii) 8055 (iii) 90 550 (iv) 9750
b Check by calculation.

4C Write each number in scientific notation.
a The payload of a space shuttle is 27 000 000 g.
b A unit in electronics is a microfarad, which is 0.000 001 farad.

5C A compact disc can store 6×10^8 bytes of information. Calculate, in standard form, the storage capacity in bytes of:
a 10 discs **b** 250 discs.

6C In 1889 a swarm of locusts weighing 550 000 tonnes, and containing 250 000 000 000 insects, crossed the Red Sea. Calculate the weight of a locust in tonnes, writing your answer in standard form.

7C This table gives the masses of some planets as fractions of the Earth's mass of 5.96×10^{24} kg. Calculate the mass of each planet in kg, in standard form.

Mercury	Venus	Mars	Jupiter	Saturn
0.054	0.81	0.107	318.4	95.3

8C Copy and complete the table to convert temperatures of 50°F, 60°F, . . . , 100°F to °C, to the nearest degree.

°F	−32	×5	÷9	°C
50	18	90		
60				

9C Giving answers in their simplest form, calculate:

a $\frac{5}{8}+\frac{1}{4}$ **b** $1\frac{2}{3}+2\frac{4}{5}$ **c** $\frac{7}{12}-\frac{1}{3}$ **d** $4\frac{3}{10}-1\frac{3}{4}$
e $\frac{3}{20}\times\frac{8}{9}$ **f** $\frac{9}{10}\div\frac{3}{4}$ **g** $4\frac{1}{5}\times 1\frac{1}{7}$ **h** $3\frac{1}{8}\div\frac{5}{12}$

10C What size of spanner is half-way between a $\frac{3}{4}$ inch and a $\frac{7}{8}$ inch one?

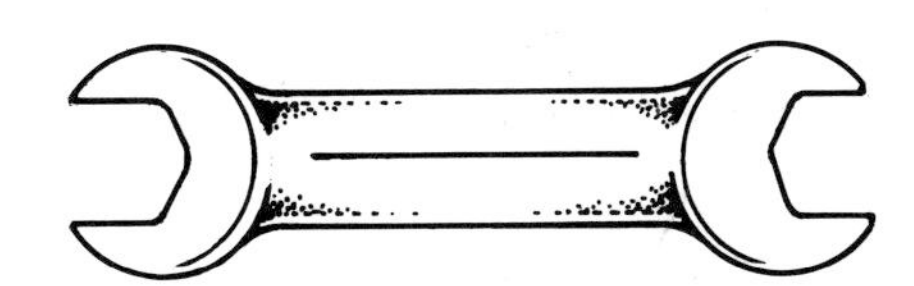

11C Calculate the area of a rectangular floor $7\frac{1}{2}$ m by $5\frac{1}{3}$ m.

12C $\frac{3}{5}$ of the students in a school are girls. $\frac{2}{3}$ of the girls and $\frac{1}{2}$ of the boys are over 13. What fraction of the students are 13 or under?

13C A rectangular carpet has length $4\frac{1}{2}$ m and area $11\frac{1}{4}$ m². Calculate its breadth.

2 SIMILAR SHAPES

LOOKING BACK

1 Calculate:
a $\frac{2}{3}$ of 15 cm **b** $\frac{3}{5}$ of 60 mm **c** 0.2 of 10 m

2 Multiply 120 by:
a $\frac{1}{5}$ **b** $\frac{2}{3}$ **c** $\frac{3}{2}$ **d** 0.1 **e** 0.7 **f** 1.6

3 Express the smaller length as a fraction of the larger, in its simplest form:

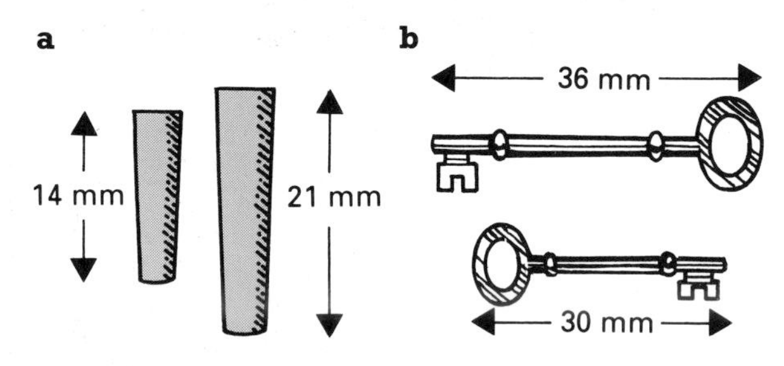

4 a Draw a rectangle 4 cm long and 1 cm broad.
b Draw two more rectangles: one with sides twice as long as the first; the other with sides half as long as the first.

5 List pairs of shapes that have:
(i) the same shape and size (are congruent)
(ii) the same shape but not the same size.

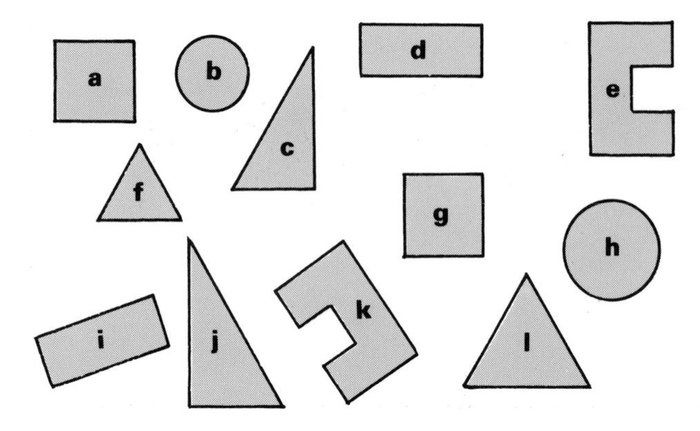

6 The model car is $\frac{1}{100}$th of the real size. If the model is 4.5 cm long, what is the length, in metres, of the real car?

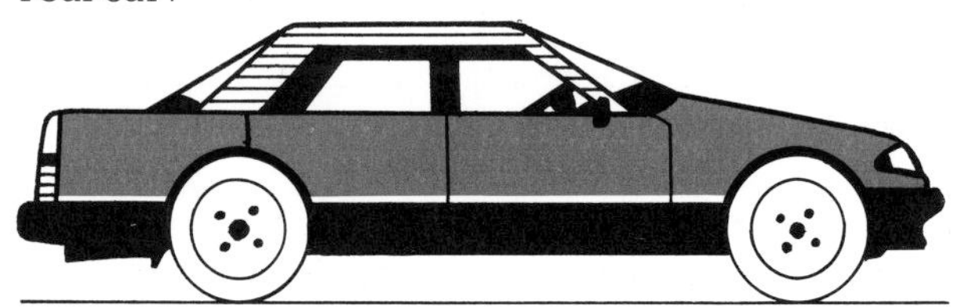

7 The snowflake is drawn 8 × actual size. The picture is 20 mm wide. What is the width of the actual snowflake?

8 Mark enters the Hall of Mirrors. Which image do you think he would prefer to see? Why?

a

b

c

d

SCALE MODELS AND SCALE FACTORS

CLASS DISCUSSION

1 The diagrams below show the real-life picture and its scale picture.

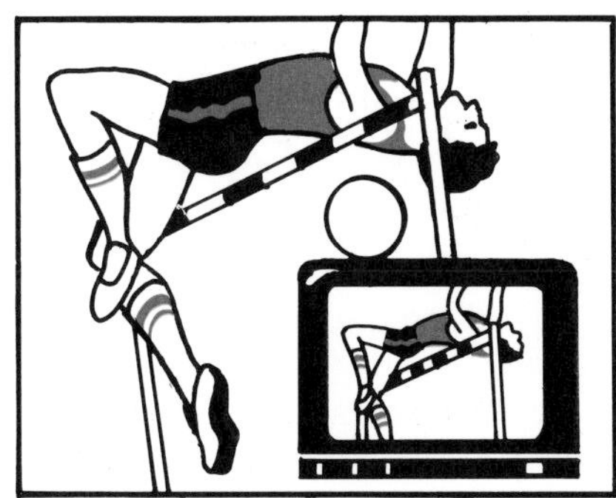

Reduction of the high-jumper | Enlargement of the silicon chip | Reduction of the sailing ship

When an object is reduced or enlarged, all the corresponding lengths are in the **same ratio**—they have the same **scale factor**.

2 This scale model of the Lazer Z fighter is $\frac{1}{100}$ full size. So every length on the model is $\frac{1}{100}$th of the corresponding length on the real aircraft.
The real fighter has a wingspan of 5 m. How many cm is this?
So what is the wingspan of the model?

Andrew's slide pictures on the screen look real. This is because in an enlargement (or a reduction):

corresponding angles are equal;
corresponding lengths are in the same ratio—
they have the same scale factor.

In an enlargement, the scale-factor is greater than 1; in a reduction, it is less than 1.

A scale factor might be given as $\frac{1}{10}$, or 1:10, or 0.1.

Example
The actual dimensions of the MIA Turbo are shown in the diagram.
On a scale model, the length is 20 cm. Calculate:

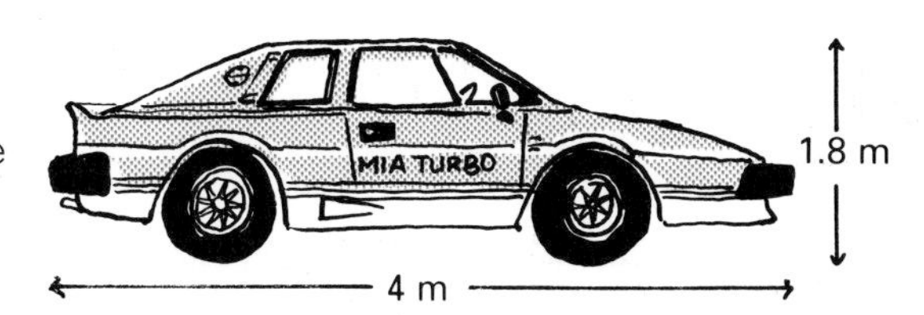

a the scale of the reduction **b** the height of the model

The scale factor $= \dfrac{\text{model length}}{\text{actual length}} = \dfrac{20\text{ cm}}{400\text{ cm}} = \dfrac{1}{20}$

The height of the model $= \frac{1}{20} \times 1.8\text{ m} = \frac{1}{20} \times 180\text{ cm} = 9\text{ cm}$

EXERCISE 1A

1 Calculate the scale factor of:
a the enlargement from (i) to (ii)
b the reduction from (ii) to (i).

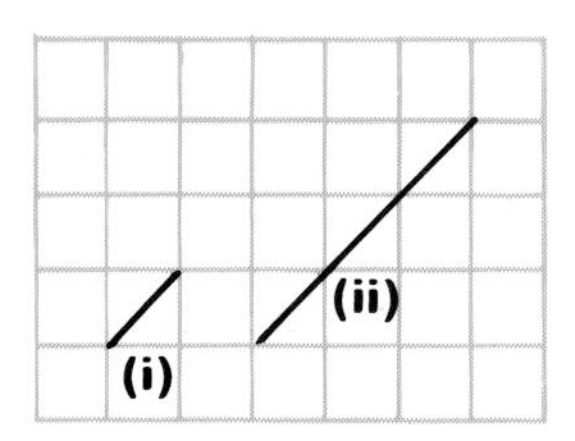

2 Calculate the scale factor of:
a the enlargements from (i) to (ii), (ii) to (iii) and (i) to (iii)
b the reductions from (ii) to (i), (iii) to (ii) and (iii) to (i).

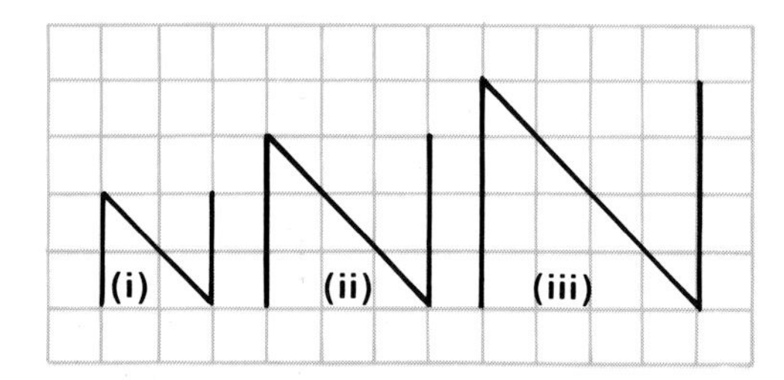

3 A line is 6 cm long. What is its length after multiplication by a scale factor of:
a 2 **b** $\frac{1}{3}$ **c** 1.5 **d** 0.5 **e** 1?

4 Write down the length and breadth of this rectangle after multiplication by:
a an enlargement factor of $\frac{3}{2}$
b a reduction factor of $\frac{1}{2}$.

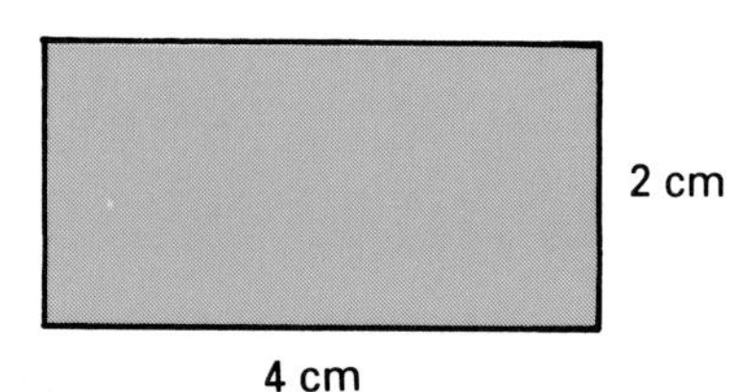

5 This film negative is enlarged by a scale factor of 4 to make a print. What size is the print:
a in mm **b** in cm?

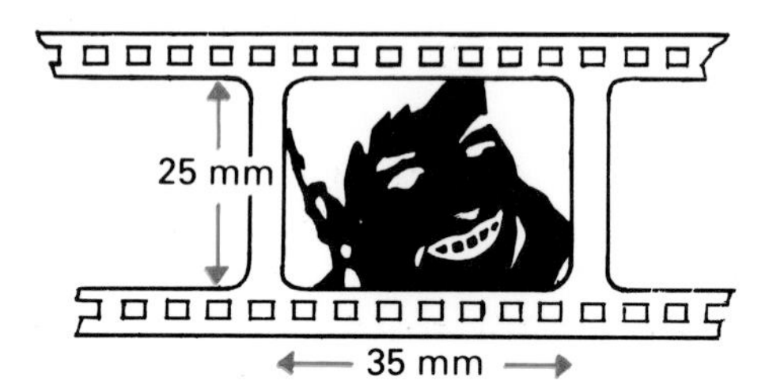

6 On squared paper, enlarge the lengths of the sides of these shapes:
A using scale factor 3, B using scale factor 2, C using scale factor 1.

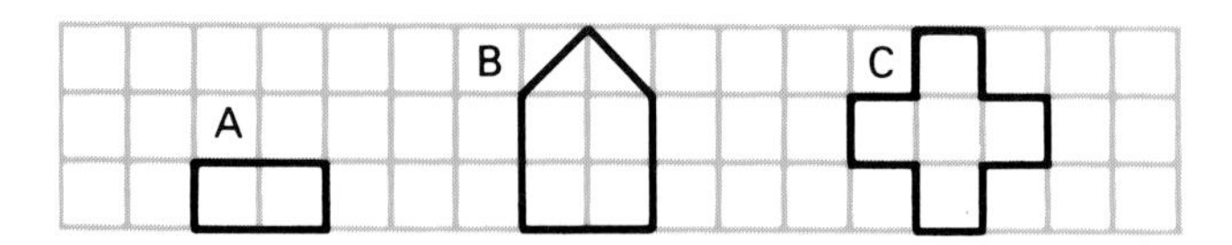

7 On squared paper, reduce the lengths of the sides of these shapes:
D using scale factor $\frac{1}{3}$, E using scale factor $\frac{1}{2}$, F using scale factor $\frac{1}{4}$.

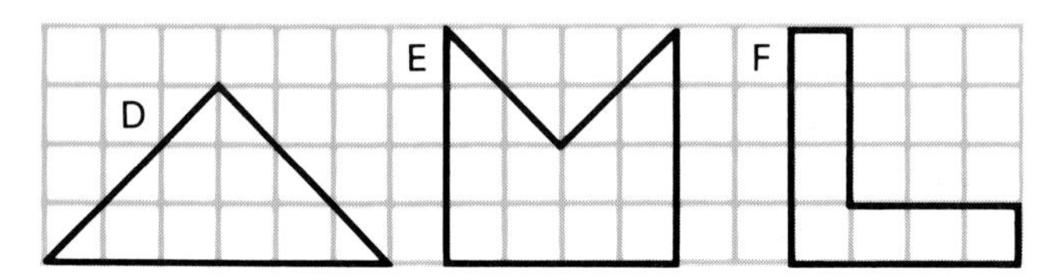

8 A model of this new rocket has to be made to a scale of 1:10.
a Write down the scale factor of the reduction as a fraction.
b Calculate the height of the model.

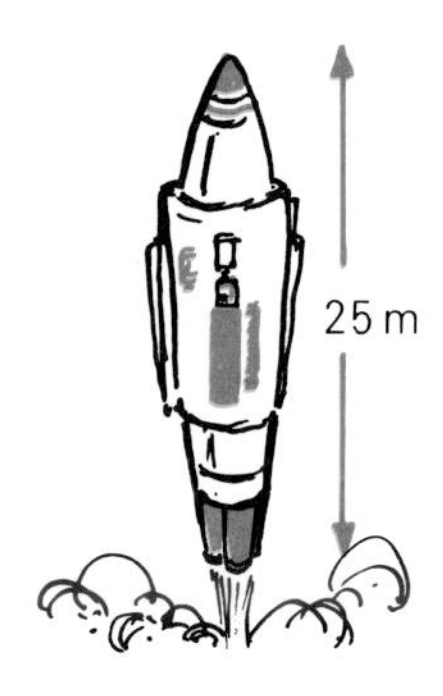

9 Fergus has a model of a yacht. The real yacht is 10 times larger (scale factor 10).

Calculate:
a the length of the real yacht
b the height of the real mast.

10 Emily's bicycle is half the size of her mother's (scale factor $\frac{1}{2}$).
a Her mother's bicycle is 80 cm high. How high is Emily's?
b The diameter of her mother's bicycle wheel is 54 cm. What is the diameter of Emily's?

11 A model engine is 20 cm long. The scale factor for the real engine is 250. How long is the real engine: **a** in cm **b** in metres?

EXERCISE 1B

1 On squared paper draw:

a enlargements of the sides of these shapes, using a scale factor of 1.5

b reductions of the sides of these shapes, using a scale factor of $\frac{1}{2}$.

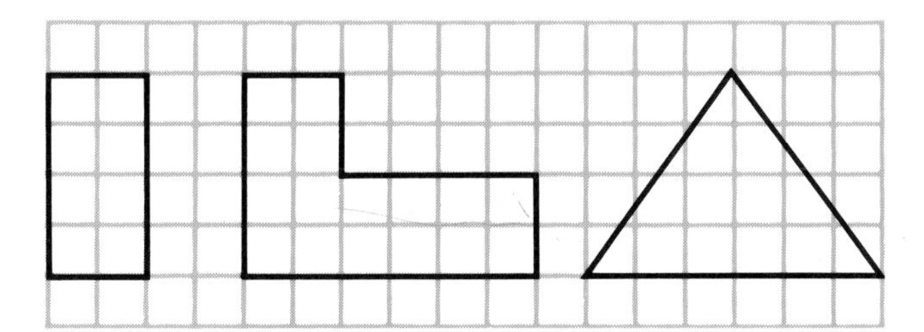

2 A company makes standard and junior playing cards. The junior pack is a scaled-down version, with scale factor 0.4. Calculate the lengths of the diagonals of the diamond (rhombus) at the centre of the junior ace.

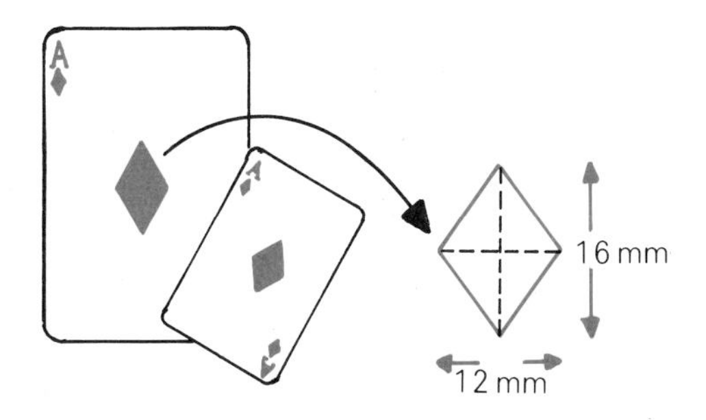

3 Mr Campbell makes two kites, one for himself and a scaled-down version for his daughter Toni. The diagonal supports of his kite are one and a half times as long as Toni's.

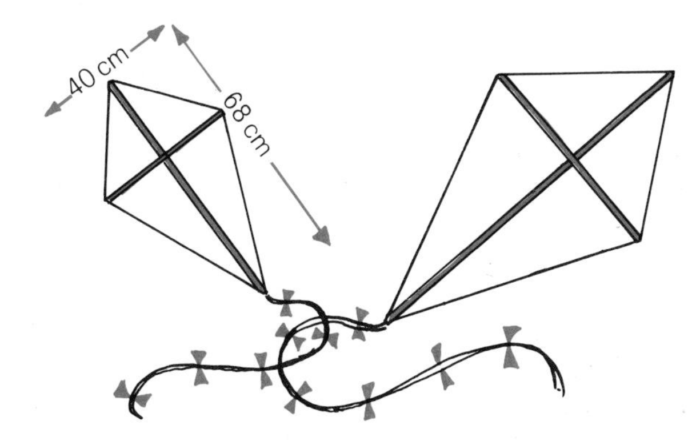

a What is the scale factor of the enlargement?

b Calculate the lengths of the diagonals of his kite.

4 One of these rugs is an enlarged version of the other.

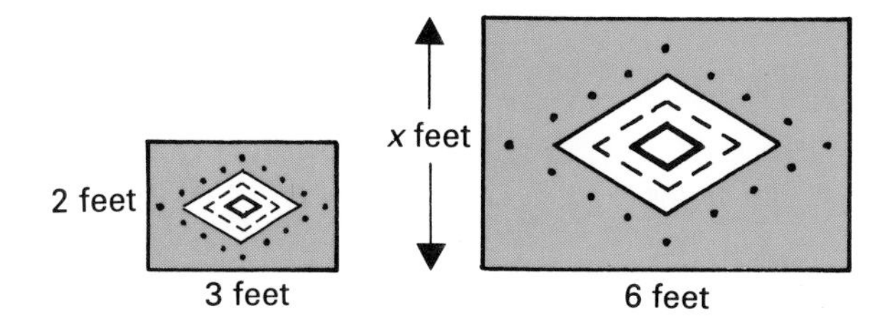

a To find x, should you enlarge or reduce?

b What is the scale factor?

c Calculate x.

5 At Tom's school the indoor football pitch is a reduction of the outdoor pitch.

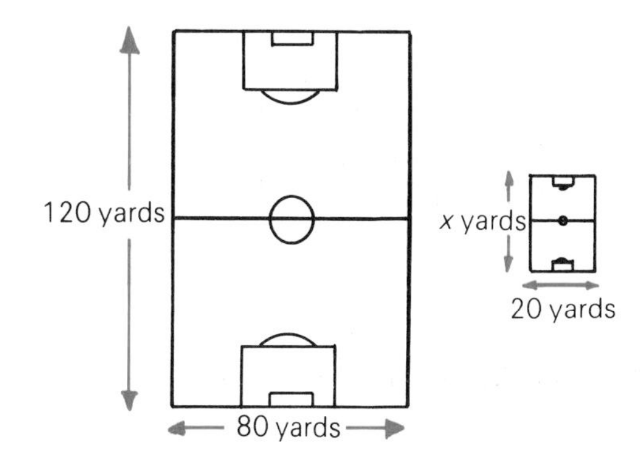

a To find x, should you enlarge or reduce?

b What is the scale factor?

c Calculate x.

6 The Clear Image Co. sell two sizes of mirror, one an enlargement of the other. Calculate:

a the scale factor of the enlargement

b x.

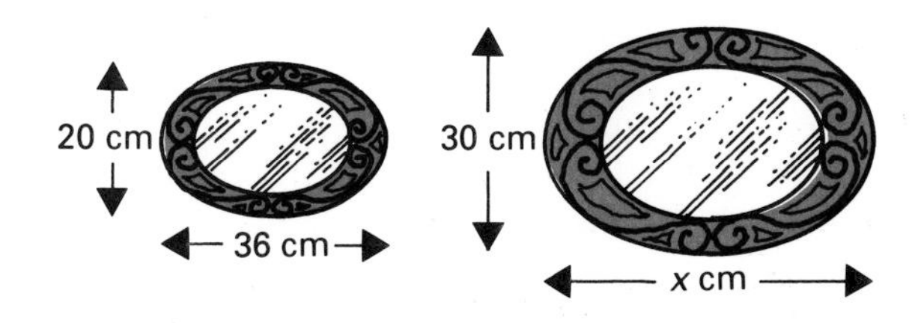

7 a The enlargement scale factor from a shape A to a shape B is $\frac{4}{3}$. What is the reduction scale factor from B to A?

b Repeat **a** for an enlargement scale factor of 1.25.

PRACTICAL PROJECT

Try one of these.

1 *Choose a suitable scale and make a scale drawing of the view from above (the plan view) of one of these:*

2 *Choose a building and make a scale model, perhaps:*

3 *As stage manager you have to make 'props' for this year's school play, Tom Thumb. To make the actors look small you have to enlarge the props! Here are some suggestions of objects to enlarge:*

What scale factor would you use to enlarge them?

SPOTTING SIMILAR SHAPES

CLASS DISCUSSION

Can you think of other everyday uses of the word 'similar'? Would you agree that to be similar, two people or things or ideas must have something in common?

Two objects are similar if they have the same shape, so that one is an enlargement of the other. If they have the same shape and size, they are congruent.

EXERCISE 2

1 Four of these Russian doll cut-outs are similar. Which is the odd one out?

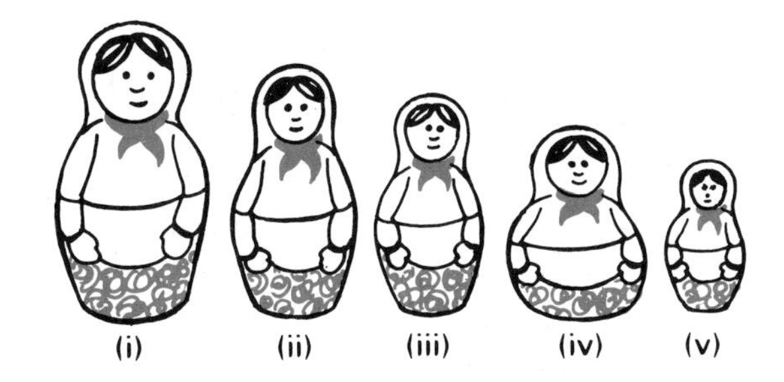

2 Three of these spanners have a similar outline. Each is an enlargement of the previous one.

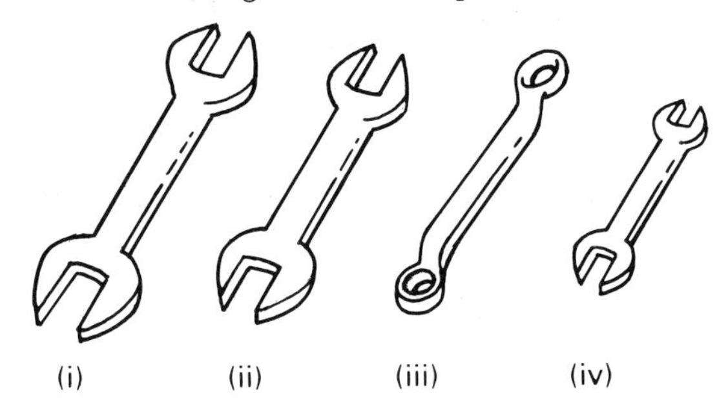

a Which is the odd one out?
b Which is congruent to the one below?

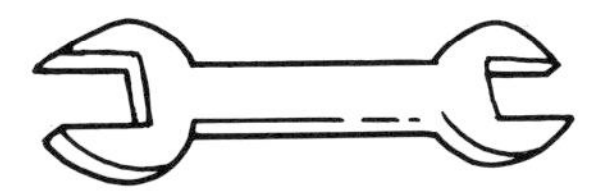

3 a Which of these traffic signs have congruent outlines?

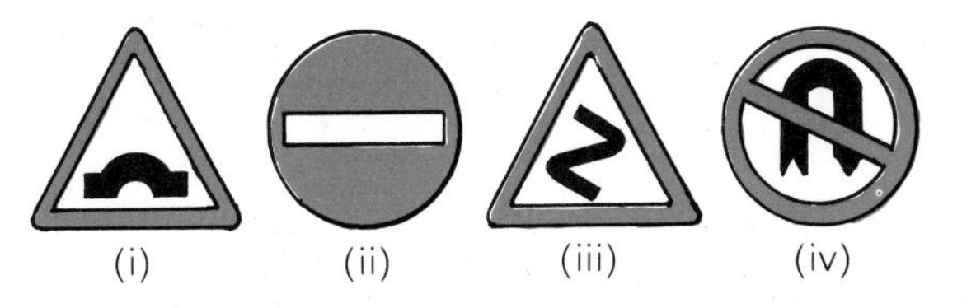

b What do the different *outlines* tell a road-user?

4 Jack is having fun in the Hall of Mirrors. Three of his images are similar. Which three?

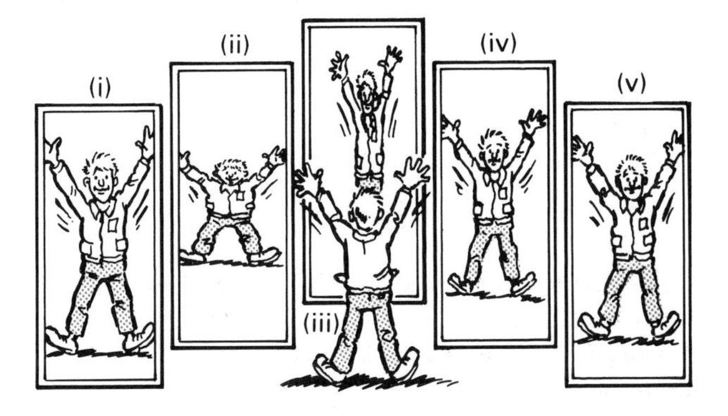

5 Dominic and his four friends win a five-a-side competition. There is a team shield and a small replica shield for the winners. The team shield is an enlargement of the small shields. Which shields are: **a** congruent **b** similar?

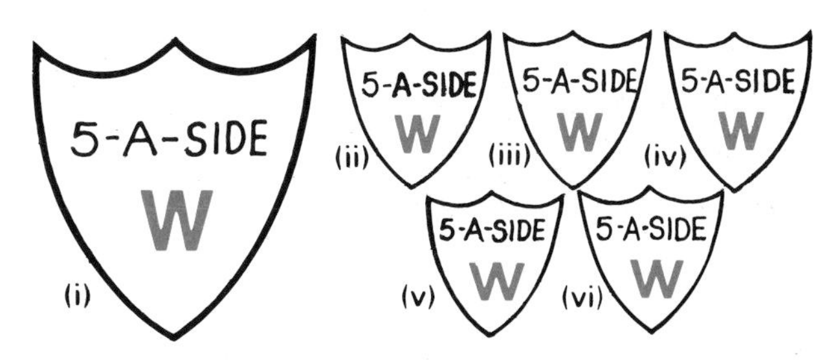

6 The top of each tier of the wedding cake is circular and each tier is a scaled up or down version of the next one, so the tops are all similar to each other. The scale factor of the reduction from (i) to (ii) is $\frac{2}{3}$, and from (ii) to (iii) $\frac{3}{4}$. If the diameter of (i) is 36 cm, calculate the diameters of (ii) and (iii).

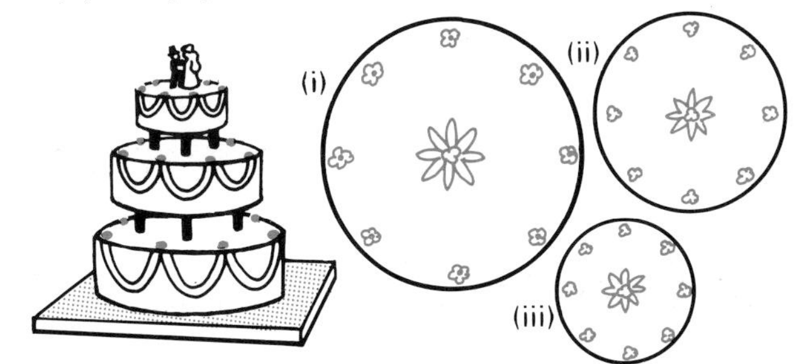

7 Are all circles: **a** similar **b** congruent? Give reasons for your answers.

A PUZZLE PICTURE

a *There are several sets of 'similar' objects hiding in the kitchen. How many can you find? Write them down.*
b *Can you think of any more sets of similar objects that you might find at home?*

CALCULATIONS WITH SIMILAR SHAPES

In geometry, two shapes are similar if one is a **scaled version** of the other.
This means that:
(i) **their corresponding angles are equal**
(ii) **their corresponding sides are in the same ratio** (have the same **scale factor**).

Example
Are the outlines of the two dominoes similar?

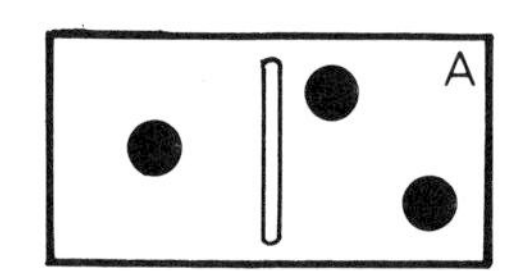

(i) *Check the angles.* All right angles, so corresponding angles are equal.
(ii) *Check ratios of sides.*

$$\frac{\text{Length of A}}{\text{Length of B}} = \frac{30\text{ mm}}{20\text{ mm}} = \tfrac{3}{2} \text{ or } 1.5.$$

$$\frac{\text{Breadth of A}}{\text{Breadth of B}} = \frac{15\text{ mm}}{10\text{ mm}} = \tfrac{3}{2} \text{ or } 1.5. \text{ So equal ratios (scale factor 1.5).}$$

The outlines *are* similar.

EXERCISE 3A

1 Newtown High School's photographer offers two sizes of photograph. Check if they are similar.

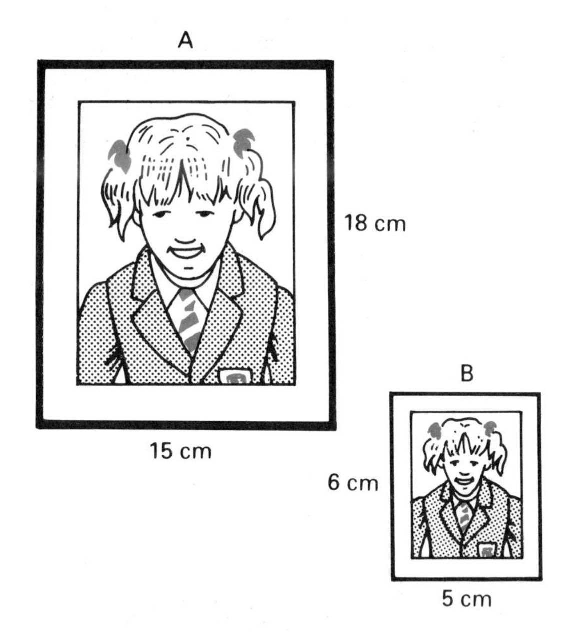

a First check corresponding angles.
b Then copy and complete:

(i) $\frac{\text{Height of A}}{\text{Height of B}} = \frac{\dots}{\dots} = \dots$

(ii) $\frac{\text{Width of A}}{\text{Width of B}} = \frac{\dots}{\dots} = \dots$

Are they similar?

2 Mike compares the school's 11-a-side and 5-a-side goals. Are they similar?

A
8 feet
24 feet
B
4 feet
12 feet

a Check corresponding angles.
b Copy and complete:

(i) $\frac{\text{Height of A}}{\text{Height of B}} = \frac{\dots}{\dots} = \dots$

(ii) $\frac{\text{Width of A}}{\text{Width of B}} = \frac{\dots}{\dots} = \dots$

Are the goals similar?

3 According to Marie, the school hockey pitch measures 54 yards by 90 yards. A full-size pitch is 60 yards by 100 yards. Are the two pitches similar? Explain.

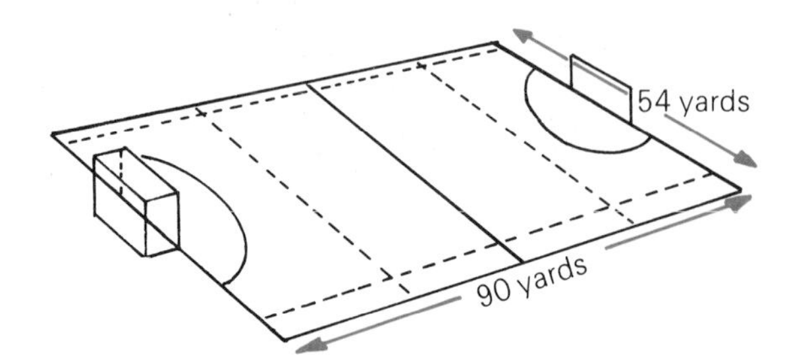

4 Measure the lengths and breadths of the windows and doors in **a** and **b**, and the fronts of **c** and **d**. Check angles and scale factors (ratios of corresponding sides) to decide which pairs of outlines are similar.

a

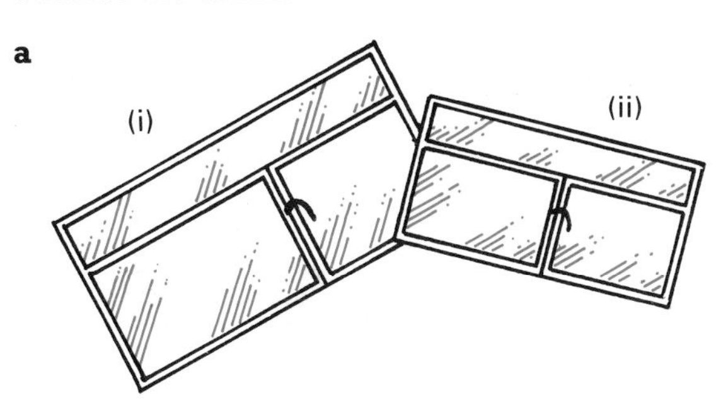

b

c

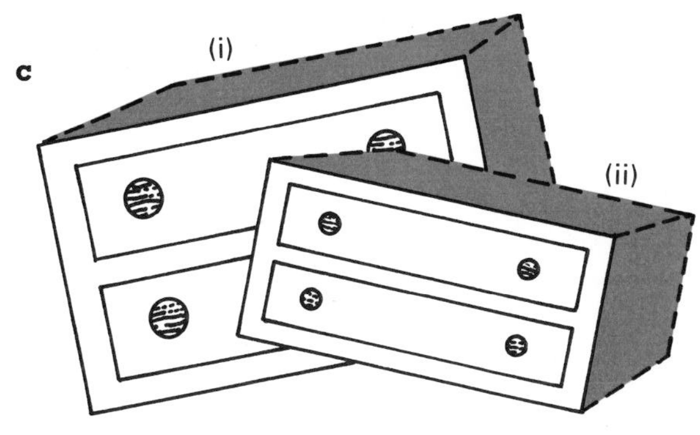

d

5 Are the inside and outside of the picture frame similar? Remember to check angles and ratios of sides.

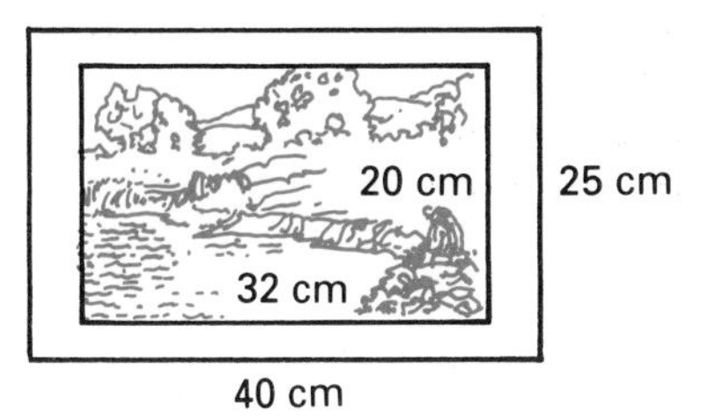

6 Is the goal area similar to the penalty area of this football pitch? Explain.

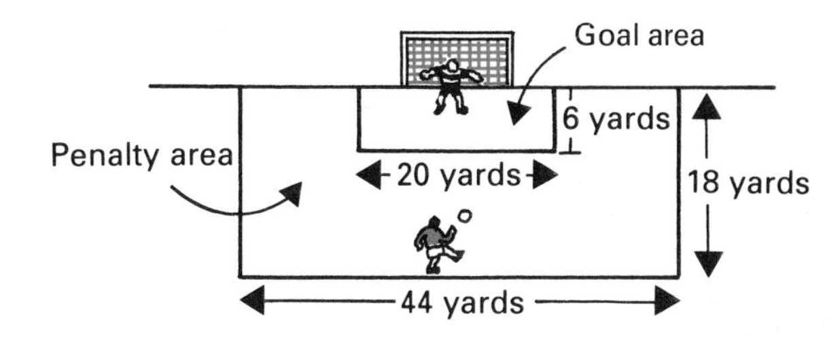

7 This toy car is $\frac{1}{6}$th life size.

a A real tyre has diameter 72 cm.
What is the diameter of a tyre on the model?

b The rear wheel has 20 equally spaced spokes.
What is the size of the angle between two spokes on:
(i) the real car (ii) the model?

8 The tyres on this toy tractor are similar, and the lengths are in millimetres. Calculate:

a the reduction scale factor

b x.

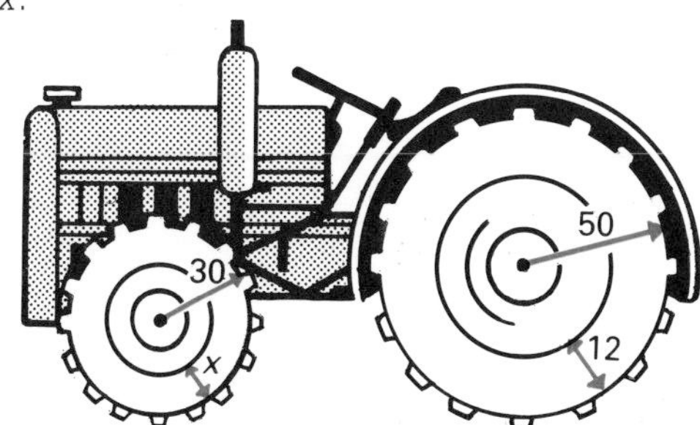

EXERCISE 3B

Example
The flat rectangular screens on the TV sets are similar. The dimensions are in centimetres.
Calculate:
a the scale factor
b the width of the smaller screen.

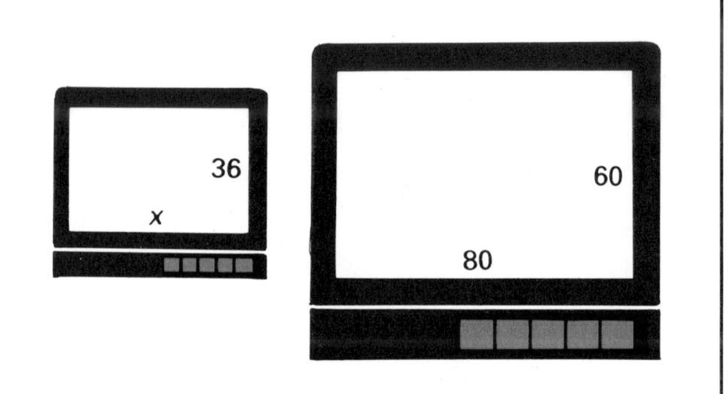

a The ratio of heights $= \frac{36 \text{ cm}}{60 \text{ cm}} = \frac{3}{5}$.

The scale factor is $\frac{3}{5}$.

b So $x = \frac{3}{5} \times 80 = 48$.

The smaller screen is 48 cm wide.

1 These two photographs are similar. One is an enlargement of the other. Calculate:
a the scale factor
b x.

2 The two bars of chocolate are similar. Calculate:
a the reduction scale factor
b x.

3 The tops of the two snooker tables are similar.
Calculate:
a the reduction scale factor
b x.

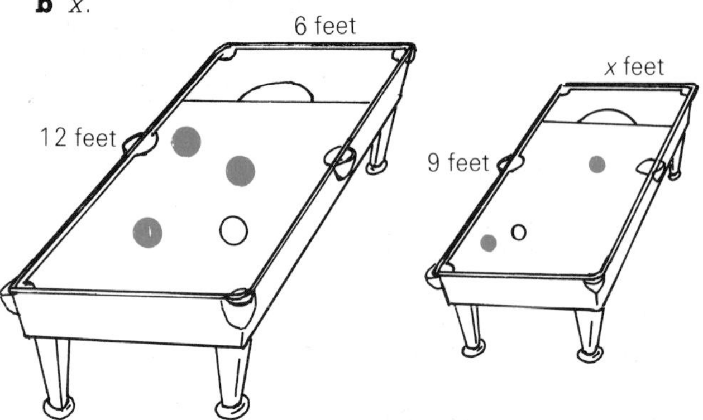

4 Maggie is making a picture frame. She wants the inside of the frame to be similar to the outside.
Calculate:
a the reduction scale factor
b x.

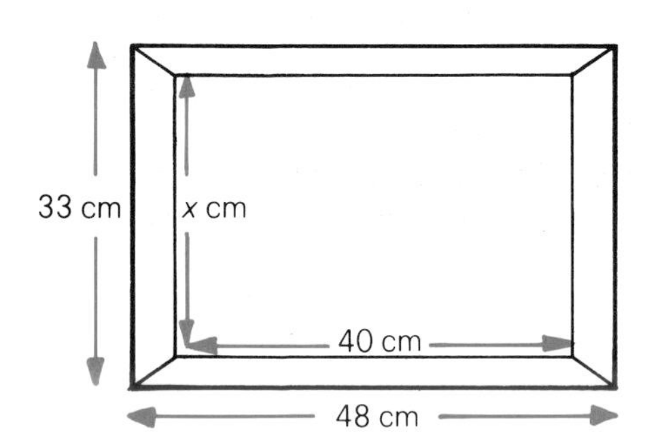

5 These two cottages are similar. Calculate:
a the reduction scale factor
b x **c** y.

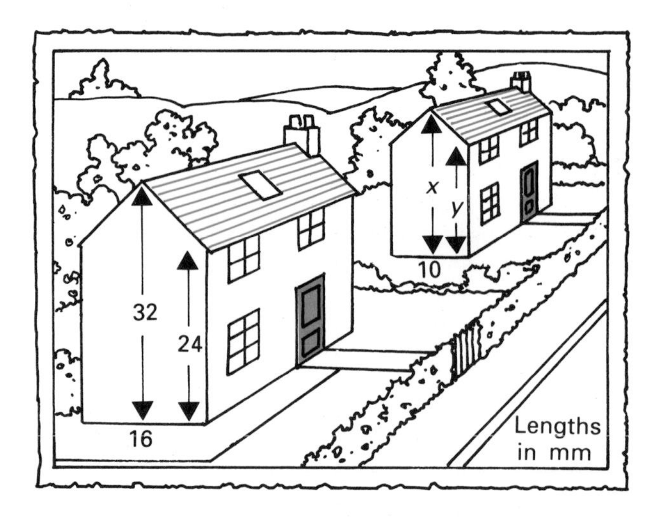

6 The flying ducks on Fenella's living-room wall are similar.

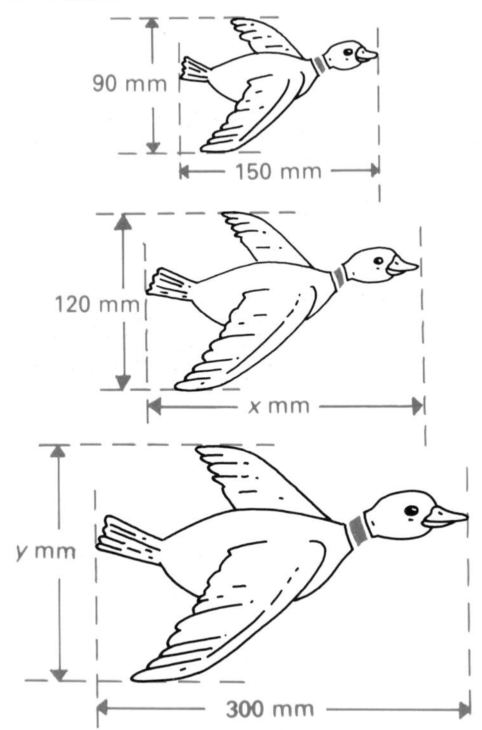

a By first calculating the scale factor, find: (i) x (ii) y.
b The smallest duck's beak is 30 mm long. Calculate the lengths of the beaks of the other ducks.

7 a What are the ratios of the corresponding sides of the two squares?
b Are the squares similar?

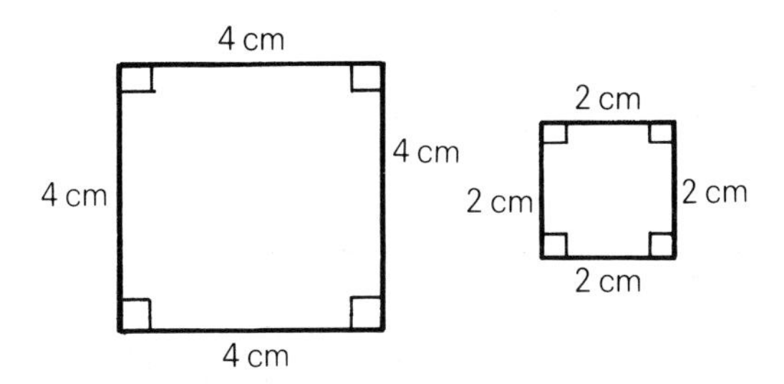

8 a What is the ratio of each side of the square to the corresponding side of the rhombus?
b Are the shapes similar? Give a reason for your answer.

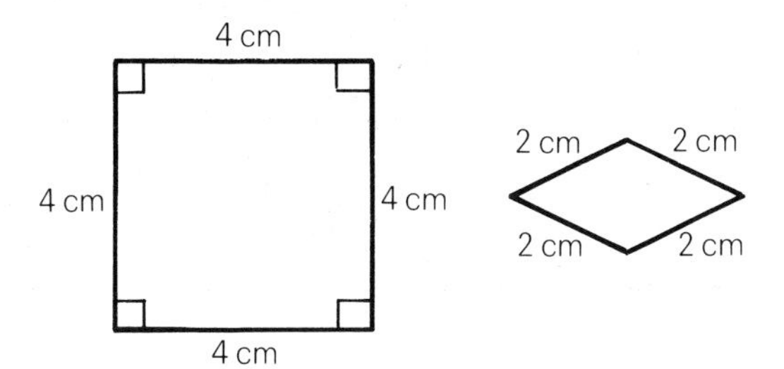

9 a The trapezium and the parallelogram have equal angles, but they are not similar to each other. Why not?

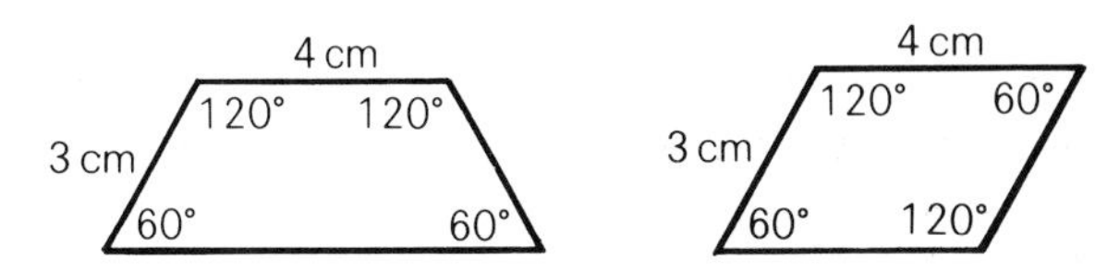

b Sketch a trapezium which *is* similar to this trapezium, and mark the lengths of its sides.
c Sketch a parallelogram which *is* similar to this parallelogram, and mark the lengths of its sides.

BRAINSTORMERS

1 *Val makes sets of three hand towels and one bath towel from strips of material.*

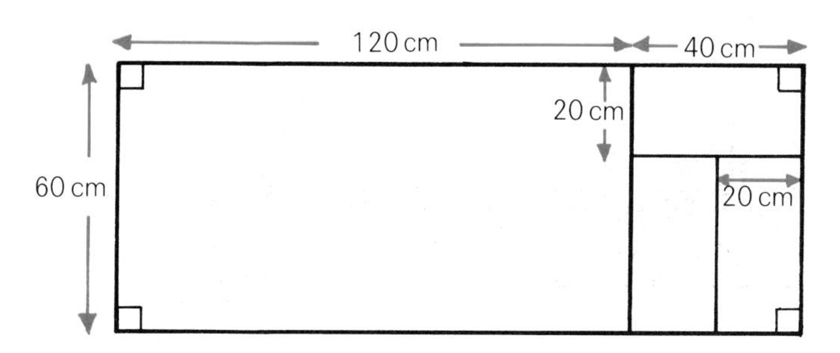

a *Are the hand towels congruent?*
b *Is the bath towel similar to a hand towel?*

2

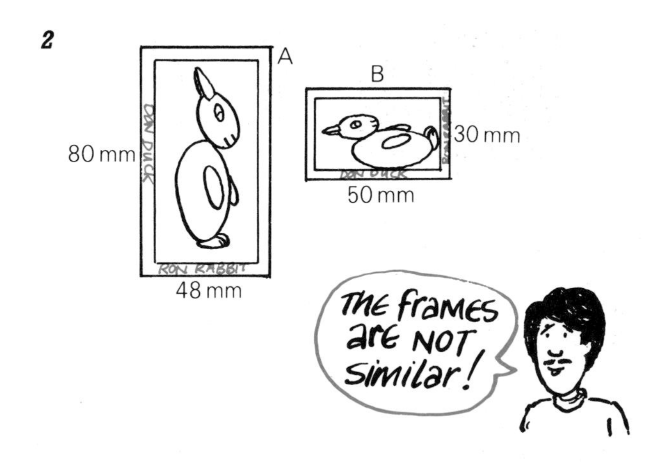

Alan calculates the ratios:

$$\frac{\textit{width of A}}{\textit{width of B}} = \frac{48\ mm}{50\ mm} = 0.96$$

$$\frac{\textit{height of A}}{\textit{height of B}} = \frac{80\ mm}{30\ mm} = 2.67$$

Is he correct? Explain.

PRACTICAL PROJECTS

1 *Measure the lengths and breadths of the fronts of a video tape case and a music cassette case. Are they similar?*

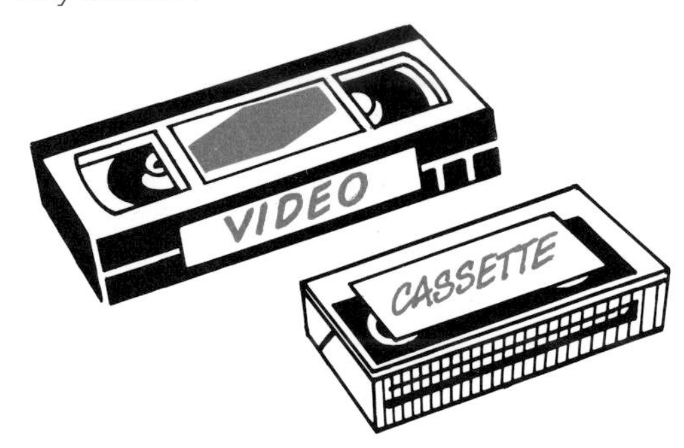

2 a *Measure the heights and widths of different television screens.*

b *Calculate $\frac{h}{w}$ for each screen.*

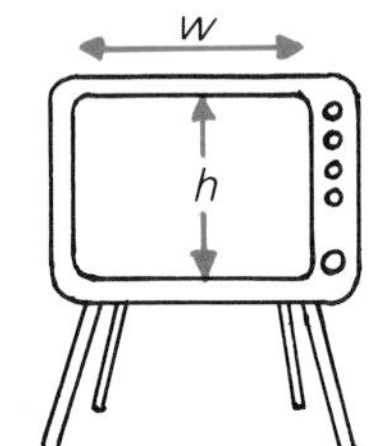

c *Are the screens all similar? Are you surprised by the answer?*

INVESTIGATION

Perspective drawing, as used by artists

a *Draw three shapes like those shown and choose a suitable point P.*

b *Draw lines, called rays, from each vertex to the point P.*

c *At some place nearer to P draw the shapes again, making sure that corresponding sides are parallel and vertices lie on the rays.*

d *Rub out the edges of the solids you have drawn that cannot be seen.*

e *Thicken the remaining lines.*

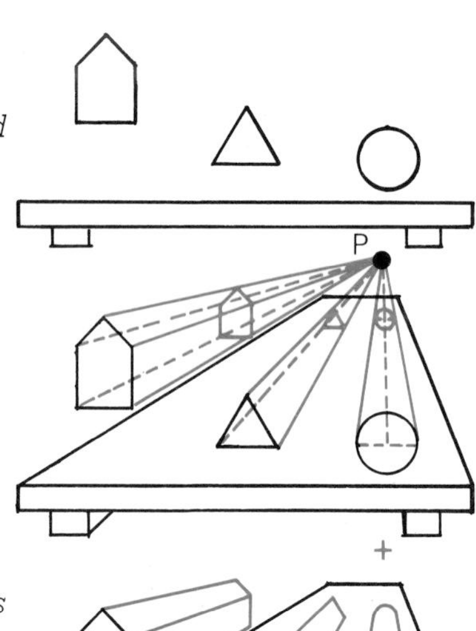

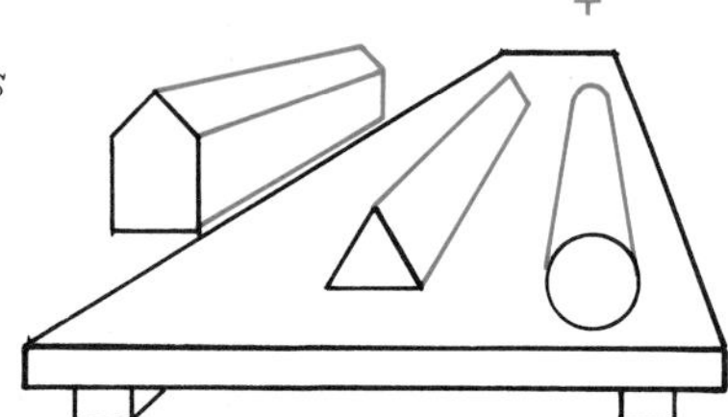

Investigate the connection between perspective drawing and similarity.

SIMILAR TRIANGLES

CLASS DISCUSSION

1 In diagram (i):

a what is true about pairs of angles in corresponding positions?

b what is the ratio (scale factor) of each pair of corresponding sides?

c are the two triangles similar?

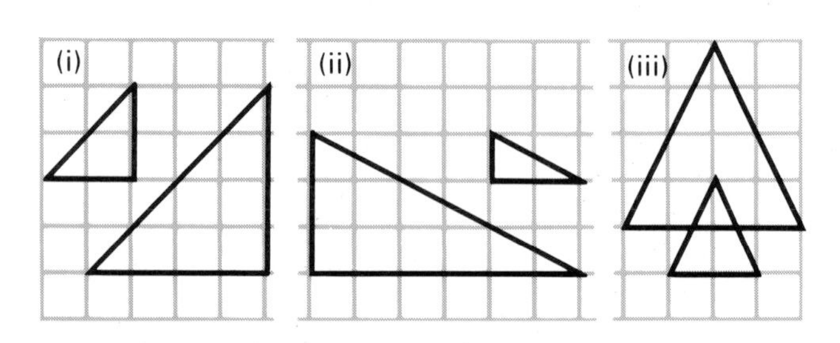

2 Repeat question **1** for diagrams (ii) and (iii).

Triangles are special.
If pairs of corresponding angles are equal, then ratios of lengths of corresponding sides are equal, and the triangles are similar.
(The converse is also true. If the ratios of corresponding sides are equal, then pairs of corresponding angles are equal, and the triangles are similar.)

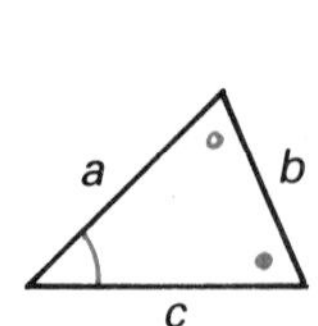

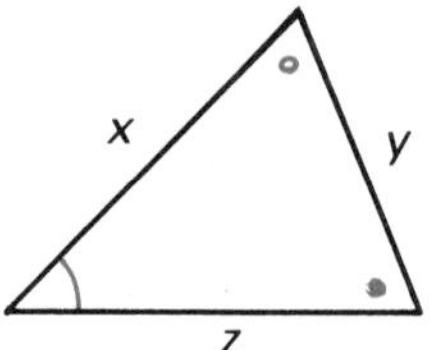

These triangles are equiangular. So their corresponding sides are in proportion (have the same ratio, or scale factor):

$$\frac{a}{x} = \frac{b}{y} = \frac{c}{z} = \text{scale factor}$$

EXERCISE 4A

1 Which of these pairs of triangles are similar?

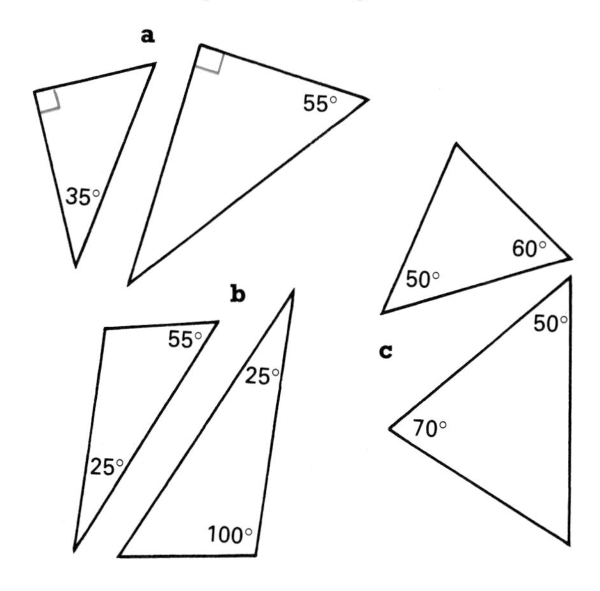

2 Ruth has a 30°, 60°, 90° set-square. Her teacher has a 30°, 60°, 90° blackboard set-square.

a Why are the two set-squares similar?

b The scale factor is 4. Calculate the lengths of the two sides about the right angle on the blackboard set-square.

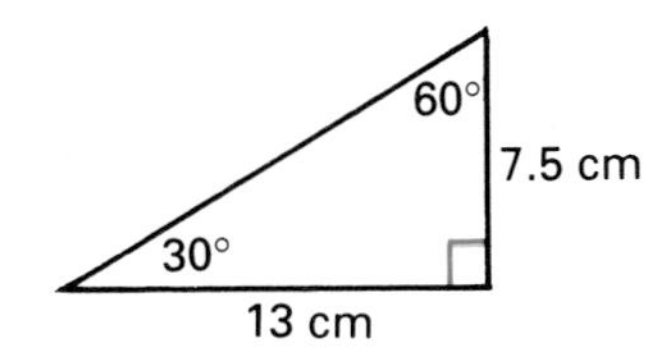

3 The triangular sides of the door wedges are equiangular. Calculate:

a their enlargement scale factor **b** x.

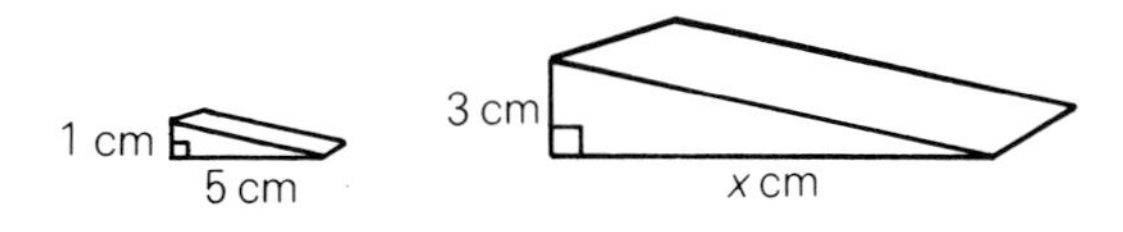

4 Mr Ford sells two sizes of equiangular brackets. Calculate:

a the reduction scale factor

b x.

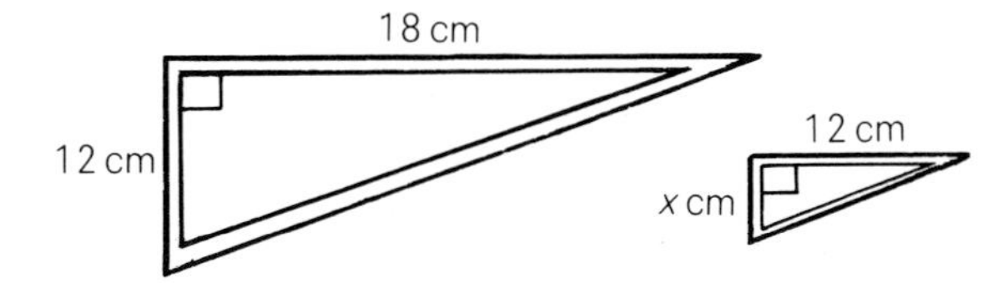

5 The corner shelves are equiangular. If AB = 20 cm, calculate the length of PQ.

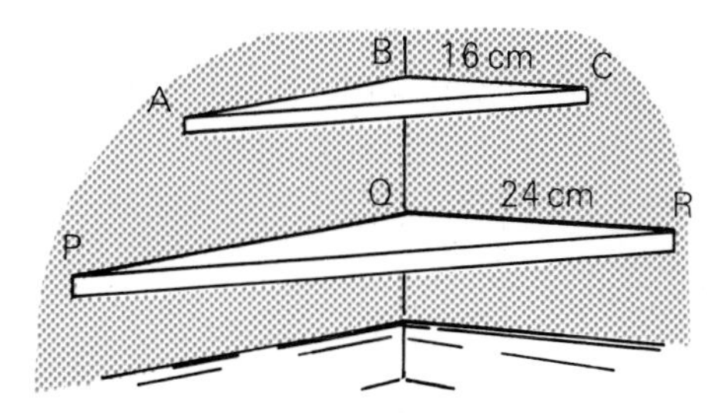

6 The Select Shed Company sells two different sizes of shed. The angle of slope of each roof is the same.

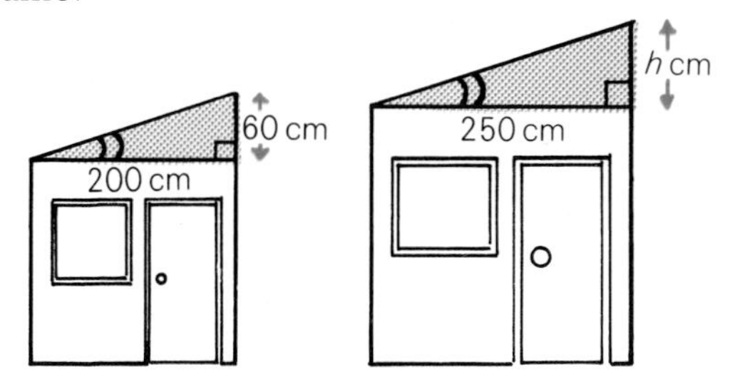

a Why are the shaded parts similar?

b Calculate h.

7 Calculate the height climbed up the hill by the car.

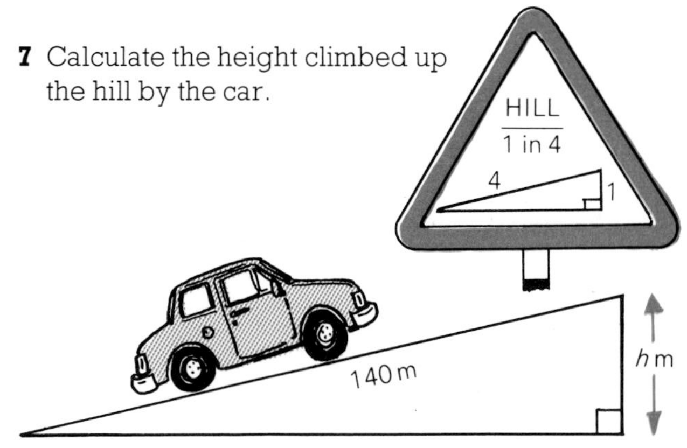

8 The angles ABC and DFE of the sails are equal.

a Why are the sails similar?

b Which edge of the smaller sail corresponds to edge EF of the larger?

c If EF = 4 m, calculate BC.

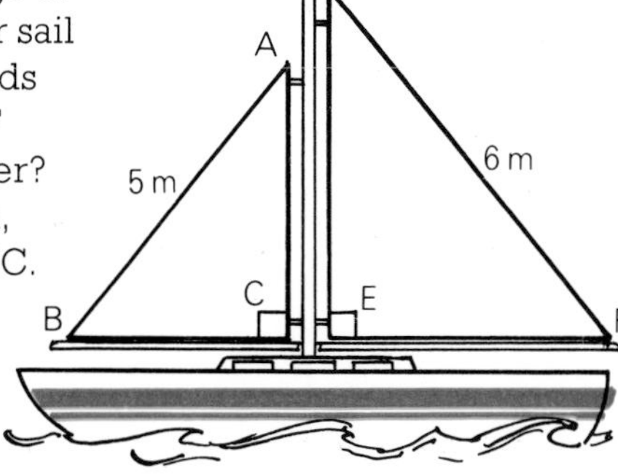

EXERCISE 4B

1 a Which pairs of triangles are equiangular, and therefore similar?

b Write down equal ratios of corresponding sides of the pairs of similar triangles. (Don't measure the sides—they are not accurate).

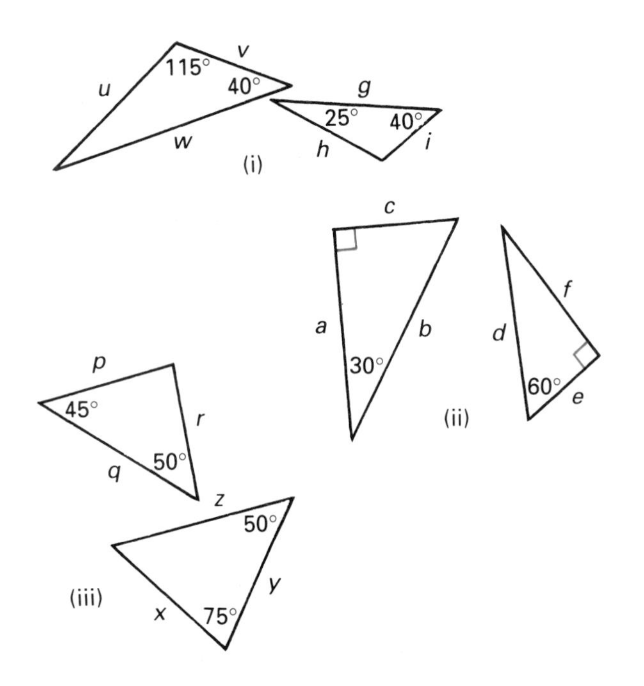

2 Which pairs of triangles have their corresponding sides in proportion, and are therefore similar? (Arrange the sides in order of size, for example, 9, 12, 15 and 6, 8, 10. Then calculate ratios.)

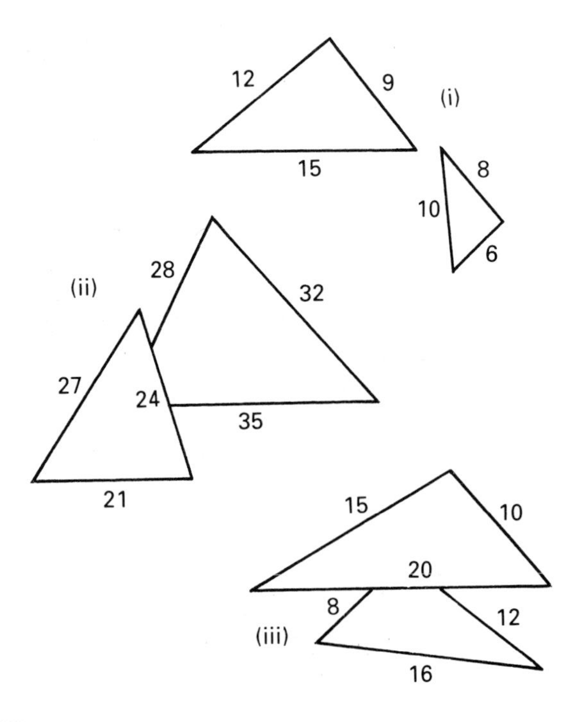

3

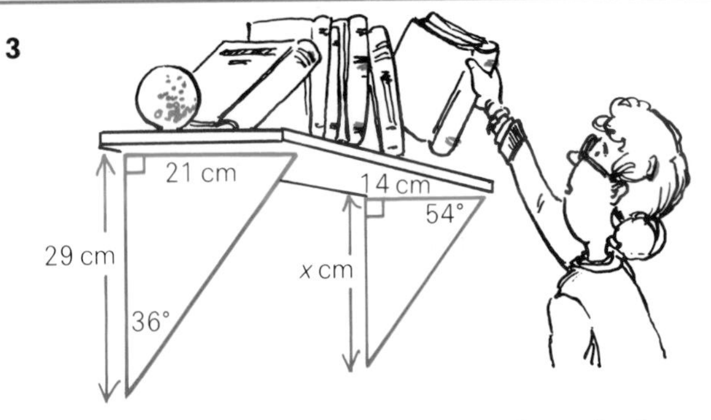

a Are the triangular brackets equiangular?

b Calculate x, correct to 1 decimal place.

4 a Are these triangular roofs equiangular?

b Calculate the length of DE.

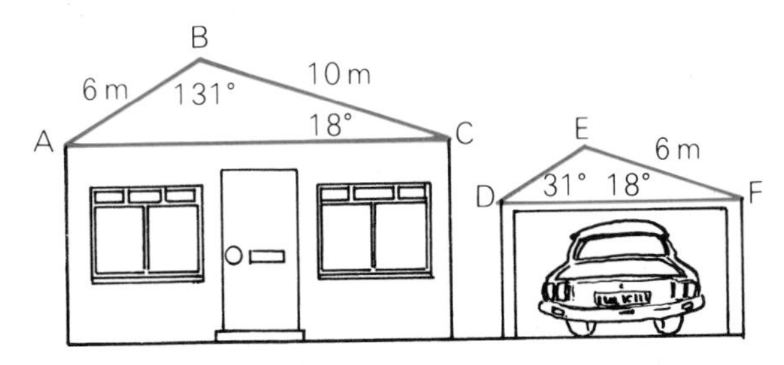

5 The triangular wing and tail are similar. Their lengths are in metres.
Calculate:
a x **b** y.

6 Both Tom and the tree cast long shadows in the sinking sun, making two similar triangles.

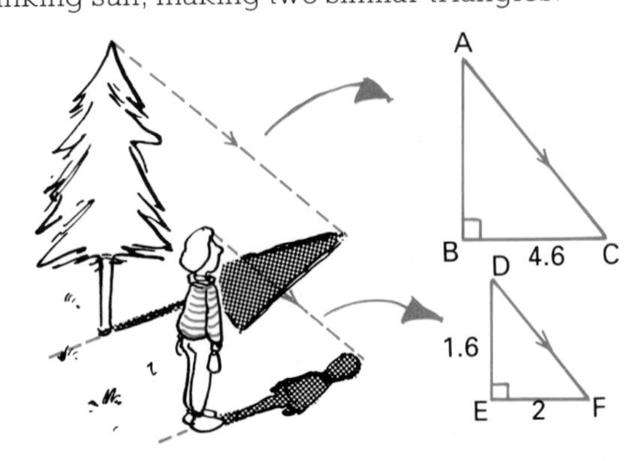

a Calculate the height of the tree. The lengths are in metres.

b Later, Tom's shadow is 3 m long. What length is the tree's shadow?

PARALLEL LINES IN TRIANGLES

BC is parallel to DE. So the corresponding angles are equal, and the $\triangle$s ABC and ADE are similar.
To calculate x, the reduction scale factor is $\frac{12}{15} = \frac{4}{5}$.

So, $x = \frac{4}{5} \times 10 = 8$

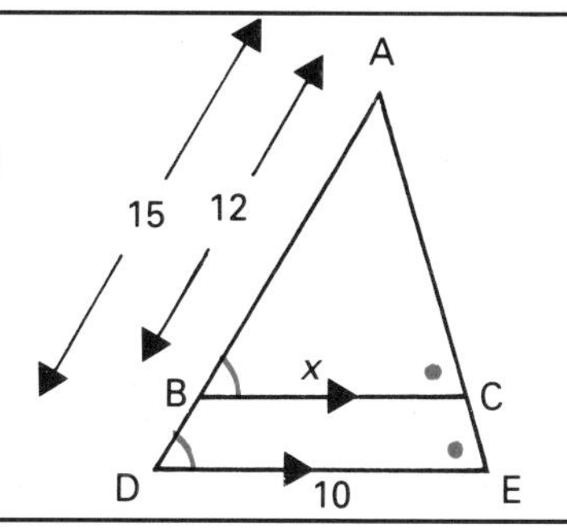

EXERCISE 5

1 In each diagram, calculate:
(i) the reduction scale factor or the enlargement scale factor to the triangle with side x
(ii) x.

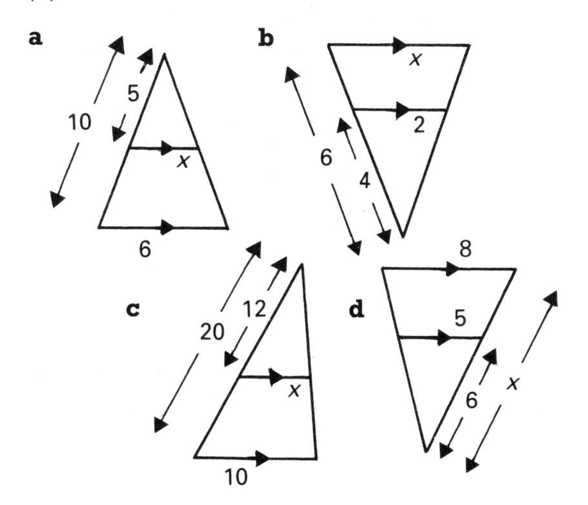

2 The Lucky Strike Gold Mine has parallel air shafts BD and CE.

a Calculate: (i) the reduction scale factor from $\triangle$ACE to $\triangle$ABD
(ii) the length of BD.

b AD = 54 m. Calculate the length, to the nearest metre, of: (i) AE (ii) DE.

3 DE and BC are horizontal. Calculate the length of:
a AC **b** EC.

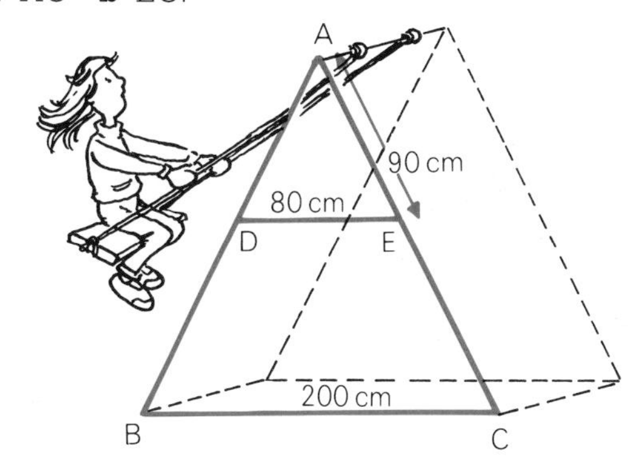

4 The observation platform at Corrieshalloch Gorge gives breathtaking views. AE and BD are horizontal.
a Calculate the height CE of the platform above the bottom of the gorge.
b If AC = 6.25 m, calculate the length of BC.

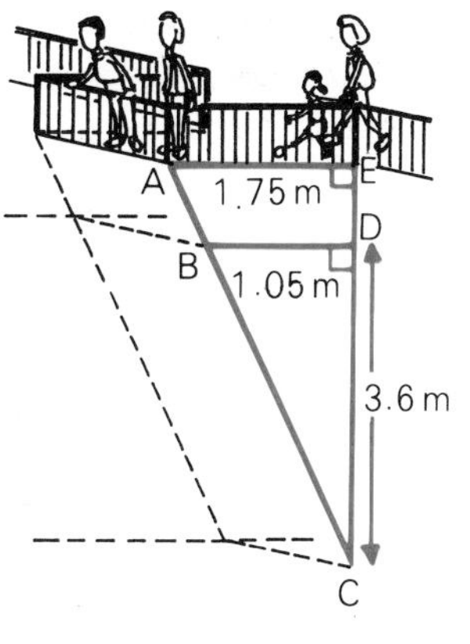

5 **a** Make a sketch of trapezium ABCD, and mark pairs of equal angles.
b Name a pair of similar triangles.
c Calculate the value of AP : PC.
d AC = 12 cm. Calculate the lengths of AP and PC.

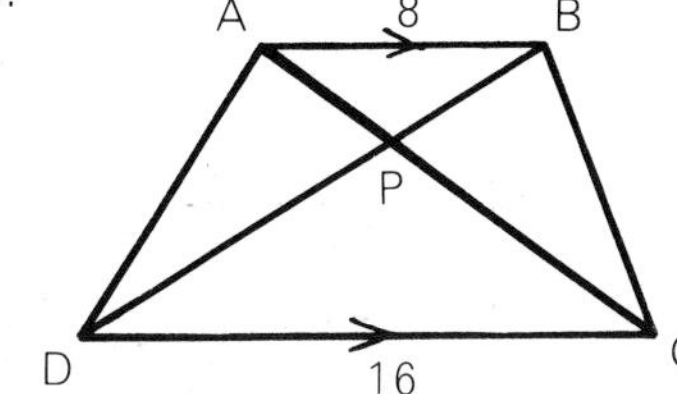

CHECK-UP ON SIMILAR SHAPES

1 On squared paper, enlarge the lengths of the sides of each shape by scale factor 2.

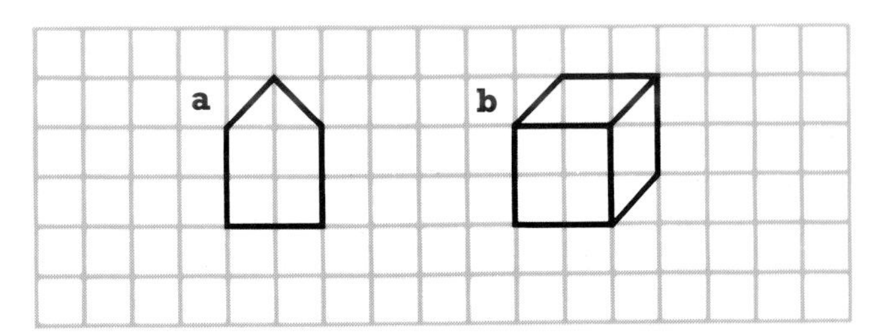

2 Calculate the scale factors, in decimal form, of:
- **a** the enlargements from flag (i) to flag (ii), and from flag (iii) to flag (iv)
- **b** the reductions from flag (iv) to flag (ii), and from flag (iii) to flag (i).

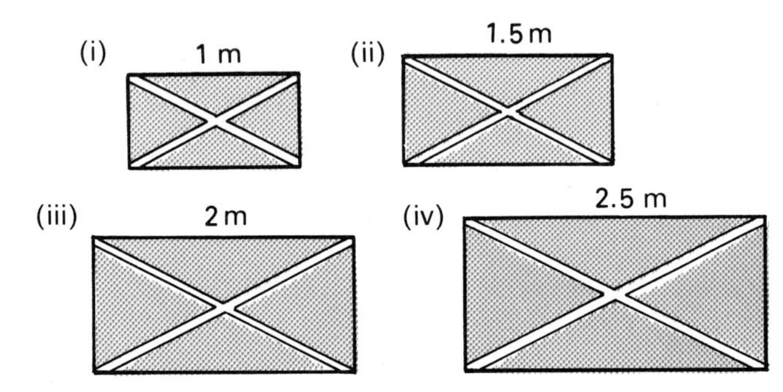

3 Eve sails her radio-controlled ferry in the local pond. Its scale factor is $\frac{1}{80}$.

- **a** If the ferry is 32 m long, what length is her model?
- **b** If the mast on her model is 10 cm high, what would the real height be?

4 One gate is a reduction of the other. Calculate:
- **a** the scale factor
- **b** x **c** y.

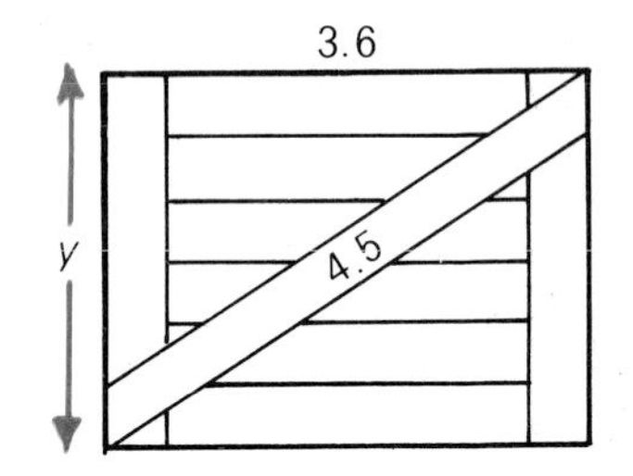

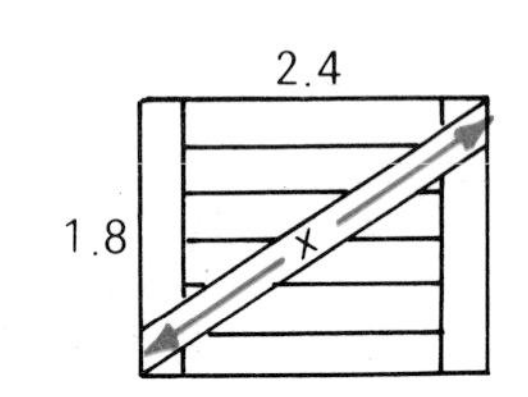

5

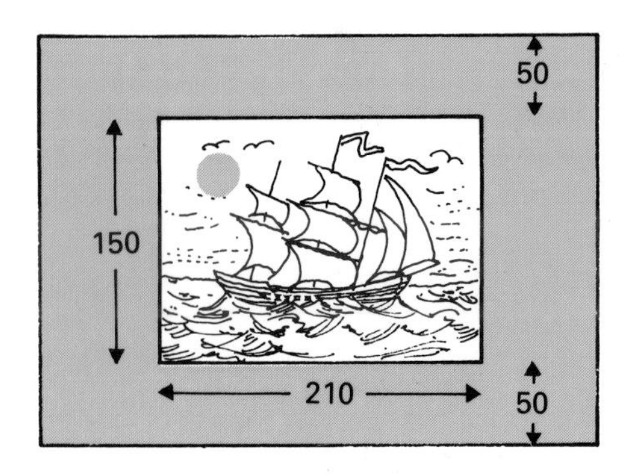

The outsides of the picture and the frame are similar. The lengths are in millimetres. Calculate:
- **a** the height of the frame
- **b** the scale factor for the enlargement
- **c** the breadth of the frame
- **d** the width of the left- and right-hand sides of the frame.

6 Which diagrams contain two similar triangles?

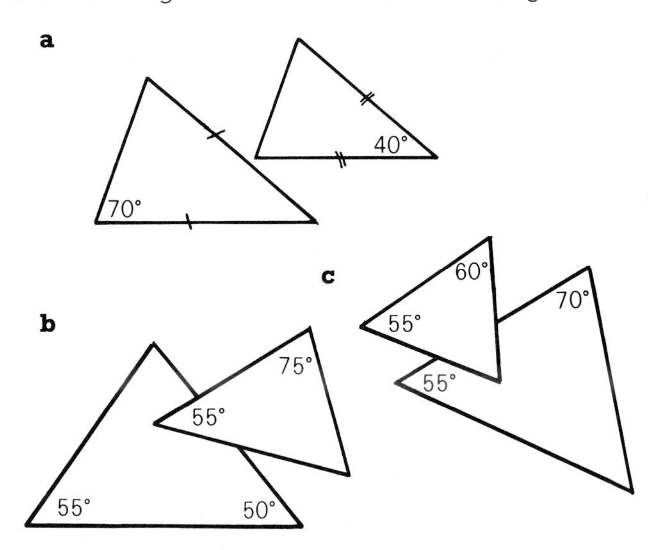

7 Calculate:
- **a** x
- **b** y

(The lengths are in metres.)

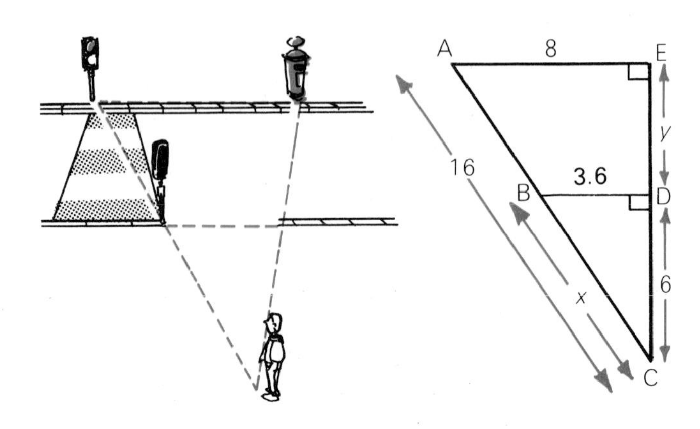

GOING PLACES

Have any of these recent record speeds been broken?

LOOKING BACK

1 This table shows the distances in straight lines, in kilometres, between the capitals.

Edinburgh			
224	Belfast		
490	391	Cardiff	
525	519	218	London

Write down the distance between:
a Edinburgh and London **b** Belfast and Cardiff.

2

How long does each of these bus journeys take?

	a	b	c	d
Depart	08 30	10 00	14 15	16 45
Arrive	09 45	11 20	15 25	18 05

3 Calculate each of these times as a decimal fraction of an hour. For example,
12 minutes = $12 \div 60$ hour = 0.2 hour.
a 30 min **b** 45 min **c** 24 min **d** 6 min

4

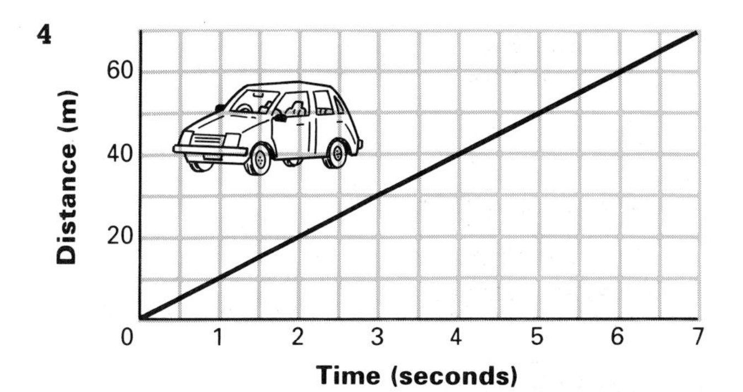

The graph shows the distance travelled by a car at a steady speed. Calculate:
a the distance travelled in (i) 2 s (ii) 7 s
b the time to travel (i) 10 m (ii) 40 m (iii) 90 m, at the same speed.

5 How long do these overnight journeys take?
a 5.55 pm to 3.19 am **b** 22 15 to 03 55 hours.

6 What is Dan's steady speed in miles per hour if he travels:
a 40 miles in 2 hours **b** 100 miles in 4 hours?

7 Estimate, then calculate, correct to: (i) the nearest whole number (ii) 2 decimal places (iii) 2 significant figures:
a 5.77×16 **b** $131 \div 19$

SPEED—HOW FAST?

Sanjay drives his van between Glasgow and London, a distance of 400 miles.
The journey takes 10 hours.

400 miles in 10 hours,
so 40 miles in 1 hour.

His speed is 40 mph, or is it?

This table shows the time he took to travel various sections of his journey. What is his speed for each part?

Distance (miles)	20	150	120	80	30
Time (hours)	1	3	3	2	1
Speed (mph)					

The speeds are all different. But the overall speed, or **average speed**, is 40 mph.

$$\text{Average speed} = \frac{\text{Distance}}{\text{Time}} \qquad \textbf{Formula: } S = \frac{D}{T}$$

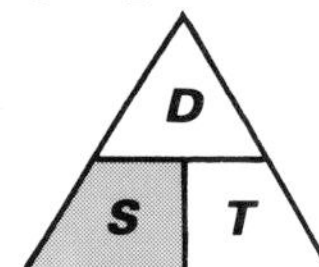

EXERCISE 1A

1 Calculate the average speed in km/h for each distance.

	a	b	c	d
Distance (km)	100	220	540	235
Time (h)	10	4	9	5

2 The best times recorded in a number of races at the school sports were:
a 100 m in 15 seconds
b 200 m in 36 seconds
c 1500 m in 5 minutes 30 seconds.
Calculate the average speed of each winner in m/s, correct to 2 decimal places.

3 Calculate the average speeds of these journeys:

a

b

4 Calculate the average speeds of these journeys. Don't forget the units: mph, km/h, m/s.
a 50 miles in 5 hours
b 300 miles in 10 hours
c 100 kilometres in 2 hours
d 4800 kilometres in 6 hours
e 50 metres in 10 seconds
f 5 metres in 10 seconds

5 Whales take 64 days to swim 6000 miles from the Mexican Coast to their feeding grounds in the Bering Sea. Calculate their average speed, to the nearest unit, in: **a** miles per day **b** mph.

6 Sanjay returned home from London to Glasgow by car in 8 hours. In the first 4 hours he drove 220 miles, and in the next 4 hours 180 miles. Calculate his average speed for:
a the first 4 hours
b the next 4 hours
c the whole journey.

7 Many aircraft can fly at speeds greater than the speed of sound.
Their speeds are often given as Mach numbers. Mach 2 means 'twice the speed of sound'.

$$\text{Mach number} = \frac{\text{speed of aircraft}}{\text{speed of sound}}$$

Calculate the Mach number, correct to 2 decimal places, for:
a a jet aircraft flying at sea-level (speed of sound 760 mph) at 1200 mph
b Concorde flying high (speed of sound 1050 km/h) at 2350 km/h.

EXERCISE 1B

1 Change the following into decimal fractions of an hour, correct to 2 decimal places if necessary.
For example, 43 minutes = $\frac{43}{60}$ hour = 0.72.
a 3 minutes **b** 40 minutes **c** 12 minutes
d 23 minutes **e** 6 h 35 min **f** 2 h 6 min

2 Action Rovers football supporters' coach travels 120 km to an away game in 1 hour 48 minutes. Calculate its average speed, to the nearest km/h.

3 Cardiff is 190 km from Exeter. The times taken for the journey by car, van and coach are:
Car 2 hours; van 3 hours 20 minutes; coach 2 hours 30 minutes.
Calculate the average speed of each vehicle.

4 An aircraft flies 1450 km in the first hour of its journey, 1650 km in the second hour and 500 km in the remaining quarter of an hour. Calculate:
a the total distance and time of flight
b the aircraft's average speed.

5 Training for the marathon, Linda ran 10 miles in 85 minutes one evening, 10 miles in 80 minutes the next evening and 6 miles in 40 minutes on the third evening. Calculate:
a the total distance she ran and the time she took
b her average speed over all three runs, correct to 1 decimal place.

6

	Depart	Distance from London (miles)
London (Euston)	09 45	0
Preston	12 05	209
Carlisle	13 15	299
Glasgow (Central)	14 45	401

a How long does the train take to travel from:
(i) London to Preston
(ii) Preston to Carlisle
(iii) Carlisle to Glasgow?
b What distance is each part of the journey in **a**?
c Calculate the train's average speed for each part, to the nearest mph.

7 Calculate the speeds of the following in km/h. Then arrange them in order, fastest first.

CHALLENGES

Clever conversions

1 km/h ↔ mph

a *Use this table, and the axes and scales shown, to draw the line joining the points (0, 0) and (80, 50).*

80 km/h ≑ 50 mph

km/h	0	80
mph	0	50

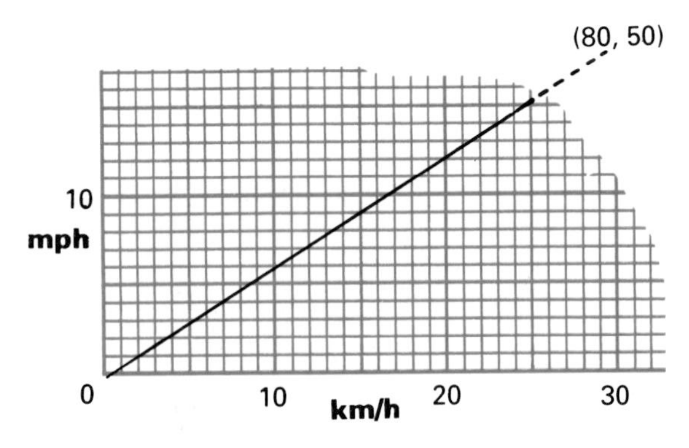

b *Use your graph to change these speed limits (mph) to km/h.*

(i) 

(ii) 

c *Use your graph to convert these speed limits (km/h) to mph (nearest unit).*

(i) 

(ii) 

2 m/s ↔ km/h

a *Study this flowchart to see how it works, then use it to change these speeds to km/h:*
(i) 10 m/s
(ii) 30 m/s
(iii) 150 m/s
(iv) 1 m/s.

b *Make a flowchart for changing km/h to m/s, and use it to change these to m/s, correct to 1 decimal place:*
(i) 72 km/h
(ii) 1 km/h.

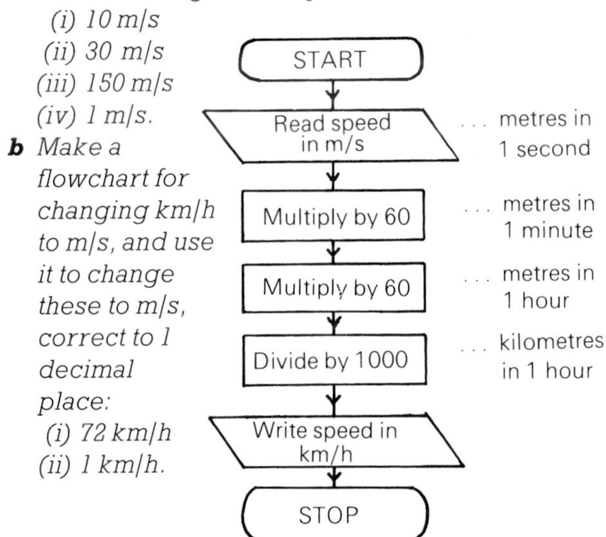

BRAINSTORMER

The Earth circles the Sun in 365 days at a distance of 9.3×10^7 miles. Show that there are 8.8×10^3 hours in 365 days, to 2 significant figures, and calculate the speed of the Earth in space, in mph.

INVESTIGATION

Windspeed is measured on the Beaufort Scale from 1 to 12. Investigate the windspeeds and related weather for the different numbers.

SPEED GRAPHS

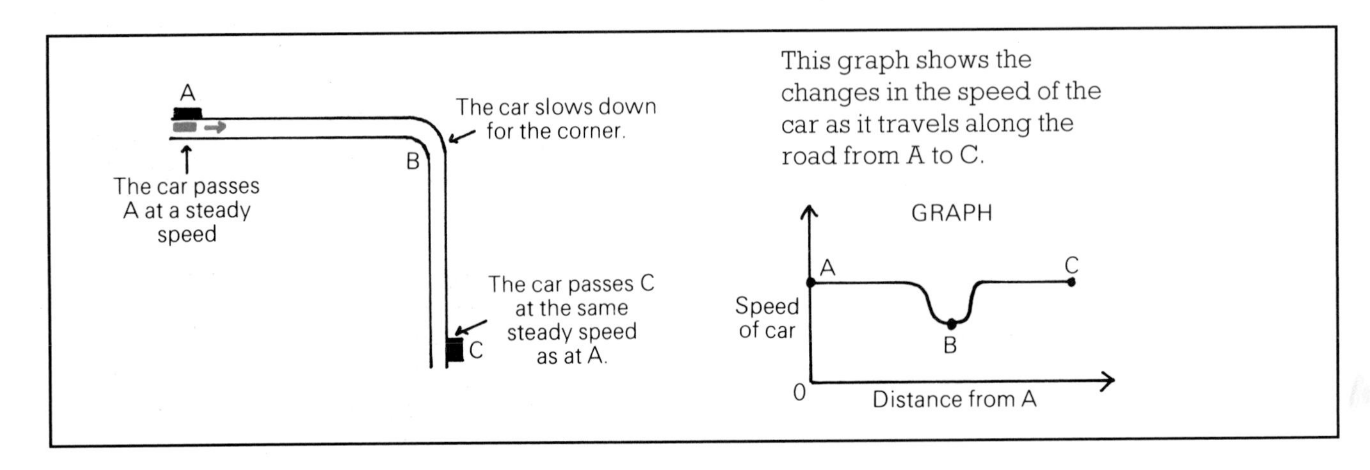

EXERCISE 2

In questions **1**—**5** copy and complete the graphs to show the changes in the speed of the car as it travels from A to C or D.

1

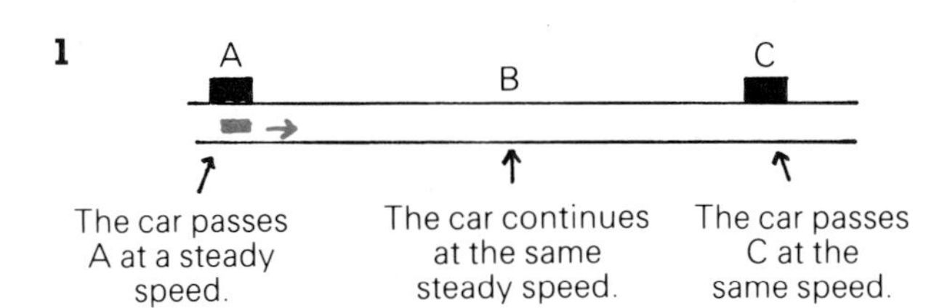

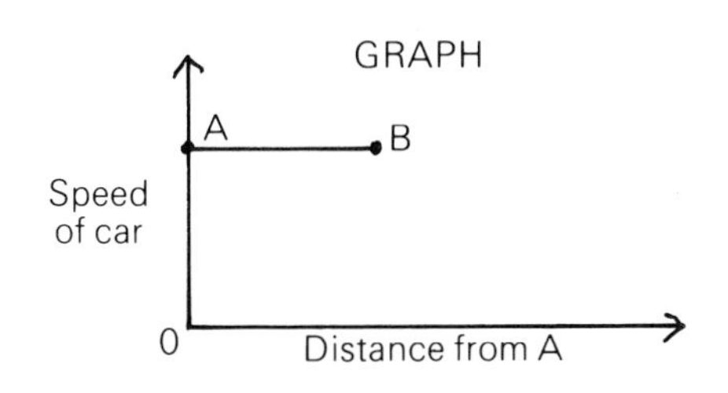

2

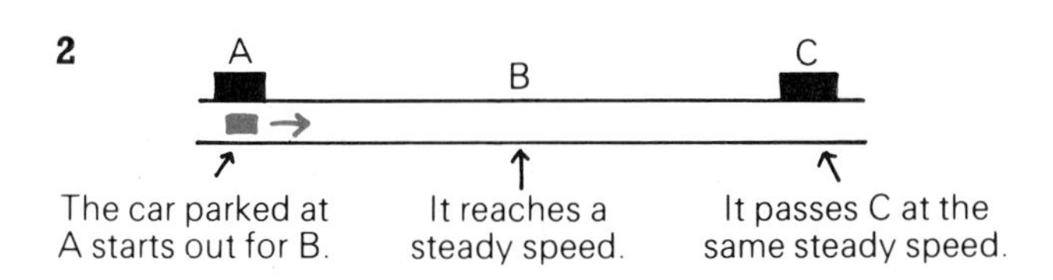

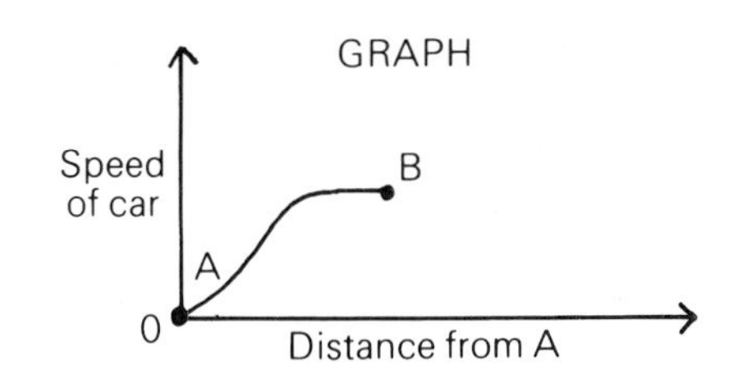

3

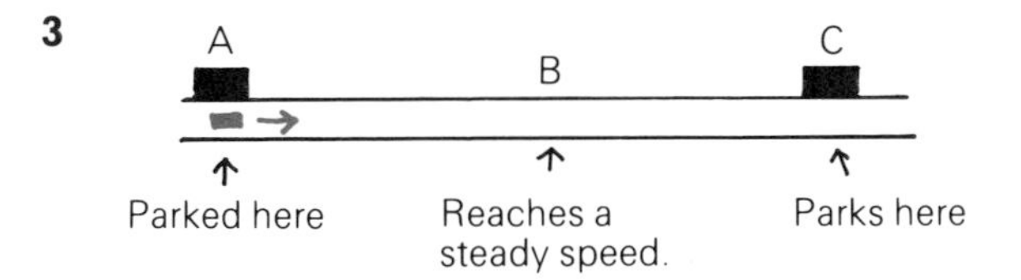

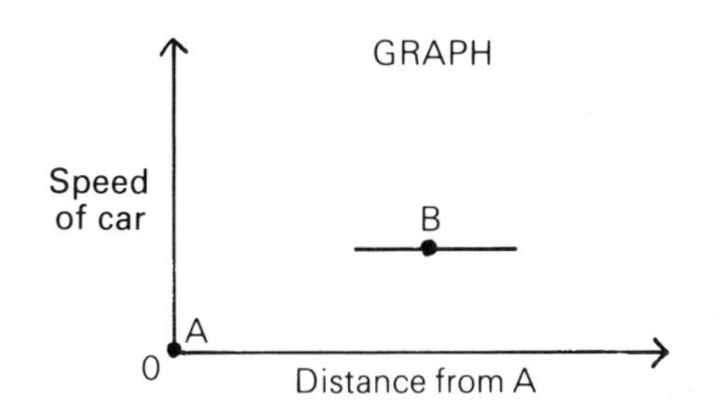

4

5

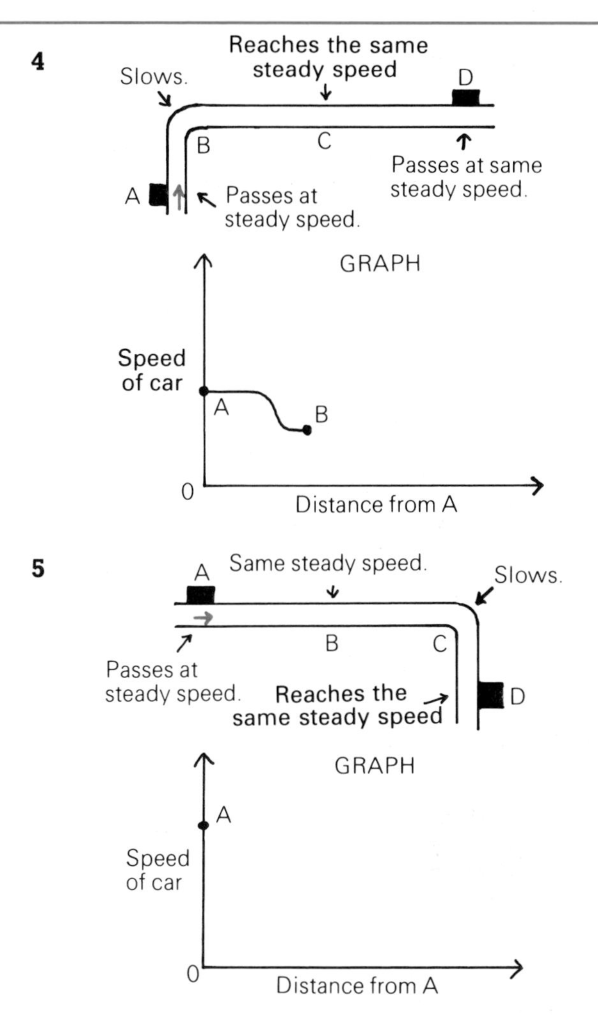

6 This graph shows the speed of a car as it travels from A to C. Can you give a reason for the shape of the graph?

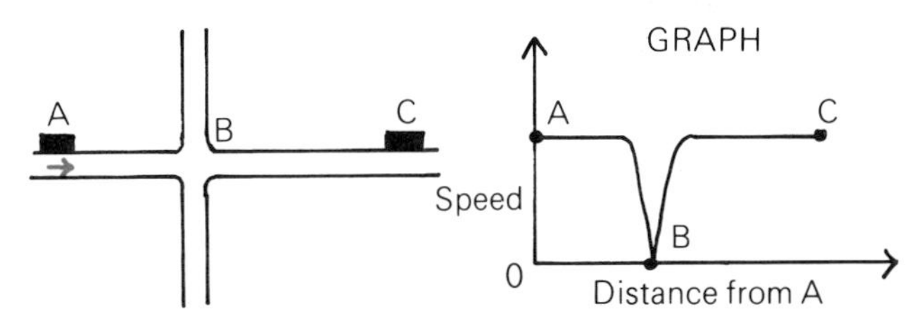

7 a There are two routes from A to D, one past B and the other past C. Which route does this graph show?

b Draw a graph for the other route, showing the speed of the car as it travels from A to D.

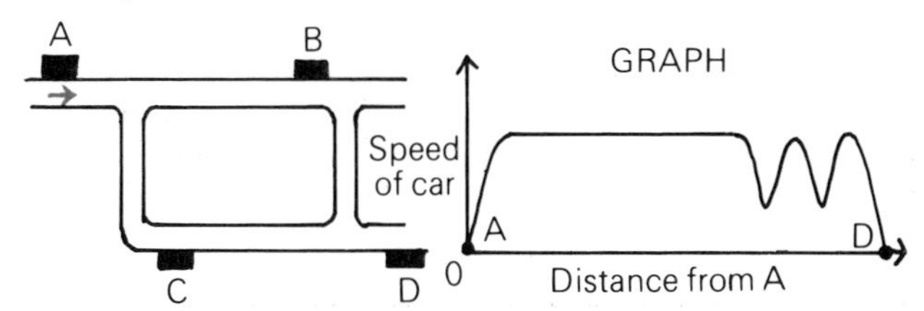

TIME—HOW LONG?

Try these.

a Distance 20 km
Speed 5 km/h
How long?

b Distance 50 miles
Speed 10 mph
How long?

$$\text{Time} = \frac{\text{Distance}}{\text{Speed}} \quad \textbf{Formula: } T = \frac{D}{S}$$

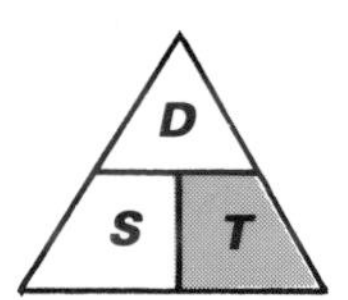

Example
Kate has to drive 450 km for a conference. She hopes to average 65 km/h. How long will the journey take?

$$\text{Time} = \frac{\text{Distance}}{\text{Speed}} = \frac{450}{65}\,\text{h} = 6.92\,\text{h} = 6\,\text{h}\ 55\,\text{min}\ (0.92 \times 60 = 55.2)$$

EXERCISE 3

1 How long does it take to travel:
a 150 miles at 25 mph **b** 1750 km at 350 km/h
c 28 miles at 8 mph **d** 100 m at 8 m/s?

2 The pilot of an aeroplane knows that he can fly 1890 km at an average speed of 840 km/h. How long will the flight take?

3 Ahmed jogs at a steady 12 km/h. How long, in hours and minutes, does he take to cover:
a 18 km **b** 6 km **c** 28 km?

4 How long for each journey in the table below? Estimate, then calculate!

	a	b	c
Distance	144 km	231 miles	1.5 m
Speed	48 km/h	42 mph	0.03 m/s

5 How long would you take to walk 1 km at a steady 8 km/h?

6 How long would these journeys take?

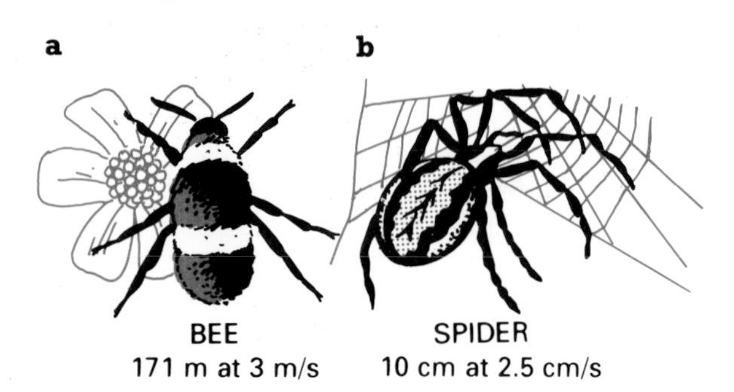

a BEE
171 m at 3 m/s

b SPIDER
10 cm at 2.5 cm/s

7 Mark's youth group plan a sponsored walk from Aberdeen to London, a distance of 900 km. They reckon that if they walk for 12 hours each day they can average 4 km/h.
a How many hours will they take?
b If they set out on Monday 10th at 9 am, on which day will they arrive?

8 On the race track, Susan cycles 4 km at an average speed of 24 km/h. How long does she take?

9 Peter thinks . . .

Can you help him out?

10 Miss Johnson is flying from Leeds to Belfast for a business meeting. The journey is 370 km. The plane averages 256 km/h. Will she be in time if she leaves home at 9 o'clock?

Home	09 00
Home to airport	1 hour
Belfast	?
Airport to meeting	20 minutes
Meeting	12 00

11 The speed of sound is about 340 m/s. Loud and clear, Andrea shouts 'Hi' towards a cliff 200 metres away. How long will it be before she hears an echo (to the nearest tenth of a second)?

12 Light travels at nearly 3×10^5 km/s. How long does light take to reach the Earth from the Sun which is 1.5×10^8 km away?

BRAINSTORMER

a *One train travelling at 112 km/h passes another going in the same direction at 85 km/h. Their lengths are 95 m and 70 m. How long does it take for the faster train to pass the other completely?*

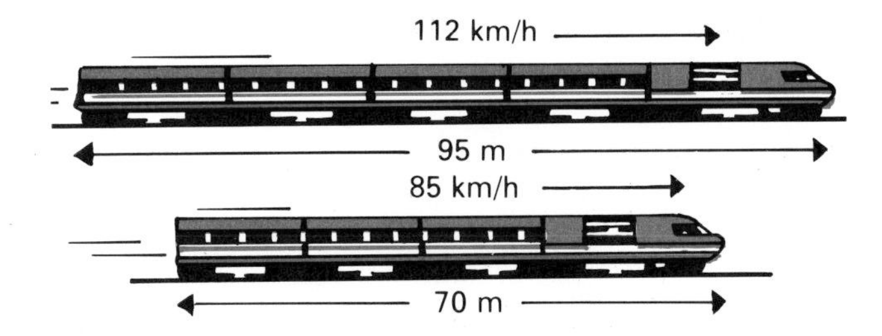

b *If the same trains were to pass each other in opposite directions, how long would it take for them to pass?*

DISTANCE—HOW FAR?

Try these.

a Speed 25 km/h
Time 2 h
How far?

b Speed 70 mph
Time 4 h
How far?

Distance = Speed × Time
Formula: *D* = *ST*

Example
Gregor thinks that he can keep up an average speed of 20 km/h on his mountain bike for $2\frac{1}{4}$ hours. How far could he travel in that time?

Distance = Speed × Time = $20 \times 2\frac{1}{4}$ km = 45 km

EXERCISE 4

1 What distance can be covered in:
a 12 hours at 35 mph **b** 4 hours at 12.5 km/h
c 9 seconds at 18 m/s **d** $2\frac{1}{2}$ hours at 1200 km/h?

2 A train travelling at an average speed of 60 km/h takes 2 hours 15 minutes for the journey from Aberdeen to Perth. What is the distance between the two towns?

3 What distance is travelled? Estimate, then calculate!

	a	b	c
Average speed	75 km/h	64 m/s	198 m/s
Time	9 h	$5\frac{1}{2}$ s	0.001 s

4 A mountain rescue helicopter left its base at 11 30 hours and reached the accident at 13 00 hours. Its average speed was 148 km/h. How far had it to fly?

5 Migrating swallows can keep up an average of 20 mph for 45 hours without stopping for food. How far is this?

6 How far would Concorde fly in $3\frac{1}{4}$ hours at an average speed of 2300 km/h?

7 At an average speed of 55 km/h the *QE2* crossed the Atlantic in 5 days 4 hours. What was the length of the crossing, to the nearest 100 km?

8 A tortoise trotted at 25 cm/minute for 2 minutes 30 seconds. How far did it go?

9 Lois and her friends cycled round the Isle of Cumbrae in 42 minutes at an average speed of 19 km/h. How far is it round the island?

10 A troop of soldiers march at a steady speed of 6.5 km/h. How far do they go in 55 minutes, to the nearest tenth of a km?

11 The new Silver Streak motorbike is being test-run. It runs for 90 minutes at 100 mph, then 3 hours at 36 mph and finally 4 hours 20 minutes at 75 mph. What distance is clocked up on the test recorder?

12 A cassette tape runs at 9.5 cm/s for 30 minutes.
a What length is the tape, in metres?
b The tape rewinds in 1 minute 30 seconds. Calculate the rewind speed.

DISTANCE/TIME GRAPHS

EXERCISE 5A

1

If you're driving in traffic you can't keep up a steady speed. You have to speed up and slow down, so the graph 'wriggles'.

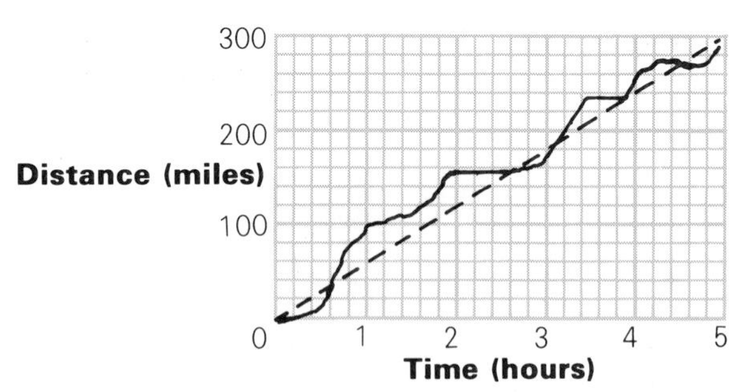

a How far does the car go in 5 hours?
b What is its average speed?
c When does the driver rest for an hour?
d What does the dotted line represent?

2

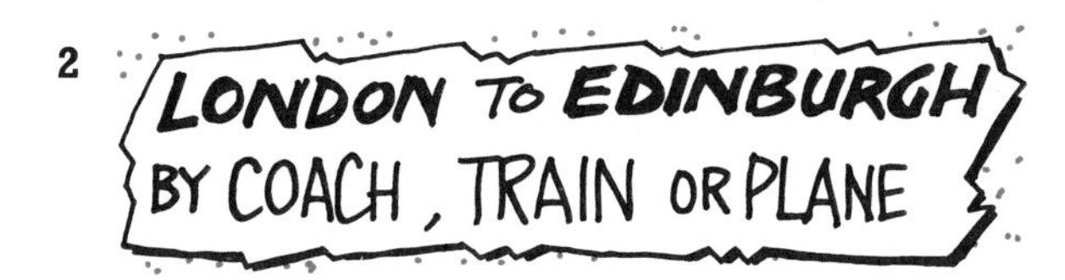

The coach takes 8 hours, the train 5 hours and the plane 1 hour.
a Copy this graph for the coach. Then plot lines for (i) the train, and (ii) the plane on the same diagram.

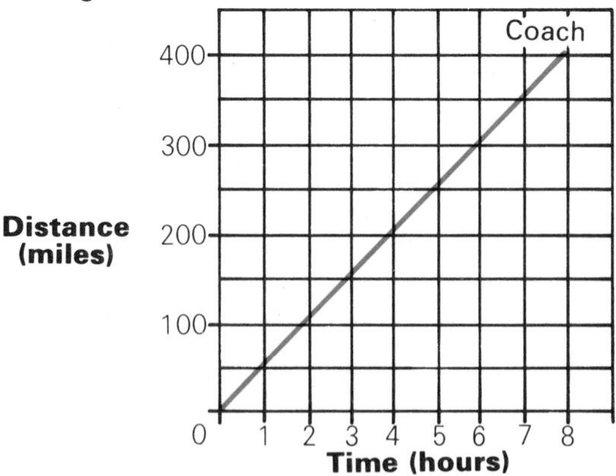

b Which line is steepest? So which is fastest?
c Calculate the average speed of each method of transport.
d Which would you choose? Write a sentence or two about the advantages and drawbacks of each.

3

The Smith family are going on holiday to the seaside. They have to travel 90 km by car. To pass the time, John decides to note the distance covered each hour.

Time	1st hour	2nd hour	3rd hour	4th hour
Distance (km)	20			
Average speed	20 km/h			

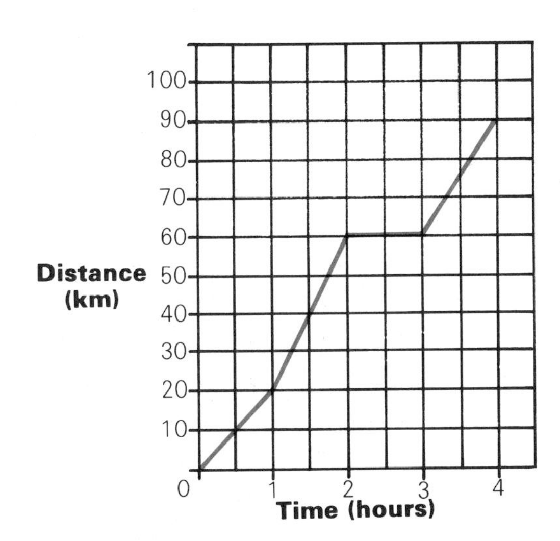

a Copy and complete the table, using the graph.
b What happens during the third hour?
c 'The greater the average speed, the steeper the graph.' Check this, using your table and graph.
d They leave home at 11.30 am. When do they arrive at the seaside?

4

Jenny cycled from Castleford to Leeds and back again.

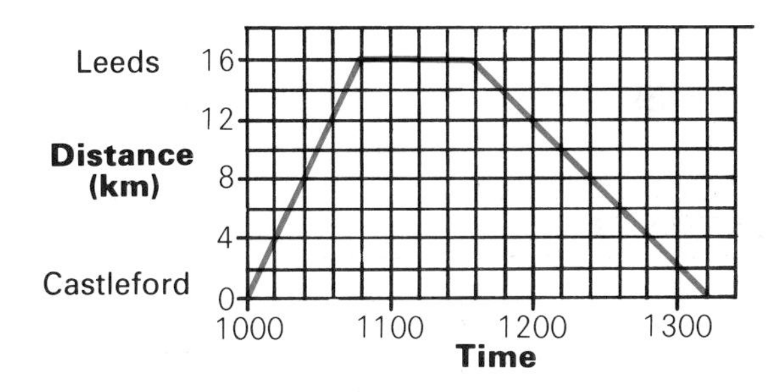

a How many minutes are in each division on the time axis?
b How long did she spend in Leeds?
c In which direction did she travel faster?
d Calculate her average speed in each direction.

5 Use the axes and scales shown to draw a graph on squared paper for Nicola's shopping trip. She has a 15 minute walk to the bus stop 2 km away, where she waits 10 minutes for a bus. The bus takes her the 8 km to the shops in a 15 minute journey. After spending 20 minutes at the shops, Nicola takes a taxi, arriving home an hour and a quarter after leaving.

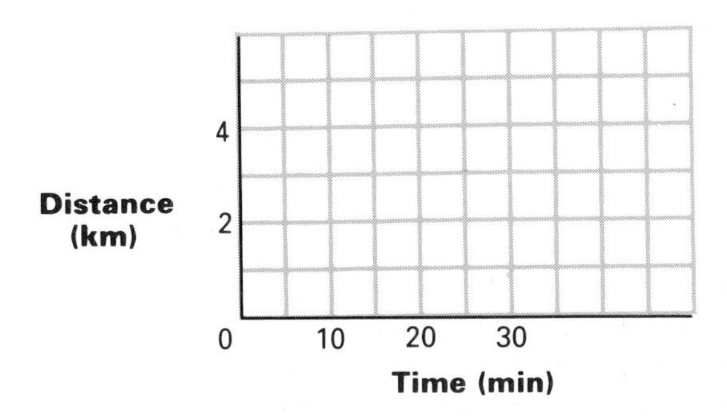

EXERCISE 5B

1 Frank and Pete decided to do some training. They ran, and they walked, then they took the bus home.

a Which parts of the graph show the running, walking and travelling by bus?

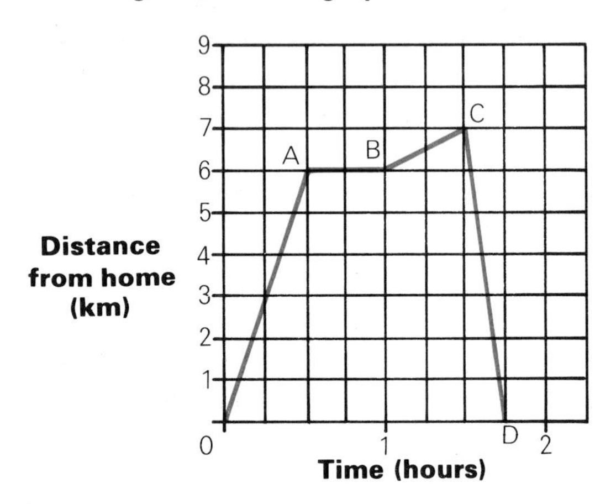

b For how long did they rest?

c Calculate the average running, walking and bus speeds.

2 The graph shows a journey from Bristol to Reading via Swindon, and back to Bristol, by car.

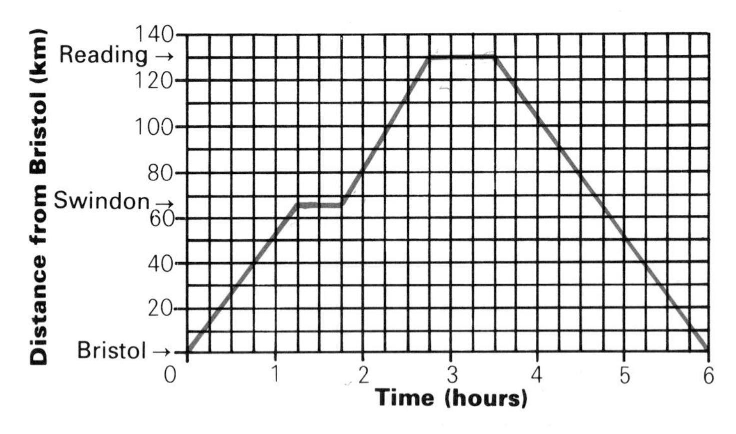

a How long did the whole journey take?

b What was the total distance covered?

c Where were stops made, and for how long?

d Calculate the average speed for each part of the journey.

3 Two salesmen leave their firm in London at the same time to drive to Gloucester, 170 km away. One drives direct, the other stops at Oxford.

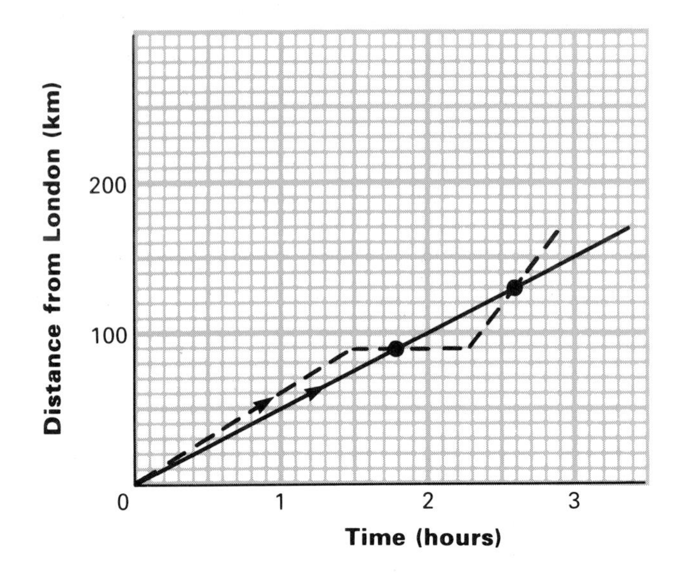

a How far is Oxford from London?

b After what times are the salesmen the same distance from London?

c How far are they from London then?

d What is the difference in their arrival times at Gloucester?

4 Look at the graph below. Paul left Newcastle at 9 am to cycle to Middlesbrough. Steve left Middlesbrough at 10.24 am to cycle to Newcastle.

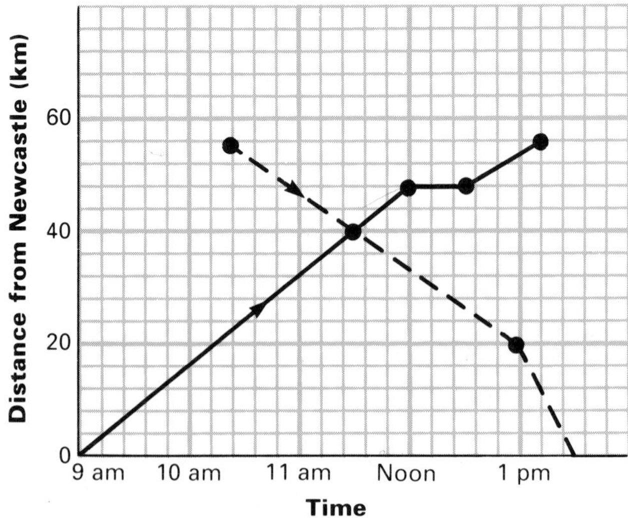

a When did they meet? How far were they from Newcastle then?

b Who had the highest average speed for any part of their journeys? What speed?

5

Describe this race between Ian and Alan in as much detail as you can.

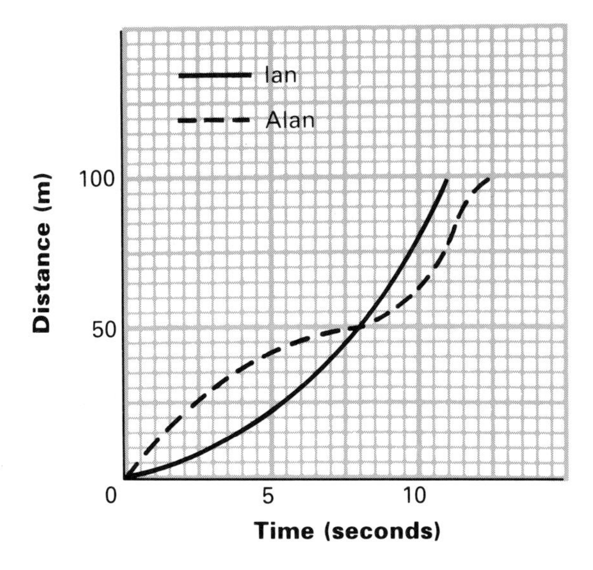

6 Joan jogs from Inverness to Nairn, and Mandy cycles from Nairn to Inverness along the same 26 km route.

a Draw a graph on the same squared paper for each girl's journey.

b How far are they from Inverness when they meet?

c When did they meet?

d Compare their average speeds.

Time	9 am	9.15	9.30	9.45	10 am	10.15	10.30	10.45	11 am
Joan's distance from Inverness (km)	0	3	6	10	13	17	20	23	26
Mandy's distance from Nairn (km)	—	—	—	0	6	13	19	26	—

PRACTICAL PROJECTS

1 *On your next car journey, note the distance travelled every 5, 10 or 15 minutes.*
When you get home draw a distance/time graph.

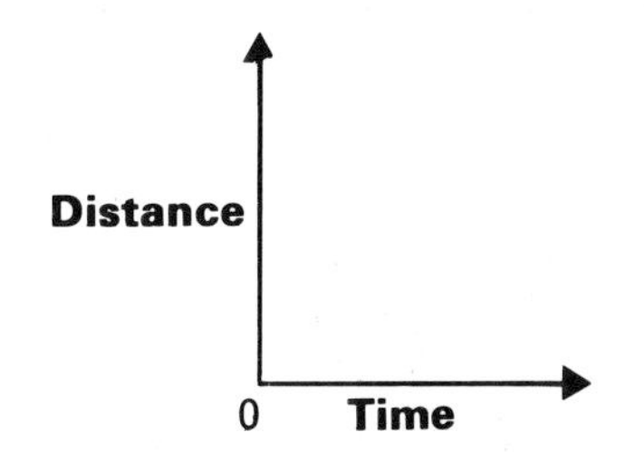

Mark the fast and slow parts, and try to remember the reasons for these. How could you have drawn a more accurate graph?

2 *Measure your time to:*
a *walk 100 m* **b** *run 100 m* **c** *cycle 100 m*
Show the results on the same distance/time graph.

INVESTIGATION

A local newspaper is claiming that 50% of the drivers in Greenleaf Avenue break the speed limit. Explain how you could test this claim. How would you check the speeds, and what data would you need?
Write a report, and describe ways of 'traffic calming', perhaps by using speed ramps.

MAKING SURE

EXERCISE 6A

1 Can you calculate speed, time and distance?

a Distance 280 km
Time 4 h
Speed?

b Distance 864 miles
Speed 72 mph
Time?

c Speed 7.8×10^3 m/s
Time 5 s
Distance?

d Distance 1 million km
Time 4 days
Speed (nearest 100 km/h)?

2 A dart flies through the air for 0.16 second. If it travels 2 metres, calculate its speed.

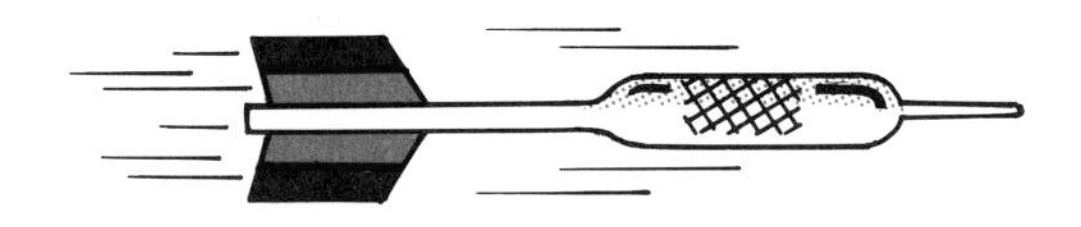

3 At 10 am, three delivery vans leave the warehouse to supply supermarkets at A, B and C.

Copy and complete the table.

Journey	A	B	C
Time		$3\frac{1}{2}$ h	
Distance	96 km		180 km
Average speed	48 km/h	42 km/h	40 km/h

4 Vince is practising his return of serve against an automatic serving machine. He stands 26 m from the machine. How long, to the nearest 0.1 s, does a ball take to reach him at:
a 15 m/s **b** 30 m/s **c** 50 m/s?

5 Melanie is hoping to cycle 80 miles in 4 hours. She averages 18 mph for the first three hours.
a How far does she travel in the first 3 hours?
b What average speed does she need to keep up in the fourth hour?

6 The average speed of traffic on a new bypass was 66 km/h. What distance did the vehicles travel in:
a 3 h **b** 30 min **c** 50 min?

7 Some British Rail Link Coach Service times between Carlisle and Galashiels are shown in the table. The distance between the towns is 65 miles.

Carlisle	09 05	12 05	13 05	15 05	17 15
Galashiels	11 10	14 10	15 10	17 10	19 15

Pam and Pat live in Carlisle, and they have arranged to meet friends in Galashiels at 6 pm.
a Which coach should they take?
b What is the average speed of their journey, to the nearest mph?

EXERCISE 6B

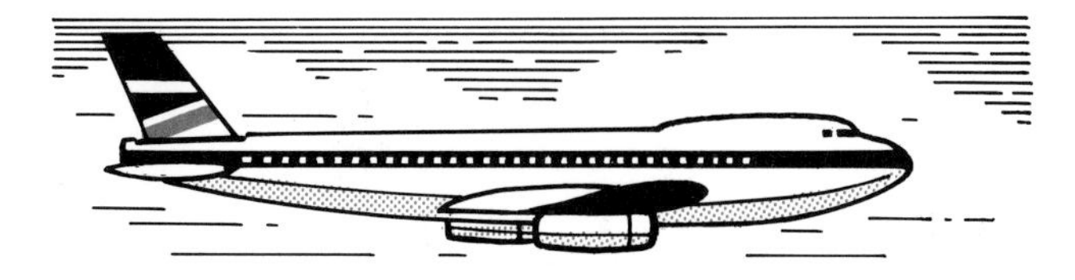

1 It's 5400 km from Glasgow to New York. BA 203 made the flight at an average speed of 900 km/h. With a tail wind, it returned at an average speed of 1000 km/h. Calculate:

a the difference in the times of the two flights

b the average speed 'there and back', to the nearest km/h.

2 In the 1992 Winter Olympics, the men's 5000 m speed skating was won in a time of 7 minutes. The winning time for the woman's 3000 m race was 4 minutes 20 seconds. Calculate the average speeds in m/s, rounded to 4 significant figures.

3 A 'spy' satellite circles the Earth 150 km above the surface at a speed of 26 500 km/h. How long does it take to complete each circular orbit, to the nearest minute? (Radius of Earth = 6350 km.)

4 In astronomy, a light year is the distance travelled by light in one year.

a Show that, if light travels at 300 000 km/second, a light year is approximately 9.46×10^{12} km.

b Convert the distances in the table to km, with the numbers in standard form ($a \times 10^n$).

Object	Distance (light years)
Prox. Centauri (nearest star)	4.2
Sirius (brightest star)	8.6
Centre of Milky Way	2.6×10^4

5 A recent world record time for the men's 100 m sprint was 9.90 seconds. Calculate the average speed in km/h, correct to 1 decimal place.

BRAINSTORMER

An escalator in a store takes 20 seconds to travel from one floor to the next. When Gita runs up the moving escalator she reaches the top in 4 seconds. How long would she take to run up if the escalator was stationary?

CHALLENGE

Mr James arranges handicaps for the 100 m event so that each competitor has the same chance of running the race in the same time. He does this by shortening the race for some competitors by giving them a 'start' of several metres.
Four athletes have reached the final. In the semi-finals their times were: Tom 12 seconds, Ian 13 seconds, Bill 12.5 seconds and Angus 12.2 seconds.
Tom has to run the full 100 metres. Calculate how far the others will have to run. Remember that each runner, theoretically, should run the race in the same time. Include a diagram of the 'staggered' start of the race.

INVESTIGATIONS

1 *Investigate the speed of travel of a clockwork toy. If you use a model train, find the scale of the model to calculate a 'realistic' speed.*

2 *Investigate the speeds, running times, lengths and costs of various cassette tapes and video tapes. Remember that graphs can sometimes give clearer information than words.*

3 *Plan a charity walk from John O'Groats to Land's End. Find out the distance, plan your route and give the names of the places you would stop each night. Detail the distance travelled each day, the total time it would take and other information you would need.*

CHECK-UP ON GOING PLACES

1 Calculate the distance and speed in the examples below.

a Speed 25 km/h
Time 6 h
Distance?

b Time 80 s
Distance 400 m
Speed?

2 Calculate the speed, time and distance to complete the table.

	a	b	c
Distance	182 km	216 km	?
Time	3 h 15 min	?	2 h 12 min
Speed	?	48 km/h	75 km/h

3 Jules Verne wrote a book called *Around the World in Eighty Days*.

a The distance is 25 000 miles. What would the average speed be, to the nearest mph?
b Do you know one form of transport that was used?

4 The record time for walking from Land's End to John O'Groats is 12 days 3 hours 45 minutes, at an average speed of 3.05 mph. How far is it, to the nearest mile?

5 Donna drove the 360 km from Stirling to Liverpool in 10 hours, and back again in 6 hours. Calculate the average speed of her journey:
a going **b** returning **c** there and back.

6 Bill, a salesman for Computerit, travelled non-stop from Manchester to Glasgow in 6 hours.
a Look at the table below and calculate his average speed, correct to 1 decimal place.

Distances in km

	Aberdeen	Glasgow	Carlisle	Liverpool	Manchester	Cardiff
Glasgow	228					
Carlisle	335	151				
Liverpool	525	341	190			
Manchester	525	341	190	56		
Cardiff	779	595	444	264	277	
London	784	634	483	317	296	248

b The following week Bill travelled from Aberdeen to London in 14 hours. What was his average speed for this journey?
c Another time Bill left Glasgow at 7 am to go to Cardiff. In addition to his driving time of 10 hours, he made stops which totalled 3 hours.
(i) At what time did he arrive in Cardiff?
(ii) What was his average speed on the road, to 2 significant figures?

7 A coach left Glasgow with holidaymakers for Blackpool. At the same time another coach left Blackpool for Glasgow, taking the same route.

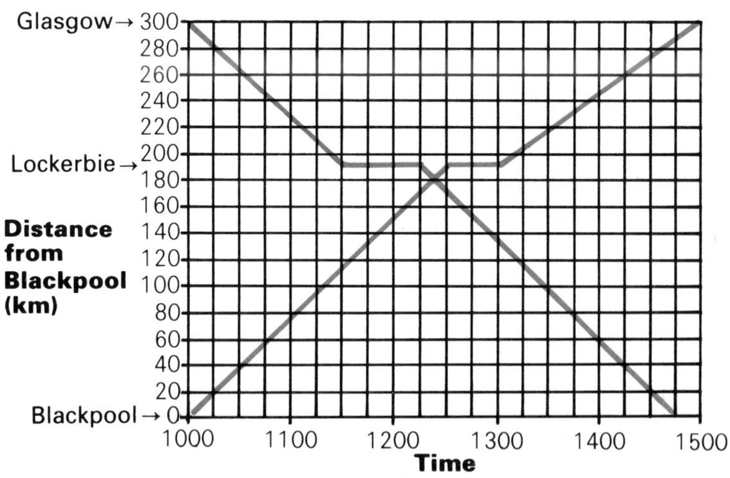

a At what time did they pass each other?
b How far from Glasgow was this?
c Calculate the average speed of the coach from Blackpool to Glasgow, excluding the stop at Lockerbie.
d For how long did the coach from Glasgow stop at Lockerbie?
e Calculate the average speed of the journey from Blackpool to Lockerbie, to the nearest km/h.

4 MONEY MATTERS - SAVING AND SPENDING

LOOKING BACK

1 What sums of money are shown on these calculator displays?

a 0.35 **b** 1.6 **c** 9.02 **d** 0.01

2 Calculate:
a $49 \times £7.05$ **b** $£1500 \div 12$ **c** $\frac{2}{3}$ of £18.18

3 Write these as decimal fractions:
a 25% **b** 50% **c** $17\frac{1}{2}\%$ **d** $117\frac{1}{2}\%$

4 Change to percentages:
a 0.5 **b** 0.75 **c** $\frac{1}{4}$ **d** $\frac{1}{5}$

5 Calculate the difference between these readings:

6 Calculate:
a 12% of £50 **b** $2\frac{1}{2}\%$ of £380

7 Calculate the sale price of these items:

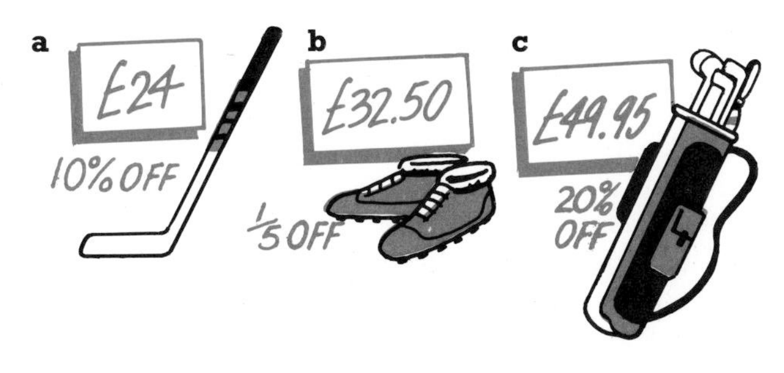

BANK of BRIT
Our interest is your interest
8% per annum

8 With £50, Lena opens a deposit account in the Bank of Brit. Calculate:
a the interest added on after one year
b the amount in her account then.

9 Calculate the cost of:
a 6 tickets in area A for Monday
b 3 tickets in area C for Saturday
c 2 tickets in area D for all six days.

Greenwood Theatre Tickets

Area	A	B	C	D
Weekday	£8	£6	£4	£2.50
Saturday	£10	£8	£6.50	£4

10 Salim buys gold rings wholesale at £40 each, and sells them for £52. Calculate his:
a profit per ring
b percentage profit, based on the wholesale price.

11 45 scientific calculators cost £348.75. Find the cost of 60 calculators.

WAGES AND SALARIES

CLASS DISCUSSION

Do you know what the terms in red mean?
Would any of these 'small ads' interest you?

Hourly rate · Annual salary · Job share · Weekly wage · Bonus · Commission · Overtime · Part-time · Self-employed

SUMMER Jobs, full time and temporary vacancies are available in wholesale import company. Duties include sales promotions, sample placement and advertising/marketing. No experience necessary, students are welcome.

WP Operator. £5.50/hr. Mornings. Temporary. Applemac Word 5 experience preferred.

COMPUTER Operator to £12K + Shift allowance + overtime. Greenfield site.

SALES Representative. £14000 + bonus + car. Selling industrial product range to end users. Candidates, aged 25-35, must have minimum 2 years industrial sales experience.

TEMP Junior Secretaries, WP Operators, legal cashiers needed. Good rates, some with overtime.

AUDIO WP Operator to £8750. Young crowd. Good overtime rates.

4 STAR HOTEL Chef de parties, required for new hotel. Live in. £11K

TELEPHONIST/Receptionist/ Clerkess required for law firm. 9am-4.30/5pm. Salary £7300.

ACCOUNT Manager. £20K+ Good Commission + Car + Car-phone + Benefits. Experience in either Communications/Computer/Office Equipment. Multi-national company.

SALES Person for busy Highland Dress shop in city centre, previous experience preferable, must be over 21. Salary £150 per week - reviewed after 3 months. Applications in writing.

EXERCISE 1

1 Elaine answered this advert, and got the job. What is her monthly salary, before tax etc?

2 Roger could sell anything, so this interested him.

The extras are £45 a week for lunches and travel. How much would he get:
a in his first year **b** each week?

3 Marion did this job for six weeks.

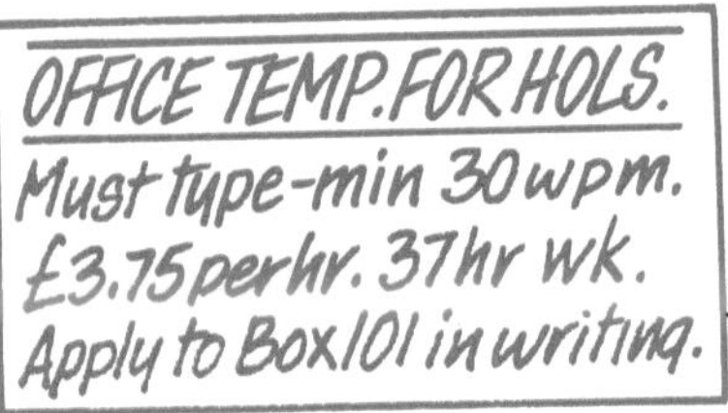

a Calculate her weekly pay.
b How much did she earn altogether?

4 Sheila took this job and agreed to do 12 hours a week.

TELE-SALES from HOME
Part-time, Phone essential.
£54 a week
Tel: 100-0001

a What is the hourly rate of payment?
b How much could she earn in 13 weeks?

5 Mr Sharma has been appointed manager of the Eversure Insurance office at a salary of £27 000 a year, with annual increases of £1200. Calculate:
a his monthly starting salary
b his annual salary and monthly salary during his fourth year as manager.

Commission and overtime

6 Mrs Patullo is in charge of a furniture department. Her monthly salary is £950 plus 2% commission on sales. In December she made sales of £6700.

a Calculate her total income in December.
b Why is commission used as part-payment?

7 Which of these companies pays its salespeople more on a month's sales of £12 500? How much more?

Sleepsound:	£900 a month + 9% sales commission
Soundsleep:	£1000 a month + 8% sales commission

8 Sleepsound decided to use this flowchart to calculate its commissions:

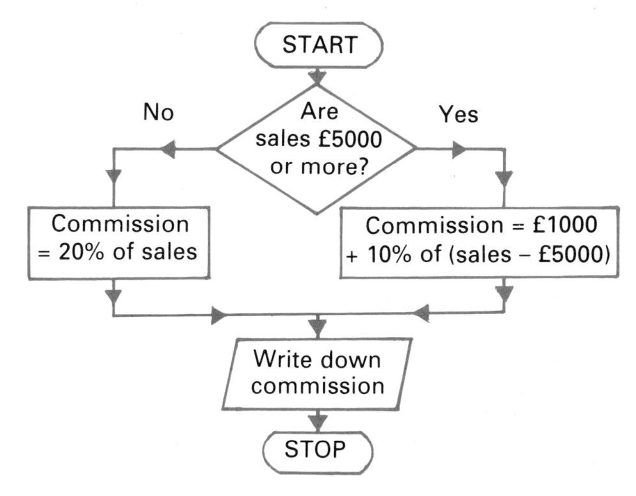

Use the flowchart to calculate the commission on sales of:

a £2500 **b** £5000 **c** £8000

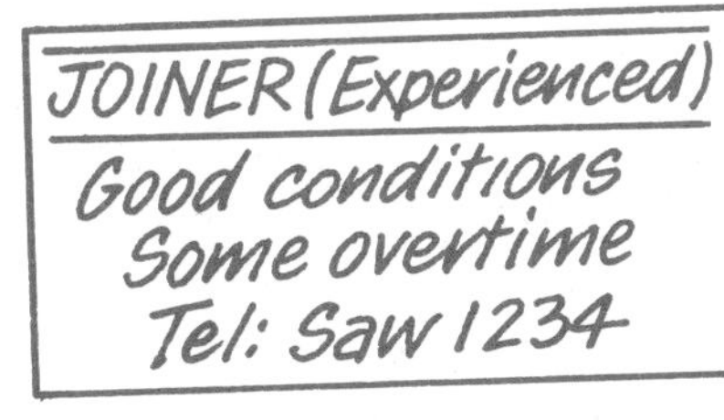

9 Dave was lucky—start next week, he was told. The pay was £9.50 an hour for a 35 hour week.

a Calculate his week's pay.
b His overtime was paid at 'time-and-a-half' (£14.25 an hour). How much would he get for 4 hours overtime?
c Why are overtime rates paid higher than normal rates of pay?

10

Graham has a holiday job at the Krazy Golf course. The pay is £3.60 an hour, plus overtime at these rates:
'double time' on Sundays (double rate of pay);
'time-and-a-half' on Saturdays and evenings after 6 pm.
How much was he paid for the following week's work?

Monday–Saturday: 10 am–1 pm and 2 pm–6 pm
Sunday: 2 pm–6 pm
Tuesday: 2 hours evening overtime

CHALLENGE

Suppose you were given this choice:

a *a salary of £12 000 a year, with annual increments of £1000, or*
b *a salary of £10 000 a year, with annual increments of 10% of the previous year's salary.*

Which would you choose? Give reasons for your choice. A graph of each salary, year by year, might surprise you.

PRACTICAL PROJECT

Look at advertisements in newspapers or Job Centres, and see how different wages, salaries, commissions and other incentives such as company cars, are offered. Make a note of some of them.

SAVINGS IN BANKS, BUILDING SOCIETIES AND POST OFFICES

EXERCISE 2

(i) Bank accounts

Billy O'Reilly decides to open a current account in the Moneymaker Bank. The bank offers all of these services. Do you know what they mean?

1 Billy pays in £50, and is given a cheque book. Here is his first cheque. Make a copy of his second cheque, number 00002, dated 28.10.94, payable to 'Hot Hi-fi', for £12.50.

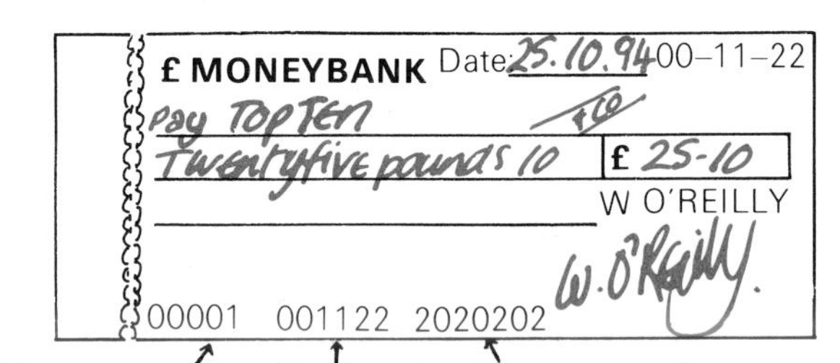

Cheque number Bank code Account number

2 Every month a bank statement comes through Billy's letterbox. It shows all the money withdrawn and paid in, and the balance.

MONEYMAKER BANK		William O'Reilly	Account 2020202	
Date	Details	Withdrawals	Paid in	Balance
20 Oct	Cash		50.00	50.00
25 Oct	Chq 00001 Top Ten	25.10		24.90
28 Oct	Chq 00002 Hot Hi-fi	12.50		12.40
30 Oct	Chq Regional Council		475.00	487.40
5 Nov	Chq 00003	100.00		

a What do the entries mean on:
(i) 20 Oct (ii) 25 Oct (iii) 30 Oct?

b How much is in his account on:
(i) 25 Oct (ii) 30 Oct (iii) 5 Nov?

3 Billy soon realised that his bank did not pay any interest on balances in current accounts. So he opened a deposit account which offered interest at 7% p.a. (per annum = yearly). How much interest would he receive in one year on:
a £200 **b** £700 **c** £1250 **d** £35?

(ii) Building Society accounts

4 Building Societies also encourage people to invest money with them. Billy's sister Maeve invests £960 with the Mushroom Building Society for 3 months.

Copy and complete:
a interest for 1 year = 8% of £960 = . . .
b interest for 3 months = . . .
c interest for 5 weeks = . . .

5 Jim Wilson, a pensioner, puts his savings of £6600 into the Money Mountain Building Society. He takes out the interest at the end of each year and uses it to buy Premium Bonds at £100 each, hoping to win the big prize. How many bonds will he have by the end of the second year?

6 The Union Jack Building Society's Special Account graph looks like this:

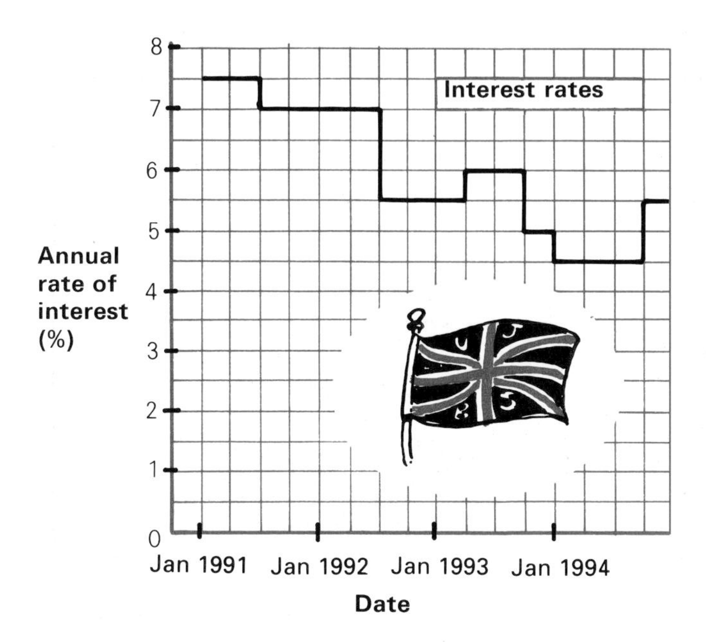

a How often has the rate changed during the four years shown?
b Calculate the interest on:
(i) £3000 for the first six months of 1992
(ii) £1200 for the first nine months of 1994
(iii) £1500 for the whole of 1991. (Careful!)

7 Invest in the Mushroom Building Society

Invest (£)	1000	10 000	20 000	50 000
After 1 year (£)	1073	10 800	21 650	54 225

a Calculate the interest rate (% p.a.) for each deposit.
b Write a sentence about the Society's policy for attracting investments.

(iii) National Savings

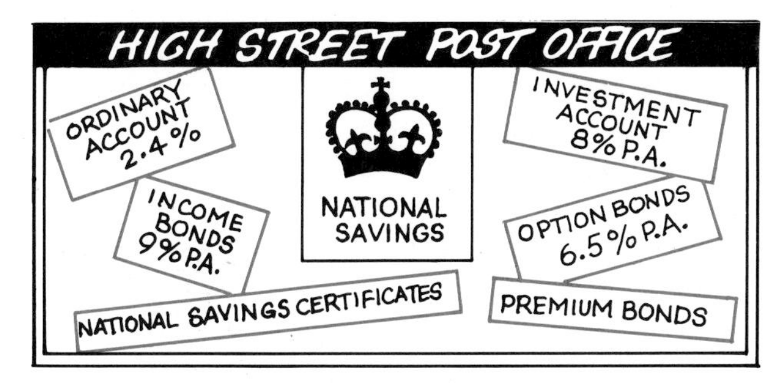

8 Joshua was browsing in the Post Office one day. He discovered that the National Savings Investment Account pays 8% p.a., before tax. (One month's notice of withdrawal is needed.)

Here is his account a year later:

INVESTMENT ACCOUNT	Joshua Simons	Account 010101				
Date	Transaction	Th	H	T	U	
31.12.93	First deposit		1	4	5	
31.12.94	Interest for 1994			1	1	60
	BALANCE		1	5	6	60
2.1.95	Twenty-five pounds 50			2	5	50
	BALANCE					

a How much is Joshua's:
(i) first deposit
(ii) first year's interest?
b Check the interest.
c Calculate the balance on January 2nd after paying in £25.50.
d Why did Joshua choose the Investment Account rather than the Ordinary Account?

9 Mrs Dixon decides to spend £500 on National Savings Certificates.

National Savings Certificates	
Interest added at end of	Interest per £100
Year 1	£4.00
Year 2	£4.58
Year 3	£6.24
Year 4	£7.75
Year 5	£9.68

a How much will they be worth in:
(i) 1 year (ii) 2 years (iii) 5 years?
b Why should she keep them for the full five years if possible?

10 Interest on some savings accounts is paid gross, that is, before income tax is deducted.
Mr Singh's National Investment Account gave him gross interest of £125 in 1992 and £230 in 1993. Calculate the net income each year after deduction of tax at:
a 20p in the £ in 1992 **b** 25p in the £ in 1993.

INVESTIGATION

Investigate the best way to invest your own savings. Banks have various offers to attract new customers, and Building Societies, National Savings, Insurance Companies, property, shares and unit trusts are other possibilities. Remember that some deduct income tax, some don't; some are risky, some are not; some will be rewarding, some may not be.

VAT (VALUE ADDED TAX)

CLASS DISCUSSION

Teachers, civil servants, police and fire brigade, nurses, . . . school books, hospitals, . . . all have to be paid for. So the Government raises money by different taxes. One of these is Value Added Tax (VAT).

SCHOOL BOOKS CHILDREN'S CLOTHES MEDICINES...

CARS, CYCLES, FUEL, RECORDS, FRIDGES, PLUMBERS, JOINERS, RESTAURANTS...

VAT IS CHARGED

FOR EACH £1 OF GOODS AND SERVICES

YOU GIVE AN EXTRA $17\frac{1}{2}$p TO THE SHOPKEEPER OR TRADESMAN ...

WHO THEN PASSES THIS TO THE GOVERNMENT

What is meant by zero-rated?
Why are some things zero-rated?
What is the present rate of VAT?

Example
A compact disc costs £7.50 plus VAT. Calculate its selling price.

Method 1
VAT = $17\frac{1}{2}\%$ of £7.50 = 0.175 × £7.50 = £1.3125 = £1.31 to the nearest penny.
Selling price = £7.50 + £1.31 = £8.81

Method 2
Selling price = $117\frac{1}{2}\%$ of £7.50 = 117.5% × £7.50 **or** 1.175 × £7.50 = £8.81

EXERCISE 3

In this exercise, take $17\frac{1}{2}\%$ as the rate of VAT.

1 Joe was in no doubt. It had to be Scotthol; it was cheaper. Teresa wasn't so sure. She did her sums. What did she find?

2 Calculate:
a the VAT on each packet of dog food
b the cost of each, including VAT.

3 Calculate this plumber's bill.

N.A. LEEK (PLUMBER)	£
3 metres plastic piping at £6.75 per metre	
4½ hours labour at £15.84 per hour	______
Subtotal	
+VAT at 17½%	______
TOTAL	______

4 Calculator in hand, Mrs Landells went round her two local 'Cash and Carry' stores, adding on the cost of VAT. Which had the lower overall cost, and by how much?

Cash and Carry (VAT to be added)	£15.00	60p	£1.00	£2.50
Locost (including VAT)	£18.00	65p	£1.15	£3.00

5 The Alroyds offered £79 000, but the Thomsons offered £82 000, and got the house. How much, to the nearest £, would the Estate Agents have charged in each case?

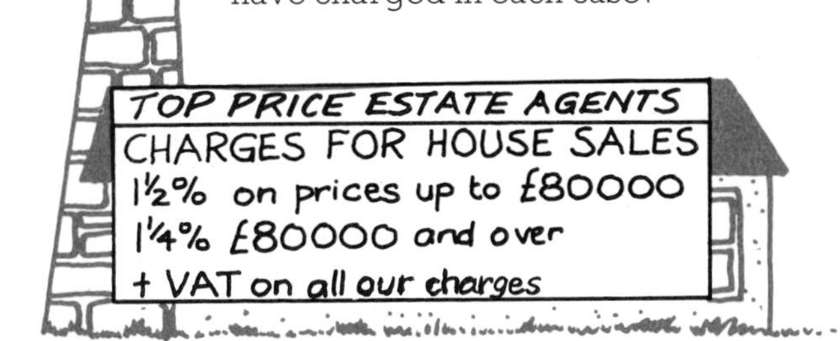

6 A TV set is advertised at £360 + VAT.

a How much more does it now cost?
b Can you find a quick way to do the calculation?

7 In 1993, gas and electricity bills were zero-rated for VAT. The following year they had 8% VAT added, and the year after that 17.5%. How much more would you have paid in the second and third years on bills of: **a** £65.50 **b** £278.95?

TELEPHONE BILLS

EXERCISE 4

Did you know that the cost of a phone call depends on these?

Study these tables:

Mon	Tue	Wed	Thu	Fri	Sat	Sun
Cheap rate: 6 pm–8 am					Cheap rate all the time	
Daytime rate: 8 am–6 pm						

LOCAL	1 min	2 min	3 min	4 min	5 min	6 min	7 min	8 min
Cheap	5p	5p	5p	10p	10p	10p	10p	15p
Daytime	5p	10p	15p	15p	20p	25p	30p	30p

1 a When is it cheapest to phone (cheap rate)?
b When is it dearest (daytime rate)?

2 What would you be charged for a local call lasting:
a 3 minutes from 2 pm on Monday
b $7\frac{1}{2}$ minutes on Saturday evening?

3 Calculate the total cost of these local calls:
5 minutes on Monday from 10 am,
3 minutes on Tuesday from 10 pm,
4 minutes on Wednesday from 3.30 pm, and
4 minutes on Sunday from noon.

4 Look at Mr Gossip's phone bill. Calculate the entries that will need to be made at A, B and C.

Phone bill for 000 111 9999

Your bill is	£ 48.88	**Call charges** £ 14.70 for direct-dialled calls of less than 10 units £ 34.18 for direct-dialled calls of 10 units and over
plus	£ 20.94	**Advance charge from 1 Mar to 31 May** £ 20.94 for the rental of your line
	£ A	Subtotal excluding VAT
plus	£ B	VAT at 17.5%
	£ C	**Total amount now due**

5 Mrs Allison's phone bill had call charges of £36.15 and advance charges of £20.94. Calculate the subtotal, VAT and the total amount she will have to pay.

INVESTIGATION

Everyone who has a telephone should have a Phone Book and a 'Guide to telephone costs and charges'. In these you can find telephone numbers and costs of local, long distance and worldwide calls. For example, the code 010 617 connects you to Brisbane in Australia.

a *Choose some faraway places and find out the costs of 5 minute calls to them.*

b *From these guides find out about other services and charges, for example calls 'via the operator', 'alarm calls' to wake you up in the morning, and 'transferred charge' calls.*

ELECTRICITY AND GAS BILLS

EXERCISE 5

1 Here is Mr Watt's electricity bill. Calculate entries A, B, C and D.

ACCOUNT ISSUED	REFERENCE	FROM	TO
15 APR		17 FEB	11 APR

METER READING Present	METER READING Previous	DETAIL OF CHARGES	AMOUNT
		Standing Charge	6.50
18223	17101	A units at 7.55p	84.71
		SUBTOTAL	B
		VAT at 17.5%	C
		TOTAL DUE	D

2 Mr Watt's next meter reading was 19 016.

a How many units had he used?

b What is the cost of these units at 7.55p each?

c Calculate the subtotal, VAT and total due on his bill. The standing charge is still £6.50.

3 Mrs Power had off-peak storage heaters installed. Here is part of her next bill:

METER READING Present	METER READING Previous	DETAIL OF CHARGES	AMOUNT
01086	00969	Domestic A units at 7.86p	B
00831	00017	Off peak C units at 3.05p	D
		STANDING CHARGE	7.15
		SUBTOTAL	E
		VAT at 17.5%	F
		TOTAL DUE	G

a Calculate entries A–G.

b What is off-peak electricity?

4 Irma Spence's gas bill looks complicated. But the cost is just based on the number of kWh (kilowatt hours) at 1.567p each. Calculate entries A, B and C.

DATE OF READING	METER READING PRESENT	PREVIOUS	GAS SUPPLIED CUBIC FEET (100s)	CUBIC METRES	kWh	CHARGES £
9 MAY	3906	3373	533	1508	16675	261.30*
STANDING CHARGE 96 days at 11.40p per day						10.94
				SUBTOTAL		A
*1.567p per kWh				VAT at 17.5%		B
			TOTAL AMOUNT DUE		£	C

5 For the next quarter, Miss Spence's bill showed 14 083 kWh at 1.605p each, and a standing charge of £11.50. How much had she to pay?

6 Calculate the entries A–E on this gas bill.

DATE OF READING	METER READING PRESENT	PREVIOUS	GAS SUPPLIED CUBIC FEET (100s)	CUBIC METRES	kWh	CHARGES £
10 SEP	9861	9388	473	1339	14873	A*
STANDING CHARGE 89 days at 11.50p per day						B
				SUBTOTAL		C
*1.62p per kWh				VAT at 17.5%		D
			TOTAL AMOUNT DUE		£	E

7 The graph shows three different schemes (A, B and C) of charging for units of energy.

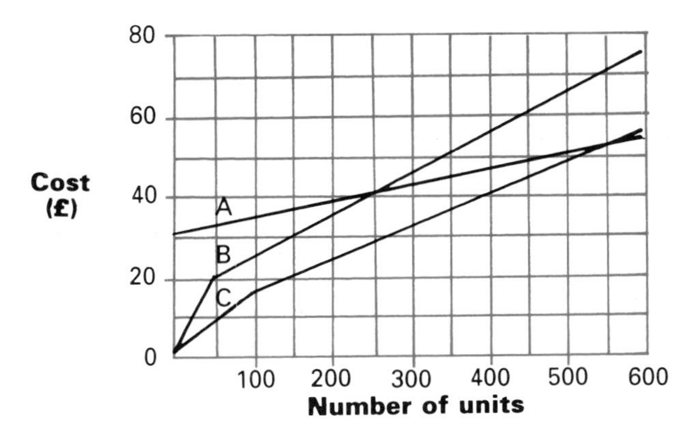

a Arrange the schemes in order of cost, cheapest first, if you used:
(i) 200 units (ii) 400 units (iii) 600 units.

b For what number of units would these two schemes cost the same:
(i) A and B (ii) A and C?

c For scheme A:
(i) what is the standing charge (before any units are used)
(ii) what is the charge per unit?

d By looking at graphs B and C, decide which has the cheaper charge per unit.

8 The Peterkins decided to pay for their gas on a monthly basis. The Gas Company estimated that the family would use about 13 000 kWh in a year. One kWh costs 1.926p, and the annual standing charge is £48.40. Calculate:
a the annual cost of gas **b** the monthly charge.

INVESTIGATIONS

1 *Ask to see some recent telephone, gas and electricity bills for your own household. Check the calculations.*

2 *Find out and compare the cost of gas and electricity for heating and cooking.*

3 *Investigate the cost of running these items by electricity: cooker, fire, colour TV, immersion water heater, lighting, iron, music centre. Compare the costs of running a 1 kilowatt fire and a 100 watt light bulb.*

HP (HIRE PURCHASE)

EXERCISE 6

1 a What is meant by hire purchase?
  b What are the advantages and disadvantages of buying goods on HP?

2

Calculate:
  a the HP price
  b the difference between the HP and the cash price.

3

Calculate:
  a the HP cost
  b the difference between the HP and the cash price.

4

How much greater is the cost by instalments?

5

How much cheaper by cash?

6 Fatima thought she could just afford the new television set in the shop window—cash price £264.96, or $12\frac{1}{2}\%$ deposit and 12 monthly payments of £24.35.
  a How much would the HP deal cost?
  b How much more than the cash price is this?

7 Muriel and Tom are setting up home, and notice a three-piece suite they like in *Homeshop's Mail Order Catalogue*. It costs £899.99 cash, or any of these terms: 26 weeks at £38.99 weekly
or 52 weeks at £21.50 weekly
or 104 weeks at £11.75 weekly.
  a Calculate the cost of each method of payment.
  b How can they decide which way to buy the suite?

8 Morna is a keen golfer. She has been saving up for a new set of clubs. How much more than the cash price would each of the three HP schemes cost?

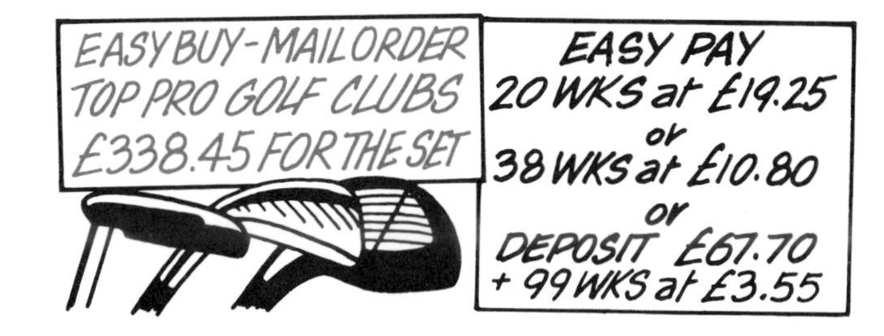

9 Allsports has a snooker table on sale at £450. The manager decides to introduce HP terms as follows: deposit 20% of cash price and 24 monthly payments; total cost 25% more than the cash price. How much will each monthly payment have to be?

INVESTIGATIONS

1 Compare the cash price with the cost of buying a television set, or guitar, or music centre, or motorbike over two or more years. Remember that if you don't pay cash you could have your money in a Building Society earning interest.

2 Imagine that you are a garage owner, trying to sell a second-hand car for £2500. Make up one or two HP offers that would attract customers, and not make a loss for the garage.

HOLIDAYS ABROAD

If you are going on holiday abroad you have to buy francs, pesetas, dollars, etc., in Britain before you go, or in the country you are visiting, once you are there. The exchange rate for buying foreign currency alters from day to day.
This table shows how much you could get for £1 on a particular day.

USA	1.50 dollars	?	229 escudos
France	8.21 francs	?	2540 lire
?	2.45 marks	?	320 drachmas
?	185 pesetas	?	2.23 Swiss francs

What countries' names are missing from the table?

EXERCISE 7A

1 Before leaving for a camp-site in Brittany, Neil and his three friends changed £80, £85, £65 and £95 into francs.
How many francs did each receive?

2 The Hill family from London went to Torremolinos on holiday. The first thing they did there was to change their traveller's cheques for pesetas. £350 was changed altogether, and the Banca d'España made a small handling charge of 1%. How many pesetas were the Hills given?

3 Matthew drew this graph before going on holiday to the Greek island of Cos. He found it useful when changing pounds to drachmas. It also let him judge how much things cost in Cos.

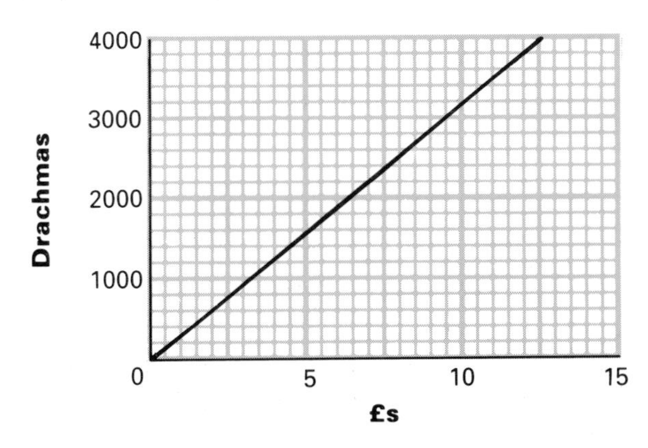

a Use the graph to find out the number of drachmas:
(i) Matthew got when he changed £10
(ii) his sister Fiona got for £5.

b How many drachmas would Matthew get for £20?

4 Matthew and Fiona went shopping, but everything was priced in drachmas. Use the graph in question **3** to convert these prices into pounds:

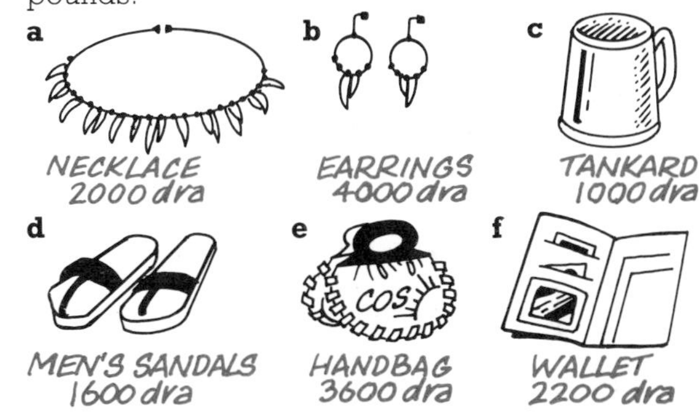

5 a Draw a pounds-to-dollars exchange graph. Take £1 = $1.50.

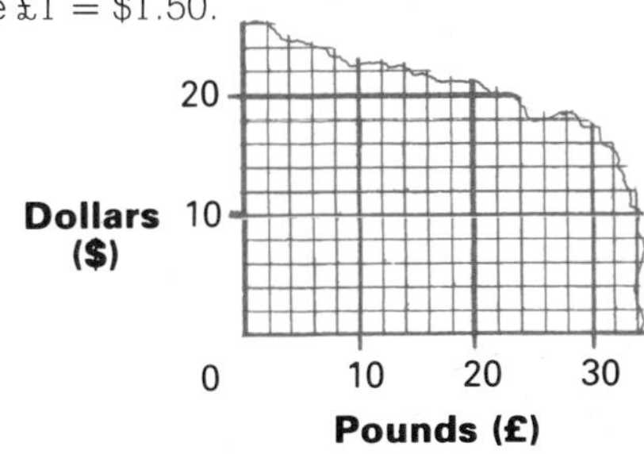

The graph is a straight line going through the points (0, 0) and (100, 150). Why?

b Copy and complete this table, using your graph:

£s	10	40	70				84	
$s				90	40	70		112

c Check the entries in your table by calculation.

EXERCISE 7B

Example

a Before leaving for a holiday in France the Kumars changed £1200 into francs at 8.19 to the £. How many francs did they get?

b They spent 9450 francs on holiday. On the way home they changed what was left to pounds, at 8.22 francs to the £. How much did they get?

a £1 ⟷ 8.19 francs
£1200 ⟷ 8.19 × 1200 francs
= 9828 francs
They got 9828 francs.

b They have 9828 − 9450 francs = 378 francs
8.22 francs ⟷ £1
1 franc ⟷ £$\frac{1}{8{\cdot}22}$
378 francs ⟷ £$\frac{1}{8{\cdot}22} \times 378$
= £45.98
They got £45.98.

1 Kate and Jenny went cycling to France.

a They changed £225 to francs at 8.24 francs to the pound. How many francs did they get?
b At the end of their holiday they had 155 francs left. How many pounds would they be given, at 8.21 francs to the pound?

2 Ali and his family went on holiday to San Francisco. They changed £750, and then another £500, into dollars at \$1.50 to the £. They spent 1750 dollars, and at Heathrow Airport changed the rest to pounds at \$1.65 to the £. How much did they receive?

3 Here are the weekly wages of five trainee chefs.

Name	Pay	Rate to the £
Andrew	£125	—
Pierre	1050 francs	8.25
Wilbur	\$180	1.55
Manuel	22 400 pesetas	190
Karl	325 marks	2.48

a Which countries do you think they come from?
b Change their wages into pounds, and put them in order, starting with the one who earns most.

4 Julie had aunts and uncles everywhere. For her birthday, her Aunt Alice in Toronto sent her 10 dollars, Uncle Ted in Paris sent 100 francs and Aunt Tina in Rome sent 10 000 lire.
Julie checked the exchange rates at the bank: 1.62 dollars, 8.30 francs, 2300 lire to the £. How much did she get altogether, in pounds?

5 Paperback books often have a list of how much they would cost in different countries.

a Use the prices shown to calculate the rates of exchange between Britain and the three other countries.
b Why might these not be accurate?

INVESTIGATIONS

1 *Investigate the rates of exchange and other details listed in some banks and building societies, or in the business sections of some newspapers, and how they change from day to day or week to week.*

2 *Compare the rates with those listed in question* **4** *of Exercise 7B. Has the £ strengthened or weakened against each currency?*

CHECK-UP ON MONEY MATTERS—SAVING AND SPENDING

1

Jane's best subject is English. She replies to the box number, and is offered the job. How much will she get, before tax etc., if she is paid:
a monthly **b** weekly?

2 An apprentice joiner had a 38-hour working week, and was paid £6.70 an hour.
a What is his weekly wage?
b How much would he receive in a week which included 5 hours' overtime at 'time and a half'?

3 Steve is paid 6% commission on sales of office equipment. What is his commission on sales of £1350?

4 Pauline has £750 to invest. Which account would give her the larger net income in one year? How much more?

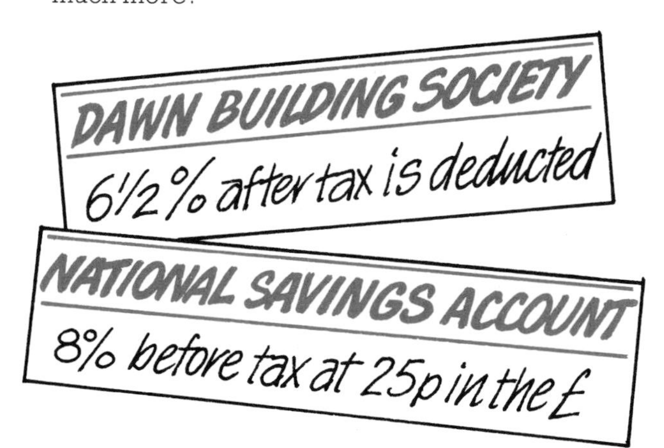

5 An investment of £1200 makes £52 interest in a year. What is the rate of interest?

6 A personal stereo costs £44.50 + VAT. If the rate of VAT is $17\frac{1}{2}$%, calculate the total cost of the stereo.

7 Here is part of a phone bill. Calculate the total amount due.

£	**Call charges** £23.09 for direct-dialled calls of less than 10 units £31.95 for direct-dialled calls of 10 units and over
£______	**Advance charges** £21.10 for the rental of your line
£	Subtotal excluding VAT
£______	VAT at 17.5%
£	**Total amount now due**

8 For these electricity meter readings, the domestic rate is 8.15p per unit, and the off-peak rate is 3.58p per unit. The standing charge is £8.95. Calculate the amount which must be paid (including VAT at 17.5%).

Meter readings	
Present	Previous
DOMESTIC	
32650	32436
OFF-PEAK	
08406	08318

9 How much more would this bike cost on hire purchase?

10 A family changed £850 into Swiss francs at 2.28 francs to the £. They spent 1750 francs abroad, and changed the rest back to £s at 2.31 francs to the £. How much of the £850 did they have left?

5 POSITIVE AND NEGATIVE NUMBERS

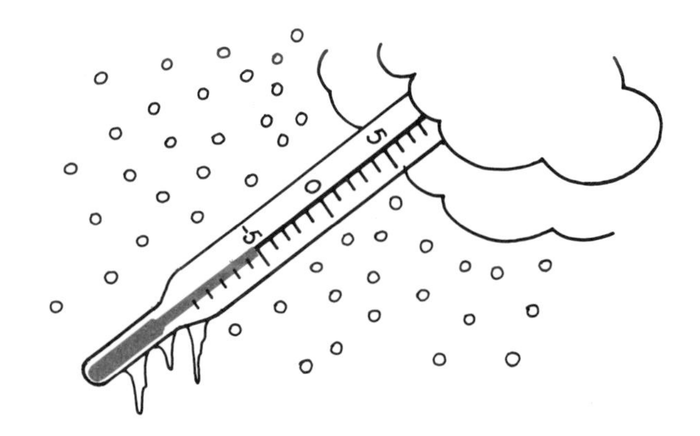

LOOKING BACK

1 Copy and label these number lines:

2 a Arrange these temperatures in order, lowest first.

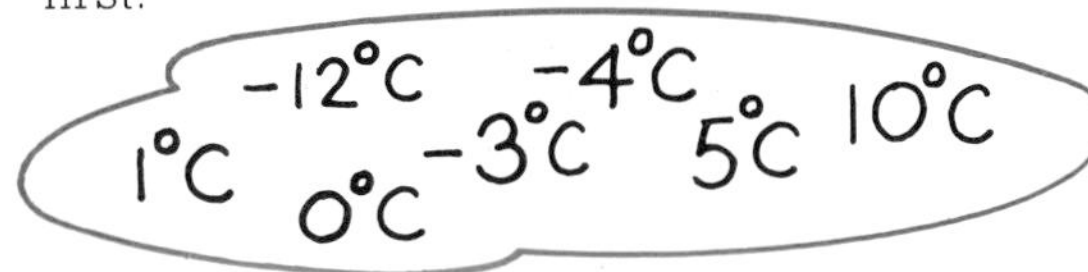

b Write down the difference, in °C, between the highest and lowest temperatures.

3 Find three more terms for each sequence below:
a 4, 3, 2, . . . **b** 10, 5, 0, . . . **c** $-6, -4, -2, \ldots$

4 a Write down the coordinates of the images of A, B and C under reflection in:
(i) the x-axis (ii) the y-axis.

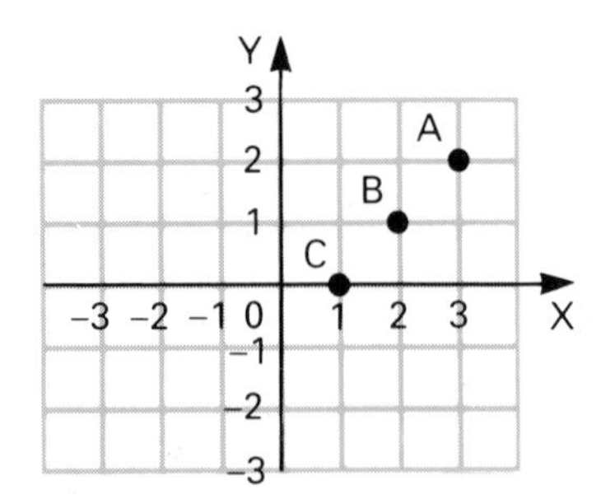

b Give the coordinates of the next three points, and the tenth point, in the sequence A, B, C, ...

5 Calculate, without using a calculator:
a $1-3$ **b** $-2-5$ **c** $-3+5$
d $-2+1$ **e** $4-(-1)$ **f** $-5-(-8)$
g $-3+2$ **h** $2+(-1)$

6 Check your answers to question **5** by calculator.

7 Arrange these bricks in pairs in three ways, with the [+] in the middle.

Calculate the three scores.

8 Solve these equations:
a $x+3=2$ **b** $x+2=-3$ **c** $x-3=-1$
d $5-x=-2$ **e** $4+x=2$ **f** $4-x=8$

9 $x=-1$, $y=-3$ and $z=5$. Find the value of:
a $x+y$ **b** $x-y$ **c** $y-z$ **d** $y+z$

10 Which of these are true and which are false?
a $3>2$ **b** $-3>-2$ **c** $0<1$ **d** $-1<0$

11 Which numbers should fill the spaces?

a

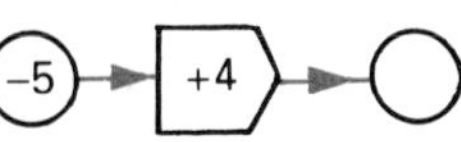

b

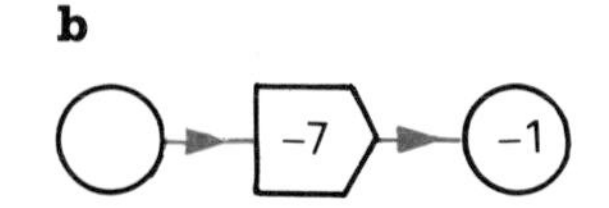

12 Simplify:
a $3x-5x$ **b** $4k+(-k)$ **c** $-3t-2t$
d $-4x-(-7x)$ **e** $-y-(-y)$ **f** $p-(-7p)$

ADDITION AND SUBTRACTION

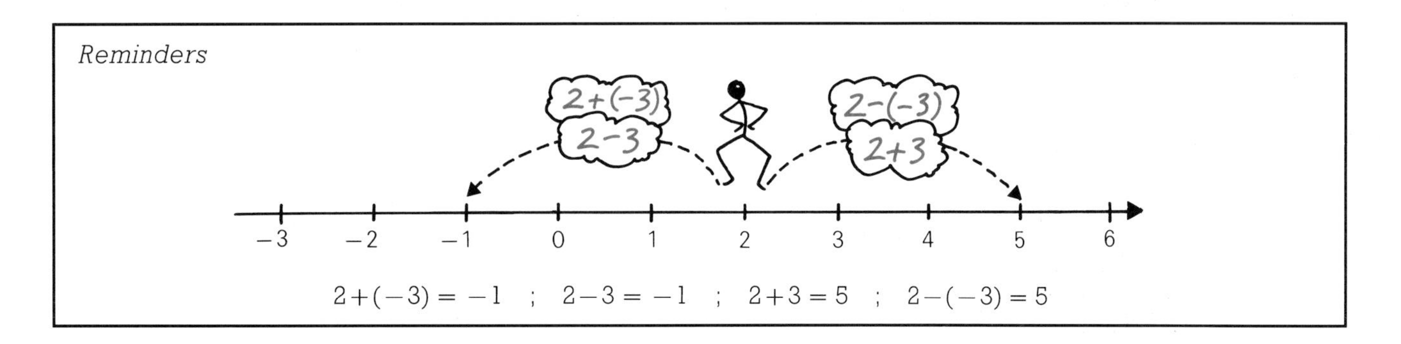

EXERCISE 1

1 Calculate these *without* using a calculator:

a $4+5$ **b** $6-4$ **c** $4-6$
d $-8+2$ **e** $-3+5$ **f** $4+(-1)$
g $5+(-5)$ **h** $-2+(-3)$ **i** $1+(-6)$
j $0+(-4)$ **k** $7-8$ **l** $-3-3$
m $-1-5$ **n** $7-5$ **o** $4-9$
p $4-(-2)$ **q** $1-(-6)$ **r** $-4-(-2)$

2 Now check your answers to question **1** *with* a calculator.

3 Again calculate *without* using a calculator.

a $-2-3$ **b** $-3-(-1)$ **c** $11-1$
d $2-9$ **e** $-1+6$ **f** $0+(-2)$
g $2-(-8)$ **h** $-7-7$ **i** $6+(-7)$
j $-1-(-1)$ **k** $15+(-9)$ **l** $-10-(-7)$
m $12-20$ **n** $-9+9$ **o** $19-(-9)$
p $-6-8$ **q** $8-(-8)$ **r** $-11+11$

4 Add as many different pairs of scores as you can.

5 Subtract as many different pairs of scores as you can.

5 -4 -1

6 Copy and complete the 'Balance' column in the bank statement. Pay-ins are marked '+' and pay-outs are marked '−'.

Date	Pay in/out (£)	Balance (£)
1 June	+25.00	25.00
4 June	−15.00	
11 June	+ 7.00	
14 June	−19.00	
20 June	+15.00	
21 June	−12.00	

7 Mr Mason runs a second-hand goods business. Each month he marks the profits (positive or negative) on a bar chart.

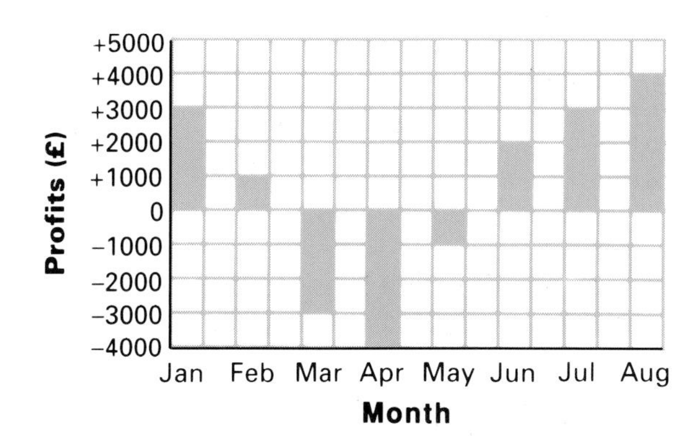

a What were his best and worst months?
b Calculate the total profit for the 8 months.
c What are 'negative profits' usually called?

8 Find three more terms for each sequence:

a 7, 5, 3, 1, . . . **b** −10, −8, −6, −4, . . .
c −10, −9, −8, −7, . . . **d** −10, −9, −7, −4, . . .

9 Hassan has challenged Moira to a round of mini-golf. He calculates his score on the strokes he takes, over and under three per hole, like this:

Hole	1	2	3	4	5	6	
Strokes	4	6	2	3	1	3	
Score	+1	+3	−1	0	−2	0	= +1

a Make a table for Moira.
She took 3, 6, 2, 3, 2, 5 strokes.
b Now make tables for the next 12 holes.
Hassan: 2, 6, 5, 3, 4, 4, 5, 3, 2, 4, 7, 4
Moira: 2, 5, 6, 1, 5, 3, 7, 4, 3, 2, 5, 3
c Who won over the 18 holes?

LETTERS AND NUMBERS

Examples

1 If $x = -2$, then:
$x+2$
$= -2+2$
$= 0$

2 If $y = -3$, then:
$5-y$
$= 5-(-3)$
$= 8$

3 Simplify:

a $x^2-(-x^2)$
$= x^2+x^2$
$= 2x^2$

b $3m-2n+m-n$
$= 3m+m-2n-n$
$= 4m-3n$

EXERCISE 2

1 Find the value of each expression using the given values of $x, y, \ldots$

a $8-x;\ x = 10$
b $10+x;\ x = -2$
c $16-y;\ y = -3$
d $y-2;\ y = -4$
e $t+5;\ t = -3$
f $t+4;\ t = 9$
g $7-k;\ k = 12$
h $9-k;\ k = -12$
i $-7-x;\ x = 0$
j $m+7;\ m = -2$
k $p-4;\ p = 1$
l $-x+7;\ x = 5$
m $-3+b;\ b = 1$
n $-5+y;\ y = -2$
o $t-10;\ t = -3$
p $a+b;\ a = 2, b = -1$
q $b-a;\ a = 4, b = 1$
r $-r-s;\ r = 1, s = -3$
s $-t-u;\ t = 4, u = -5$

2 The height h metres of a hang glider above the top of a cliff-edge after t seconds is $h = t-20$.

a Find the height after:
(i) 30 seconds (ii) 20 seconds (iii) 5 seconds

b Explain what each answer in **a** means.

3 Hi-fly Airline calculates its profit (£P) per flight by using the formula $P = 100n-12\,000$, where n is the number of passengers.

a Calculate the profit when the number of passengers is:
(i) 50 (ii) 100 (iii) 150 (iv) 200.

b What is the 'break even' number?

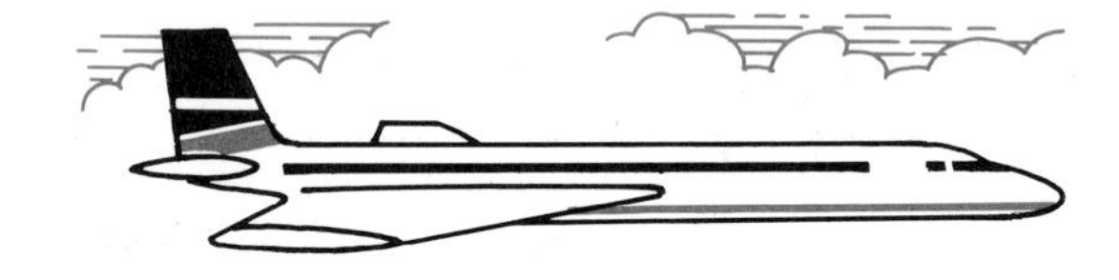

4 A ball is thrown up from the edge of a cliff 12 metres above the sea.

The height of the ball above the cliff, h metres, after t seconds, is given by the formula $h = 4t-t^2$.

a Copy and complete this table:

t	0	1	2	3	4	5	6	7
h	0	3				-5		

b Using axes and scales as shown, draw a graph on squared paper of the flight of the ball.

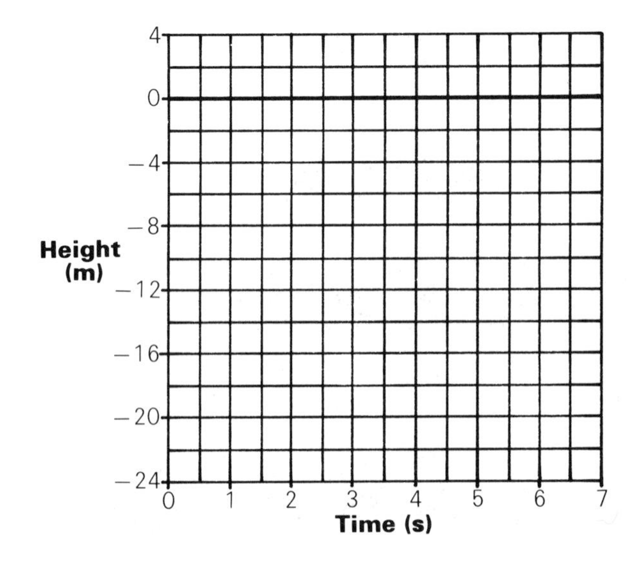

c How long does it take for the ball to reach the sea?

5 Copy and complete these lists.

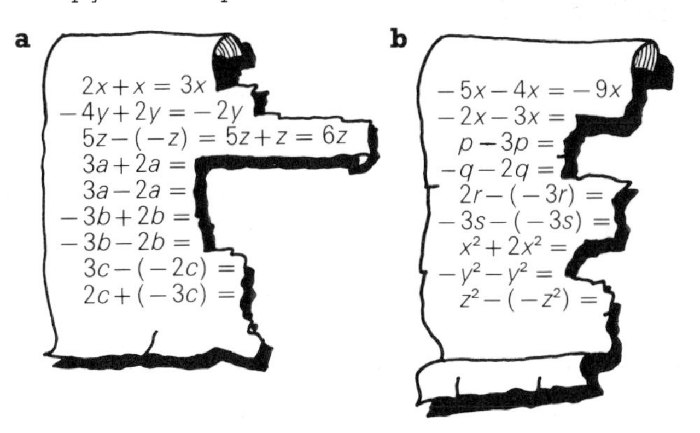

6 Simplify:

a $4a+3a-2a$
b $3b-2b-b$
c $c+2c-3c$
d $2d+2d-(-2d)$
e $-e-e-e$
f $4f-5f+f$
g $2g-2g-2g$
h $-h-3h+2h$
i $2x^2-x^2+3x^2$
j $x^2-(-x^2)+x^2$
k $-y^2+4y^2-y^2$
l $-z^2-5z^2+9z^2$
m $x+2x+y-2y$
n $3p-p+q-3q$
o $3m+n-4m-n$
p $s-t-2s-t$
q $-5u-v+u+2v$
r $a^2+b^2-2b^2-a^2$
s $2x^2-y^2-4x^2+3y^2$
t x^2-2x+x^2-2x
u $ab-a^2+ab+3a^2$
v $x^2-y^2+z^2-x^2-y^2-z^2$

INVESTIGATION

Calculate the possible differences between pairs of dice in each set. How many different values can you find for each set?

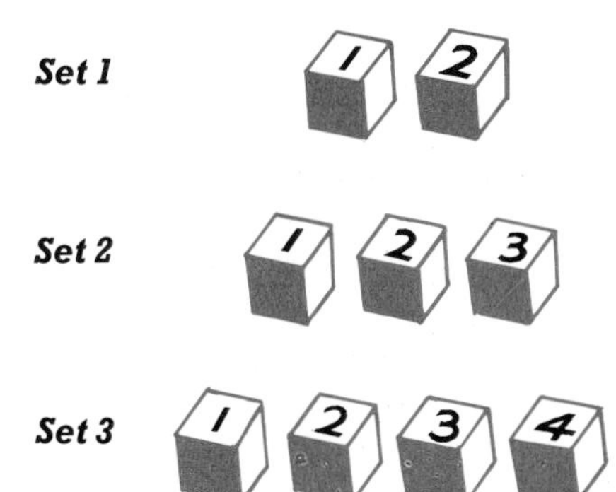

Can you find a formula for set n?

Set n

CHALLENGES

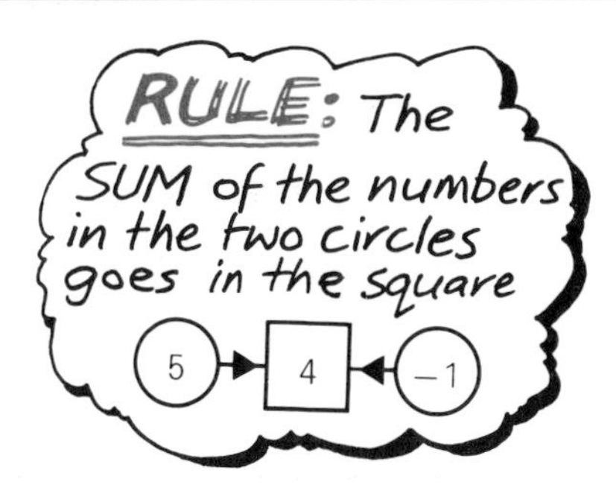

1 *Copy and complete:*

a

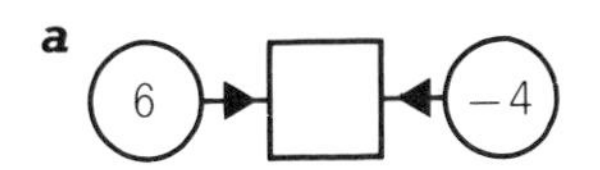

b **c**

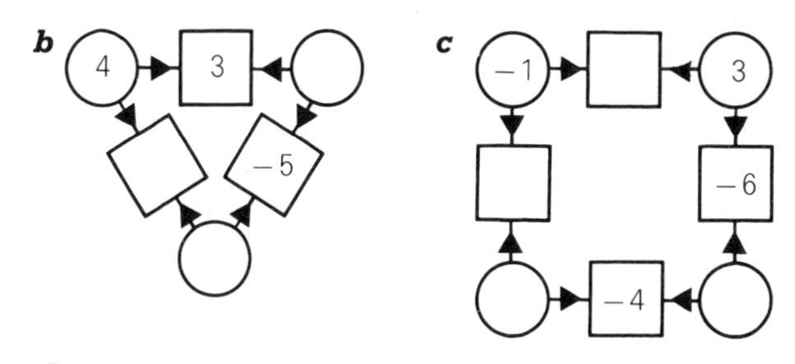

d

x — 1 — ○ — 2 — ○ — 3 — ○ . . .

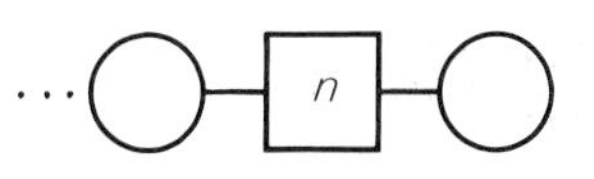

(Different entries exist for these last two circles for n even and n odd.)

2 *Copy and complete this obstacle course.*

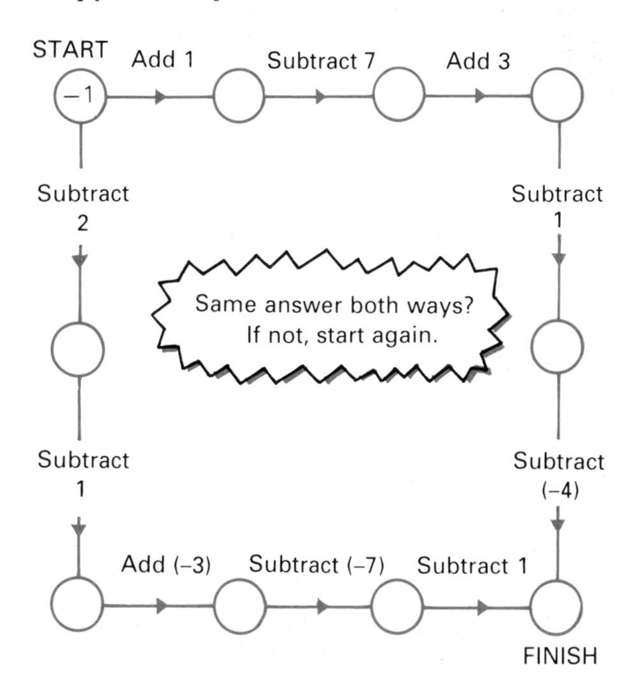

MULTIPLICATION OF POSITIVE AND NEGATIVE NUMBERS

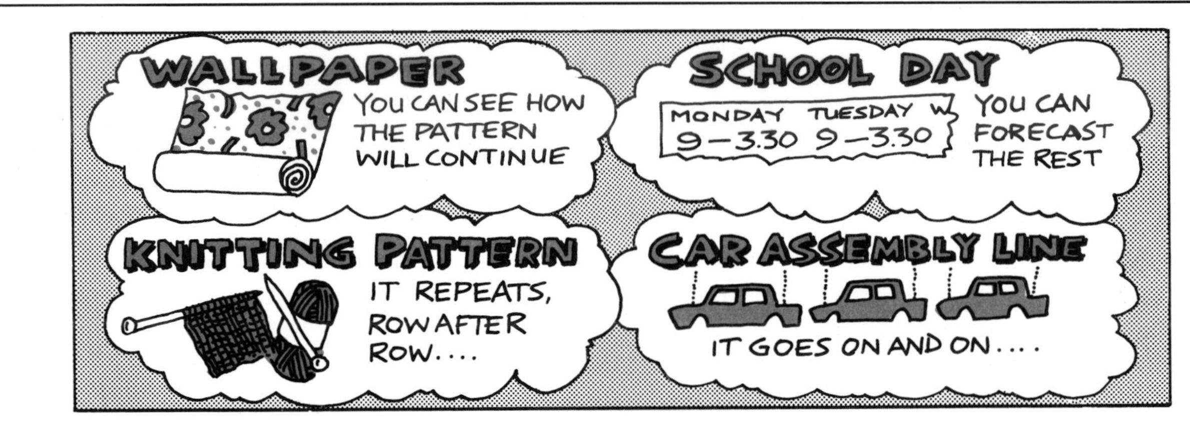

There are lots of patterns in the world around you, and there are patterns too in mathematics. The pattern in the sequence 9, 6, 3, 0, . . . can be given by the rule 'subtract 3', leading to $-3, -6, -9, \ldots$. Sequences like this are very useful, as you'll find in the next exercise.

EXERCISE 3A

1 a Copy this table:

×	3	2	1	0	−1	−2	−3
3	9	6	3	0			
2	6	4	2	0			
1	3	2	1	0			
0	0	0	0	0			
−1							
−2							
−3							

b Use the sequences of numbers along the rows and down the columns to complete the table.

2 Use your table to write down the values of these:
a $3\times(-2)$ **b** $2\times(-1)$ **c** $1\times(-3)$
d -1×1 **e** -2×2 **f** -3×3

3 Check your answers to question **2** by calculator.

> Can you see that:
>
> **a positive number × a negative number = a negative number?**

4 Use your table to write down the values of:
a $-1\times(-1)$ **b** $-1\times(-2)$ **c** $-1\times(-3)$
d $-2\times(-1)$ **e** $-3\times(-2)$ **f** $-3\times(-3)$

5 Check your answers to question **4** by calculator.

> Can you see that:
>
> **a negative number × a negative number = a positive number?**

> *Summary*
>
> For a product of two numbers:
>
> **s**ame **s**igns, **p**roduct is **p**ositive
>
> *Examples:* $3\times2=6$
> $-3\times(-2)=6$
>
> different signs, product is negative
>
> *Examples:* $3\times(-2)=-6$
> $-3\times2=-6$

6 Calculate, without using a calculator:
a 5×7, $5\times(-7)$, -5×7, $-5\times(-7)$
b 4×6, $4\times(-6)$, -4×6, $-4\times(-6)$
c 9×2, $-9\times(-2)$, $9\times(-2)$, -9×2
d $7\times(-7)$, 7×7, -7×7, $-7\times(-7)$
e $-5\times(-10)$, $5\times(-10)$, -5×10, 5×10

7 ALL MIXED UP

Calculate, then *check* by calculator:
a 1×5 **b** $2\times(-5)$ **c** -3×5
d $-4\times(-5)$ **e** 5×0 **f** $7\times(-3)$
g $-1\times(-1)$ **h** -4×9 **i** 6×7
j $6\times(-6)$ **k** $-3\times(-6)$ **l** -3×8
m $8\times(-3)$ **n** -4×0 **o** 5×5
p $-2\times(-2)$ **q** $(-3)^2$ **r** $(-4)^2$
s $(-5)^2$ **t** $(-1)^2$ **u** 9×7
v -9×7 **w** $-9\times(-7)$ **x** $9\times(-7)$

8 For each set of dice (**a**–**f**), multiply three pairs of numbers to get three different answers.

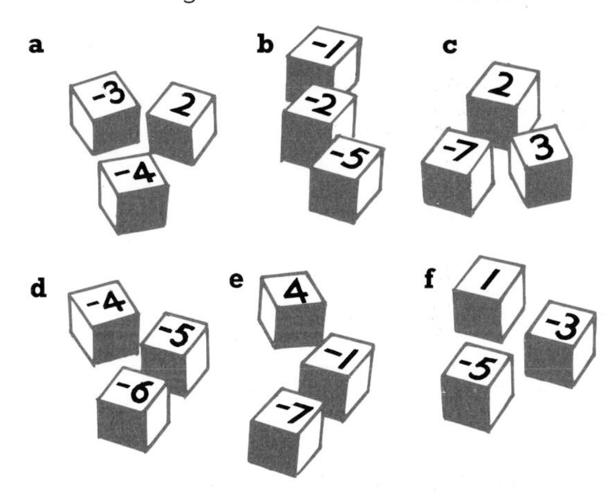

9 The Blue Bus Company and the Red Bus Company are in competition. Calculate their total profit or loss for this day.

	No. of routes making profit		Day's profit (£100s)	
Profit (£100s)	Blue Co.	Red Co.	Blue Co.	Red Co.
−5	2	3	−10	−15
−4	3	1		
−3	5	6		
−2	4	1		
−1	1	2		
0	10	8		
1	5	6		
2	8	9		
3	12	10		

EXERCISE 3B

No calculators in this exercise!

1 Calculate:

a 5×9 **b** -5×9 **c** $5\times(-9)$
d $-5\times(-9)$ **e** 5×0 **f** $-4\times(-1)$
g -6×4 **h** $8\times(-3)$ **i** $0\times(-5)$
j $1\times(-1)$ **k** 6×9 **l** $-9\times(-6)$
m $10\times(-9)$ **n** -7×7 **o** $-8\times(-7)$
p -3×7 **q** $7\times(-7)$ **r** $-6\times(-8)$
s $-1\times(-1)$ **t** $-10\times(-100)$ **u** -1×0

2 Give a rule for each sequence below, then write down three more terms:

a 1, −2, 4, −8, . . .
b 5, −5, 5, . . .
c −8, 4, −2, . . .

3 Calculate:

a $(-1)^2$ **b** $(-3)^2$ **c** $(-2)^3$
d $-1\times 2\times(-3)$
e $3\times(-2)\times(-1)$
f $4\times 3\times(-2)$
g $(-3)^4\times 0\times(-1)$
h $-2\times(-5)\times(-6)$
i $-3\times(-2)^2$

4 Calculate:

a $2\times 3+(-1)$ **b** $-2\times(-1)+4$
c $3+2\times(-1)$ **d** $-1+2\times(-1)$
e $(-4)^2-3$ **f** $-10+(-2)^2$
g $-2\times(-3)+(-1)\times(-3)$
h $(-3)^2+(-2)^2$
i $(-6)^2-(-3)^2$
j $4\times(-2)-(-1)^2$

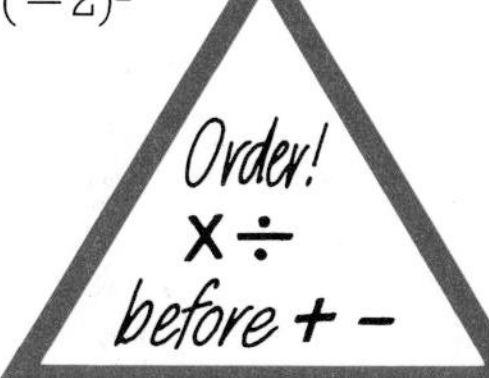

5 Calculate:

a $2(-1+3)$ **b** $(-2+4)^2$
c $-3(5-2)$ **d** $-2(1-4)$
e $6-(2-3)$ **f** $-3-(5-3)$
g $4-2(3-7)$ **h** $12-2(5-8)$

Brackets come first

TARGET NUMBERS

−3	2	1	0
5	−4	−2	3
6	2	−5	7
4	−1	−6	−7

*Choose any three numbers, in order, in a row, column or diagonal. Multiply the first two, then add or subtract the third. This gives you a **target number**. For example, here are five ways of getting the target number 2:*

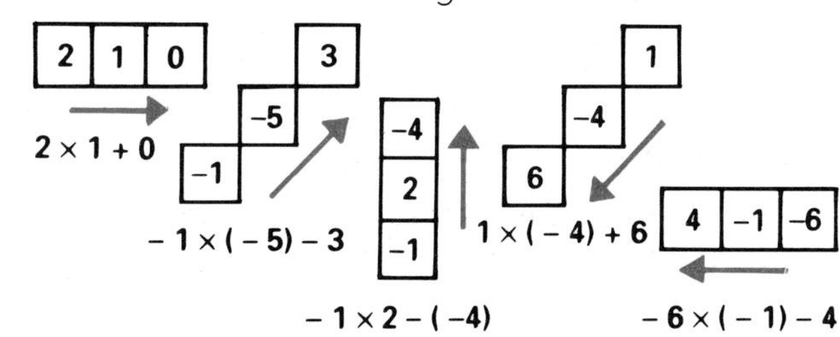

a *There are four more ways of getting the target number 2. Can you find them?*

b *Find one solution for each of these target numbers:*
(i) 5 (ii) −2 (iii) 7 (iv) −10 (v) 10

c *Find:*
(i) the highest possible target number
(ii) the lowest target number
(iii) the only target number from −10 to 10 that can't be found!

LETTERS AND NUMBERS

Examples

a Given $x = -3$,
$5-2x$
$= 5-2(-3)$
$= 5+6$
$= 11$

b Given $y = -5$,
$6-2y^2$
$= 6-2\times(-5)^2$
$= 6-2\times 25$
$= 6-50$
$= -44$

c $4x\times(-2x)$
$= -8x^2$

d $(-t)^2$
$= -t\times(-t)$
$= t^2$

EXERCISE 4A

1 Given $x = -1$, calculate the values of:
a $4x+2$ **b** $7x-3$ **c** $2x+10$
d x^2+1 **e** $2x^2$ **f** $2(x+1)^2$

2 Given $y = -2$, calculate the values of:
a $16-2y$ **b** $22-5y$ **c** $6y+2$
d y^2 **e** $1-y^2$ **f** $3y^2$

3 Given $t = -3$, calculate the values of:
a $9-5t$ **b** $4t+7$ **c** $3+t$
d $1-t$ **e** $3+3t$ **f** t^2+1
g $6t^2$ **h** $4+2t^2$ **i** $3-3t^2$

4 $a = 2$, $b = -3$ and $c = -4$. Find the values of:
a ab **b** $ab+c$ **c** $a+bc$
d $a-bc$ **e** $8ab$ **f** $ac+ab$

5 An oil pipe-line comes up from the sea, and through a pumping station at a constant slope. Each joint has a number (n), and its height (h m) above sea-level is found from the formula $h = 30-10n$.

a Calculate the height of joint number:
(i) 2 (ii) 6 (iii) -1 (iv) -6.
What does your answer to (ii) mean?
b Calculate the difference in height between joints:
(i) 5 and 8 (ii) -2 and -5 (iii) -4 and 7.

6

The train is slowing down steadily.
Ron starts his stop-watch as it passes him, travelling at speed u m/s.
The formula Ron uses to calculate the train's speed (v m/s) t seconds after passing him is $v = u-5t$.
a The train passes Ron at 30 m/s. Calculate its speed:
(i) 2 seconds later (ii) 4 seconds later.
b How many seconds after passing Ron will the train stop?
c (i) What does $t = -3$ mean?
(ii) What was the speed of the train at this time?
d The train started braking 7 seconds before reaching Ron. What was its speed then?

7 This flask is used to store very cold liquids, but the label states that liquids should not be stored in it if their boiling point is below -100°F. The table of boiling points is in °C.

	Liquid	Boiling point
a	Ether	35.0°C
b	Sulphurous Anhydride	-10.0°C
c	Chlorine	-33.6°C
d	Ammonia	-33.5°C
e	Carbon Anhydride	-78.2°C
f	Nitrous Oxide	-87.9°C

Convert each to °F, and then say whether or not the flask is suitable for storing that substance. The conversion formula is: $F = 1.8\times C+32$

EXERCISE 4B

1 $a = 1$, $b = 0$ and $c = -3$. Calculate the values of:

a abc **b** $a+b+c$ **c** $a^2+b^2+c^2$
d $ab+bc+ca$ **e** $a+b-c$ **f** $2a^2+2c^2$
g a^3+c^3 **h** a^2-c^2 **i** $(a+c)(a-c)$

2 Copy and complete:

a
$3 \times x = \ldots$
$2 \times x = \ldots$
$1 \times x = \ldots$
$0 \times x = \ldots$
$-1 \times x = \ldots$
$-2 \times x = \ldots$

b
$-3 \times (-x) = \ldots$
$-2 \times (-x) = \ldots$
$-1 \times (-x) = \ldots$
$0 \times (-x) = \ldots$
$1 \times (-x) = \ldots$
$2 \times (-x) = \ldots$

3 Simplify:

a $u \times v$ **b** $u \times (-v)$ **c** $-u \times v$
d $-u \times (-v)$ **e** $a \times a$ **f** $2b \times 2b$
g $(3c)^2$ **h** $d^2 \times d$ **i** $(-k)^2$
j $(-k)^3$ **k** $3p \times 4q$ **l** $-3p \times (-4q)$
m $(ab)^2$ **n** $a^2 \times b^2$ **o** $-a^2 \times b^2$
p $a^2 \times (-b^2)$

4 Simplify:

a $3a-(-a)+2a$ **b** $5b-6b+b$ **c** $2c \times 3c \times 4$
d $t \times 2t \times 3t$ **e** $m \times m + n \times n$ **f** $ab+ba$
g $pq-2pq+qp$ **h** $c^2-4c^2+5c^2$ **i** $ab \times (-ab)$
j $2 \times 3x \times (-4x^2)$

5 Multiply out, to remove brackets:

a $a(b+c)$ **b** $p(q-r)$ **c** $y(2m-n)$
d $x(2a+3b)$ **e** $a(a-b)$ **f** $n(n^2-m)$

6 (i) Use these number machines to calculate the y-coordinates of the points with x-coordinates -3, -2, -1, 0, 1, 2, and 3.
(ii) Plot the seven points on an XOY diagram, and join them up with a straight line.

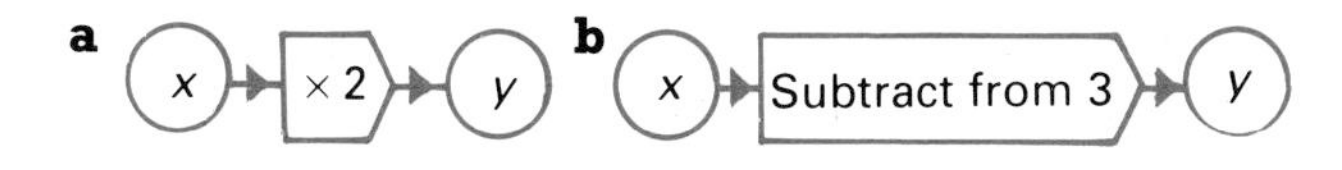

CHALLENGE

Use each of the numbers 1, 2, 3 and some of the symbols +, −, × and ( ) to produce each whole number from 0 down to −8. For example, $(2-1) \times (-3) = -3$.

INVESTIGATIONS

1 **a** *(i) If $x = 4$, then $x^2 = 16$. True or false?*
(ii) If $x = -4$, then $x^2 = 16$. True or false?
b *(i) If $x^2 = 16$, $x=4$.*
(ii) If $x^2 = 16$, $x = -4$.
(iii) If $x^2 = 16$, $x = 4$ or -4.
Only one of these is true. Which one?

2 **a** *Investigate possible solutions of the equation $x^n = 1$, where n is a whole number.*
b *What happens when $x^n = -1$?*

3 *Study the values of $(-1)^2, (-1)^3, (-1)^4, (-1)^5, \ldots$*
a *Investigate the value of $(-1)^n$, where n is a whole number.*
b *Investigate whether a^n is positive or negative, where a is a positive or negative whole number, and n is a positive whole number.*
c *Write up your investigation.*

DIVISION OF POSITIVE AND NEGATIVE NUMBERS

Use your calculator to check that: (i) $\frac{6}{3} = 2$ (ii) $\frac{-6}{-3} = 2$ (iii) $\frac{-6}{3} = -2$ (iv) $\frac{6}{-3} = -2$.

The same pattern of signs appears as for products of two numbers. For any quotient:

same signs, positive quotient
different signs, negative quotient

EXERCISE 5

Only use your calculator to *check* some of your answers.

1 Say whether each of these is positive or negative:

a $\frac{-3}{1}$ **b** $\frac{5}{4}$ **c** $\frac{7}{-2}$ **d** $\frac{-8}{-3}$

e $\frac{-1}{8}$ **f** $\frac{-6}{-1}$ **g** $\frac{12}{-13}$ **h** $\frac{-12}{-13}$

2 Calculate, then check by calculator:

a $8 \div 2$ **b** $8 \div (-2)$ **c** $-8 \div 2$
d $-8 \div (-2)$ **e** $-8 \div 4$ **f** $21 \div (-7)$
g $-25 \div (-5)$ **h** $9 \div (-1)$ **i** $-9 \div (-1)$

3 Calculate:

a $\frac{16}{-4}$ **b** $\frac{-12}{3}$ **c** $\frac{-1}{-1}$ **d** $\frac{30}{-2}$ **e** $\frac{-9}{-3}$

f $\frac{-100}{10}$ **g** $\frac{36}{9}$ **h** $\frac{13}{-1}$ **i** $\frac{-27}{-3}$ **j** $\frac{54}{18}$

4 Calculate the mean of these scores:
5, −2, 3, 1, −1, 4, −2, 1, 0

5 Temperatures (°C) around the world one day in January were:
−3, 0, 2, −7, −10, 1, 4, 0, −18, −1, 3, −1.
Calculate the mean temperature.

6 Use the formula $F = \frac{9}{5}C + 32$ and $C = \frac{5}{9}(F - 32)$ to convert:

a to Celsius
(i) 41°F (ii) 23°F (iii) 5°F

b to Fahrenheit
(i) −10°C (ii) −15°C (iii) −150°C.

7 Given $p = 8$, $q = -4$ and $r = -2$, calculate the values of:

a $\frac{p}{q}$ **b** $\frac{p}{r}$ **c** $\frac{q}{r}$ **d** $\frac{pq}{r}$

e $\frac{qr}{p}$ **f** $\frac{pr}{q}$ **g** $\frac{p-q}{q-r}$ **h** $\frac{q+r}{p+r}$

8 Robin found a way to calculate the coordinates of the midpoint of a line. Here it is:

$$M\left(\frac{-6+2}{2}, \frac{1+(-3)}{2}\right) = M(-2, -1).$$

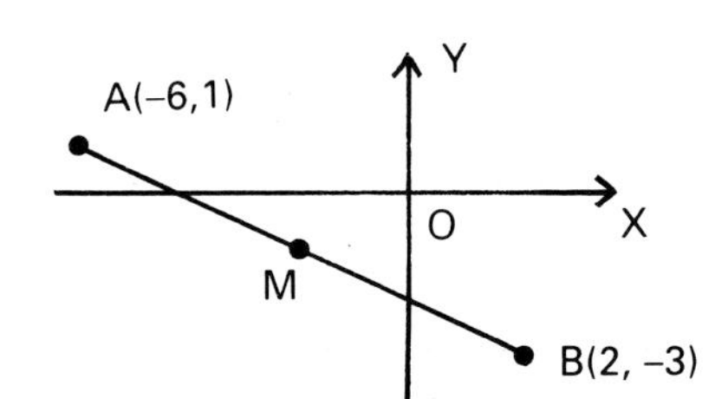

Use his method to find the midpoint of the line joining:

a A(−1, −6), B(3, 8) **b** P(−5, 5), Q(−1, 1)
c O(0, 0), T(−10, 8) **d** R(−3, −5), S(−3, 1)

9 You can find the time of day at any place P in the world from the formula:

$$\text{Time at P} = \text{Time at London} + \frac{\text{Longitude of P}}{15}.$$

West longitude is negative, so 20°W means (−20).

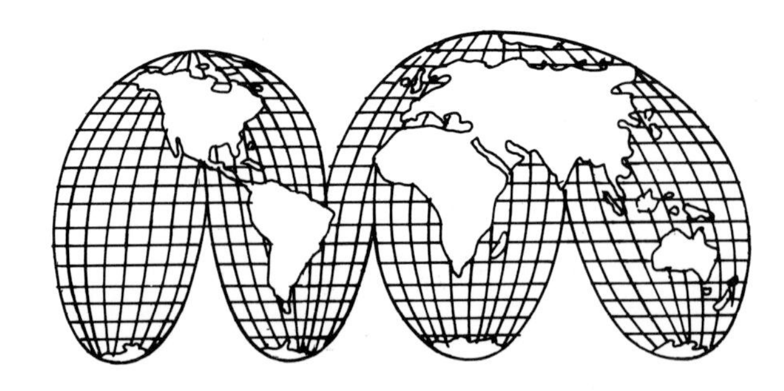

a Find the time, to the nearest hour, in these places when it is 6 pm in London:
(i) Las Palmas, 15°W (ii) St Petersburg, 30°E
(iii) Adelaide, 135°E (iv) San Francisco, 120°W.

b Comparing times with noon in London, find the time gap between Cairo USA, 90°W and Cairo Egypt, 30°E.

EQUATIONS

Reminders

$a = b$

$a+2 = b+2$

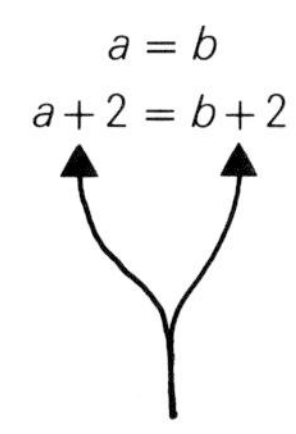

Add the same number to each side

$a = b$

$a-3 = b-3$

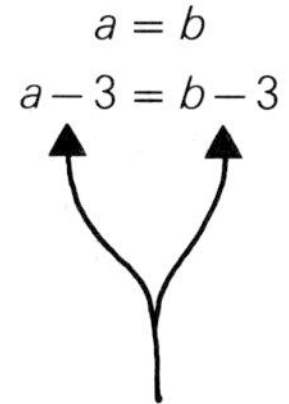

Subtract the same number from each side

$a = b$

$3a = 3b$

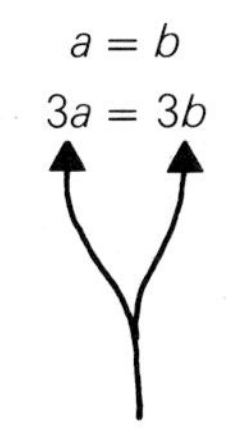

Multiply each side by the same number

$a = b$

$\frac{a}{4} = \frac{b}{4}$

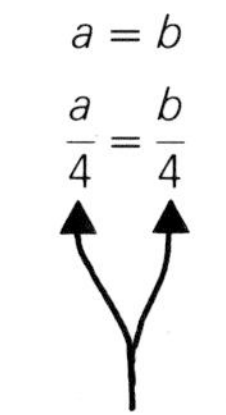

Divide each side by the same non-zero number

Example Solve these equations.

a
$$\begin{aligned} 14-x &= 8 \quad \text{(Subtract 14 from each side.)} \\ -14 \quad & \quad -14 \\ -x &= -6 \quad \text{(Multiply each side by } -1.) \\ x &= 6 \end{aligned}$$

b
$$\begin{aligned} 3x-4 &= -10 \quad \text{(Add 4 to each side.)} \\ +4 \quad & \quad +4 \\ 3x &= -6 \quad \text{(Divide each side by 3.)} \\ x &= -2 \end{aligned}$$

EXERCISE 6A

1 Solve the equations. Check with the solutions below.

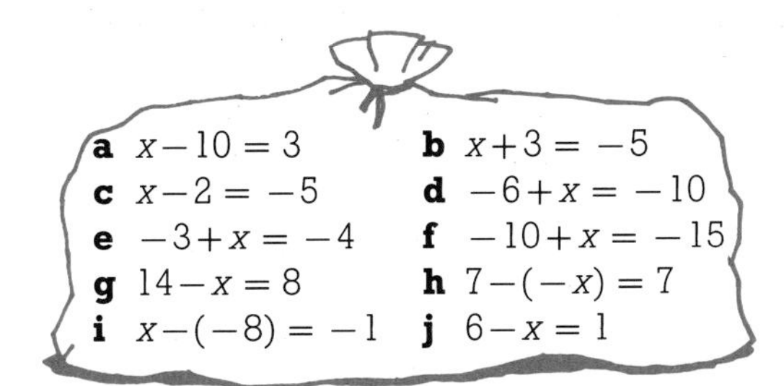

Solutions (not in order)
$-9, -8, -5, -4, -3, -1, 0, 5, 6, 13$

2 Now solve these equations.

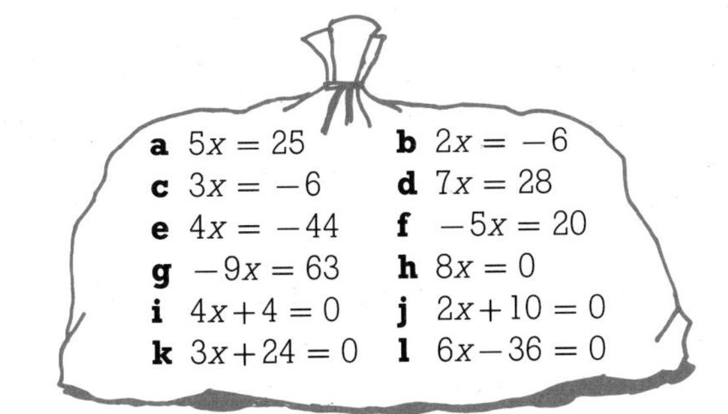

Solutions (not in order)
$-11, -8, -7, -5, -4, -3, -2, -1, 0, 4, 5, 6$

3 Solve these equations.

a $x-8 = 3$ **b** $y-5 = -2$
c $z+6 = -4$ **d** $x+7 = 5$
e $x-10 = 9$ **f** $t+3 = -5$
g $x-6 = -1$ **h** $n+9 = 4$
i $-5+k = -6$ **j** $-3+x = 2$
k $-8-m = -12$ **l** $12-p = 7$
m $8-(-x) = 8$ **n** $a-(-2) = 1$
o $14-x = 16$

4 Solve these equations.

a $2y-1 = 3$ **b** $3y-2 = -8$
c $4t+1 = -11$ **d** $5t+3 = 18$
e $2p-11 = 7$ **f** $3q-5 = -11$
g $4r-7 = -7$ **h** $-2s-7 = 3$
i $1-3t = 4$ **j** $3-4u = 7$
k $5 = 1+2x$ **l** $8 = 2-3v$

EXERCISE 6B

Example

Solve: $4x-2=7x-11$ (Add 2.)
$4x=7x-9$ (Subtract $7x$.)
$-3x=-9$ (Divide by -3.)
$$\frac{-3x}{-3}=\frac{-9}{-3}$$
$x=3$

or

$4x-2=7x-11$ (Interchange sides.)
$7x-11=4x-2$ (Add 11.)
$7x=4x+9$ (Subtract $4x$.)
$3x=9$ (Divide by 3.)
$x=3$

1 Solve these equations.

a $3x=x-4$ **b** $5x=2x+6$
c $6x=x-10$ **d** $5x=x-12$
e $2y-9=-y$ **f** $5y+8=y$
g $y-5=2y$ **h** $y+6=3y$
i $2x+4=x+2$ **j** $5y-3=y-19$
k $10t+4=12t+2$ **l** $2-2x=14+2x$
m $4a+2=a-13$ **n** $6b-1=2b-17$
o $3c+4=8c-1$ **p** $-5d-1=2d-8$

2

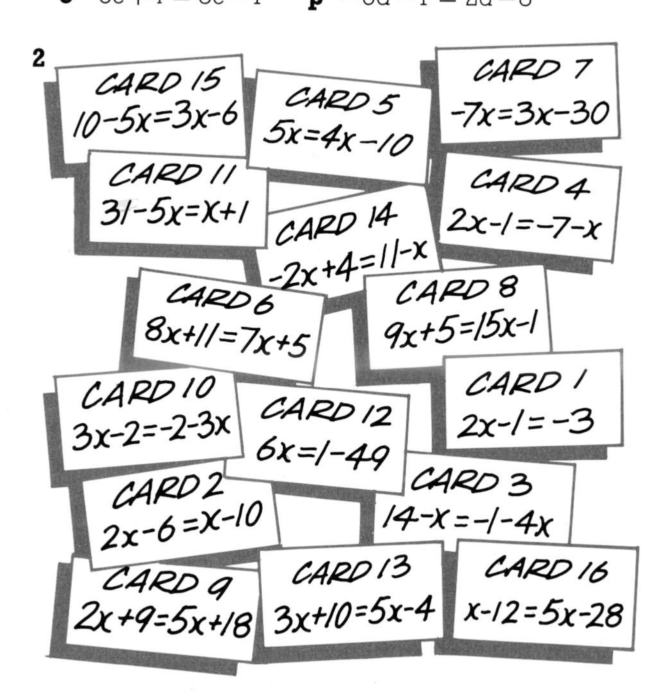

Copy the blank 4 × 4 grid. Then complete it to show where the cards go.

3	−8	−1	−7
−4	7	1	5
4	−5	0	−10
−3	−6	2	−2

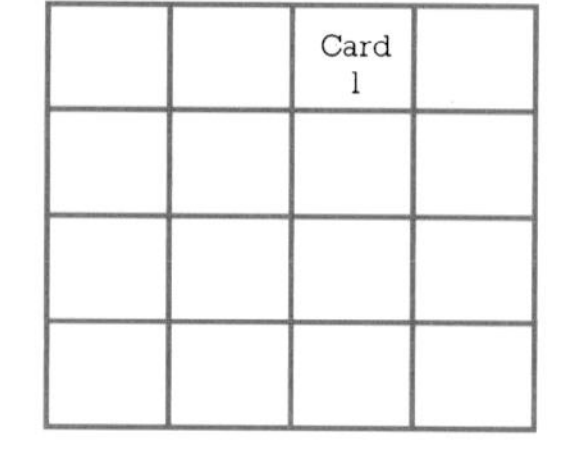

When you have completed the grid correctly you will have a magic square. (Be sure to check this!)

BRAINSTORMER

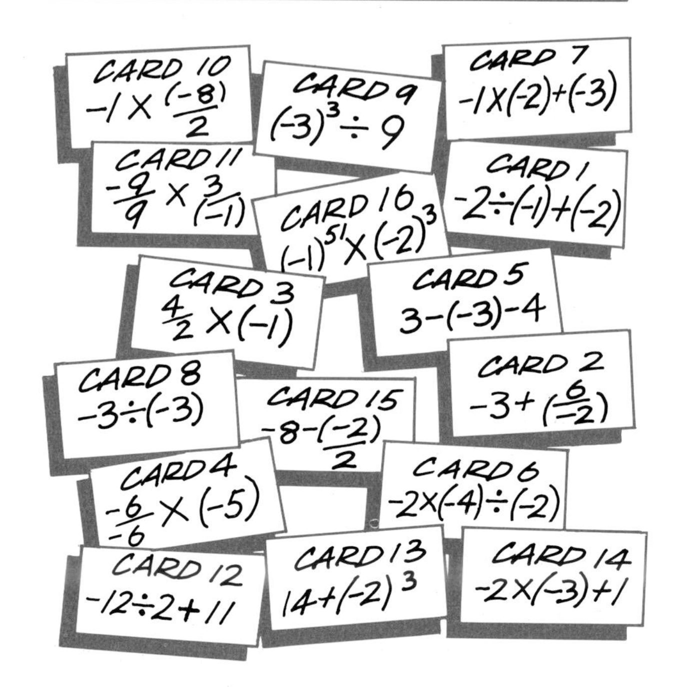

Copy this blank 4 × 4 grid. Then complete it to show where the cards go.

0	1	−7	4
7	3	−5	2
5	6	−4	−2
−1	−6	−3	8

Card 1			

When you have completed the grid correctly you will have a magic square—check this!

CHECK-UP ON POSITIVE AND NEGATIVE NUMBERS

1 Simplify:
a $-4+5$ **b** $-4-5$ **c** $-4-(-5)$
d $2x-6x$ **e** $3a-(-3a)$ **f** $6c-2c+(-4c)$

2 Calculate the IN and OUT numbers:

a

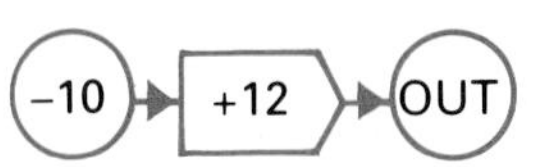

b

3 Find three more terms in each sequence below:
a 15, 10, 5, . . . **b** $-1, 3, -9, \ldots$ **c** 10, 5, 1, . . .

4 Find the entries in the third row of this table of temperatures (°C):

At 4 pm	8	0	-2	2	-1	1
At 4 am	3	-3	-3	-2	-11	-1
Fall (°C)						

5 Choose pairs of numbered cubes in **a** and **b**, and calculate all possible scores:

a

b

6 Use the three values of x to obtain three values of each expression:

Values of x
1 -1
-2

Expressions
a $x-3$ **b** $14-x$
c $x-10$ **d** $-10-x$

7 Simplify:
a -4×2 **b** $-4\times(-2)$ **c** $4\times(-2)$
d $3\times(-2t)$ **e** $-1\times(-k)$ **f** $-2a\times(-2a)$

8 Simplify:
a $16-6t$, given $t=-4$ **b** $4r^2+1$, given $r=-1$
c $2a^2-5a^2$ **d** $x^2-(-x^2)$

9 a Copy and complete Lauren's 18 hole golf score. She counts 1 over par $(+1)$ and 1 under par (-1).

Number of holes	Score		Total
1	-2	$1\times(-2)$	
6	-1		
5	0		
5	1		
1	2		
	Total for round		

b Her partner Joanne kept her score in the same way. What is her total?

Number of holes	7	6	2	3
Score	-1	0	1	2

c Who won? By how many strokes?

10 Simplify:
a $8\div(-2)$ **b** $-10\div(-10)$ **c** $\dfrac{4\times(-3)}{-6}$

11 Solve these equations.
a $4x=-12$ **b** $3x+27=0$
c $16-4y=36$ **d** $6c+11=2c+7$
e $5x-2=8x+4$

12 Simplify:
a $a-2a-3a$ **b** $b\times 2b\times 3b$
c $2c^2-c^2-3c^2$ **d** $2x\times(-x)\times(-x)$

PYTHAGORAS

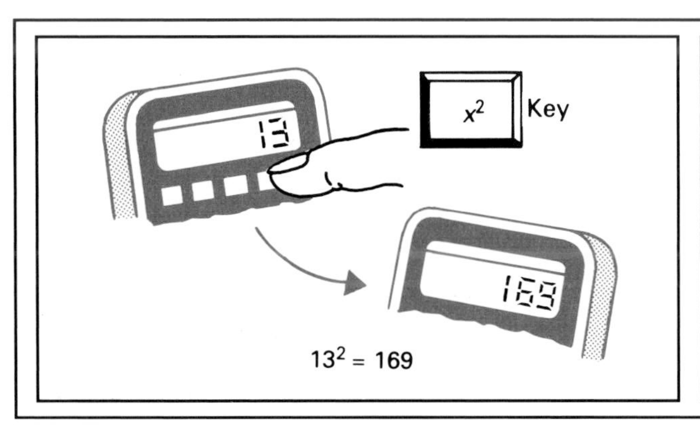

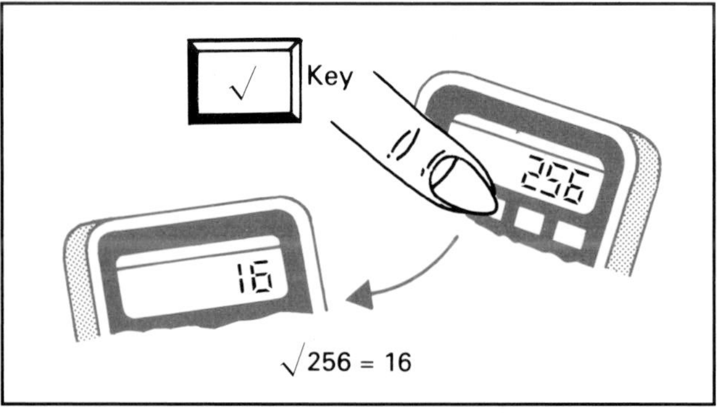

LOOKING BACK

1 How many square slabs are needed to make each of these square paved areas?

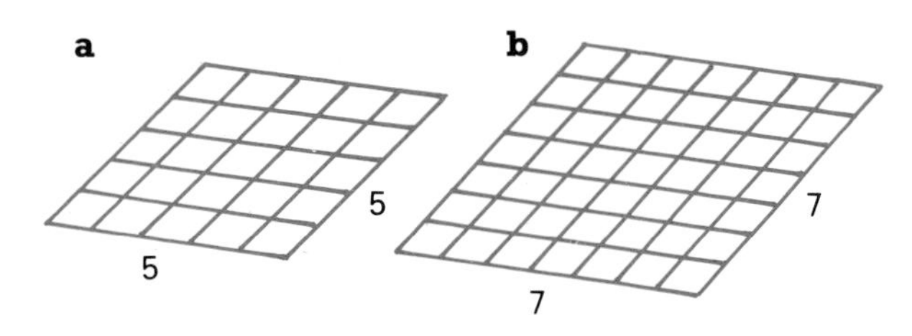

2 How many slabs are there in square paved areas containing:
a 8 rows of 8 slabs **b** 10 rows of 10 slabs?

3 How many rows of square slabs with 1 metre sides are needed for these square paved areas?

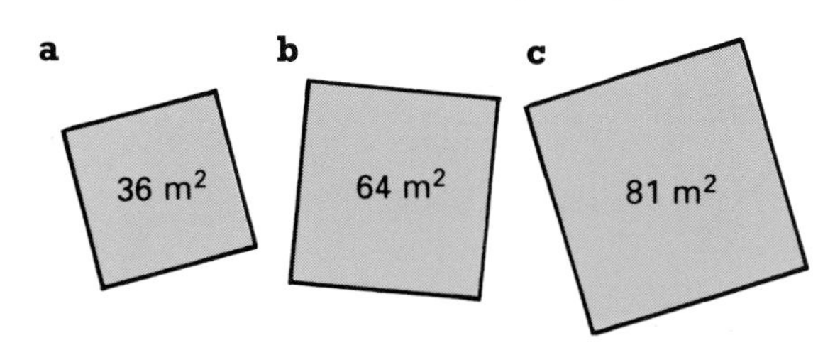

4 How many rows of slabs with 1 metre sides are needed for a square paving which has an area of: **a** 16 m² **b** 144 m² **c** 4 m²?

5 Write down the value of:
a 3^2 **b** 4^2 **c** 9^2 **d** 1.5^2 **e** 0.2^2

6 Find the value of:
a $\sqrt{100}$ **b** $\sqrt{121}$ **c** $\sqrt{12.25}$ **d** $\sqrt{1296}$

7 Calculate the area of each square below.

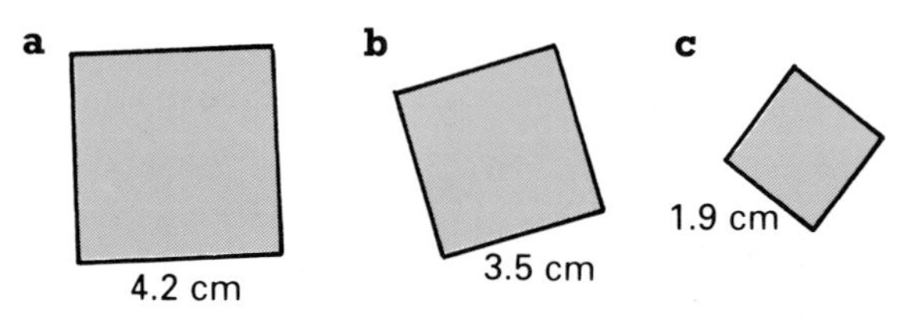

8 $a = 9$ and $b = 12$. Find the value of:
a a^2 **b** b^2 **c** a^2+b^2 **d** $\sqrt{(a^2+b^2)}$

9 How many right-angled triangles can you find in these diagrams (**b** is a cuboid)? Name them.

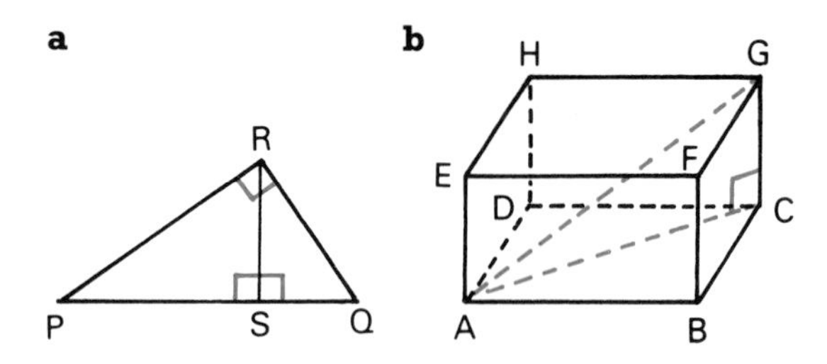

10 The diagram shows a square in a square.

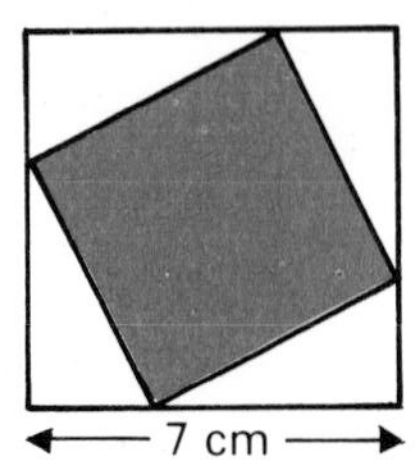

The area of each triangle is 6 cm². Calculate:
a the area of each square
b the length of a side of the smaller square.

EXERCISE 1/CLASS DISCUSSION

1 a How far up the wall will the ladder reach?

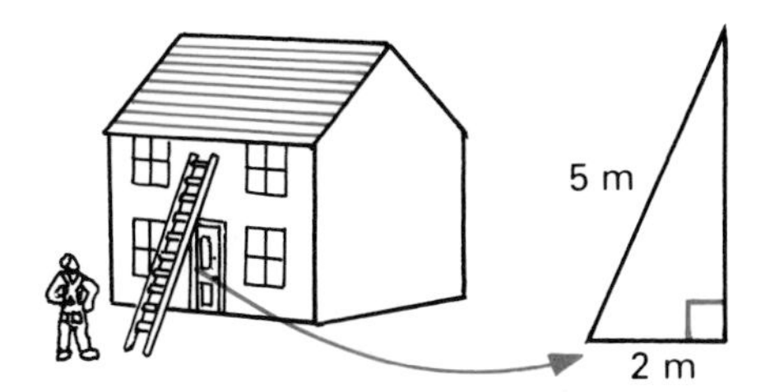

b What is the length of the car ramp?

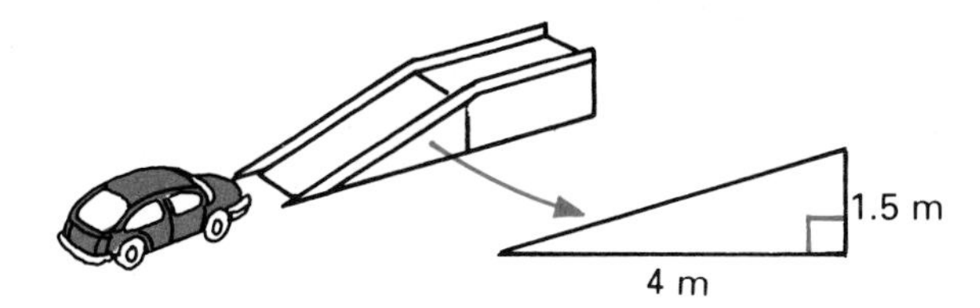

c How long is the sloping edge of the roof?

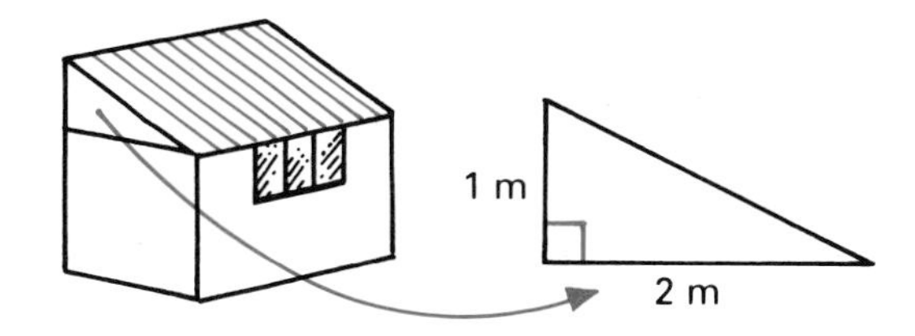

Estimate the answers to the questions above. Scale drawings *could* be used to find the lengths, but a quicker and more accurate way is to use the Theorem of Pythagoras. Pythagoras was a Greek mathematician who lived in Sicily about the sixth century BC.

2 Squares are drawn on the sides of a right-angled triangle.

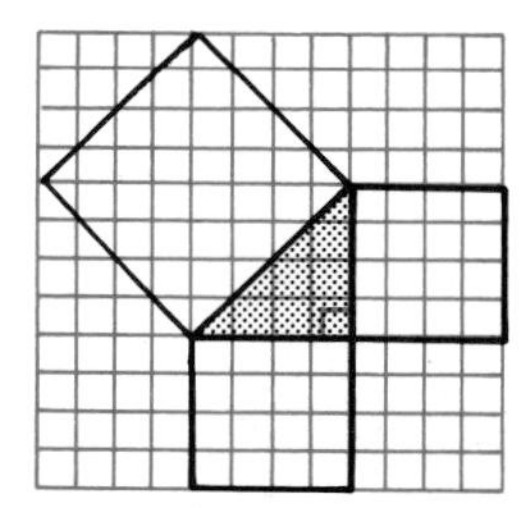

a Count the number of small squares in each one. (Two half squares make one square.)

b Check that the area of the square on the hypotenuse (the side of the triangle opposite the right angle) is equal to the sum of the areas of the other two squares.

3 a Draw a right-angled triangle with the sides about the right angle 3 cm and 4 cm long.

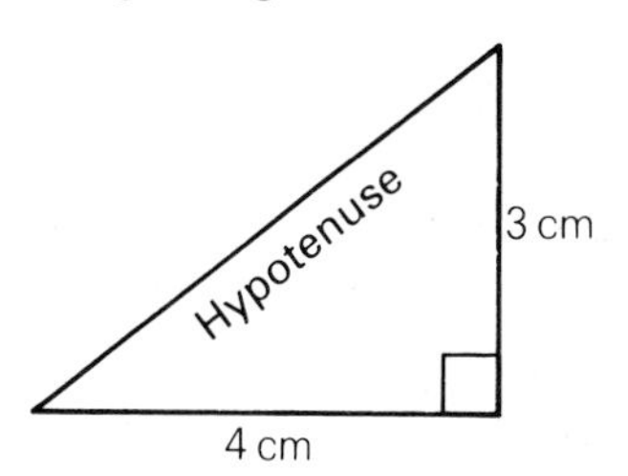

b Draw a square on each side, and divide each one into small squares of side 1 cm.

c By counting the 1 cm squares, check that the area of the square on the hypotenuse is equal to the sum of the areas of the squares on the other two sides.

4 a Draw these right-angled triangles as accurately as you can.

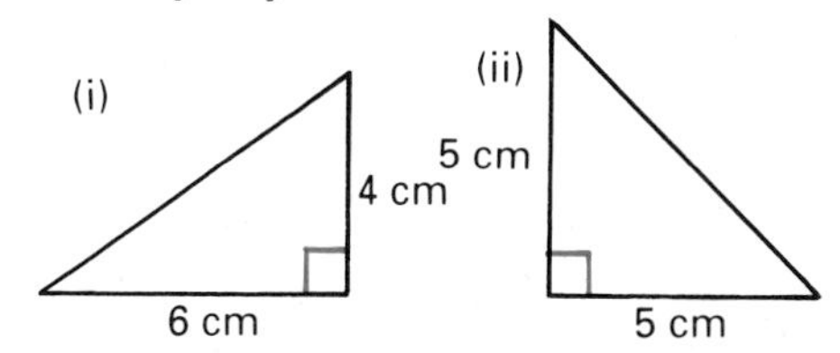

b Measure the length of each hypotenuse to the nearest 0.1 cm.

c *Calculate* the areas of the squares on all three sides of each triangle, then check that the area of the square on the hypotenuse is equal to the sum of the areas of the squares on the other two sides.

5 Repeat question **4** for some right-angled triangles of your own choice.

PRACTICAL PROJECT

a *Draw squares on the sides of a right-angled triangle.*

b *Draw lines through the centre O of one square, parallel and perpendicular to the hypotenuse.*

c *Cut out parts 1, 2, 3, 4 and 5. You should be able to cover square 6 completely with them. What does this tell you about the areas of the three squares?*

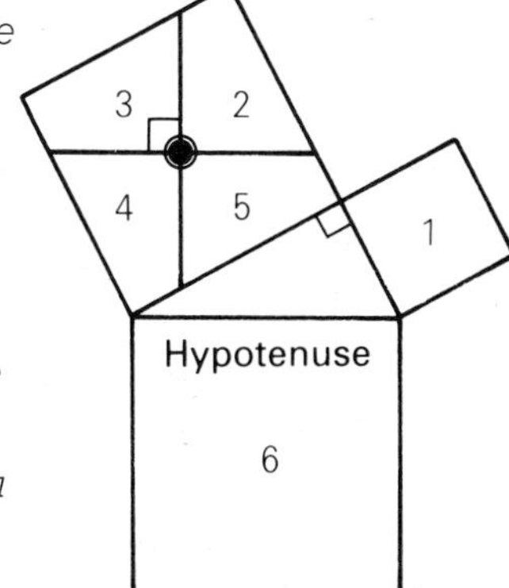

THEOREM OF PYTHAGORAS

The Theorem of Pythagoras

In a right-angled triangle, the square on the hypotenuse (the side opposite the right angle) is equal to the sum of the squares on the other two sides. In the diagram, $a^2 = b^2 + c^2$.

Pythagoras may have worked out the proof of his theorem while examining the mosaic on a temple floor.

The pattern, or mosaic, could have been made up of pairs of large congruent squares, marked off in the lengths of the sides of the right-angled triangles, like this:

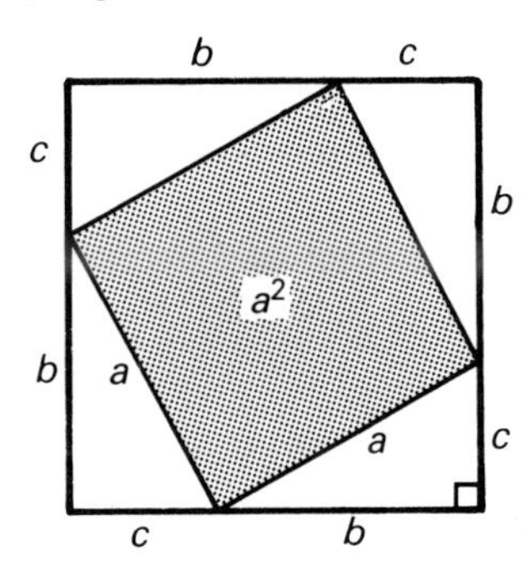

and

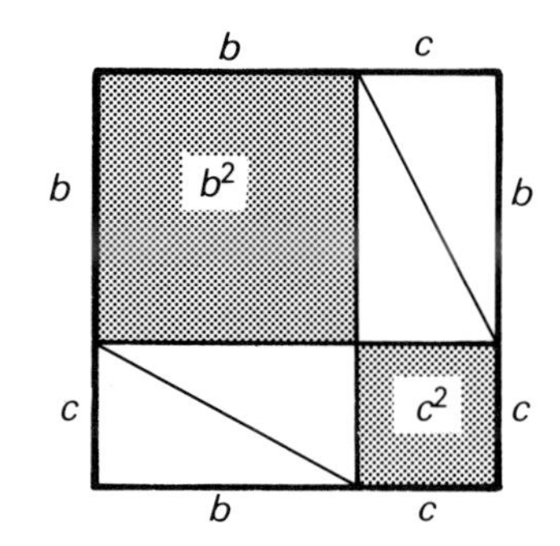

Area of large square
= area of shaded square + 4 triangles

Area of large square
= area of 2 shaded squares + same 4 triangles

So $\boldsymbol{a^2 = b^2 + c^2}$

Although the theorem is about areas, it is also very useful for calculating lengths.

Example
Calculate a.

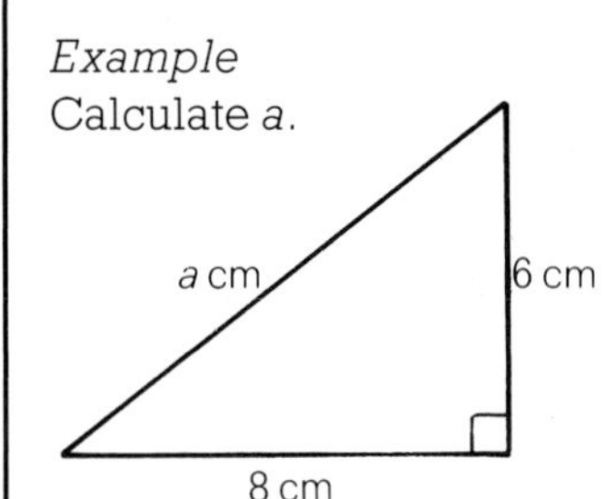

The triangle is right-angled. So by the Theorem of Pythagoras,

$$a^2 = 6^2 + 8^2 = 100$$
$$a = \sqrt{100} = 10$$

Note Throughout these exercises units of length are only marked in practical situations.

EXERCISE 2A

1 This is a right-angled triangle.
Using Pythagoras' Theorem,

$$a^2 = b^2 + c^2.$$

Write equations for the triangles **a**—**d**.

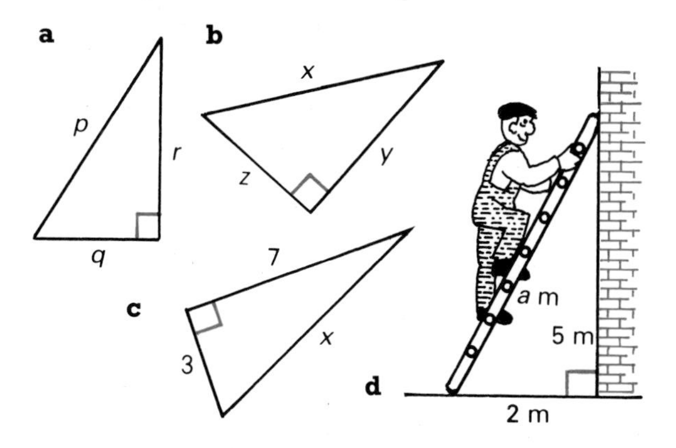

2 (i) Use Pythagoras' Theorem to write equations for these triangles.
(ii) Use your calculator to check that the measurements are correct, for example that $5^2 + 12^2 = 13^2$.

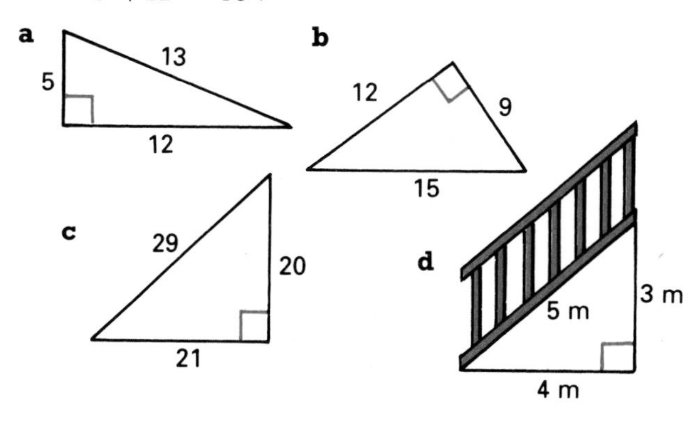

3 Calculate the length of the hypotenuse in each triangle below.

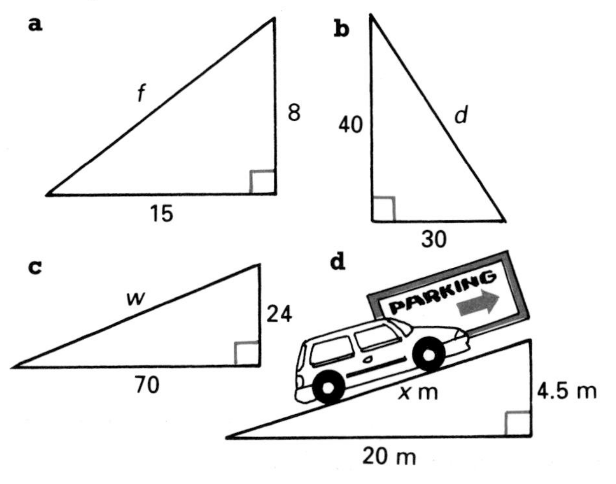

4 Calculate x, correct to 1 decimal place.

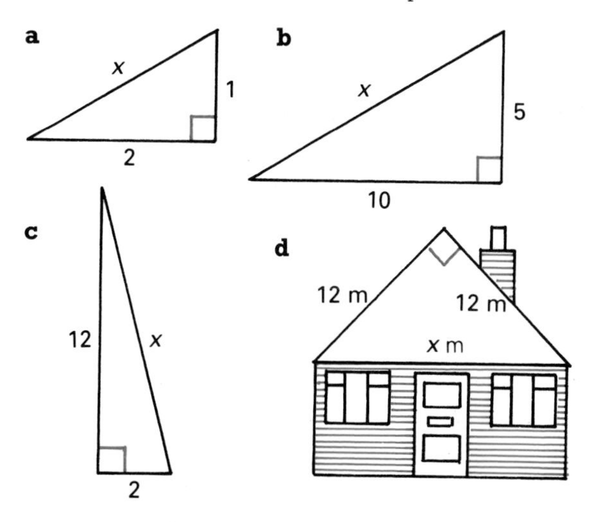

To calculate one of the smaller sides in a right-angled triangle.

Example
Calculate y.

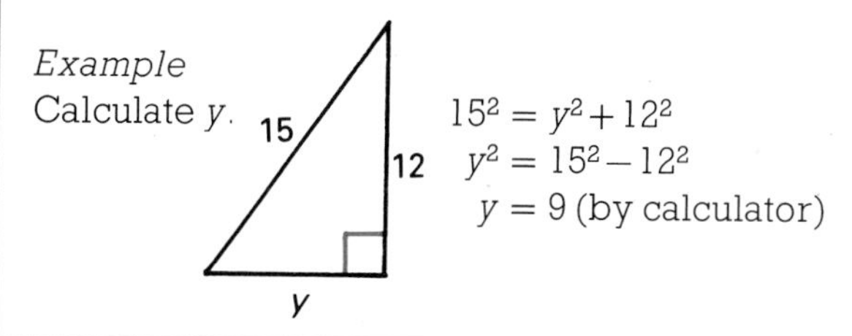

$$15^2 = y^2 + 12^2$$
$$y^2 = 15^2 - 12^2$$
$$y = 9 \text{ (by calculator)}$$

5 Calculate x, y, z.

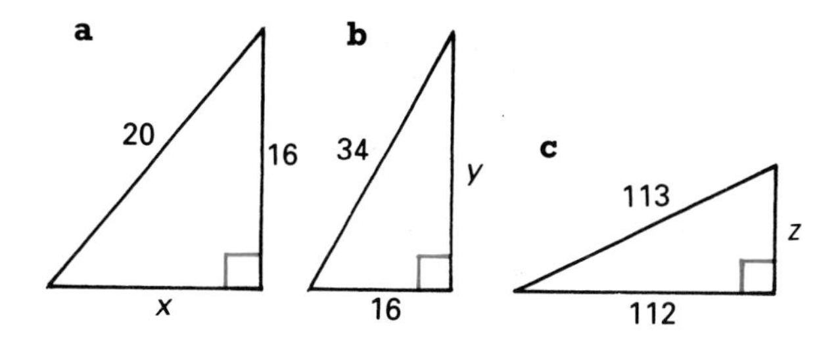

6 Calculate x, correct to 1 decimal place, in each triangle below.

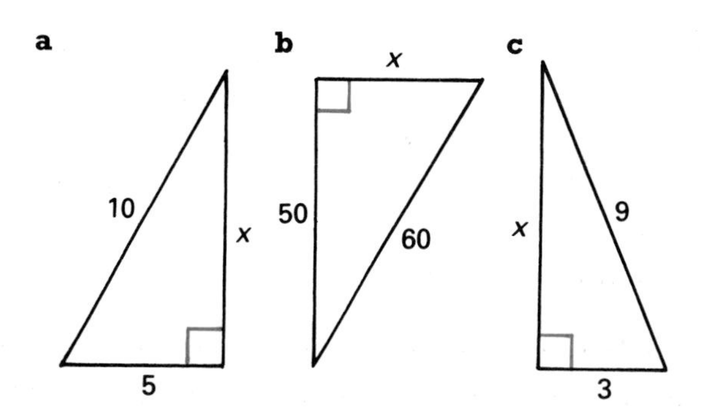

7 Calculate the length of the third side in each triangle, but take care! Sometimes it's the hypotenuse, sometimes it's not.

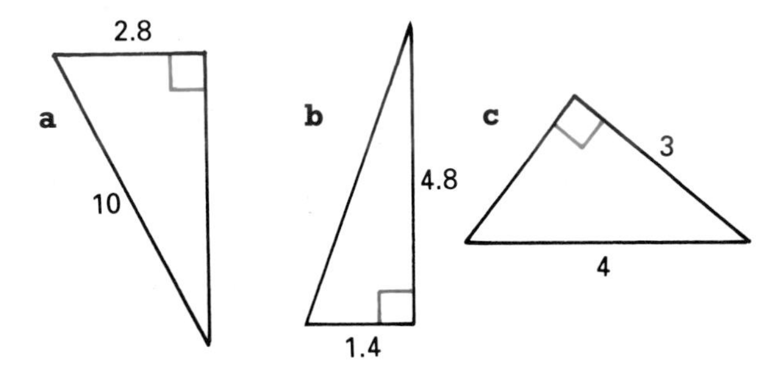

8 Find the right-angled triangle in each diagram, and calculate x, y, z and p.

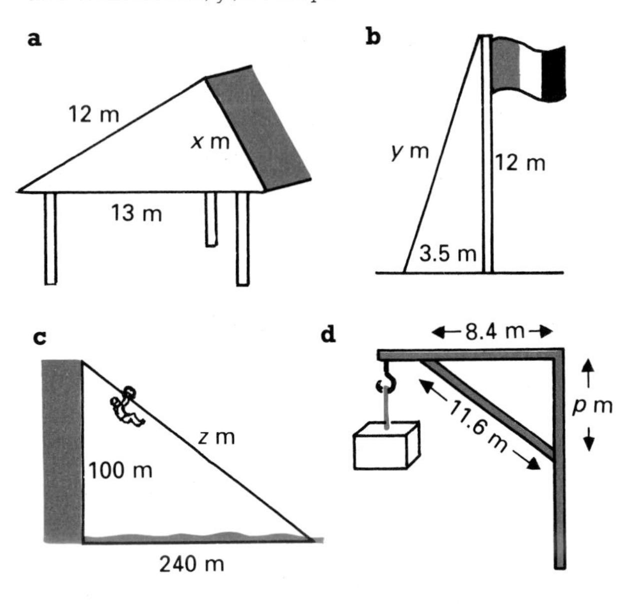

9 Use Pythagoras' Theorem to find x, then y.

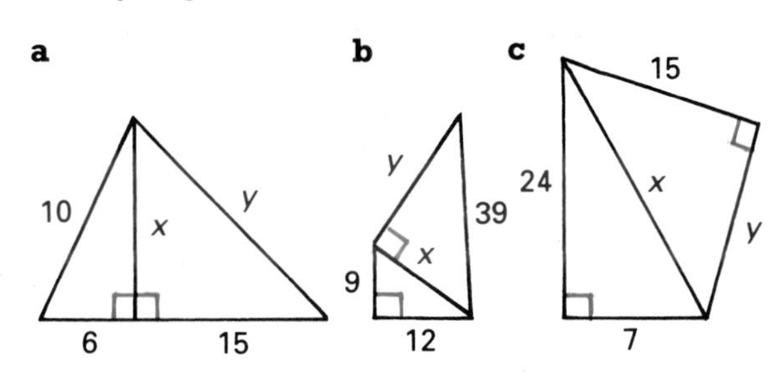

10 Calculate the length of the slide.

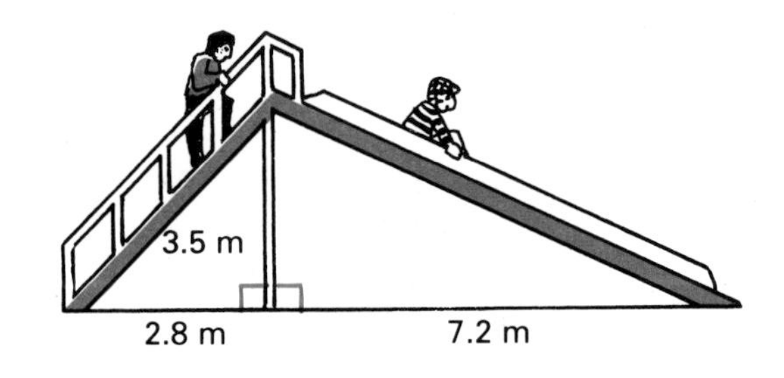

11 a How much shorter would it be to go across this rectangular field from corner to corner than around the sides?

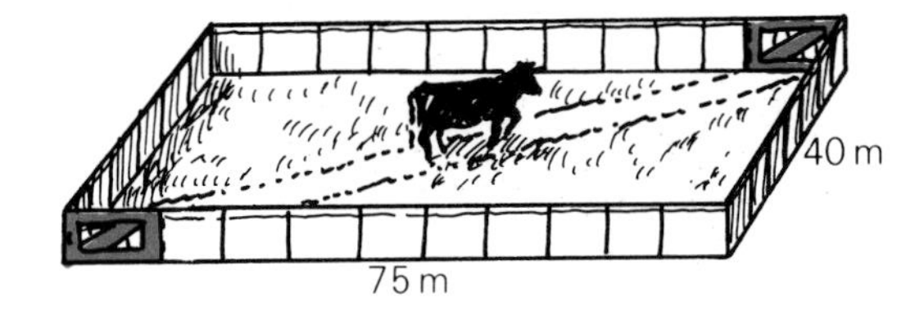

b Which way would you go?

EXERCISE 2B

1 Each diagram has an axis of symmetry in an isosceles triangle.

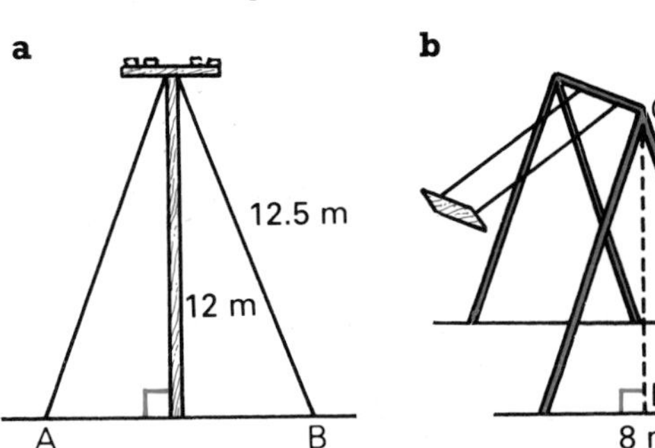

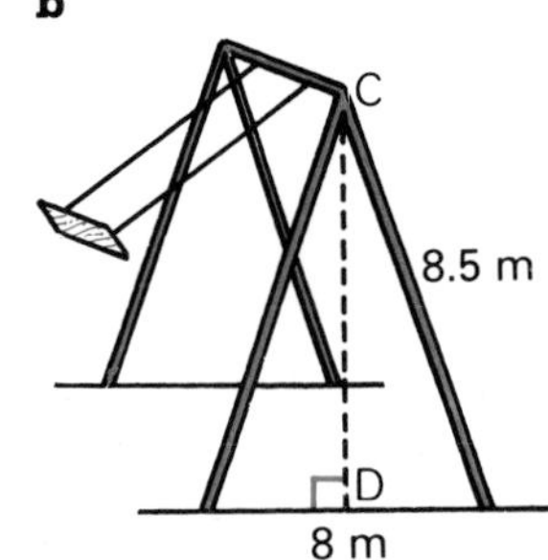

a Find the distance between the feet of the guy ropes (AB).

b How high above the ground is the cross-bar of the swing (CD)?

2 Calculate x in each diagram.

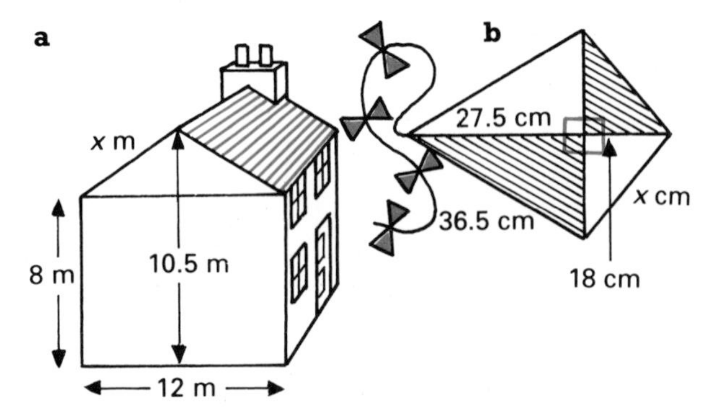

3 Fairy lights are strung across a river as sides of a series of isosceles triangles with base lengths 60 m.

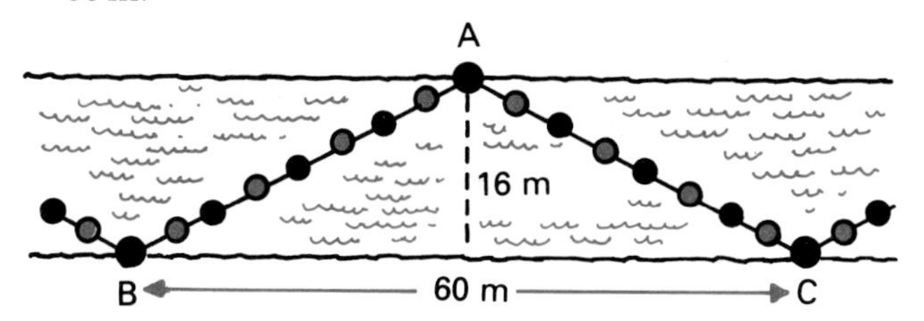

a Calculate the length shown from B to A to C.
b What length of lights would be needed for a riverbank distance of 1080 m?
c What total length, to the nearest metre, is saved if this new pattern of lights is used instead?

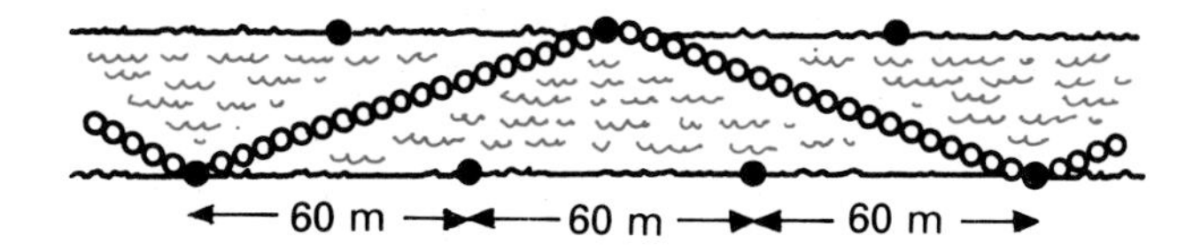

4 The tank in Alf's milk tanker is a cylinder with radius 2 m. He checks the milk level after he visits each farm.

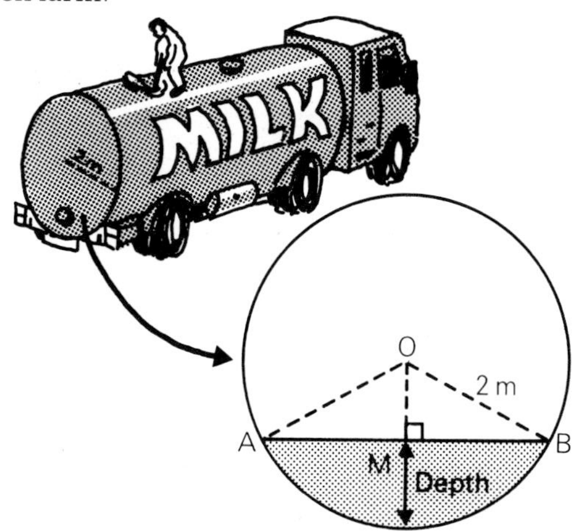

a What is the depth when the tank is full?
b How far is O above the milk surface when the depth is 1 m?
c Calculate, to the nearest cm, the width of the milk surface (AB) when the depth of milk is:
(i) 1 m (ii) 1.5 m (iii) 3.25 m.

5 An aircraft is flying from A to B. It is nearest the control tower X when it reaches N. Calculate the distances AN and NB, to the nearest 100 m.

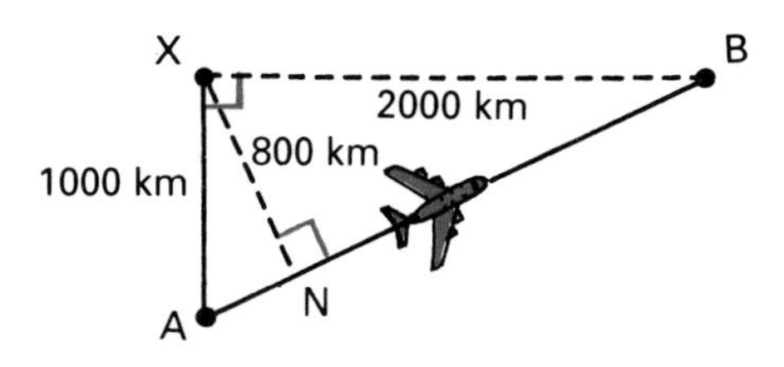

6 Natalie's pram has equal, overlapping wheels. AC = 30 cm and BD = 40 cm.

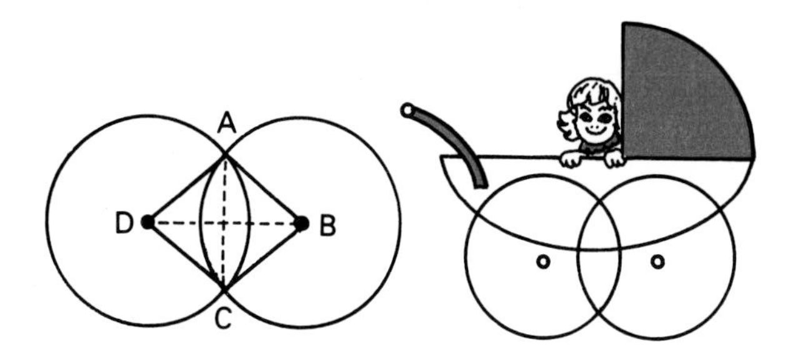

a What shape is ABCD?
b Calculate the length of the spokes.

7 Amy's pram has unequal, overlapping wheels. Show this in a diagram.
a What shape is ABCD? (Lettering as in question **6**.)
b If AC = 60 cm, and the radii of the wheels are 50 cm and 60 cm, calculate (to the nearest cm) the distance (BD) between the centres of the wheels.

CONSTRUCT AND CALCULATE

a *Construct a whirl of right-angled triangles like this:*
(i) Draw OA = 2 cm.
(ii) With set-square or protractor, make a right angle at A.
(iii) Draw AB = 2 cm and join OB.
(iv) Make another right angle at B.
(v) Draw BC = 2 cm and join OC.
(vi) Continue the whirl as far as you can.

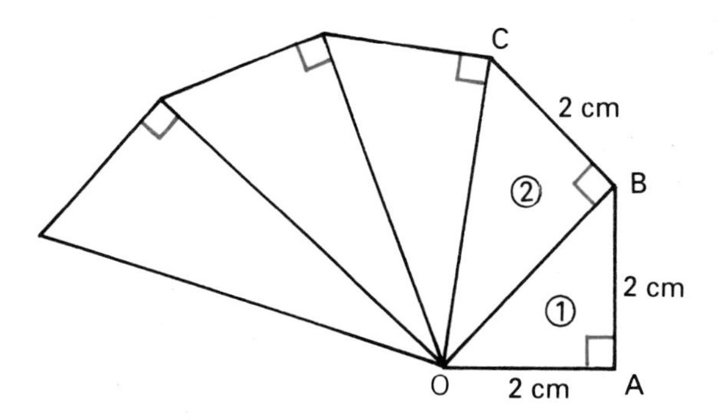

b *OA is 1 unit long. So AB = BC = . . . = 1 unit.*
(i) Show that OB = $\sqrt{2}$ units.
(ii) Copy and complete the table.

Triangle	1	2	3	4	10	n
Hypotenuse (units)	$\sqrt{2}$					

EXERCISE 3

A 3-dimensional example

Calculate the length of the space diagonal AD of this cube.

To make right-angled triangles, join AC.

In right-angled $\triangle ABC$, $AC^2 = AB^2 + BC^2 = 8^2 + 8^2 = 128$.

In right-angled $\triangle ACD$, $AD^2 = AC^2 + CD^2 = 128 + 8^2 = 192$.

So AD = 13.9 cm, correct to 1 decimal place.

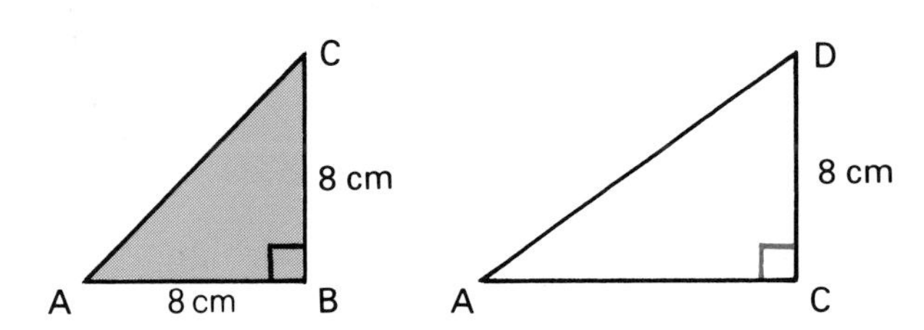

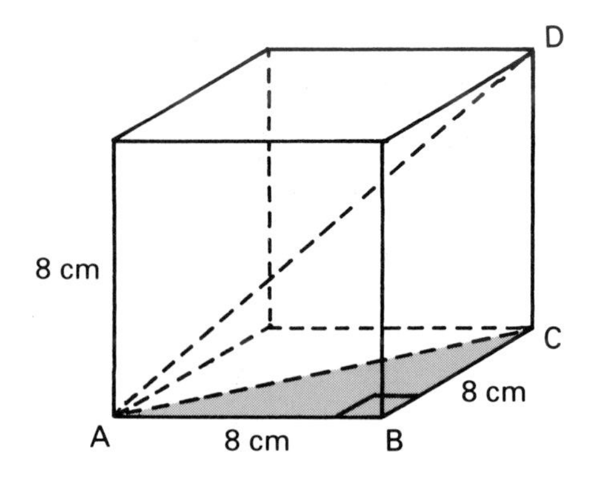

1 Follow the method used in the worked example above to calculate the length of a space diagonal of a cube with edges 6 cm long.

2 Calculate the length of the dotted space diagonal in each cuboid.

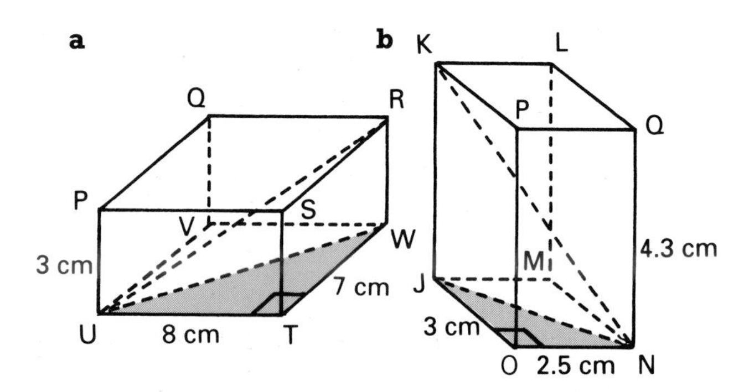

3 An oil rig in the North Sea has a square horizontal deck of side 60 m. The vertical legs stand 20 m above the sea. Together they form a cuboid.

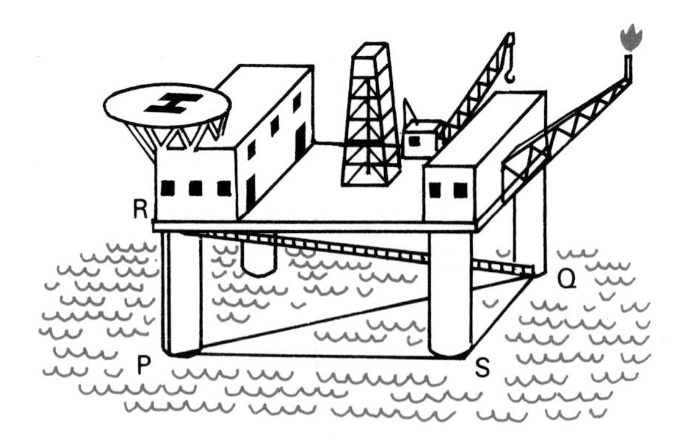

Calculate, to the nearest metre:

a the distance from P to Q

b the length of the walkway QR.

4 Wires are fixed near the top of a vertical television mast at T, 60 m from the foot. The wires are anchored in horizontal ground at X, Y and Z. Calculate the length of:

a TX **b** TY **c** ZF

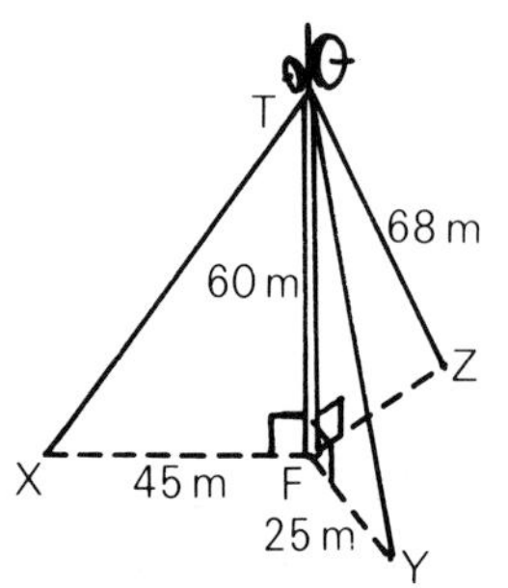

5

H G 18 cm E D F C 7 cm A 24 cm B

ABCDEFGH is a cuboid. In each triangle listed below, name a right angle, then calculate the length of the given line:

a $\triangle ABC$, AC

b $\triangle ABF$, AF

c $\triangle AFG$, space diagonal AG, correct to 1 decimal place.

6 Andrew Service runs this delivery van. A firm of heating engineers asks him to deliver a pipe 5.75 m long. Will it fit into his van?

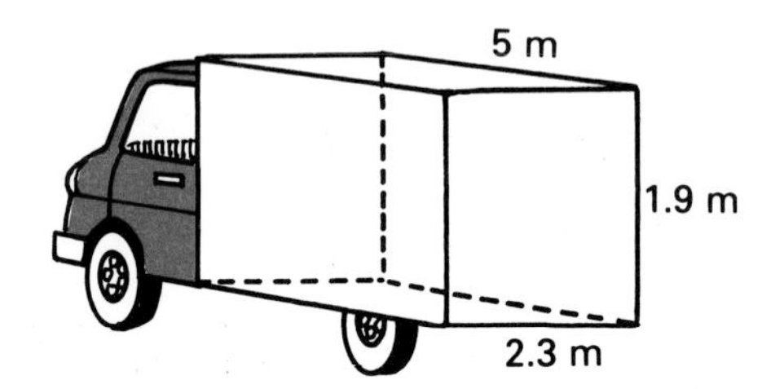

7 The card table is a square of side 1 metre. The legs form the four space diagonals of a cube.

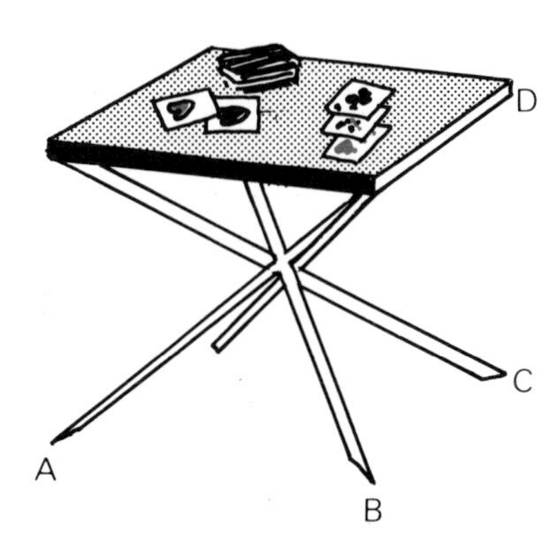

Calculate the length of a leg of the table, correct to the nearest centimetre.

BRAINSTORMER

A small powerful fan heater has to be placed so that it heats the room as efficiently as possible. It is put somewhere on AB.

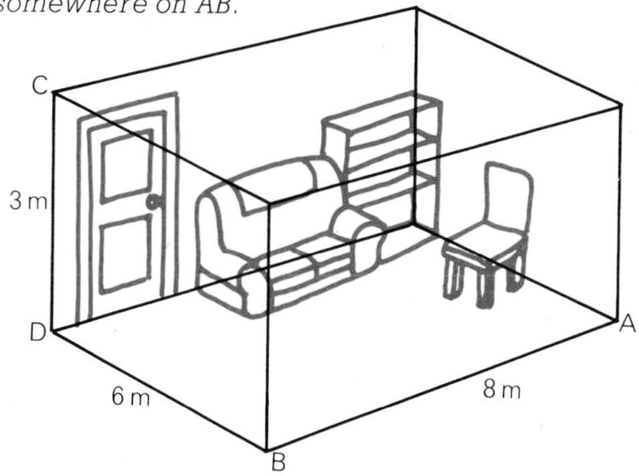

a *How far is it from C, correct to 2 decimal places, when placed at: (i) B (ii) A?*

b *What is the best point for it on AB? Explain why.*

CALCULATING DISTANCES USING COORDINATES

How can you find the distance between these two electric sockets in the wall if no measuring tape is available?

Choose origin O, and take the side of a tile as the unit of length. The sockets are at the points (1, 1) and (5, 4). Plot these points on a grid, and join them.

By Pythagoras' Theorem, $d^2 = 4^2 + 3^2$

so $d = 5$.

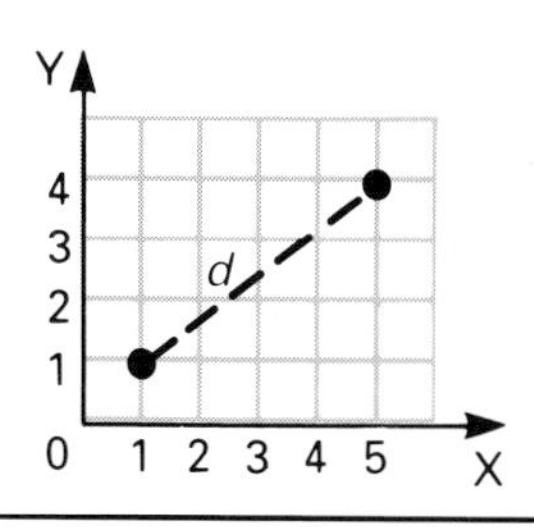

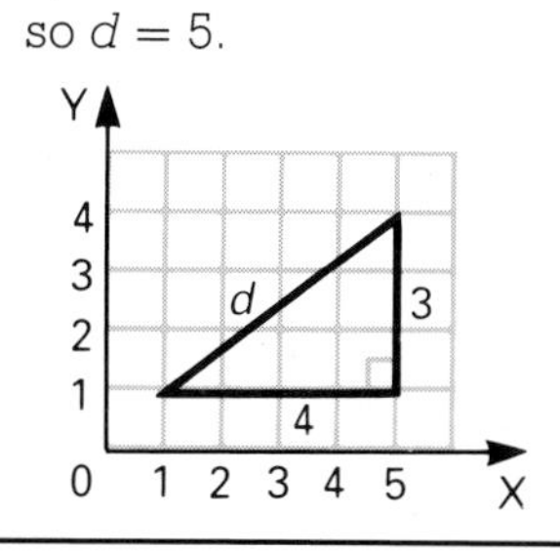

The sockets are 5 tile units apart.

EXERCISE 4A

1 Calculate the lengths of AB, CD, EF and GH, correct to 1 decimal place.

a

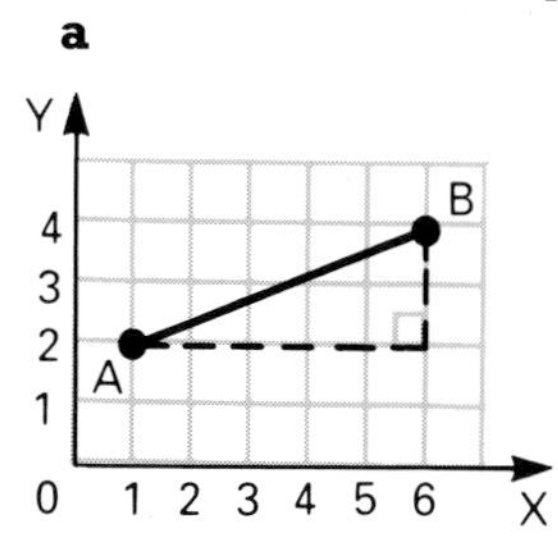

b

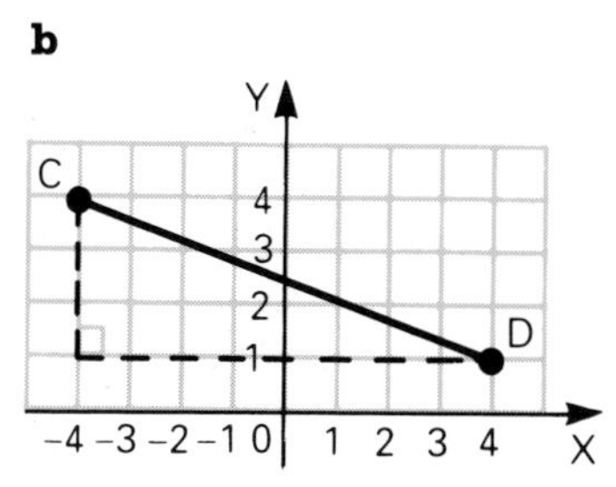

c

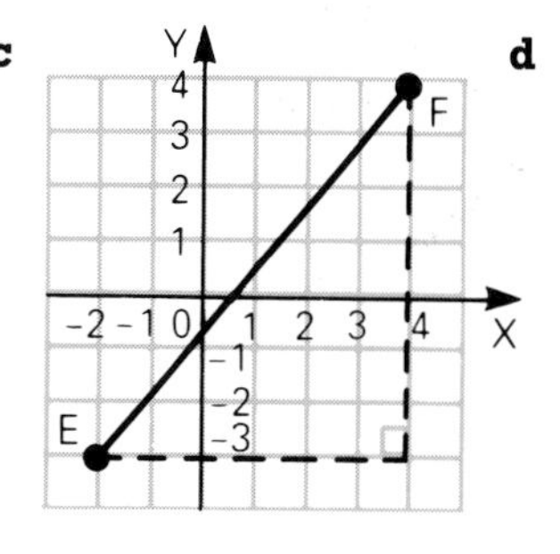

d

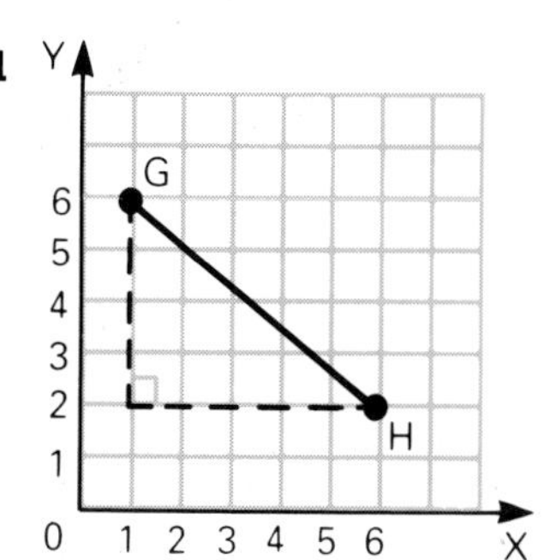

2 Calculate the distance between each pair of points below.

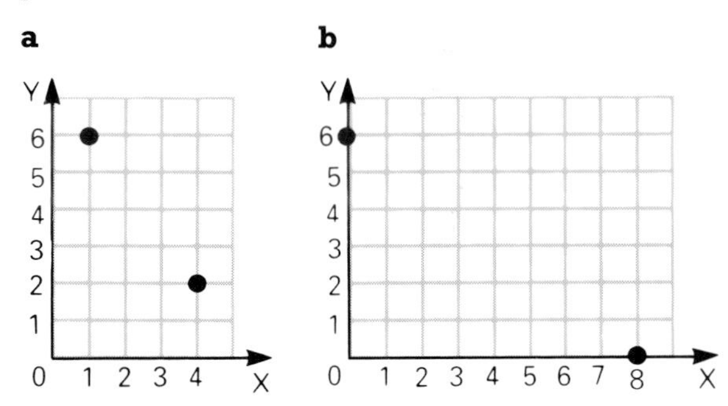

3 The classroom ceiling is covered with square tiles of side 1 metre. Use Pythagoras' Theorem to calculate the distance between the lamps A and B, to the nearest centimetre.

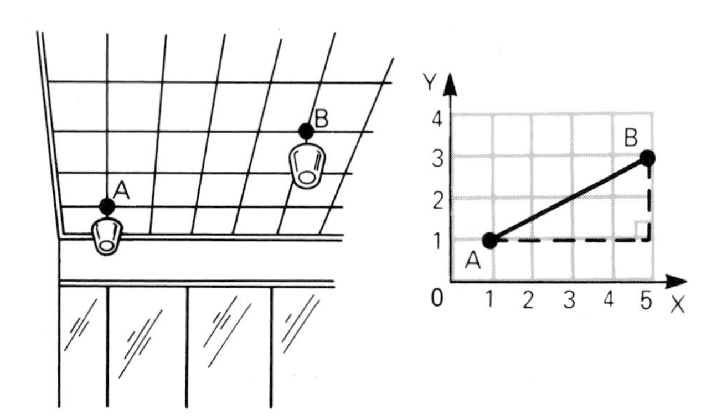

4 Plot these pairs of points on squared paper, and calculate the lengths of the lines joining them:
a O(0, 0) and M(8, 6) **b** P(1, 2) and Q(9, 8)
c R(3, 6) and S(8, −6) **d** T(−1, −1) and U(2, 3)

5 A computer simulates sports events. It models the shot putt by setting up a grid with the competitor at the origin and giving the coordinates of the points where the shot lands.

Five competitors' throws land at A(8, 1), B(7, 4), C(6, 5), D(5, 7) and E(3, 7), the unit being the metre. Calculate, to the nearest cm, the length of each throw. Which competitor won?

6 Calculate the lengths of the sides of this pentagon, correct to 1 decimal place.

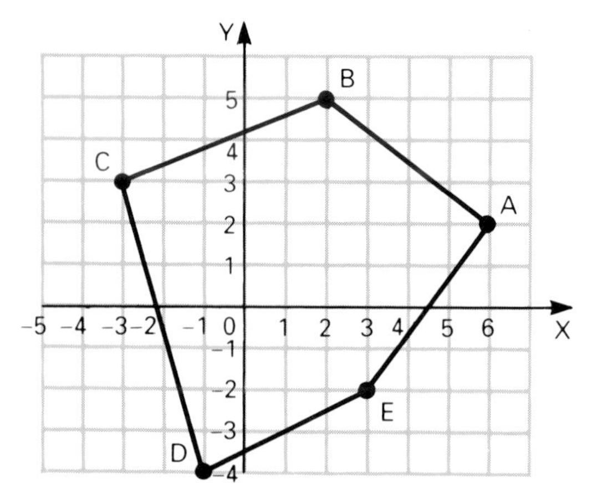

7

The only way to get from A to B in this mountainous area is to go by the pass CD. The dotted line shows the route of a proposed tunnel.

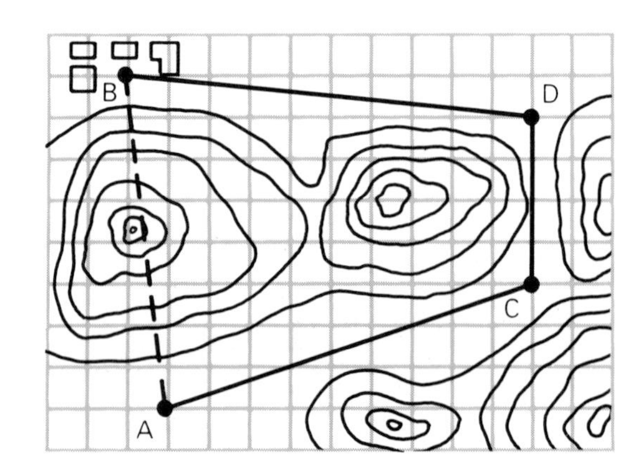

The side of each square is 1 km. Calculate, correct to 2 decimal places:
a the length of the tunnel route, AB
b the length of the road route A to C to D to B
c the saving in distance if the tunnel is built.

EXERCISE 4B

1 A(2, 2) and B(10, 8) are ends of the diameter of a circle.

a Find the coordinates of the centre of the circle.
b Calculate the length of its radius.
c Is the point (9, 1) inside, on, or outside the circle?

2 a Prove that △OAB is isosceles.
b Write down the coordinates of M, the midpoint of AB.
c Calculate the area of △OAB.

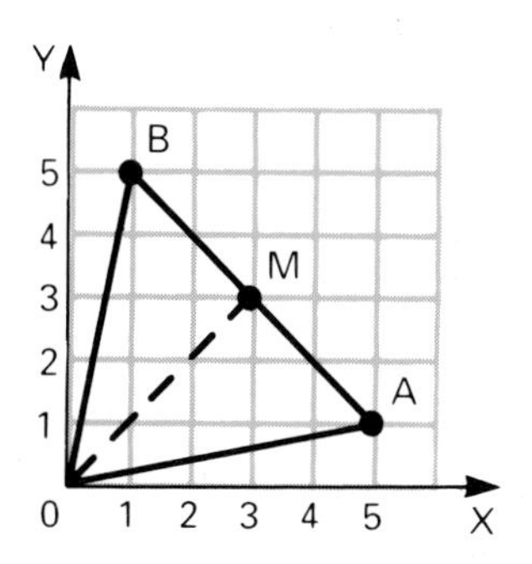

3 The vertices of a quadrilateral are A(−2, 1), B(1, 3), C(4, 1) and D(1, −2).

a Draw ABCD, and its diagonals, which cross at E.
b By calculating lengths, find four isosceles triangles in your diagram.
c Which line is an axis of symmetry of ABCD?

4 Young trees are planted a metre apart. The ones at (3, 2) and (8, 14) died of a disease which can spread up to 10 metres from a diseased plant.

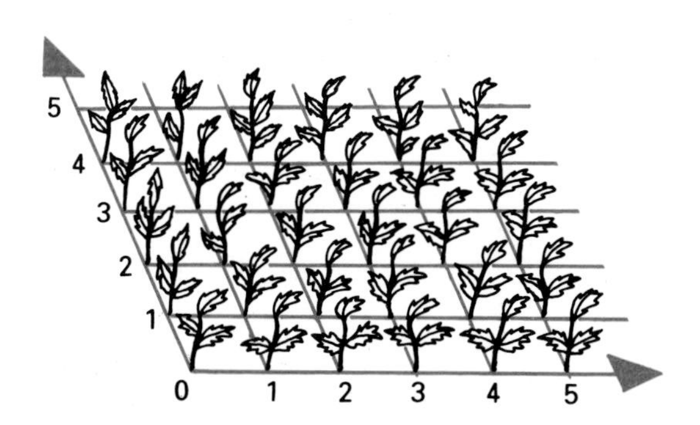

a Could the disease have passed from one of these two trees to the other?
b Another tree at (12, 9) is diseased.
Could this explain how the first two trees were infected?

5

This map has a reference grid on it, made up of 100 m squares. A geographer would call the point A 224575. A mathematician would call A (22.4, 57.5). Calculate, correct to 2 decimal places:

a the distance from A to B 284522
b the length of Holehouse Brae, which runs from 267613 to 308546.

BRAINSTORMER

Into 3 dimensions with coordinates

Axes OX, OY and OZ are at right angles to each other. A is the point (4, 2) and B is (7, 6) in the plane XOY. Point C in space needs a third coordinate, in the z-direction.
C is the point (7, 6, 5), A is now (4, 2, 0) and B(7, 6, 0). Give lengths correct to 1 decimal place where necessary.

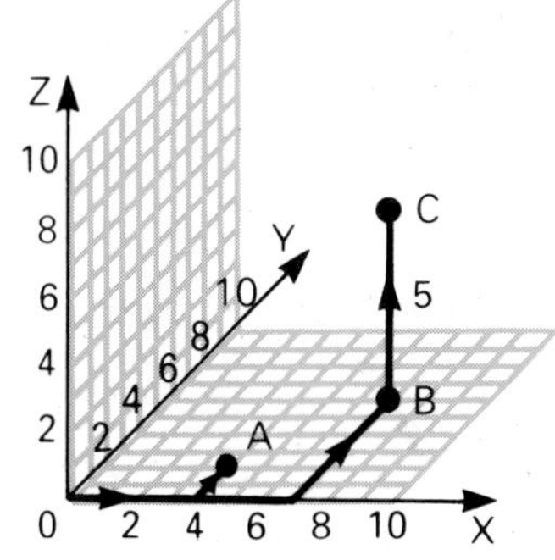

1 *Calculate the length of:*
a *AB* **b** *AC* **c** *OB* **d** *OC*

2 *Draw axes OX, OY and OZ, and number them as in the diagram. Plot the points D(1, 0, 0), E(7, 8, 0) and F(7, 8, 4). Calculate the length of:*
a *DE* **b** *DF* **c** *OE* **d** *OF*

3 *Draw the axes again, and plot P(0, 5, 5), Q(0, 5, 0) and R(5, 5, 0). Calculate the length of:*
a *OP* **b** *OR* **c** *PR*

THE CONVERSE OF THE THEOREM OF PYTHAGORAS

Theorem

If $\triangle ABC$ is right-angled at A, then $a^2 = b^2 + c^2$.

Converse

If in $\triangle ABC$ $a^2 = b^2 + c^2$, then angle A is a right angle.

We can show that the converse is true, like this:

(i)

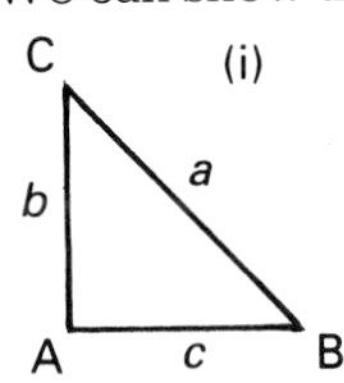

Suppose that in this triangle, $a^2 = b^2 + c^2$

(ii)

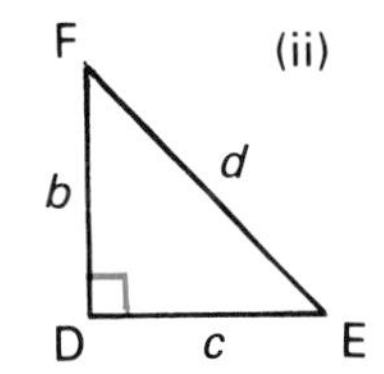

Suppose that in this triangle, $\angle D = 90°$, so that $d^2 = b^2 + c^2$

From the results of (i) and (ii), $a^2 = d^2$, so $a = d$.
The triangles must be congruent as their three sides are equal.
So $\angle A = \angle D = 90°$; the converse is true.

Examples

a Andy is checking that the walls are at right angles.
He has to measure AB, BC and CA, to make his report.

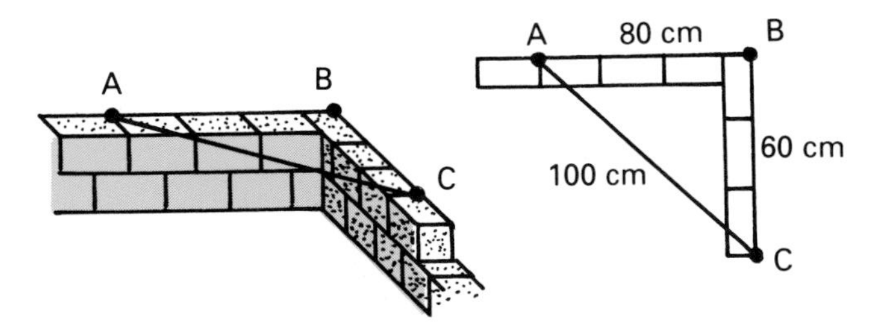

Shorter sides: $AB^2 = 80^2$
$BC^2 = 60^2$

$AB^2 + BC^2 = 10\,000$

Longest side: $AC^2 = 100^2 = 10\,000$

$AB^2 + BC^2 = AC^2$.

So $\angle B = 90°$, by the converse of Pythagoras' Theorem.

b Tom is trying to check that the lamp-post is at right angles to the pavement. His report gives the result of his measurements.

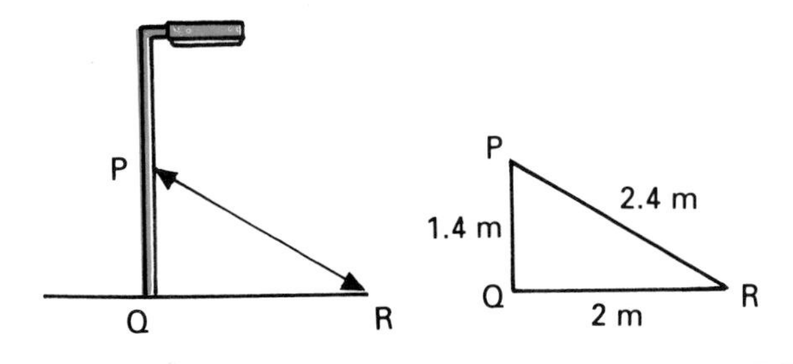

Shorter sides: $PQ^2 = 1.4^2$
$QR^2 = 2^2$

$PQ^2 + QR^2 = 5.96$

Longest side: $PR^2 = 2.4^2 = 5.76$

$PR^2 \neq PQ^2 + QR^2$.

So $\angle Q$ is **not** $90°$.

PRACTICAL PROJECT

Long ago builders made right angles using a rope with 12 evenly-spaced knots in it. Why does this give a right angle? Many bricklayers use the same trick today to check their work. Take a piece of string and mark it in twelve equally spaced places with a pen. Check how accurately you can make a right angle with the marked string.

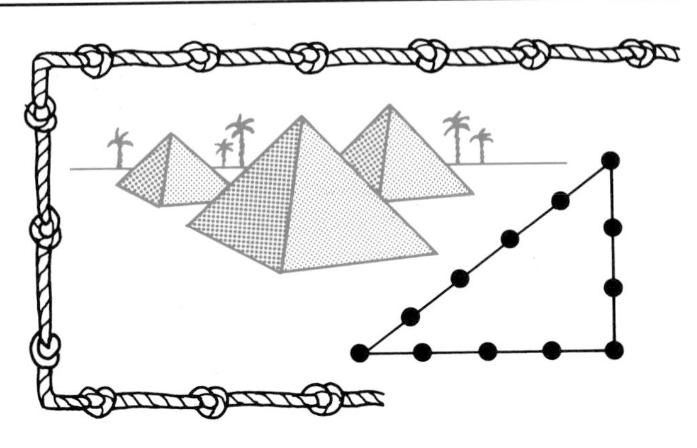

EXERCISE 5A

1 Use the converse of Pythagoras' Theorem to decide which of these triangles are right-angled.

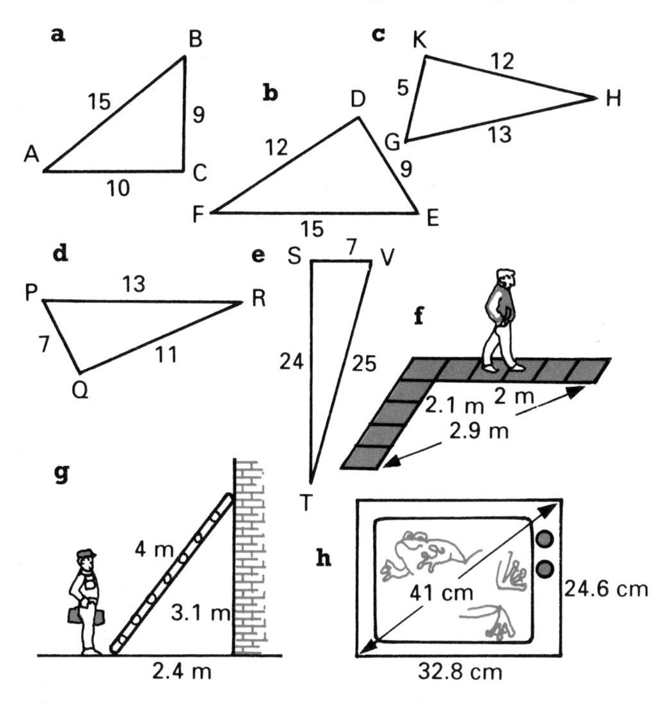

2 Many angles which look like right angles are not. Which ones in this picture are right angles, and which are not? All the measurements are in millimetres.

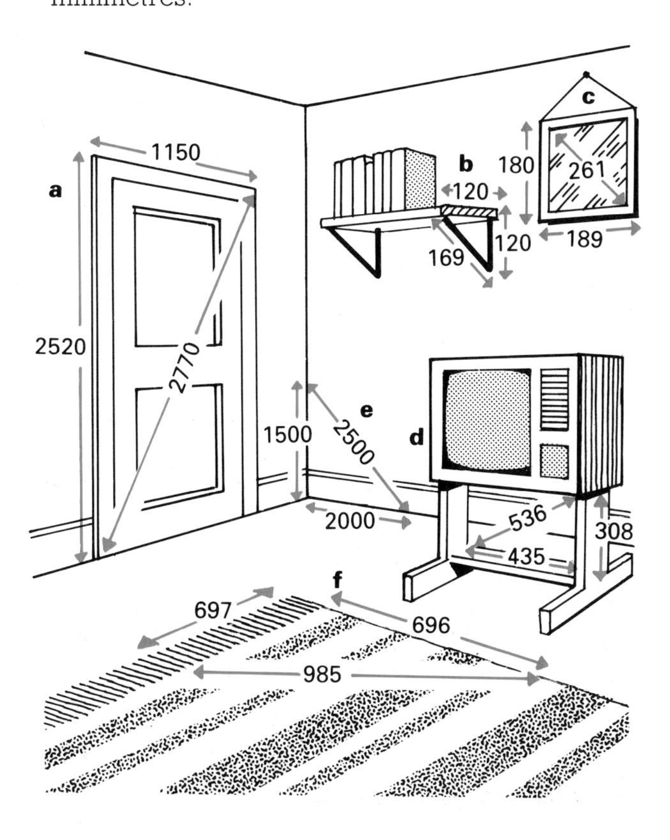

3 a Calculate the length of:
(i) AC (ii) AB (iii) BC.

b Is $\triangle$ABC right-angled?

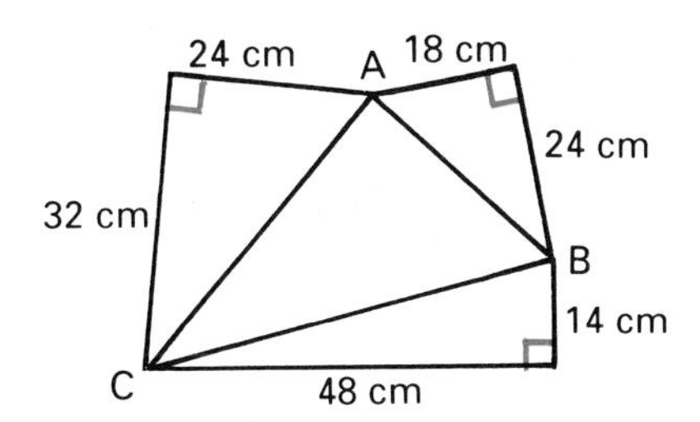

4 a Calculate the length of:
(i) PQ (ii) QR.

b Prove that $\triangle$PQR is right-angled.

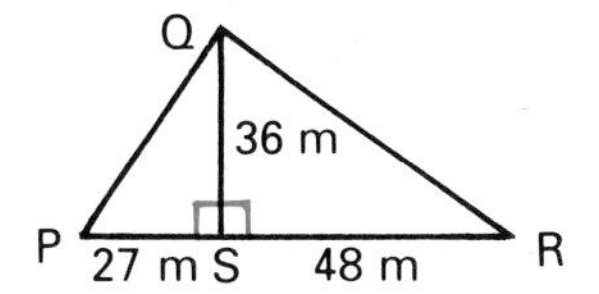

5 This diagram shows part of a bridge.

a Calculate the length of each sloping girder.

b Prove that $\angle$PSR between these girders is a right angle.

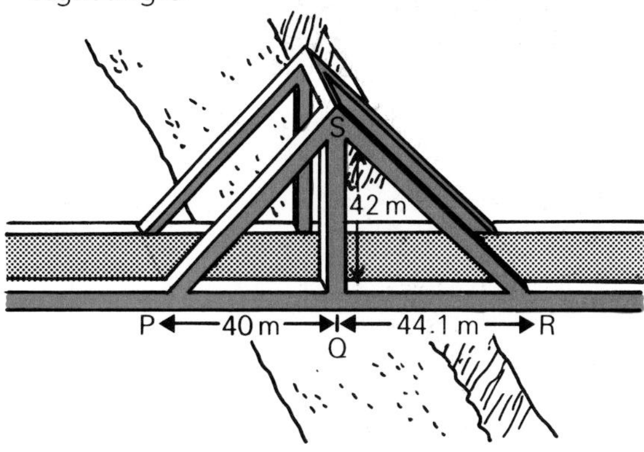

6 The model bridge swings open to let a toy boat through. It has three rectangular parts. Those on either side are fixed, and the centre part swings open.

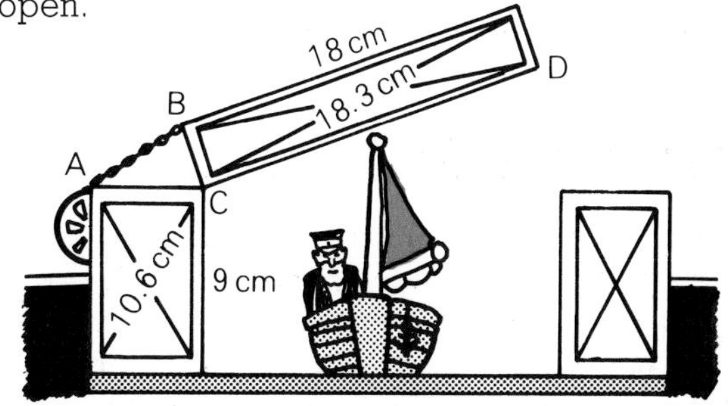

a Calculate: (i) AC (ii) BC.

b Show that when chain AB is 6.5 cm long, CD is horizontal.

EXERCISE 5B

1 You'll need Pythagoras' Theorem and its converse to prove that each of the largest triangles in the diagrams below is right-angled.

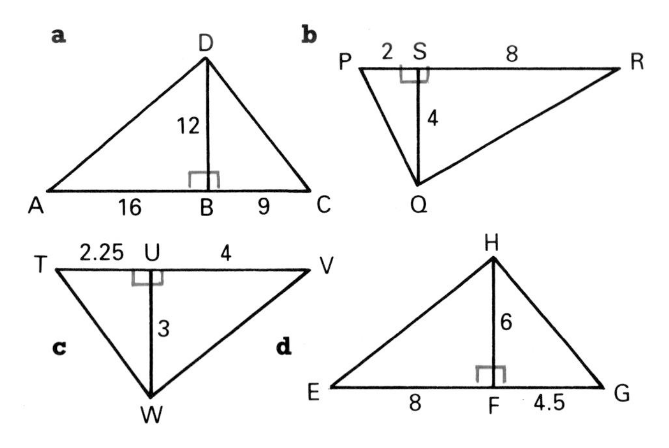

2 Prove that $\triangle CEF$ is right-angled, given that ABCD is a rectangle.

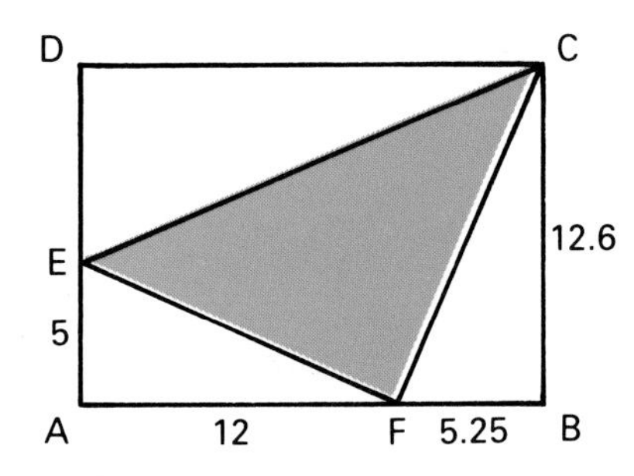

3 Prove that the triangle formed by each set of points is right-angled, and name each right angle.

a A(3, 1), B(1, 4), C(9, 5)

b P(−2, −2), Q(−5, 1), R(2, 2)

4 Prove that the rectangle formed by joining the points A(−1, 4), B(2, 5), C(4, −1) and D(1, −2) is a rectangle.

5 A jet aircraft at J is 12 km east and 5 km north of an airport A, at a height of 1000 m. A helicopter H is 5 km from the airport, also at a height of 1000 m, and 14 km from the jet.

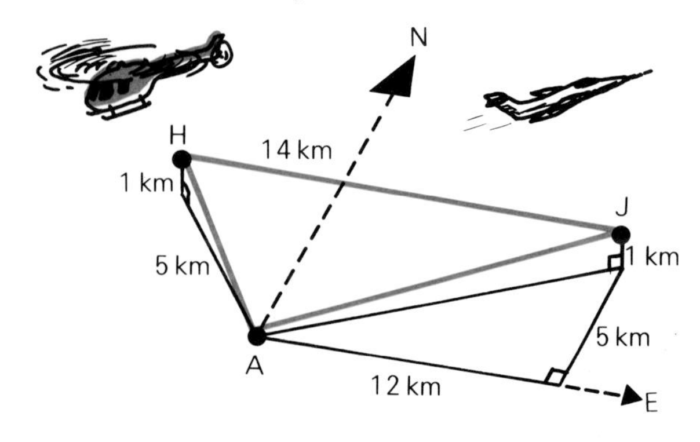

Prove that the flight paths HA and JA are at right angles to each other.

INVESTIGATIONS

Pythagorean triples

1 a *Choose two numbers a and b, with a > b.*

b *Use them to calculate x, y and z where:*

$x = a^2 + b^2$

$y = a^2 - b^2$

$z = 2ab$

c *Check that, with these values, $\triangle ABC$ is right-angled.*

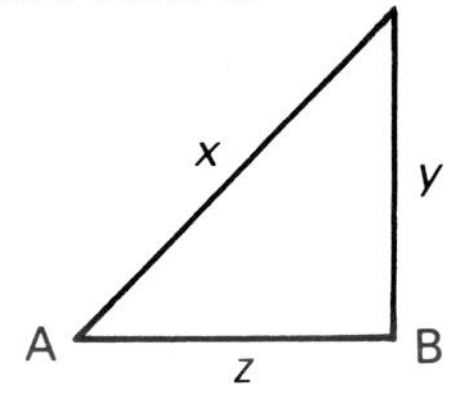

2

a	b
2	1
5	2
12	5
29	12

These pairs of starting numbers produce triples in which the two smaller numbers differ by 1. Investigate:

a *the sequence 1, 2, 5, 12, 29, . . . used to form them*

b *how to use the sequence of triples to find fractions which are close to the value of $\sqrt{2}$.*

3 *If you have access to a spreadsheet, the following will generate the triples for you:*

	A	B	C	D	E
1	2	1	= A1*A1+B1*B1	= A1*A1−B1*B1	= 2*A1*B1
2	= IF(A1−B1 = 1, A1+1, A1)	= IF(A1−B1 = 1, 1, B1+1)	= A2*A2+B2*B2	= A2*A2−B2*B2	= 2*A2*B2
3					
4					

Copy line 2 down as far as you like

CHECK-UP ON PYTHAGORAS

1 Copy and complete:
a $r^2 = \ldots$ **b** $p^2 = \ldots$ **c** $q^2 = \ldots$

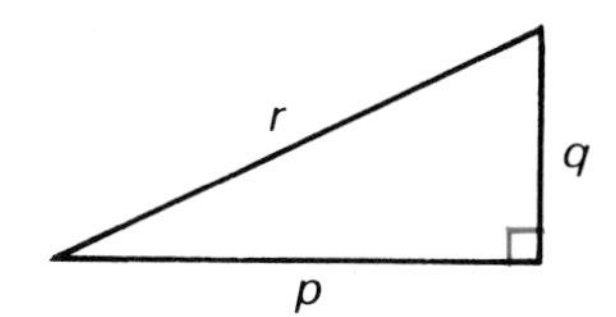

2 Calculate x in each triangle below.

a

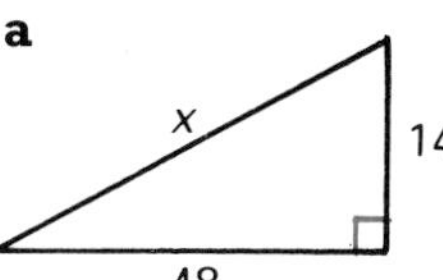

b

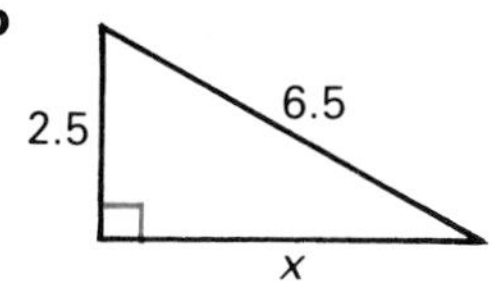

3 The diagonals of the rhombus are 16 cm and 12 cm long. Calculate its perimeter.

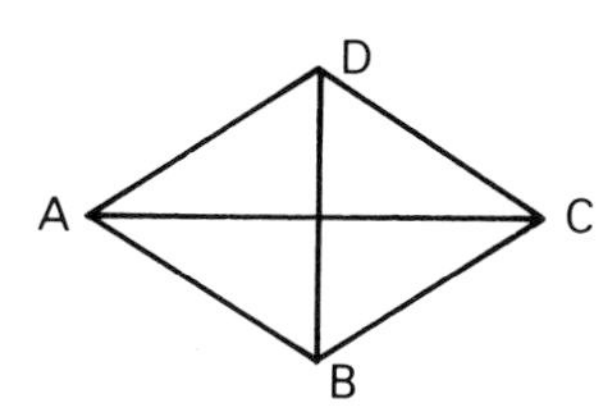

4 In the kite PQRS, calculate:
a the length of diagonal QS
b the area of the kite.

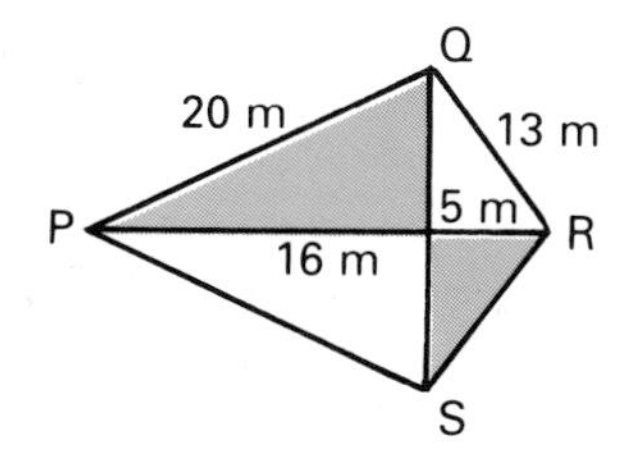

5 Calculate the length of AB, correct to the nearest cm. Each tile is 10 cm long.

6 Plot these pairs of points, then calculate the distance between the points in each pair.
a P(1, 1), Q(9, 16) **b** R(−3, 1), S(6, −11)

7 This container is in the shape of a cuboid.

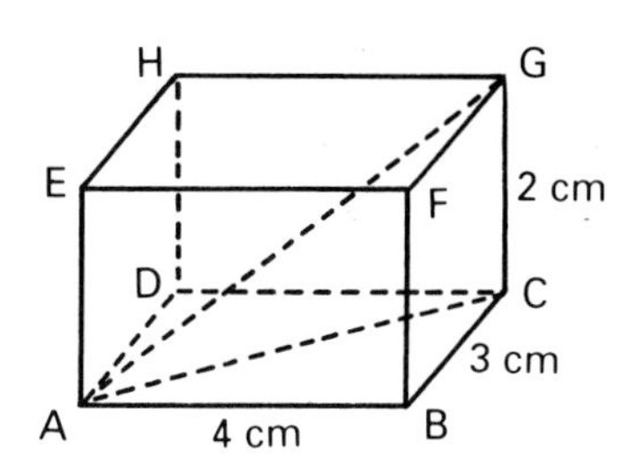

Calculate the length of:
a face diagonal AC
b space diagonal AG, correct to 1 decimal place.

8 **a** If $\angle A = 90°$ in $\triangle ABC$, then $a^2 = b^2 + c^2$. Write down the converse statement.
b Are any of these triangles *not* right-angled?

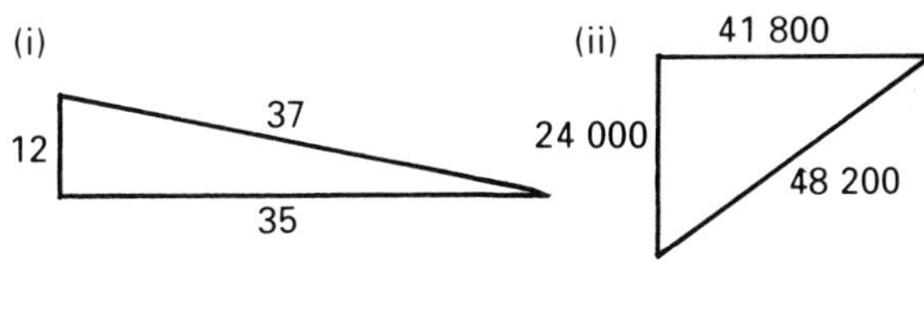

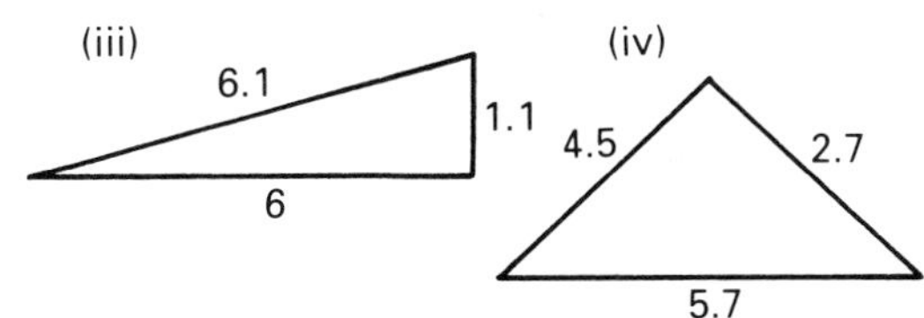

9 Triangles ABC and ADE are isosceles.
a Calculate BC. (Copy the diagram and use right-angled triangles.)
b Is $\triangle ABC$ right-angled? Explain your answer.

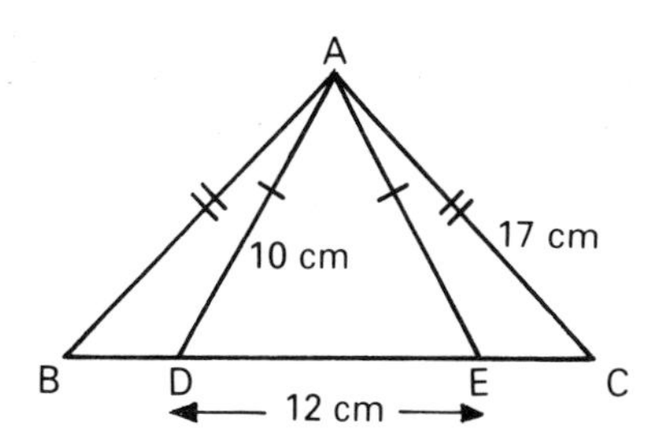

10 Prove that $x^2 - y^2 = a^2 - b^2$.

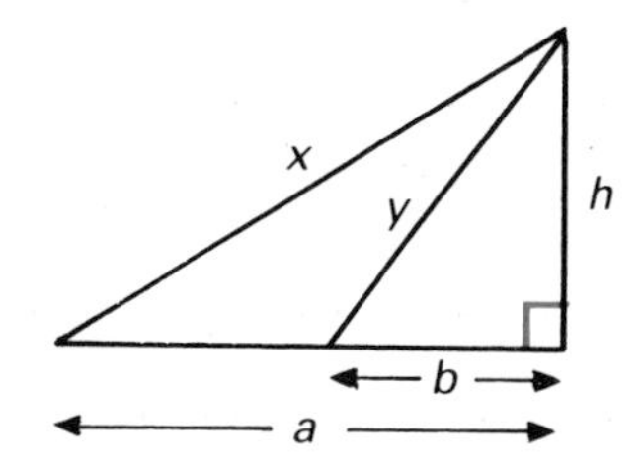

7 BRACKETS AND EQUATIONS

LOOKING BACK

1 Calculate, without using a calculator:

a $2+4$ **b** $2-4$ **c** $-2+4$
d $-2-4$ **e** 2×4 **f** $2\times(-4)$
g -2×4 **h** $-2\times(-4)$ **i** $\frac{2}{4}$
j $\frac{2}{-4}$ **k** $\frac{-2}{4}$ **l** $\frac{-2}{-4}$

2 Check your answers to question **1** with a calculator.

3 Simplify:

a $3x-x$ **b** $4y-8y$ **c** $t-(-t)$
d $a\times a$ **e** $2b\times b$ **f** 7^2
g $(-5)^2$ **h** $x+3+2x$ **i** $4+6y-2y-4$

4 Write these without brackets:

a $8(x+2)$ **b** $3(1-x)$ **c** $2(3x+1)$
d $5(3-2x)$ **e** $x(x-1)$ **f** $y(y+x)$

5 Solve these equations:

a $5x=5$ **b** $2x=-8$ **c** $-3x=6$
d $-2x=-4$ **e** $5y+15=0$ **f** $7-3y=1$
g $2k-4=k$ **h** $7x-4=3x+12$

6 Solve:

a $2(x+1)=10$ **b** $3(y-2)=12$

7 a

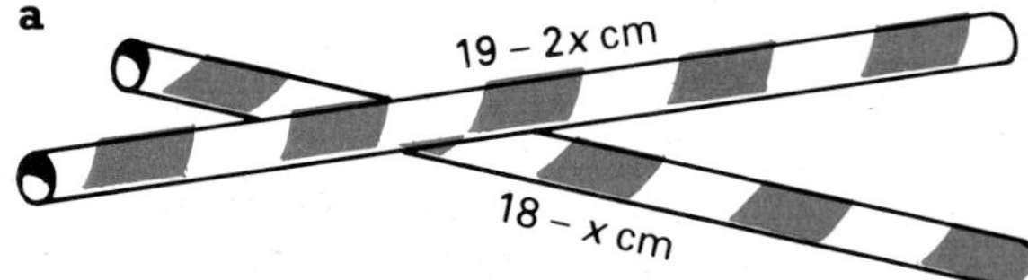

The straws have equal lengths. Make an equation, solve it, and find the length of the straws.

b

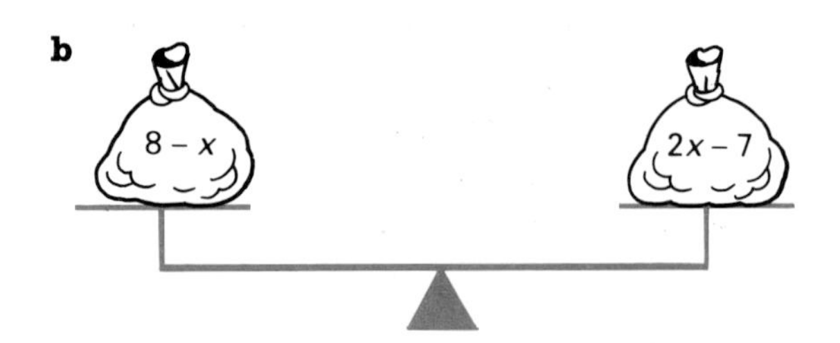

Make an equation, solve it, and find the number of £1 coins in each bag.

8 a Write down an expression for the area of this rectangle.

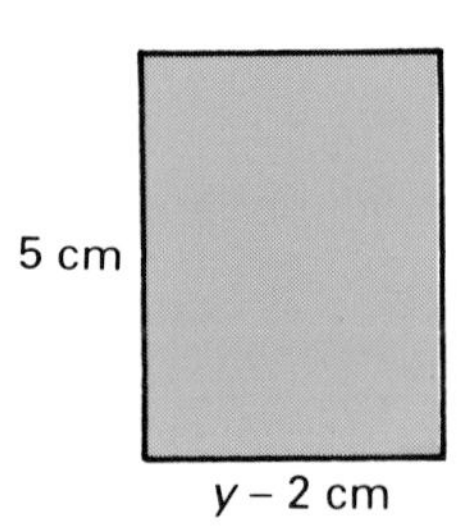

b If the area is $3y$ cm^2, make an equation and find y.

9 In **a** and **b** below find the lengths of straws (i) and (ii) in terms of x. The lengths are in cm.

a

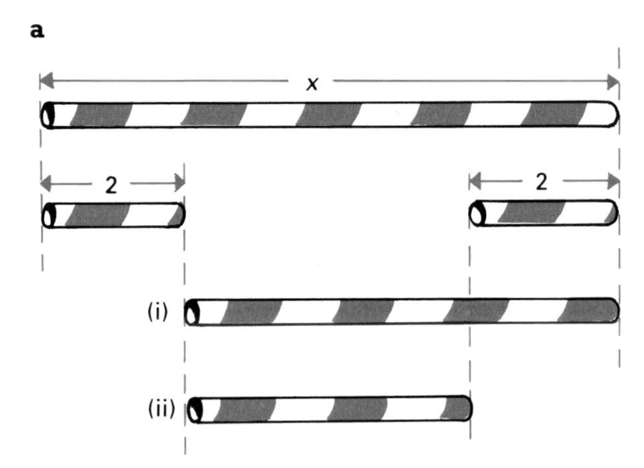

b

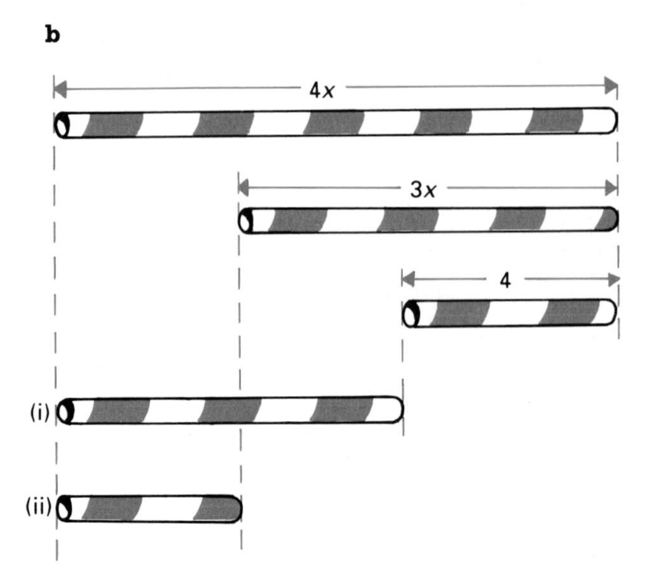

REMOVING BRACKETS

Reminders

$a(b+c) \ldots a(b+c) = ab+ac$

$a(b-c) \ldots a(b-c) = ab-ac$

Examples

(i) $3(x+y) = 3x+3y$

(ii) $2(n-3) = 2n-6$

(iii) $4(2-t) = 8-4t$

EXERCISE 1A

1 Remove the brackets:

a $2(x+5)$ **b** $3(x+2)$ **c** $4(y+1)$
d $5(y+3)$ **e** $6(b+2)$ **f** $3(x-2)$
g $2(x-5)$ **h** $4(y-1)$ **i** $6(m-2)$
j $8(n-1)$ **k** $2(3+x)$ **l** $3(1+y)$
m $4(5-a)$ **n** $7(2-b)$ **o** $6(1-c)$
p $5(a+b)$ **q** $3(x-y)$ **r** $9(5-x)$
s $10(t+w)$ **t** $11(y-1)$ **u** $1(1-x)$

2 For each picture below, give the total number of coins in two different ways (with brackets, then without brackets).

a

5 bags

b

4 bags

c

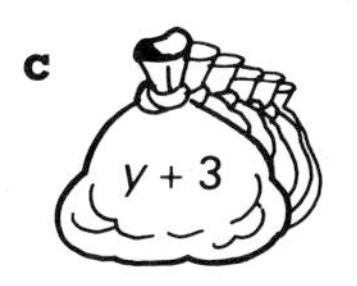

6 bags

d

t − 3

3 bags

e

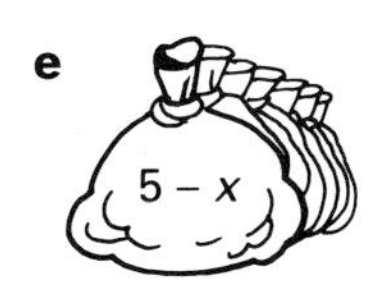

7 bags

f

2 bags

g

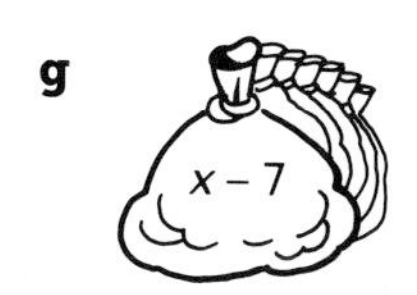

8 bags

h

9 bags

i

4 bags

3 An orange weighs $y+3$ grams. Write down the weight of five oranges in two different ways.

4 The area of each sticker is $x+1$ cm^2. Write down the total area of the sheet in two different ways.

5 Each glass holds $12-j$ ml of juice. Write down the total amount in the three glasses in two different ways.

6 Each car is $t-11$ metres long. Write down the total length of the cars in two different ways.

EXERCISE 1B

1 Remove the brackets:

a $2(3t+4)$ **b** $2(5x-3)$ **c** $3(1+2x)$
d $4(2-2y)$ **e** $6(2p-1)$ **f** $a(x+3)$
g $b(y+2)$ **h** $c(k-1)$ **i** $d(m-5)$
j $e(n+2)$ **k** $a(b+c)$ **l** $a(x+y)$
m $b(x-y)$ **n** $x(x+y)$ **o** $a(a-b)$
p $r(r-4)$ **q** $x(1-x)$ **r** $y(y-4)$
s $x(x+2y)$ **t** $y(3y-4)$ **u** $x(3x-5)$
v $y(8x-7y)$ **w** $t(2t-7)$ **x** $w(15-3w)$

2 Remove the brackets:

a $2(x+y+5)$ **b** $5(x+y+3)$
c $3(x-y-4)$ **d** $4(a+2b+c)$
e $6(2a-b+c)$ **f** $7(3x-4y+2)$
g $x(x^2-1)$ **h** $y(y^2+1)$
i $a(a^2-5)$ **j** $b(b^2+b)$
k $c(c^2-c)$ **l** $t(t^3-3t^2)$

A NEGATIVE MULTIPLIER

The 'distributive law' $a(b+c) = ab+ac$ is true for all real numbers, positive, negative and zero.

Examples

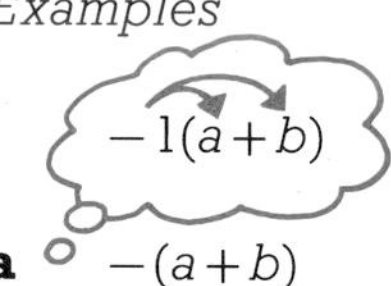

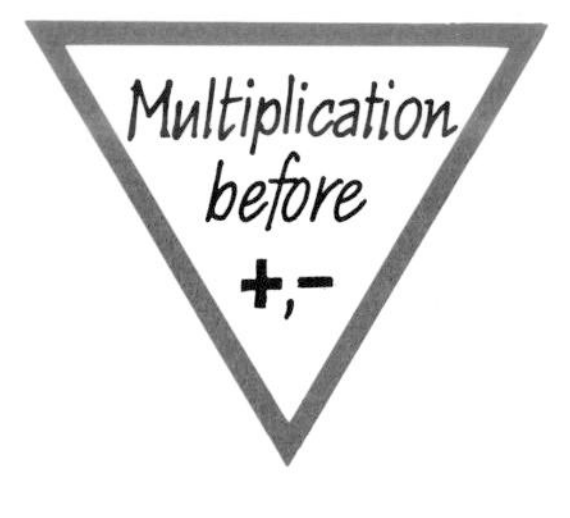

$-3(x-2)$ $\qquad$ $4-3(x-2)$

a $-(a+b)$
$= -a-b$

b $-3(x-2)$
$= -3x+6$

c $4-3(x-2)$
$= 4-3x+6$
$= 10-3x$

EXERCISE 2A

1 Remove the brackets:

a $-2(x+3)$ **b** $-3(x-4)$ **c** $-5(x+1)$
d $-4(x-2)$ **e** $-6(1+y)$ **f** $-2(3-m)$
g $-7(4+n)$ **h** $-10(1-k)$ **i** $-2(p+q)$
j $-5(p-q)$ **k** $-1(m+n)$ **l** $-(m+n)$

2 Simplify, but make sure to multiply before adding or subtracting:

a $5+2(x+1)$
$= 5+2x+2$
$= \ldots\ldots\ldots$

b $1-3(y-2)$
$= 1-3y+6$
$= \ldots\ldots\ldots$

c $6-4(x+1)$
$= 6-4x-\ldots$
$= \ldots\ldots\ldots$

d $4+5(w+3)$ **e** $4+2(x+1)$ **f** $10+4(x-2)$
g $3+3(x-1)$ **h** $4+(x+3)$ **i** $6-2(y+1)$
j $5-3(y+2)$ **k** $7-4(y+1)$ **l** $8-(y+2)$
m $5+(n-1)$ **n** $8-2(m-2)$ **o** $11+5(k-3)$
p $3-7(x+1)$ **q** $3+2(a-2)$ **r** $6-3(b+2)$
s $3-(c-3)$ **t** $4+8(w-1)$

3 Copy and complete:

Area of whole garden $= 10y\,\text{m}^2$
Area of flower bed $= 8(y-2)\,\text{m}^2$
So area of path $= 10y-8(y-2)\,\text{m}^2$
$= \ldots\ldots$
$= \ldots\ldots$

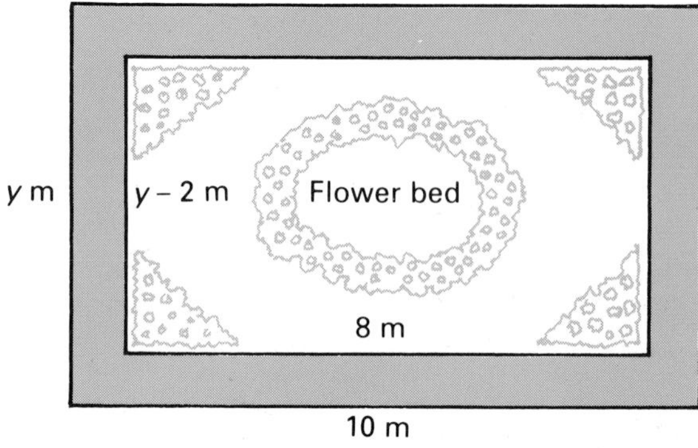

4 Repeat question **3** for these gardens. The path is 1 m wide in each one.

a

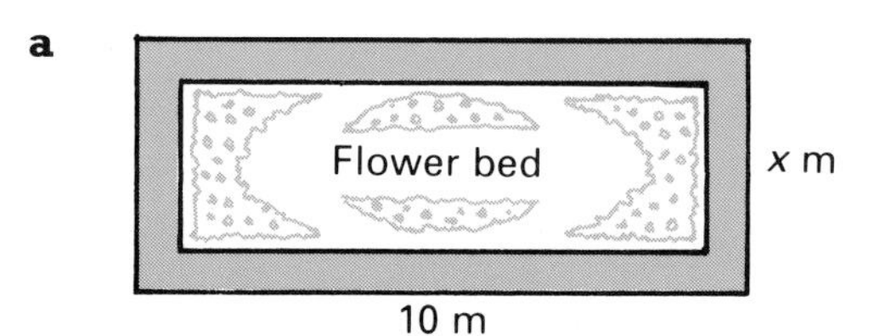

b

Flower bed
t m
7 m

c

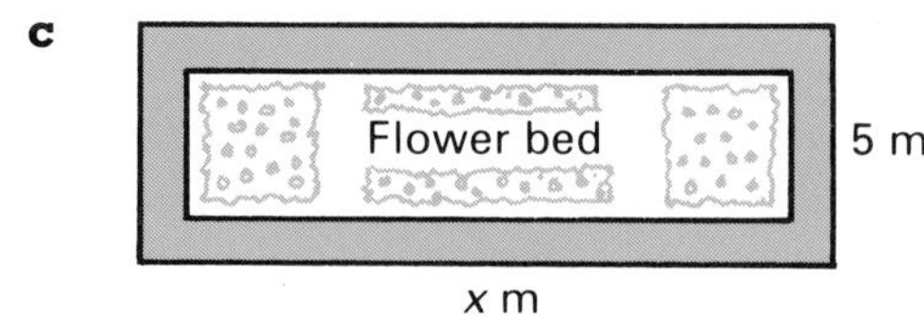

d

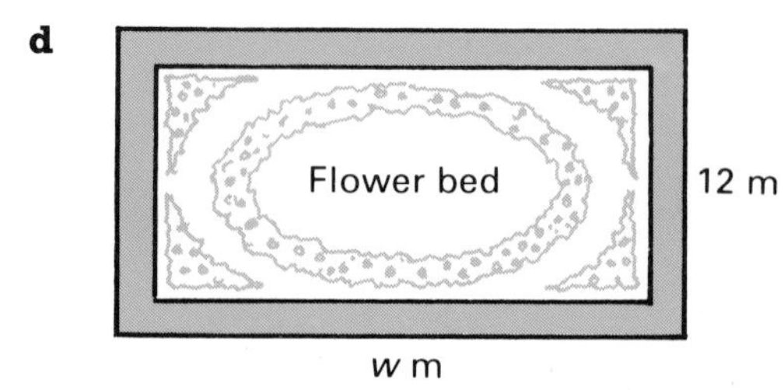

5 A flower bed is 8 m long and x m wide. It has a path 1 m wide round the outside. Make a sketch, and find an expression for the area of the path.

EXERCISE 2B

1 In **a**—**d** below write down expressions for the areas of the rectangles and their differences, area (i) − area (ii), then simplify the difference in each. (Lengths are in cm.)

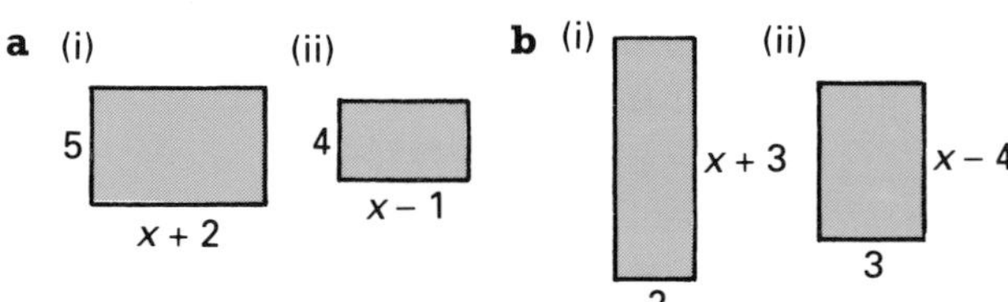

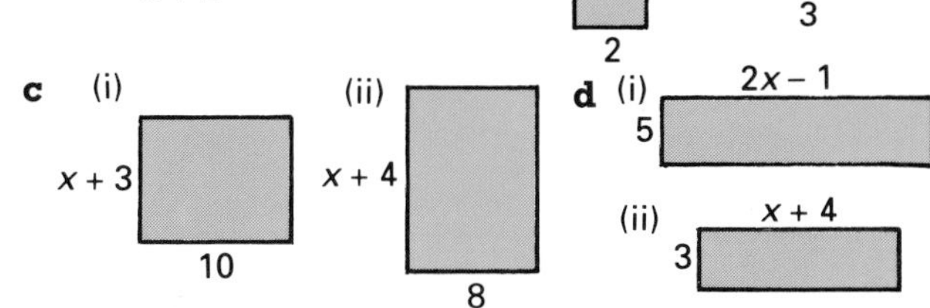

2 How many more coins are there in the first load of bags in each diagram below than in the second?

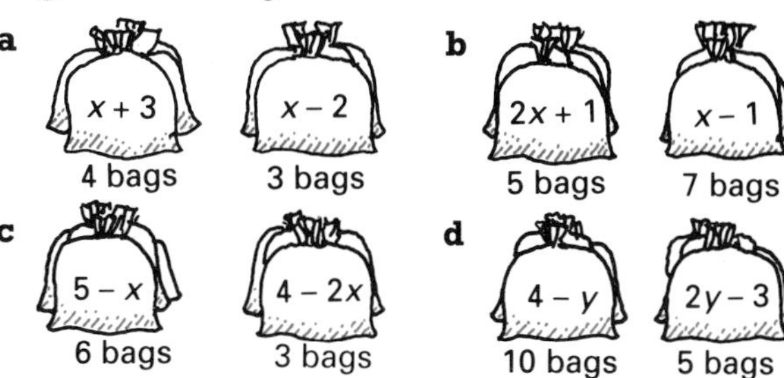

3 Simplify:

a $3(x-2)-2(x+1)$ **b** $5(y+3)-3(y-2)$
c $2(t-2)+3(t-3)$ **d** $2(w+4)-(w+5)$
e $3(x-2)-(2x+1)$ **f** $5(2y-3)-(4y+2)$
g $3(4-2x)-2(5-x)$ **h** $x(x-2)-x(x+1)$

EQUATIONS WITH BRACKETS 1

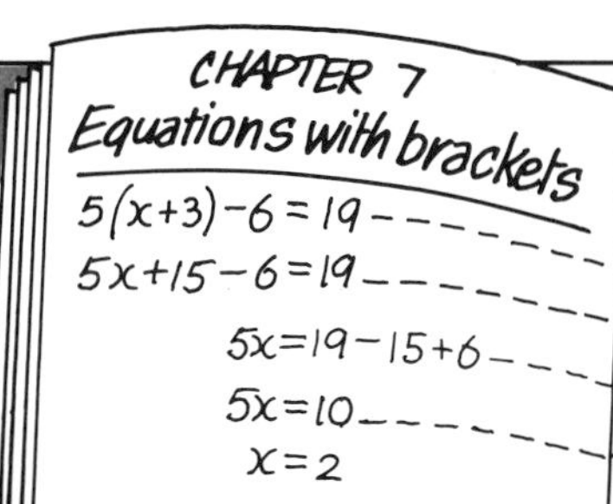

EXERCISE 3A

1 Solve, by first multiplying out brackets:

a $2(x+3)=12$ **b** $2(x-3)=12$
c $2(x+1)=10$ **d** $2(x-5)=8$
e $3(y+1)=15$ **f** $3(y-1)=15$
g $3(y-4)=6$ **h** $3(y+5)=15$
i $2(x+4)=12$ **j** $3(x-2)=15$
k $4(y+1)=8$ **l** $5(y-1)=20$

2 Solve:

a $3(2x-1)=15$ **b** $2(2x+1)=10$
c $6(2y+4)=36$ **d** $5(3y-1)=10$
e $4(y+1)-8=0$ **f** $5(u+1)+3=18$
g $8(v-1)-4=20$ **h** $3(3w-1)-11=40$

3 Four of these gold bars

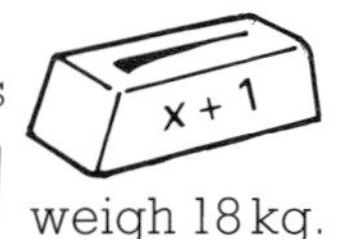

and one of these [2] weigh 18 kg.
So $4(x+1)+2=18$.
Solve the equation, and then write down the weight of the larger bar.

4 Make an equation for each picture and solve it. Then write down the weights of the eight gold bars.

a

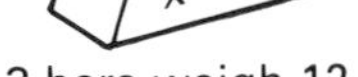

3 bars weigh 12 kg

b

4 bars weigh 20 kg

c

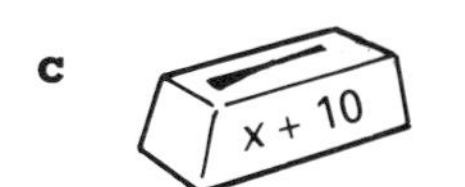

6 bars weigh 90 kg

d

5 bars weigh 35 kg

e

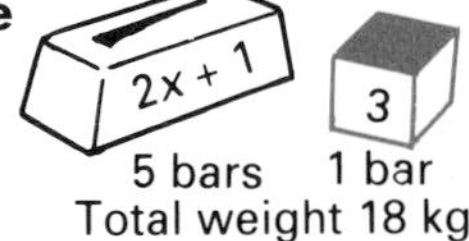

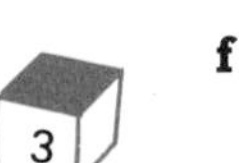

5 bars 1 bar
Total weight 18 kg

f

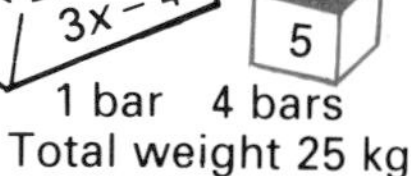

1 bar 4 bars
Total weight 25 kg

g

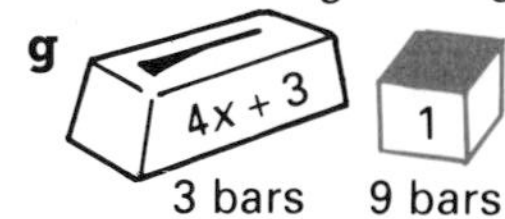

3 bars 9 bars
Total weight 18 kg

h

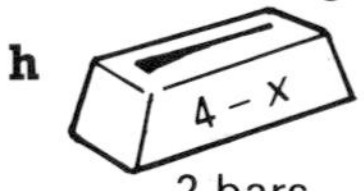

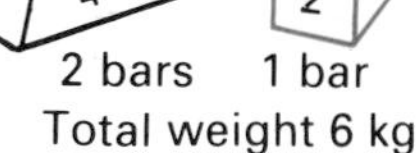

2 bars 1 bar
Total weight 6 kg

EXERCISE 3B

1 Solve:

a $5(x-1)=3(x+1)$ **b** $3(y+1)=2(y+2)$
c $5(k-3)=3(k-1)$ **d** $4(p+2)=2(p+6)$

2 In each picture the number of coins in (i) is the same as the number in (ii). Make an equation for each picture and solve it, then write down the number of coins in each bag.

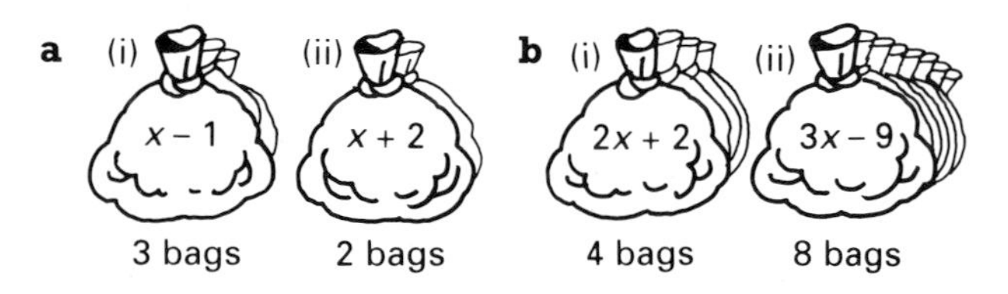

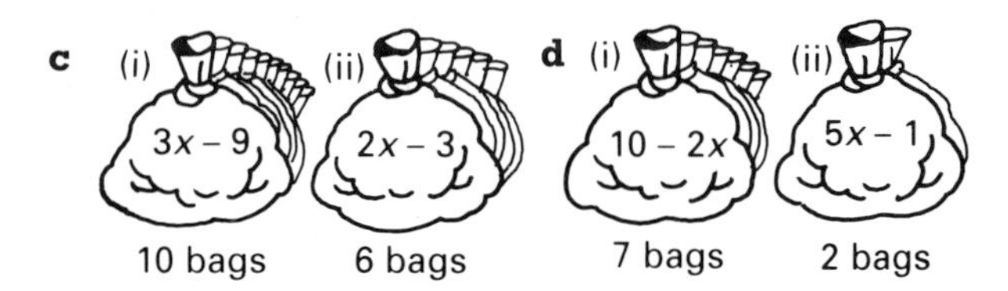

3 A rectangle has length $2x-1$ cm and breadth 4 cm. Its area is 36 cm^2. Calculate x.

4 A car travels for two hours at an average speed of s km/h, and for three hours at an average speed of $(s-2)$ km/h. Altogether it travels 204 km. Make an equation, and calculate the two speeds.

5 Repeat question **4** for a train which travels for two hours at an average speed of s km/h, and then for two hours at an average speed of $(s-5)$ km/h, covering a distance of 330 km.

6 Solve:

a $5-2(x+1)=x$ **b** $8x=4-7(2-x)$
c $3x-1=7-(8+3x)$ **d** $5(6-p)+2=p+2$
e $2(m+3)-2=7+3m$ **f** $x-3(x-2)=3(2-x)$
g $4-(y-7)=3$ **h** $1-2(k-3)=3$
i $4+7(2x+1)=81$ **j** $6(x-1)-5=4-(x+1)$
k $2-3(1-y)=2(y+1)$ **l** $1-2(2x+1)=1-(x-1)$

PUZZLE CORNER

Majid set Debbie a puzzle.

Debbie was puzzled. She tried 1, 2, 3, . . . no good!
Majid said to try x, like this:

$$x \to 2x \to 2x+3 \to 4(2x+3)$$

$$\begin{aligned} \text{Then}\quad 20-4(2x+3) &= 28 \\ 20-8x-12 &= 28 \\ -8x+8 &= 28 \\ -8x &= 20 \\ 8x &= -20 \\ x &= -2\tfrac{1}{2} \end{aligned}$$

The number was $-2\frac{1}{2}$.
Debbie thought this was unfair!

1 *Majid was good at inventing puzzles. Here are two more of them for you to try:*

a *Think of a number.*
Multiply it by 3 and add 1.
Multiply the answer by 2.
Take this answer from 16.
The number left is 38.
What was the original number?

b *Think of a number.*
Multiply it by 4 and subtract 1.
Multiply the answer by 3.
Take this answer from 15.
21 is left.
What was the number?

2 *Make up some number puzzles like these, and ask a friend to solve them.*

MULTIPLYING PAIRS OF BRACKETS

1 Extending the $x(x+4) = x^2+4x$ pattern,

a $(x+6)(x+4)$
$= x(x+4)+6(x+4)$
$= x^2+4x+6x+24$
$= x^2+10x+24$

b $(x-1)(x+2)$
$= x(x+2)-1(x+2)$
$= x^2+2x-x-2$
$= x^2+x-2$

2

A SHORT-CUT

$(a+b)\ (c+d)$

① ( )( )
② ( )( )
③ ( )( )
④ ( )( )

$(x-1)(x+2)$

① x^2 Firsts
② $+2x$ Outsides
③ $-x$ Insides
④ -2 Lasts
$= x^2+x-2$

Mnemonic FOIL

EXERCISE 4A

1 Multiply these brackets, using either method shown above.
a $(x+3)(x+1)$ **b** $(x+2)(x+1)$
c $(x+4)(x+5)$ **d** $(a+2)(a+6)$
e $(b+4)(b+4)$ **f** $(y+3)(y+4)$
g $(m+1)(m+1)$ **h** $(t+2)(t+3)$

2 Check your answers to question **1** by marking areas in sketches like this, to show that $(x+6)(x+4) = x^2+10x+24$.

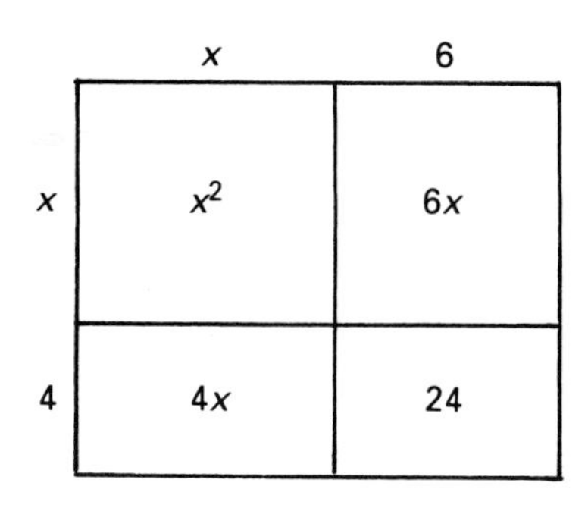

3 Multiply:
a $(y-2)(y-1)$ **b** $(w-3)(w-2)$
c $(u-5)(u-6)$ **d** $(b-1)(b-5)$
e $(c-3)(c-3)$ **f** $(x-2)(x-3)$
g $(n-2)(n-2)$ **h** $(s-3)(s-4)$

4 Multiply:
a $(x+5)(x-2)$ **b** $(p+4)(p-1)$
c $(t+1)(t-3)$ **d** $(x-2)(x+1)$
e $(y-3)(y+2)$ **f** $(z-4)(z+5)$
g $(m+6)(m-6)$ **h** $(n-1)(n+1)$

5 Multiply:
a $(x+6)(x+1)$ **b** $(m+2)(m-4)$
c $(r-7)(r+3)$ **d** $(t-5)(t-2)$
e $(y+9)(y+9)$ **f** $(x-6)(x+10)$
g $(y+7)(y-2)$ **h** $(n+1)(n+3)$
i $(q+6)(q-2)$ **j** $(t-4)(t-1)$
k $(x+10)(x+10)$ **l** $(y-7)(y+11)$
m $(n+12)(n+12)$ **n** $(a-9)(a+9)$
o $(b-8)(b+3)$ **p** $(c-5)(c-5)$

6

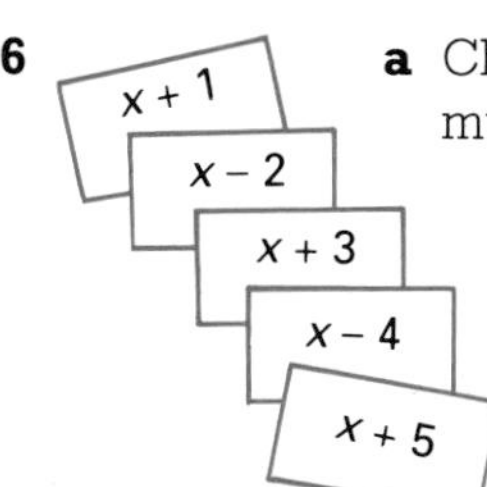

a Choose a pair of cards, and multiply the expressions.
b Then repeat **a** for other pairs of cards. You should be able to find ten different pairs altogether.

EXERCISE 4B

You have a choice:

$(2x-1)(3x+2)$
$= 2x(3x+2)-1(3x+2)$
$= 6x^2+4x-3x-2$
$= 6x^2+x-2$

or

$(2x-1)(3x+2)$
$= 6x^2 \quad +4x-3x \quad -2 \quad = 6x^2+x-2$

Firsts **O**utsides **I**nsides **L**asts

1 Multiply these brackets:

a $(2a+1)(2a-1)$ **b** $(3x-1)(3x+1)$
c $(2t-3)(2t+3)$ **d** $(2b-3)(2b-2)$
e $(2c-4)(3c-2)$ **f** $(s+2)(2s-1)$
g $(t-2)(4t+1)$ **h** $(w-1)(3w+4)$
i $(2m+3)(2m-3)$ **j** $(2z+4)(3z+2)$
k $(3n-5)(2n-3)$ **l** $(4x-2)(5x+3)$
m $(2-x)(1-x)$ **n** $(5-y)(3+2y)$
o $(1-4t)(1-5t)$

2 Pick pairs of cards, and multiply the two expressions together. There are six possible products. Try to find them all.

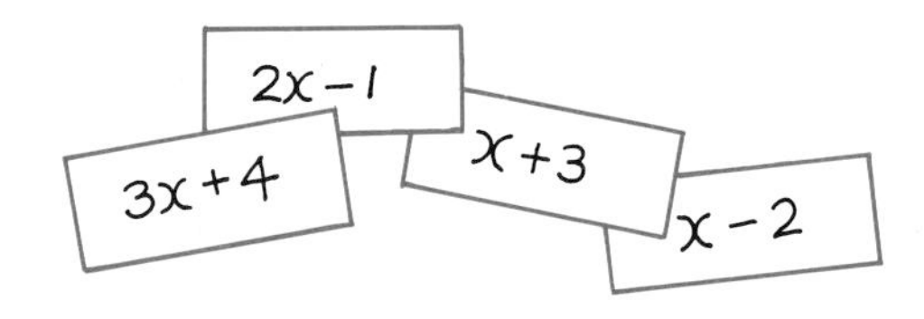

3 Copy and complete:

$(x+3)(x^2-2x+3)$
$= x(x^2-2x+3)+3(x^2-2x+3) = \ldots\ldots$

4 Multiply out, and simplify:

a $(x+2)(x^2-3x+1)$
b $(x-3)(x^2+x-3)$
c $(x-1)(x^2-2x-4)$

5 A rectangular metal plate is x mm long and y mm broad. When clamped and heated it expands. Its length becomes $(x+a)$ mm and its breadth $(y+b)$ mm.

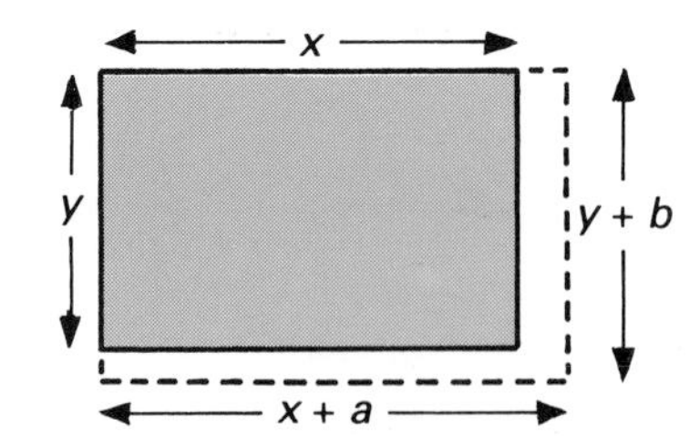

a Write down expressions for the areas of: (i) the cool plate (ii) the hot plate.
b Find the simplest expression you can for the increase in area of the plate.

6 A knitted square is pulled sideways into a rectangular shape. (Lengths are in cm.) Which area is greater, and by how much?

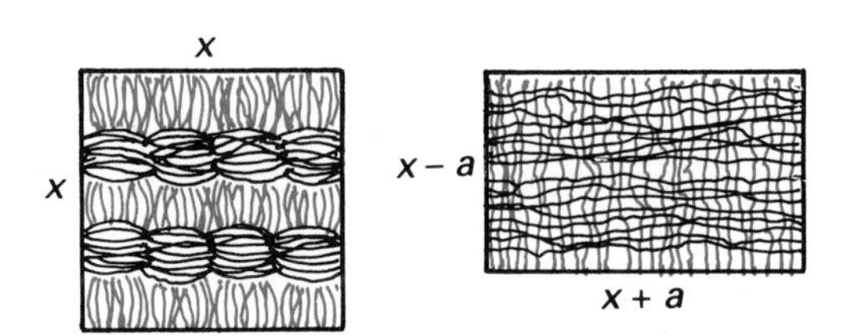

INVESTIGATIONS

1 a *Find the sequences of numbers you get when you put $n = 1, 2, 3, \ldots$ in $2n+6$ and $2(n+3)$.*
b *Explain why the sequences are the same.*
c *Find the value of n when $2(n+3) = 102$.*

2 *Repeat parts **a** and **b** of question **1** for:*
(i) n^2+n and $n(n+1)$ (ii) $(n+1)(n-1)$ and n^2-1.

3 a *By multiplying out, show that:*
(i) $(n-1)(n+1) = n^2-1$
(ii) $(n-1)(n^2+n+1) = n^3-1$
(iii) $(n-1)(n^3+n^2+n+1) = n^4-1$
b *Write down the next part of the pattern, and check it by putting $n = 1, 2, 3, \ldots$*

TWO USEFUL SQUARES

$(a+b)^2$
$= (a+b)(a+b)$
$= a^2+ab+ba+b^2$
$= a^2+2ab+b^2$

$(a+b)^2 = a^2+2ab+b^2$
$(a-b)^2 = a^2-2ab+b^2$

Examples $(t-1)^2 = t^2-2t+1$
$(2a+3b)^2 = 4a^2+12ab+9b^2$

$(a-b)^2$
$= (a-b)(a-b)$
$= a^2-ab-ba+b^2$
$= a^2-2ab+b^2$

EXERCISE 5A

1 Multiply out:

a $(x+y)^2$ **b** $(m+n)^2$ **c** $(u+v)^2$ **d** $(s+t)^2$
e $(a+b)^2$ **f** $(w+z)^2$ **g** $(x-y)^2$ **h** $(u-v)^2$
i $(p-q)^2$ **j** $(c-d)^2$ **k** $(e-f)^2$ **l** $(r-s)^2$
m $(x+5)^2$ **n** $(y+3)^2$ **o** $(a+1)^2$ **p** $(b+10)^2$
q $(w+2)^2$ **r** $(y+4)^2$ **s** $(c-4)^2$ **t** $(d-1)^2$
u $(e-6)^2$ **v** $(f-8)^2$ **w** $(x-5)^2$ **x** $(y-9)^2$

2 Find expressions for the areas of these squares. Multiply out, and simplify. Lengths are in cm.

a $x+1$

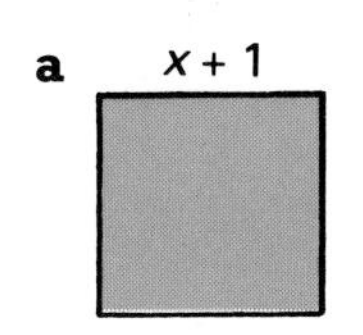

b $y-3$

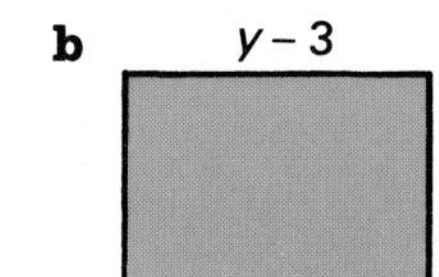

c $w+2$

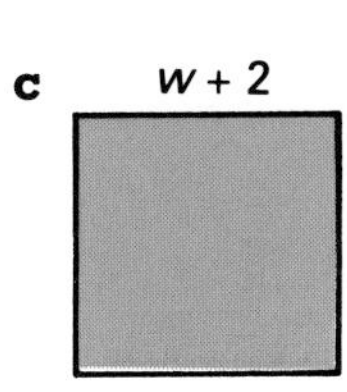

3 $(x+1)^2 = x^2+2x+1$
$(x+2)^2 = x^2+4x+4$
$(x+3)^2 = x^2+6x+9$
. .
. .
. .

a Write down the next five lines in this list of squares.
b Describe any number patterns that you can see.
c Write down the nth line.

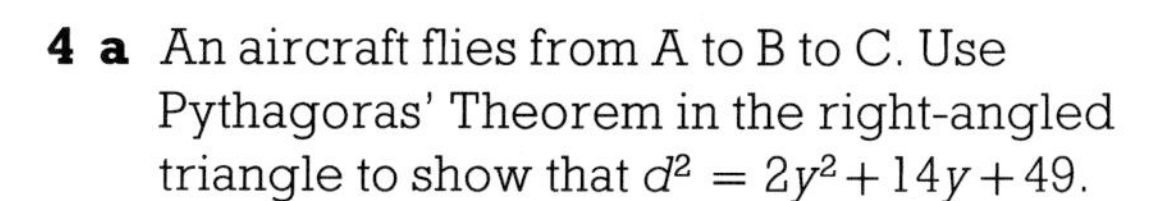

4 a An aircraft flies from A to B to C. Use Pythagoras' Theorem in the right-angled triangle to show that $d^2 = 2y^2+14y+49$.

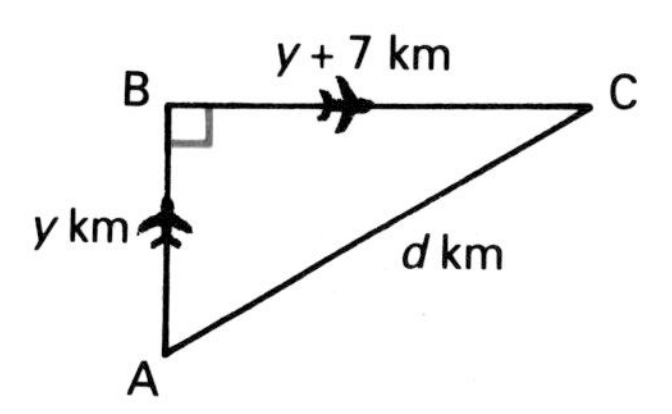

b Calculate d, correct to 1 decimal place, when $y = 10$.

5 a Use Pythagoras' Theorem to show that $h^2 = 10x+25$.
b Calculate h when $x = 20$.

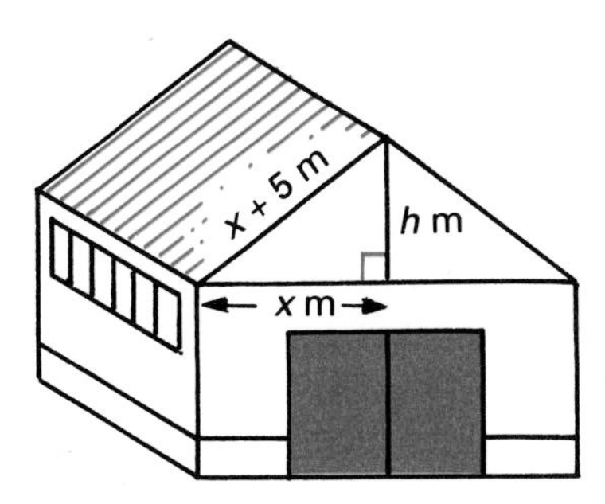

EXERCISE 5B

1 Multiply out:

a $(2x+3)^2$ **b** $(3y+1)^2$ **c** $(3a-4)^2$
d $(5b+2)^2$ **e** $(2c-3)^2$ **f** $(3a-5)^2$
g $(2x+y)^2$ **h** $(x-3y)^2$ **i** $(2a+3b)^2$
j $(4c-2d)^2$ **k** $(5p+2q)^2$ **l** $(4u-5v)^2$

2 Multiply out:

a $\left(x+\frac{1}{x}\right)^2$ **b** $\left(y-\frac{1}{y}\right)^2$ **c** $\left(2m+\frac{1}{2m}\right)^2$

3 a Copy and complete:

$(3x-1)^2-(x-1)^2$
$= 9x^2-6x+1-(x^2-2x+1) = \ldots$

b Multiply out and simplify:
(i) $(x+3)^2+(x+1)^2$ (ii) $(x+3)^2-(x+1)^2$
(iii) $(x-2)^2+(2x-1)^2$ (iv) $(5x-1)^2-(5x-2)^2$

4 Find expressions for:
(i) the sum (ii) the difference
of the areas of the squares in each pair.

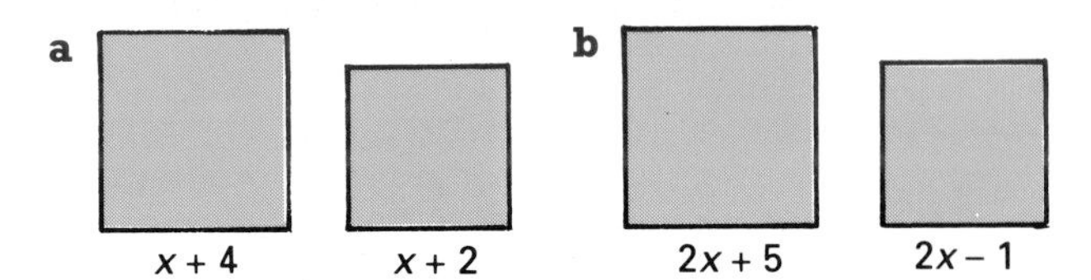

5 A square metal plate is 1 m long. When heated, the plate expands by x m as shown.

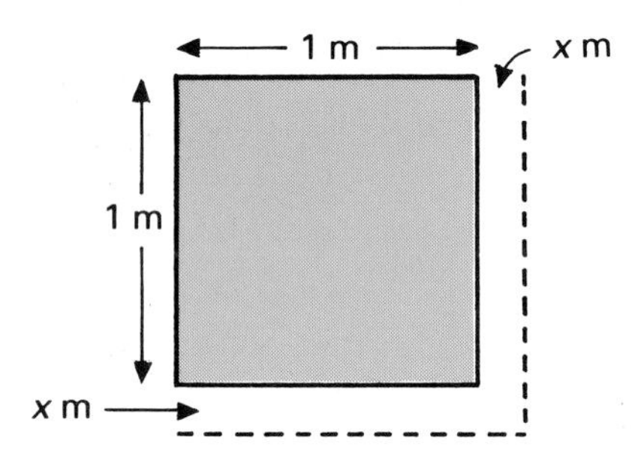

a Find an expression for the increase in area.
b Prove that, for an increase in length of 1 centimetre, the increase in area is $0.02\,\text{m}^2$, correct to 2 decimal places.

6 The span of a bridge is $2a$ metres and the rise of the arch is h metres. Find the radius of the arch, r metres, as follows.

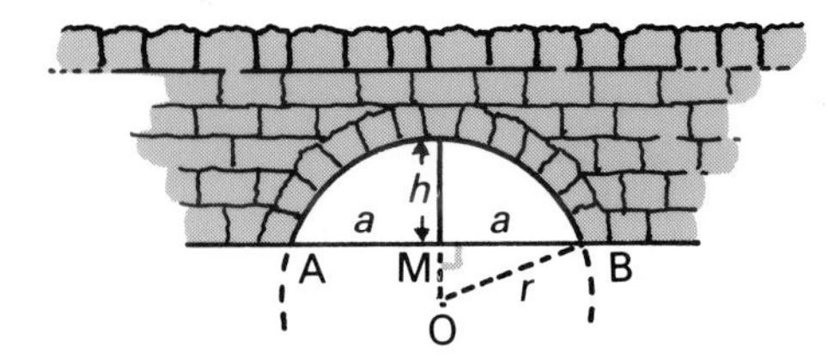

a Express OM in terms of r and h.
b Use Pythagoras' Theorem in $\triangle$OMB to find an equation in r, a and h.
c Hence show that $2rh = a^2 + h^2$.
d Find r when $a = 20$ and $h = 10$.

INVESTIGATIONS

1 *Investigate ways of illustrating the expansions of $(x+y)^2$ and $(x-y)^2$ by the areas of squares and rectangles.*

2 *Multiply out $(x+1)^2$ and $(x+1)^3$, and arrange the terms in order, highest powers of x first.*
Now try $(x+1)^4$ and $(x+1)^5$. Can you write down expansions for $(x+1)^6$, $(x+1)^7$, . . . ?
Have you seen this pattern of numbers before?

EQUATIONS WITH BRACKETS 2

The gable end of the solar roof is right-angled, with dimensions as shown. Calculate its actual dimensions.

Using Pythagoras' Theorem, $(x+1)^2 = x^2+4^2$

$$x^2+2x+1 = x^2+16$$
$$2x+1 = 16$$
$$x = 7\tfrac{1}{2}$$

So the dimensions of the gable end are 4 m, 7.5 m and 8.5 m.

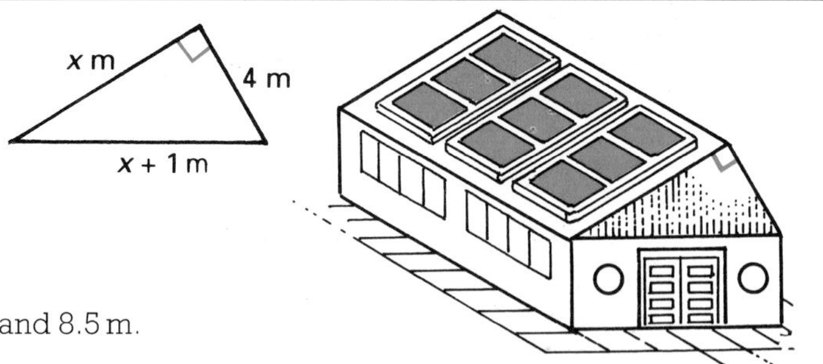

EXERCISE 6A

1 Solve:
a $x(x+3) = x^2+6$
b $x(x+2) = x^2-4$
c $x(2x+3) = 2(x^2-3)$
d $(x+4)(x-3) = x(x-2)$
e $x(x+5) = (x+3)(x-1)$
f $(y+2)^2 = y(y+2)$
g $(y-3)^2 = y(y+3)$
h $(y+6)^2 = y(y-6)$

2 The ladder just reaches the top of the wall.
a Use Pythagoras' Theorem to make an equation.
b Solve the equation, to find the length of the ladder.

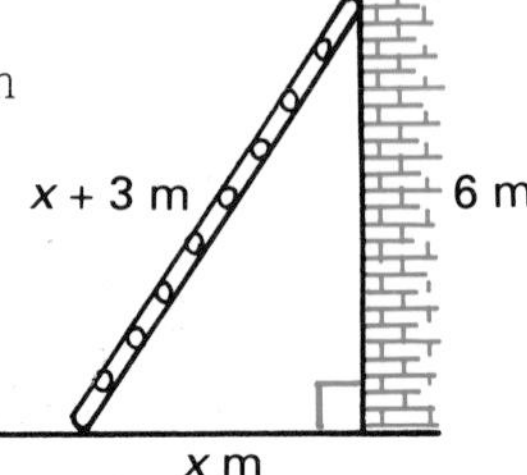

3 Use Pythagoras' Theorem to help you find the lengths of the sides of each triangle (all metres).

a

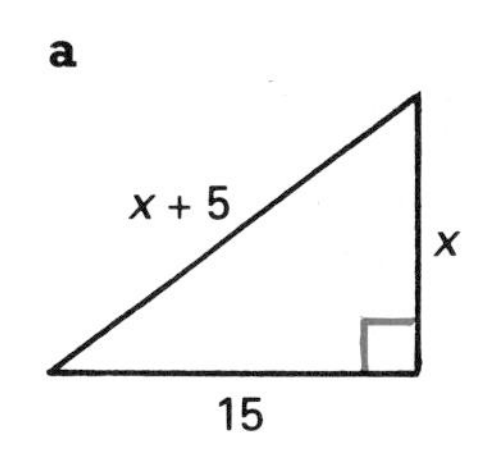

b (triangle with sides x, 10, $x + 2$)

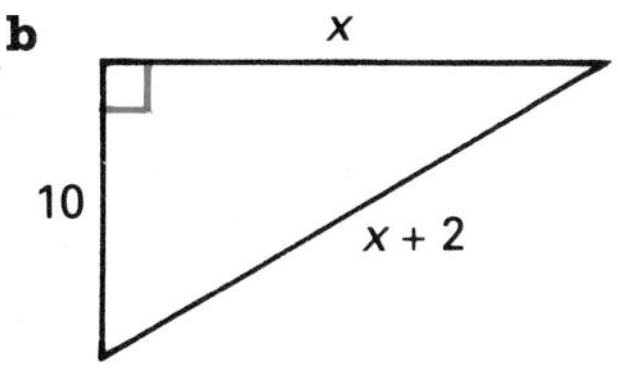

4 The bridge straddles a ravine. Find the length of:
a the bridge **b** the support struts.

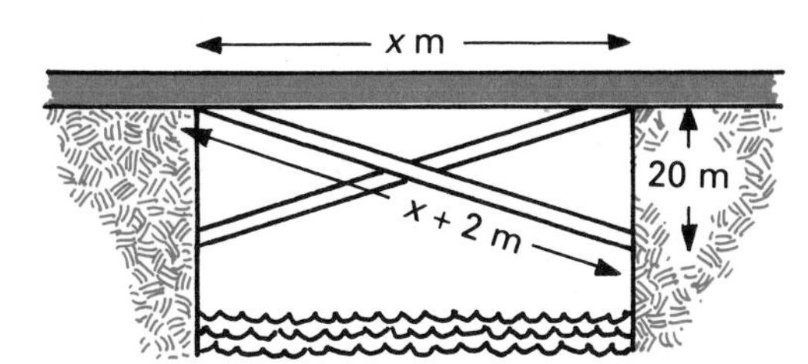

EXERCISE 6B

Example
The areas of these two pictures are equal. Find the actual dimensions of the pictures.

The areas are $x(x+10)\ \text{cm}^2$ and $(x+4)^2\ \text{cm}^2$.

So $x(x+10) = (x+4)^2$
$x^2+10x = x^2+8x+16$
$10x = 8x+16$
$2x = 16$
$x = 8$

The dimensions are 8 cm by 18 cm, and 12 cm by 12 cm.
Check The area of each picture is $144\ \text{cm}^2$.

1 The paintings in each pair below have the same area. Make an equation for each pair, and find the lengths and breadths of the paintings. All lengths are in cm.

a

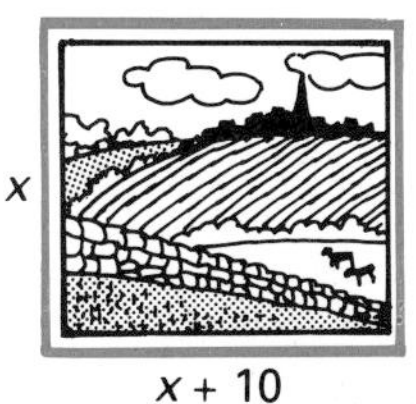

b

c

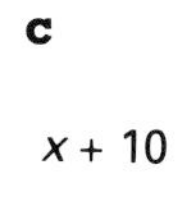

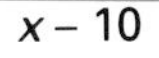

d

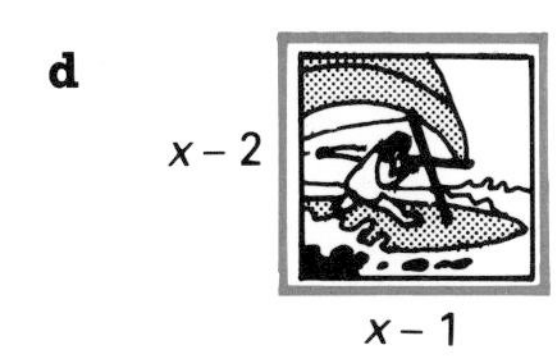

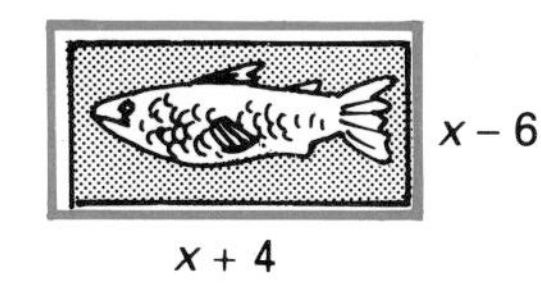

2 Solve:
a $x^2-x(x-1) = 2$
b $x^2-x(x+3) = 6$
c $x^2-x(5+x)+10 = 0$
d $(t+3)^2 = (t+1)^2$
e $(t+4)^2 = (t+2)^2$
f $(t-5)^2 = (t-1)^2$
g $(p+2)(p-2) = (p-1)^2$
h $(p+4)(p-4) = (p-2)^2$
i $(4p+1)(p-1) = (2p-1)^2$
j $y^2-(y-4)^2+8 = 0$
k $y^2-(y-10)^2 = 0$
l $y^2-(y+5)^2-20 = 0$

BRAINSTORMER

The diagram shows a right-angled triangle in a rectangle. Show that $y = x+1$.

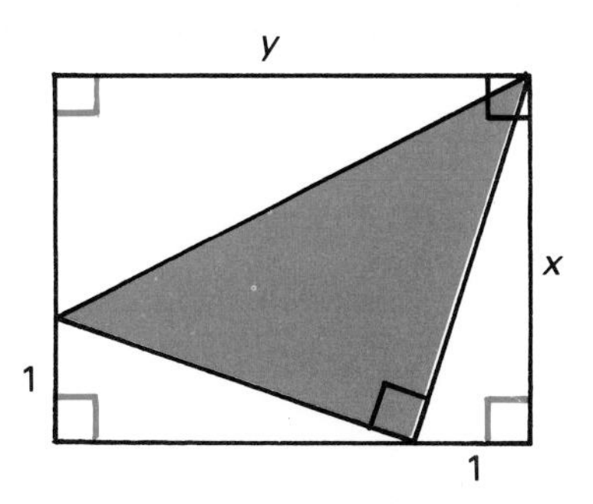

CHECK-UP ON BRACKETS AND EQUATIONS

1 Remove the brackets:

a $5(x+2)$ **b** $3(y-4)$ **c** $4(2k+1)$
d $7(1-n)$ **e** $a(b+c)$ **f** $x(x-y)$
g $-2(t+1)$ **h** $-3(t-1)$ **i** $-5(x+y)$

2 Multiply out, then simplify:

a $3+2(x-1)$ **b** $5-2(x+3)$
c $1+2(x-1)$ **d** $6-3(w-2)$

3 Solve:

a $4(x-3) = 16$ **b** $2(y+3)+3 = 15$

4 Make an equation for the collection of bags in each picture and find the number of coins in each bag.

a

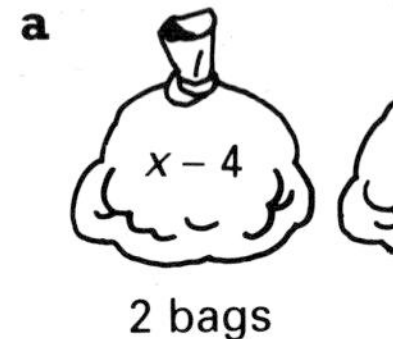

2 bags 3 bags

Total coins: 35

b

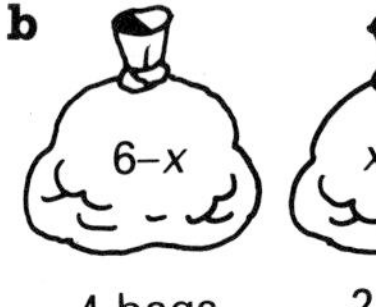

x + 2

4 bags 2 bags

Total coins: 22

5 A rectangle 8 cm long and $2x+1$ cm broad has an area of 40 cm^2. Make an equation, solve it for x and find the breadth of the rectangle.

6 Three people can lunch at the Café Royal for the same price as two at the Bistro.

a Make an equation in x, and solve it.
b How much is lunch at:
(i) the Café Royal
(ii) the Bistro?

7 Multiply out:

a $(x+1)(x+3)$ **b** $(y-2)(y-1)$
c $(n-5)(n+2)$ **d** $(2m+3)(3m+2)$
e $(4n-3)(5n-1)$ **f** $(u+5)^2$
g $(a-4)^2$ **h** $(2w+3)^2$

8 Solve:

a $(x+2)^2 = x^2+12$
b $(y+4)^2 = (y+3)(y+6)$
c $(2m+2)^2 = 4m(m+1)$
d $t^2-(t-9)^2 = 9$

9 Use Pythagoras' Theorem to form an equation, then solve it to find the lengths of the sides of the triangle.

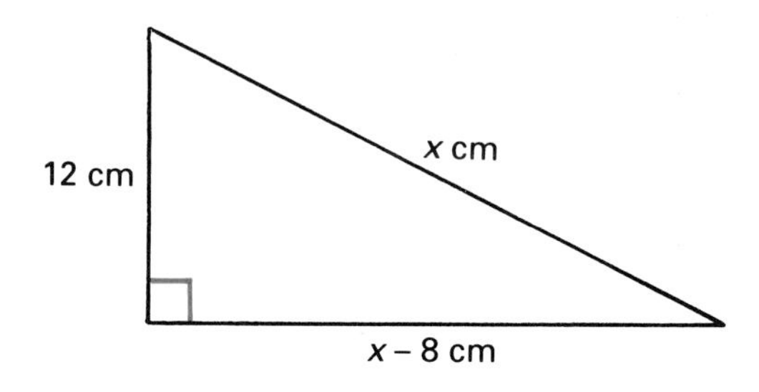

10 These two paintings have the same area. Make an equation, and find their dimensions.

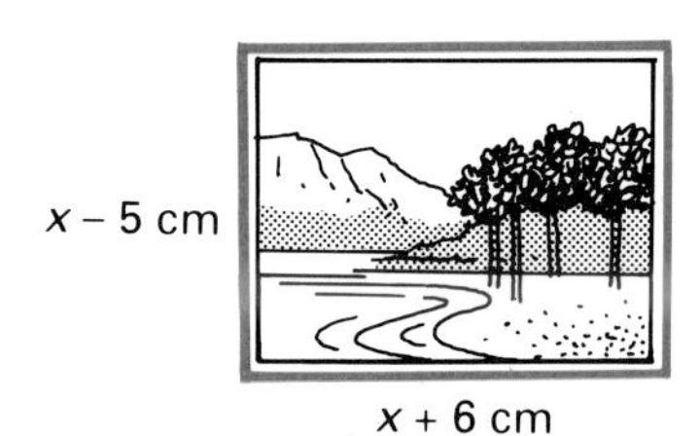

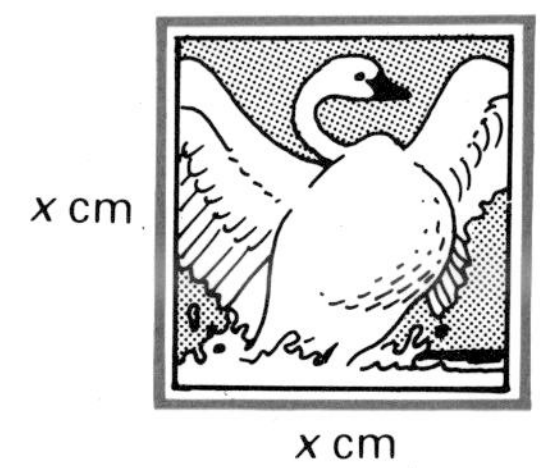

11 a Find the sequence of numbers when you put $n = 1, 2, 3, 4, \ldots$ in $(n+2)(n+4)-2(3n+4)$.
b What special numbers do you get? Explain why, by multiplying out the brackets.

12 Repeat question **11** for the expression $(3n+2)^2-(2n+3)^2$.

8 STATISTICS

LOOKING BACK

1 Starting at 10 am, a weather station records the windspeed at the top of Ben Nevis every $\frac{1}{2}$ hour.

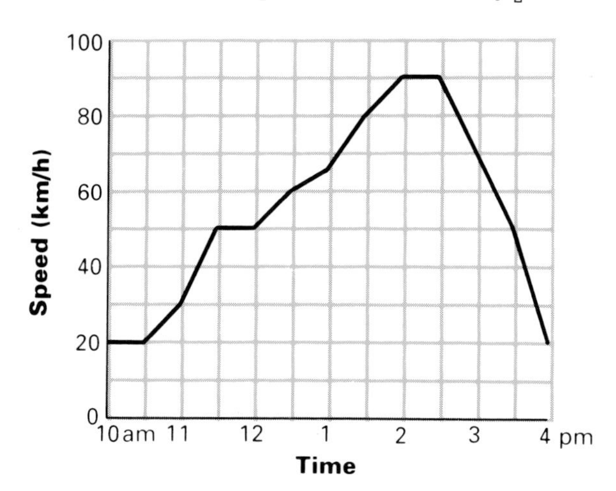

a What is the highest speed shown on the graph, and at what time is it recorded?
b For how long is the wind speed 50 km/h or more?
c In which 30 minute period does the wind:
(i) increase most quickly
(ii) decrease most quickly?

2 a Make a frequency table for this set of marks, using a class interval of 5, beginning 1–5.

15	17	20	10	16	25	8	22	12	19
23	18	19	20	13	21	21	14	18	20
17	19	15	21	14	20	5	10	15	22

b Show the results in a frequency diagram.
c What is the modal class interval?

3 Annette made a survey of car ages and recorded the results in a table.

Age (years)	0	1	2	3	4	5	6	7	8
Frequency	10	11	12	11	8	6	5	6	3

a Write down:
(i) the mode (ii) the range, of the ages.
b Calculate:
(i) the median age (ii) the mean age.
c Draw a frequency polygon of the ages.

4

Weather recordings at Sunny Beach holiday resort for two weeks in June produced these scatter diagrams.

a

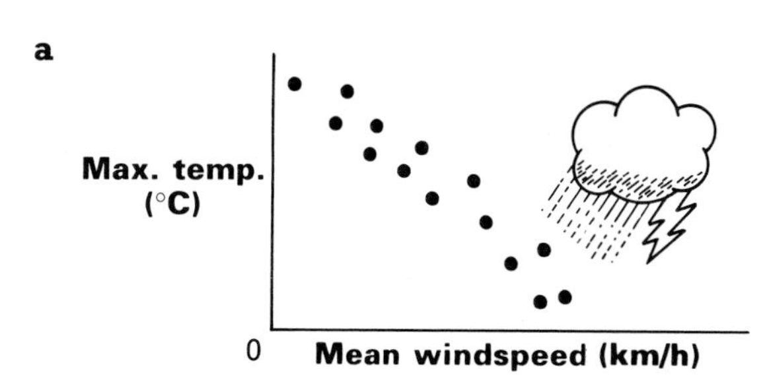

b

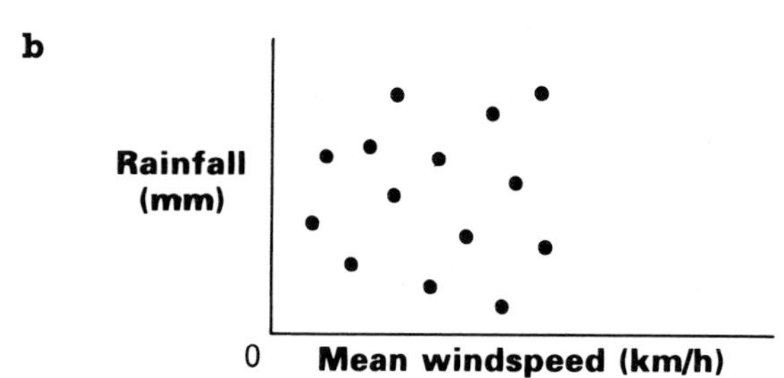

c

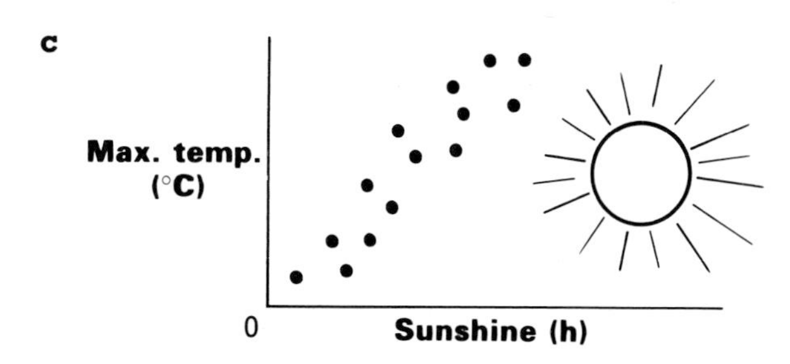

Describe the kind of correlation in each. Would you have expected these correlation results?

INTERPRETING AND COMPARING STATISTICS

EXERCISE 1

Throughout the following exercises, give approximate answers correct to 1 decimal place.

1 This bar chart shows the result of a survey (in 1992) of the number of adults who listened to Radios 1 to 5 for at least five minutes each week.

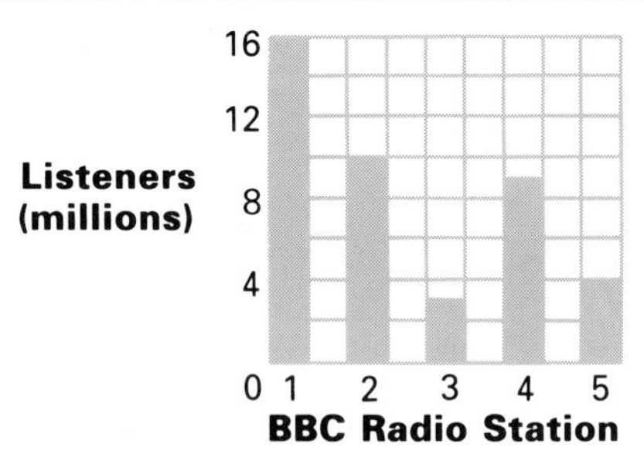

a Which station is:
(i) most popular (ii) least popular?

b Why is the condition 'at least five minutes' unfair to some stations?

2 The pie charts indicate the energy consumption of different fuels in the UK.

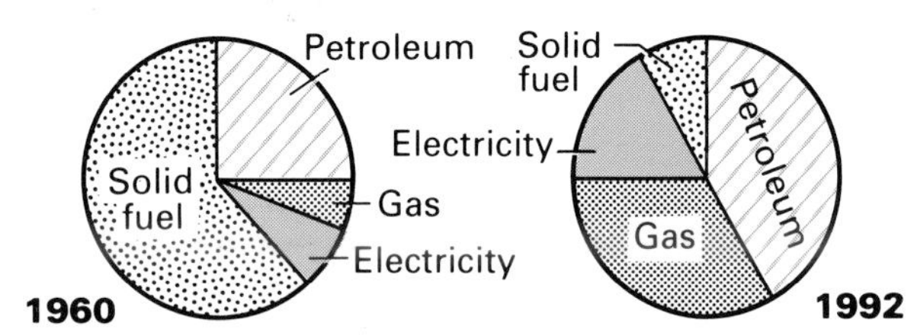

a What changes strike you most?

b Estimate the percentage share of solid fuel in:
(i) 1960 (ii) 1992

3 This diagram shows the rainfall (to the nearest 5 mm) last June in two holiday resorts.

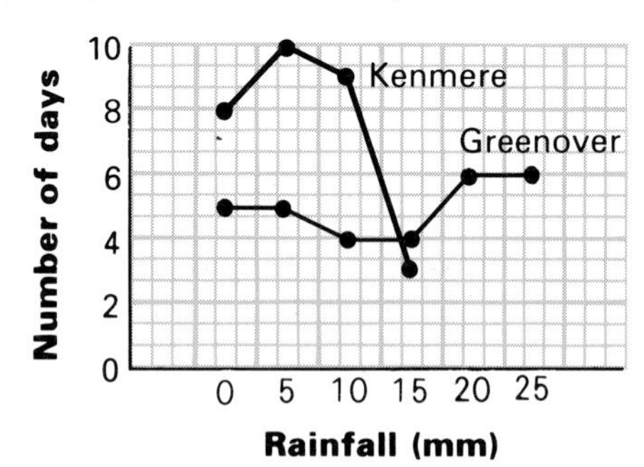

a On how many days did each place have 5 mm of rain?

b Which place had the heaviest rainfall in one day? How much?

4 15 members of Greenside Golf Club hold a competition in which each plays two rounds. The table gives their net scores.

Round 1	80	74	76	72	81	78	73	82	72	73	79	86	75	84	73
Round 2	78	74	74	70	77	77	75	77	69	70	76	78	72	75	73

a Calculate the mode, median, mean and range for the scores in each round.

b In which round was the weather better for golf?

c Which player won?

d Which average do you think best describes each round?

5 Roger asks his friends 'On average, how many hours a week do you spend watching TV, and how many doing homework?' He made this table of replies:

TV	16	18	10	5	11	9	9	4	13	17	11	7
Homework	7	4	10	14	9	12	11	15	8	6	8	12

a Calculate the mean number of hours his friends spent:
(i) watching TV (ii) doing homework.

b (i) Draw a scatter diagram.
(ii) Plot the point M giving the mean number of hours TV and homework, and draw the best-fitting straight line through it.

c Roger spends 12 hours a week watching TV. Estimate the time that he might spend on homework.

6 a Would you expect a positive or negative correlation between the heights of a group of students and their shoe sizes?

b Calculate the mean shoe size and the mean height for this data.

Shoe size	5	11	7	10	$8\frac{1}{2}$	6	7	$7\frac{1}{2}$	$6\frac{1}{2}$	8	6	9
Height (cm)	158	179	166	173	170	160	164	165	161	170	163	172

c (i) Draw a scatter diagram.

(ii) Plot the mean shoe size against the mean height, and draw the best-fitting straight line through the point. Was your answer to **a** correct?

d Jim's height is 178 cm. Estimate his shoe size.

INVESTIGATION

Statistics are used in the media to advertise, and to support arguments and reports. Examine some examples of these critically, and write a report on what you find.

MODE, MEDIAN AND MEAN FROM A FREQUENCY TABLE WITH CLASS INTERVALS

You may remember the question in Book 2 where 150 students at Action Academy were tested on the number of sit-ups they could do in $1\frac{1}{2}$ minutes.

We can now extend the frequency table of results, and calculate the mean number of sit-ups, like this:

Number of sit-ups	Mid-value	Frequency	Mid-value × frequency
40–44	42	11	462
45–49	47	20	940
50–54	52	28	1456
55–59	57	36	2052
60–64	62	24	1488
65–69	67	19	1273
70–74	72	12	864
	TOTAL	150	8535

Taking the mid-value to represent each class interval, the PE staff calculated the mean number of sit-ups per student.

$$\text{Mean} = \frac{\text{Total number of sit-ups}}{\text{Frequency (no. of students)}}$$

$$= \frac{8535}{150}$$

$$= 56.9$$

The modal class interval is 55–59 sit-ups.

The median score is half-way between the 75th and 76th scores. Counting the frequencies, $11+20+28+36 = 95$. The 75th and 76th scores are in the class interval 55–59 sit-ups.

Calculator Your calculator can help you with these calculations if it has a statistics mode. If it has, key all the 'mid-value × frequency' products, then use the

[n] key to display the total frequency

[Σx] key to display the total 'mid-value × frequency'

[$\bar{x}$] key to display the mean.

EXERCISE 2A

1 Irma counts the number of letters in the words in a paragraph of a newspaper.

Number of of letters	Mid-value	Frequency	Mid-value × frequency
1–3	2	57	
4–6		46	
7–9		28	
10–12		6	
13–15		3	
	TOTAL		

a Copy and complete the table.
b Which is the modal (most popular) class interval?
c In which class interval is the median word length?
d Calculate the mean word length.

2 3A's French exam marks are grouped in this table.

French mark	1–5	6–10	11–15	16–20	21–25	26–30	31–35	36–40	41–45	46–50
Frequency	1	3	7	18	25	27	19	12	10	2

a Write down:
(i) the modal class interval
(ii) the class interval which contains the median score.
b Calculate the mean mark.

3 This table shows a week's production of eggs at Happy Valley farm.

Size	Weight (g)	Mid-value	Frequency (dozen)	Mid-value × frequency
1	70 and over	Use 72	147	
2	65–69	67	278	
3	60–64		332	
4	55–59		240	
5	50–54		173	
6	45–49	47	61	
7	<45	Use 42	39	
		TOTAL		

a Calculate the mean weight of an egg.
b Find the median size of egg.

4

The table shows data on the ages of competitors in the Coast and Country half marathon. Calculate the mean age of the competitors.

Age (years)	Mid-value	Frequency	Mid-value × frequency
10–17	13.5	26	
18–29		64	
30–45		88	
46–59		110	
60–75		72	
	TOTAL		

EXERCISE 2B

1 Here is a list of the number of hours of sunshine in December at 48 places in the UK.

15	15	11	36	13	42	35	37	28	28	22	30
22	41	4	23	27	27	38	20	34	32	26	18
29	22	18	28	44	19	22	18	32	29	3	16
24	31	28	25	19	12	25	24	26	44	4	9

a Choose your own class intervals, and make a frequency table.
b Use your table to calculate the mean number of hours of sunshine.
c Find:
(i) the modal class interval
(ii) the class interval which contains the median.

2 Moira records the heights, to the nearest cm, of boys and girls in her year.

Height (cm)	Girls	Boys
151–155	III	I
156–160	𝍸 III	II
161–165	𝍸 IIII	𝍸 II
166–170	𝍸 𝍸 𝍸 I	𝍸 𝍸
171–175	𝍸 𝍸 II	𝍸 𝍸 IIII
176–180	II	𝍸 𝍸 IIII
181–185		𝍸
186–190		𝍸
191–195		II

a Make a frequency table for each group.
b Calculate the mean height of each group, to the nearest cm.
c Draw frequency polygons for both groups on the same diagram.
d Write a short report comparing the heights of the girls and boys based on the means and frequency polygons.

3 Steve spends a day training for the discus, and records his throws in a diagram like this.

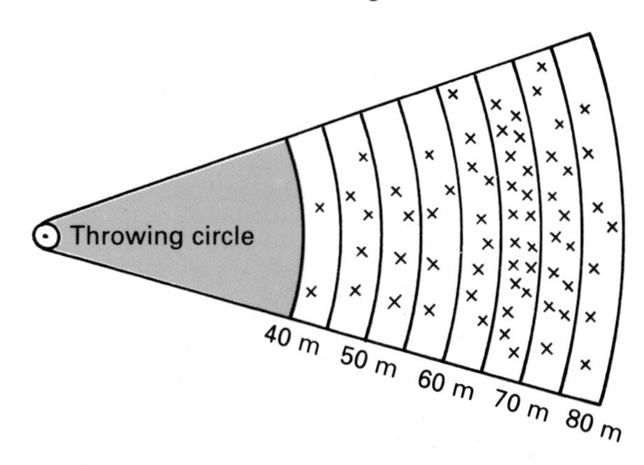

a Make a frequency table, using 5 m class intervals ($40 < d \leqslant 45$) etc.
b Find the modal class interval.
c In which class interval is his median throw?
d Calculate his mean throw.

CHALLENGES

1 *There are six possible ways of listing the three averages in ascending order of size. For example, mean, median, mode; mean, mode, median, etc. Write them all down, and then try to find sets of numbers which will fit each triple arrangement. For example, to make the mean the greatest and the mode the least you could choose the four numbers 1, 1, 3, 11. (Check this). Explain your methods.*

2 *Averages are meant to be typical of the data. Give advantages and disadvantages of the mode, median and mean, and find examples when you would choose one rather than another. For example, the average length of 100 raspberry canes, all different—which is the easiest average to use?*

INVESTIGATIONS

1 *Investigate the effect of choosing different class intervals. Go back to question **1** of Exercise 2B, and try different class intervals. What conclusions do you reach?*

2 *Check the lengths of words used in different newspapers. Collect data, analyse it and write a short report.*

CUMULATIVE FREQUENCY DISTRIBUTIONS

The PE staff at Action Academy wanted a way to tell at a glance how many pupils managed 'up to 49 sit-ups', 'up to 54', and so on. So they made a **cumulative frequency table**, and drew a **cumulative frequency curve**.

Cumulative frequency table

Number of sit-ups	Frequency	Cumulative frequency
40–44	11	11
45–49	20	31
50–54	28	59
55–59	36	95
60–64	24	119
65–69	19	138
70–74	12	150
TOTAL	150	

In the table, they added the frequencies, line by line. For example, $11 + 20 = 31$, which shows that 31 pupils managed only 40–49 sit-ups.

Cumulative frequency curve

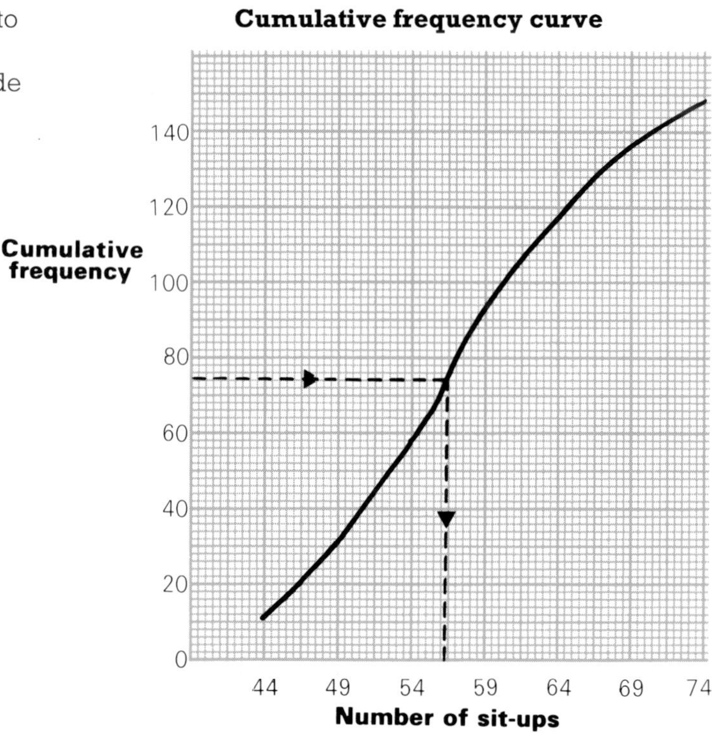

In the graph, they used the upper end of each class interval on the horizontal axis. For example, (44, 11) and (49, 31) are points on the curve.
To find the median number of sit-ups they read up the cumulative frequency axis until they came to the 75th sit-up. Then they followed the dotted lines. Check that the median is 56.5.

EXERCISE 3A

1 The rainfall was measured every day for a month at a weather station on the west coast of Scotland.

a Copy and complete the table.

Rainfall per day (mm)	Frequency	Cumulative frequency
0–4	6	6
5–9	9	15
10–14	8	
15–19	5	
20–24	2	
25–29	1	

b Draw the cumulative frequency curve.

c Estimate the median rainfall per day.

2 Over the year the weekly sales of singles in a record department were recorded.

a Copy the table and add a cumulative frequency column.

b Draw the cumulative frequency curve.

c Estimate the median weekly sales.

d Calculate the number of weeks when sales exceeded 330, and give this as a percentage of the year.

Number sold	Frequency
251–270	2
271–290	5
291–310	12
311–330	16
331–350	8
351–370	6
371–390	3
TOTAL	52

3 a Using the information in the frequency diagram, make a cumulative frequency table of the number of repeated programmes on TV in the week shown.

b Construct a cumulative frequency curve.

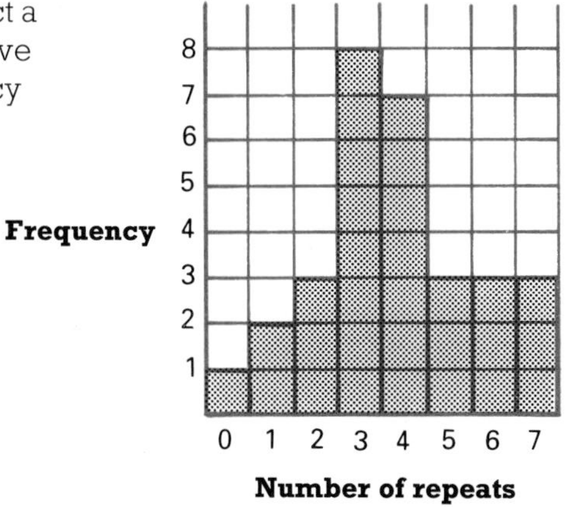

4 The table opposite shows the distribution of marks in History and Geography tests for a group of pupils.

Mark	Frequency	
	History	Geography
21–30	3	0
31–40	16	14
41–50	22	23
51–60	39	25
61–70	43	35
71–80	24	34
81–90	13	21
91–100	0	8

a Draw two cumulative frequency curves on the same squared paper.

b Amy missed both tests, but her teachers thought she would have done as well as her twin sister Alison who was 18th in History and 39th in Geography. Estimate Amy's marks.

c Any mark over 70% was given an A. What percentage of pupils got an A in:
(i) History
(ii) Geography?

INTERQUARTILE RANGE

Data about Danny's golf drives are shown in this cumulative frequency table and curve.

Distance (yards)	Frequency	Cumulative frequency
221–240	5	5
241–260	12	17
261–280	20	37
281–300	25	62
301–320	33	95
321–340	17	112
341–360	8	120

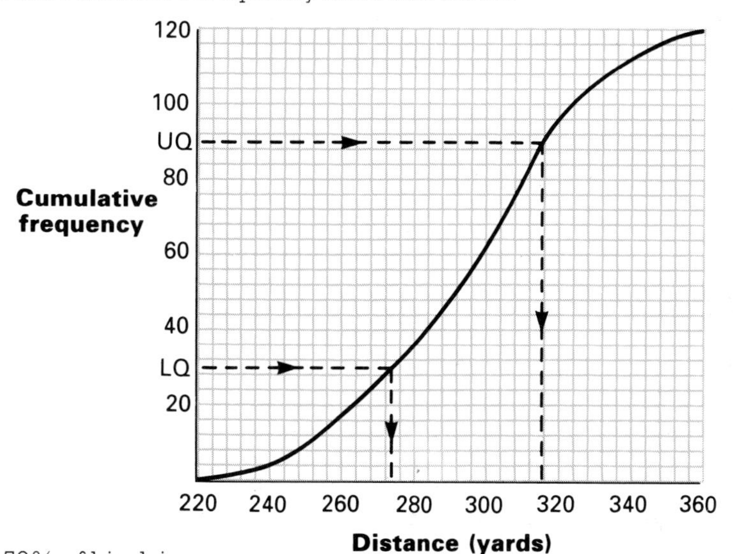

Danny wants to find the range of the middle 50% of his drives.
25% of his 120 drives = 30 drives. How far does his 30th drive go? From the cumulative frequency curve, the distance is 276 yards. This is the **lower quartile** (**LQ**).
75% of his 120 drives = 90 drives. How far does his 90th drive go? From the curve, the distance is 316 yards. This is the **upper quartile** (**UQ**).

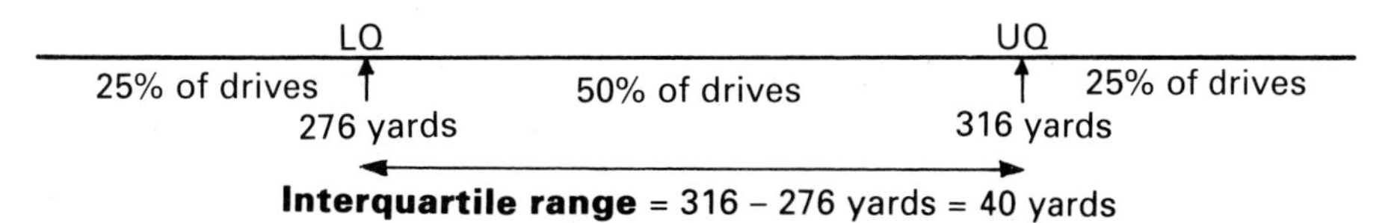

The middle 50% of his drives range from 276 yards to 316 yards.

Summary
Each of the following gives a concise picture of statistical data in terms of:
(i) its average (ii) its spread:

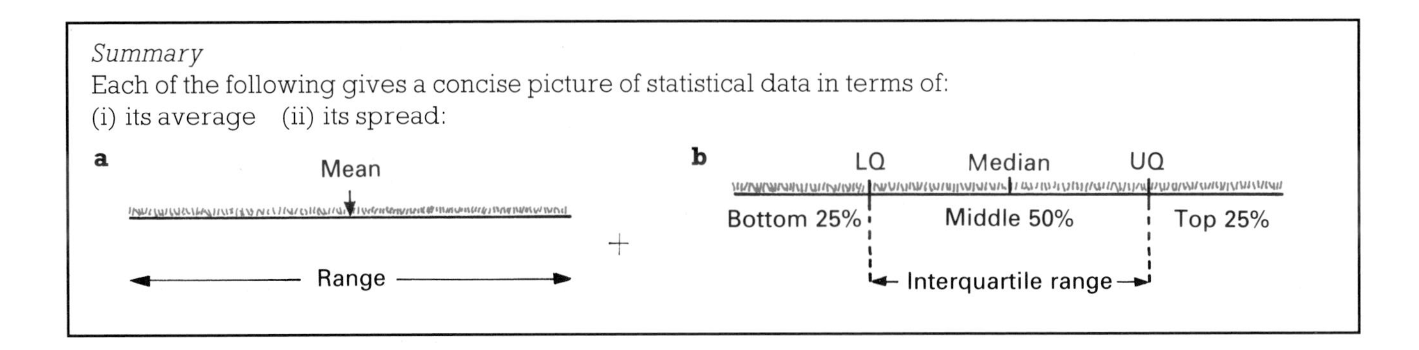

EXERCISE 3B

1 All 12–13 year olds in Action Academy were tested in spelling.

Mark	11–20	21–30	31–40	41–50	51–60	61–70	71–80	81–90	91–100
Frequency	7	10	18	24	28	29	22	18	4

a Calculate the mean mark.
b Write down the modal interval.
c Make a cumulative frequency table and curve.
d Find:
(i) the interquartile range
(ii) the median mark.
e What kind of results should a well-designed test produce? Comment on this test.

2 a Use the data in question **2** of Exercise 2B to make cumulative frequency tables and curves (on the same diagram).
b Calculate the interquartile ranges of the heights. What does this tell you?

3 Helen and Sally are competing in an archery competition. Each ring on the target is 12 cm wide.

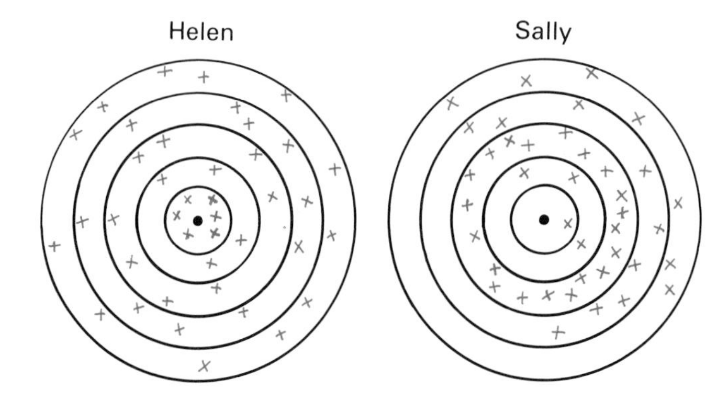

a Make a cumulative frequency table for each girl.

Distance from centre (cm)	Frequency	Cumulative frequency
$0 < d \leqslant 12$	6	6
$12 < d \leqslant 24$	4	10

b Draw cumulative frequency curves on the same diagram.
c Find the interquartile range for each one.
d Calculate the mean distance from the centre for each girl's arrows.
e Who is the better archer, Helen or Sally? Give your reasons.

PROJECT

Make cumulative frequency tables and curves of the heights, weights or armspans of students in your year. Analyse the data, and write a report.

INVESTIGATION

A day trip

The pupils in Mrs Adams' Class, 4C, are planning a day trip. There are five suggestions for the outing, and the pupils list their preferences in order, 1 to 5.

1 *Two groups of four friends have exactly the same preferences. What are their initials?*

2 a *Make a tally chart for each suggestion, like this one for the seaside.*

Preference	Tally	Seaside frequency
1	~~IIII~~ ~~IIII~~ I	II
2		
3		
4		
5		

b *Which destination has been given most:*
(i) first choices
(ii) last choices?

3 *From your statistics try to reach an agreement on which place to visit.*

4 *Mrs Adams has to decide on the destination. How can she make the fairest choice?*

5 *In the end she decides that the form will split up and go to two different places.*

a *Is it possible for all the pupils to have:*
(i) their first choice
(ii) their first or second choice?

b *Which two places does she choose? List the initials of the pupils going to each one.*

Pupil	Seaside	London	Zoo	Concert	Funfair
J.A.	1	4	2	5	3
W.A.	5	1	3	4	2
R.B.	1	2	5	3	4
T.C.	3	5	2	1	4
S.E.	5	1	3	4	2
J.F.	1	5	2	3	4
T.F.	5	4	2	1	3
I.H.	2	1	5	3	4
J.M.	1	2	5	4	3
P.M.	4	1	5	3	2
R.M.	1	4	2	5	3
S.N.	5	2	3	1	4
A.P.	5	1	4	3	2
J.R.	4	2	5	1	3
T.R.	1	5	2	3	4
W.R.	5	1	3	4	2
K.S.	1	5	2	4	3
N.S.	4	2	5	1	3
C.T.	5	1	3	4	2
G.T.	1	2	5	3	4
S.T.	1	5	2	3	4
A.W.	5	4	2	1	3
K.W.	5	4	2	1	3
P.Y.	1	5	2	3	4
R.Y.	1	5	2	4	3
T.Y.	5	2	3	1	4

PRACTICAL PROJECTS

*Choose one of the following topics of research. Make some claim, or state some fact you think is true about your topic (a **hypothesis**). Collect, display and analyse relevant data to see if they confirm or contradict your hypothesis. State your conclusions carefully and consider whether the research has opened any new avenues for you to explore. It may be possible to design and input data by computer. Software may be available to form spreadsheets and databases, and to draw and print out graphs.*

1 Questionnaires

Topic	***Data you might collect***
Holidays	*Where, when, how often?*
Shopping	*Type of goods, how far to travel?*
School meals	*Food bought, cost, healthy diet?*
Student budget	*Income, earnings, saving, spending?*
Musical tastes	*Preference, amount spent, concerts?*

2 Physical data collection

Topic	***Data you might collect***
Weather	*Sunshine, rain, wind, temperature?*
Absences	*Which day, which year/group, reasons?*
Traffic	*Movements at intersection, age of cars?*
Average contents	*Crisps (weigh), matches in box?*

3 Newspaper or other media claims

Many news articles contain claims of one sort or another. Examples:

- *10% of voters will support the Green Party at the next election.*
- *Most packaging in a supermarket is yellow because this attracts most customers.*
- *5% of cars on the road have an out-of-date tax disc on display.*

Look out for a claim like this. Then collect and analyse as much data as you can to see if the claim is true.

CHECK-UP ON STATISTICS

1 In a science experiment the resistance in a circuit was changed and the currents were measured to produce this scatter diagram.

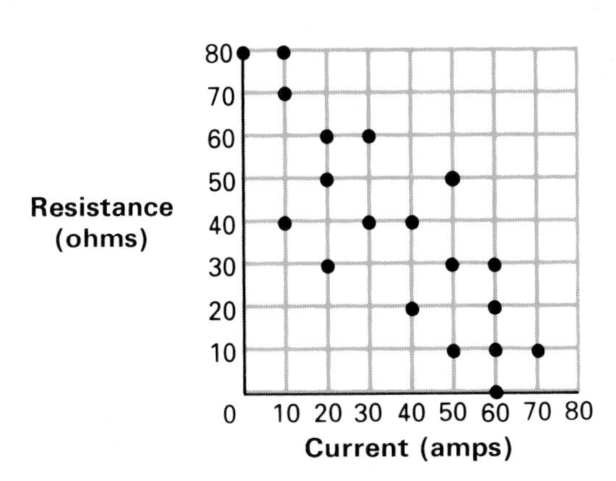

Hold your ruler along the best-fitting line and estimate the resistance for a current of:
a 55 amps **b** 25 amps

2 The results of the long jump and the 100 m in a decathlon are shown in the table:

Long jump (m)	7.60	7.40	7.15	7.90	7.45	7.50	7.90	7.70	7.25	7.70
100 m (s)	10.8	11.0	11.3	10.5	11.1	10.9	10.6	10.7	11.2	10.8

a Calculate the mean and range for the long jump, to 2 decimal places.
b Calculate the mode, median and mean for the 100 m race.
c (i) Draw a scatter diagram, and a best-fitting line.
(ii) Estimate the long jump distance of a competitor who runs the 100 m in 10.65 s.

3 Naomi is a student nurse. She tests the pulse rates of some of her friends.

Pulse rate (beats/minute)	Frequency
51–55	5
56–60	10
61–65	14
66–70	27
71–75	36
76–80	34
81–85	26
86–90	14
91–95	3
96–100	1

a Which is the modal class interval?
b In which class interval is the median rate?
c Calculate the mean pulse rate.

4 Look again at the data in question **3**.
a Draw a frequency diagram and a frequency polygon.
b Make a cumulative frequency table, and draw a cumulative frequency curve.
c Use your graph to find: (i) the median (ii) the upper quartile (iii) the lower quartile (iv) the interquartile range of pulse rates.
Show the results for **c** in a sketch like the one in the summary on page 100.

9 TRIGONOMETRY

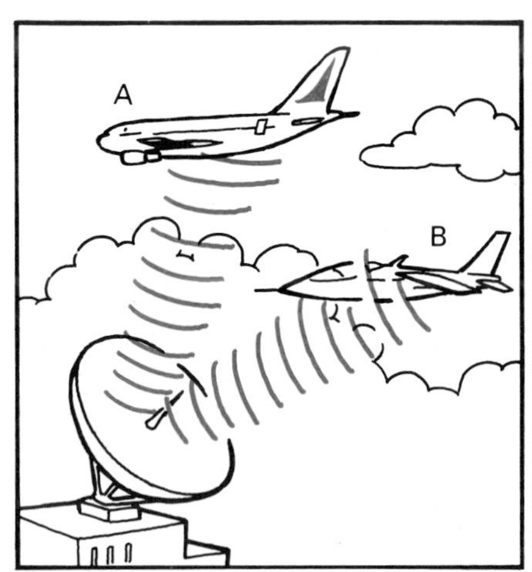

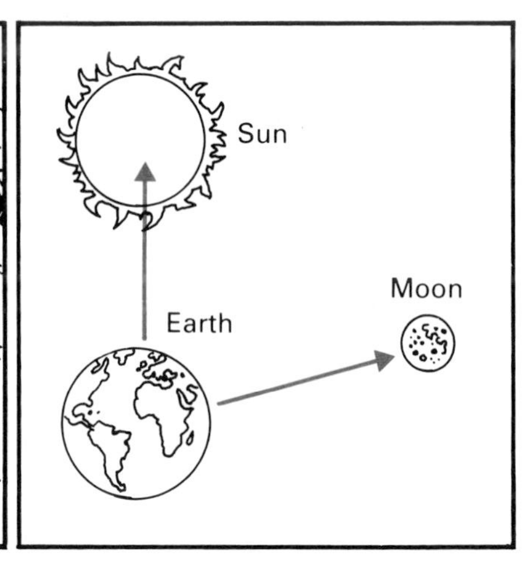

LOOKING BACK

1 Use your calculator to find x, correct to 2 decimal places.

a $x = 15 \times 0.123$ **b** $x = \dfrac{9}{0.58}$

2 Calculate n in each triangle below:

a

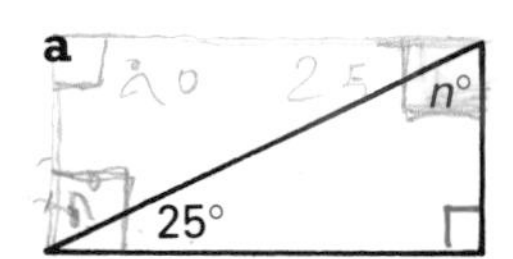

b

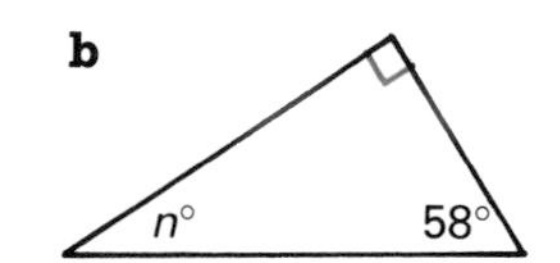

3 Write down, in their simplest form, the ratios of lengths:

a $\dfrac{AD}{OD}$ **b** $\dfrac{BE}{OE}$ **c** $\dfrac{CF}{OF}$

C
B
A
3
2
1
O 2 D 2 E 2 F

4 In right-angled triangle $\triangle PQR$:

a name the hypotenuse

b use Pythagoras' Theorem to calculate PR, correct to 1 decimal place.

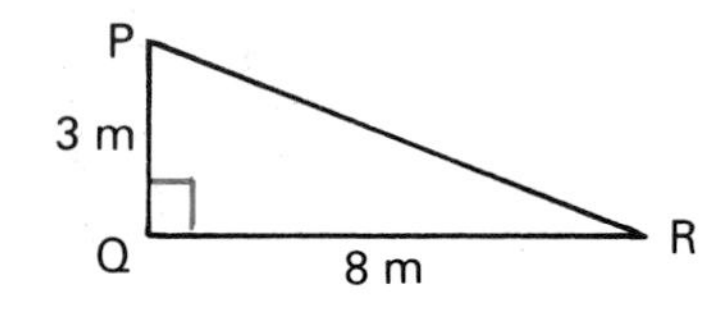

5 a Name:

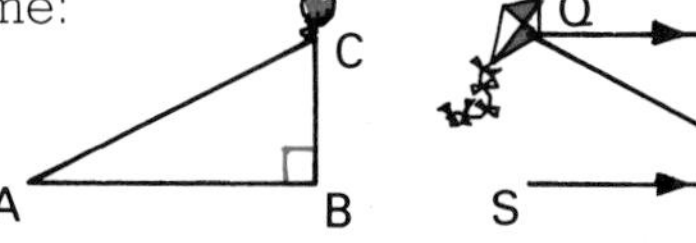

(i) the angle of elevation of C from A.

(ii) the angle of depression of R from Q.

b If $\angle PQR = 38°$, what size is $\angle QRS$?

6 Use the 'cover up' method to solve these equations:

a $\dfrac{x}{3} = 4$ **b** $\dfrac{x}{6} = 7$ **c** $\dfrac{x}{8} = 1.5$

7 (i) Solve $\dfrac{x}{4} = 5$ by multiplying each side by 4.

(ii) Use this method to solve the equations in question **6**.

8 a Make a scale drawing to help Frances to find the height of the school, to the nearest metre. Remember to add on her height.

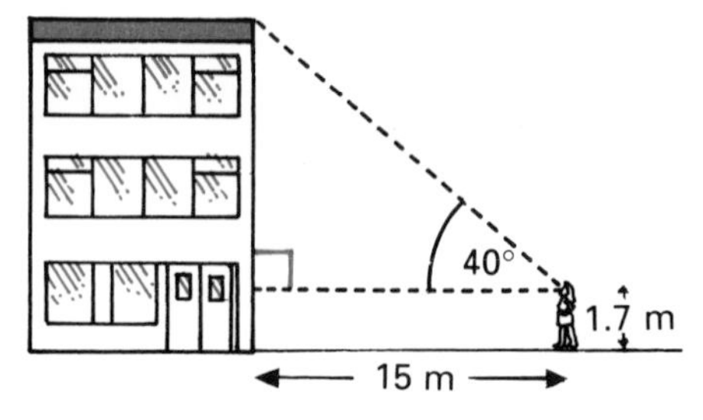

b Compare your answer with others in your class. Then list some drawbacks of scale drawings.

FROM SCALE DRAWING TO CALCULATION

EXERCISE 1/CLASS DISCUSSION

1 The diagram below shows the flight-path of an aircraft climbing in a straight line at an angle of $a°$ to the ground.
At A, its height is 50 m, and its horizontal distance from O is 100 m.

The ratio $\dfrac{\text{height}}{\text{horizontal distance}}$ is $\dfrac{AM}{OM}$.

$$\frac{AM}{OM} = \frac{50\text{ m}}{100\text{ m}} = 0.5$$

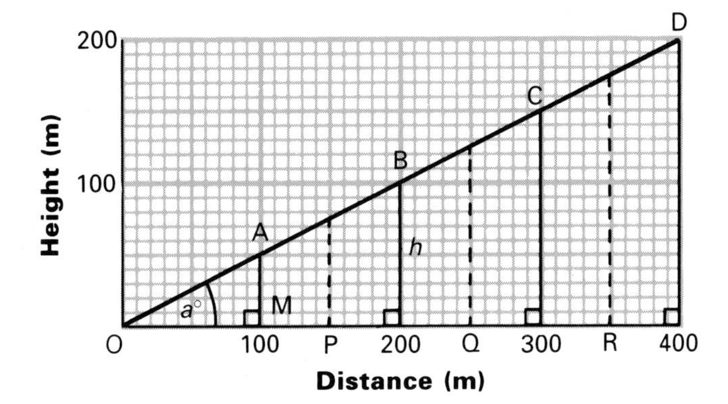

a Calculate the value of $\dfrac{\text{height}}{\text{horizontal distance}}$ at points B, C and D.
b What do you find?

> The ratio $\dfrac{\text{height}}{\text{horizontal distance}}$ is the same for every right-angled triangle on the flight-path as long as angle $a°$ is fixed. All these triangles are **similar**.
> The ratio is called the tangent of the angle at O.
> We say **the tangent of a° is 0.5**, and write **tan a° = 0.5**.

You can now calculate the height, h metres, of any plane on the flight-path.

At B, $\tan a° = \dfrac{h}{200}$

so $h = 200 \times \tan a° = 200 \times 0.5 = 100$.
Check this on the graph.

2 Now calculate the height of the plane at:
a P, 150 m from O **b** Q, 250 m from O
c R, 350 m from O **d** S, 1000 m from O.

3 a Calculate the tangents of the angles $a°$, $b°$ and $c°$ of the three flight-paths in this diagram.

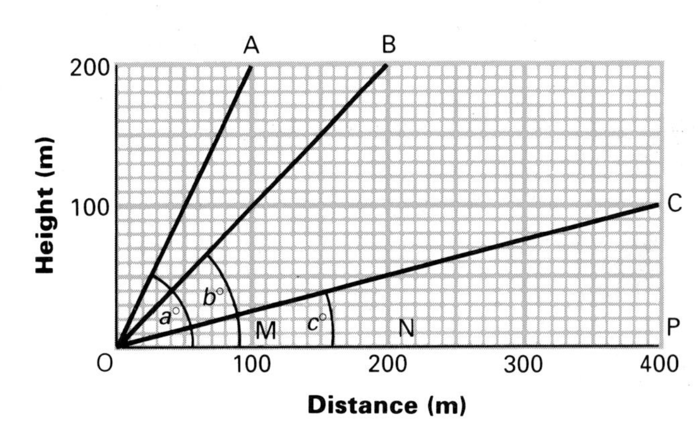

b What is the value of tan 45°?
c Use tan 45° to find the height of the plane on the flight-path of 45° when it is at a horizontal distance from O of:
(i) 50 m (ii) 150 m (iii) 3000 m.

4 Calculate tan 10°, tan 20° and tan 30°, correct to 1 decimal place for these flight-paths.

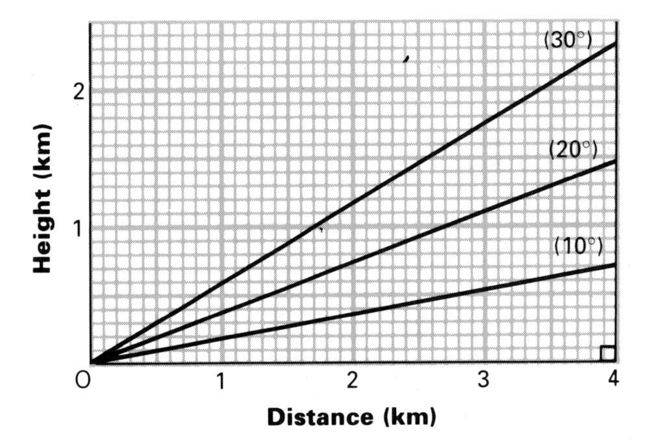

THE TANGENT OF AN ANGLE

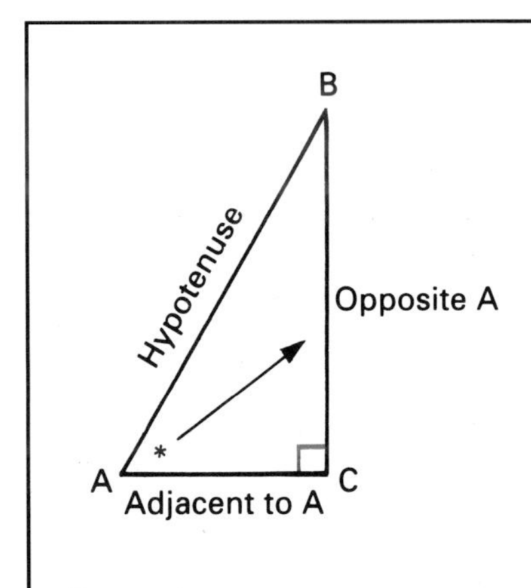

In right-angled △ABC,
AB is the **hypotenuse**.
From angle A,
BC is the **opposite side**
and AC is the **adjacent side**
('adjacent' means 'next to').

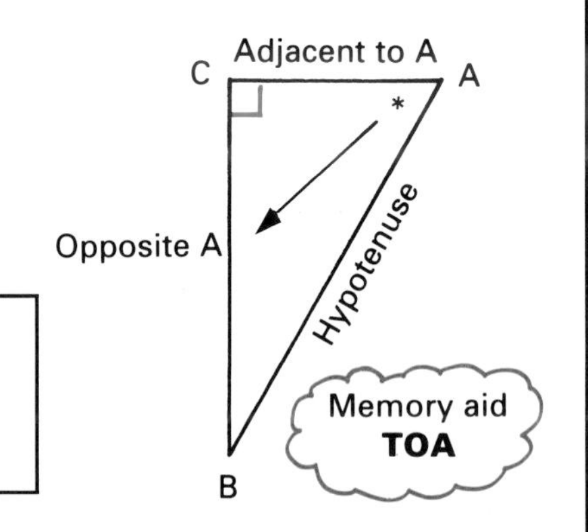

> The tangent of angle A is defined as
>
> $$\textbf{T}\text{an A} = \frac{\textbf{O}\text{pposite side}}{\textbf{A}\text{djacent side}}$$

Examples

a

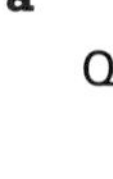

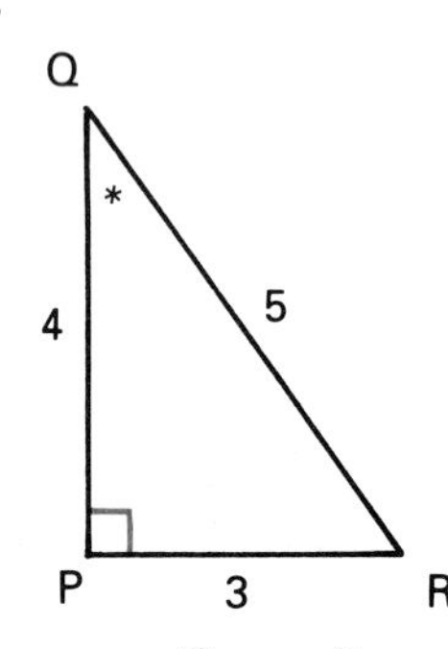

$$\text{Tan Q} = \frac{\textbf{O}\text{pposite}}{\textbf{A}\text{djacent}}$$
$$= \frac{3}{4}$$
$$= 0.75$$

b

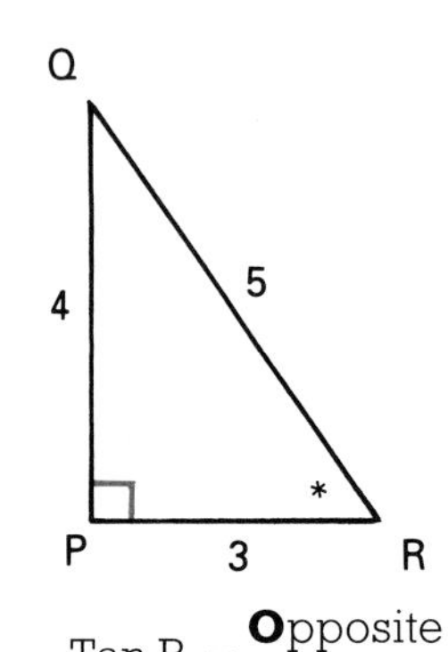

$$\text{Tan R} = \frac{\textbf{O}\text{pposite}}{\textbf{A}\text{djacent}}$$
$$= \frac{4}{3}$$
$= 1.33$, correct to 2 decimal places

c

G, 5, 12, F, 13, H

$$\text{Tan F} = \frac{\textbf{O}\text{pposite}}{\textbf{A}\text{djacent}}$$
$$= \frac{12}{5}$$
$$= 2.4$$

EXERCISE 2

1 In this right-angled triangle,

$$\tan A = \frac{\text{Opposite}}{\text{Adjacent}} = \frac{1}{2}.$$

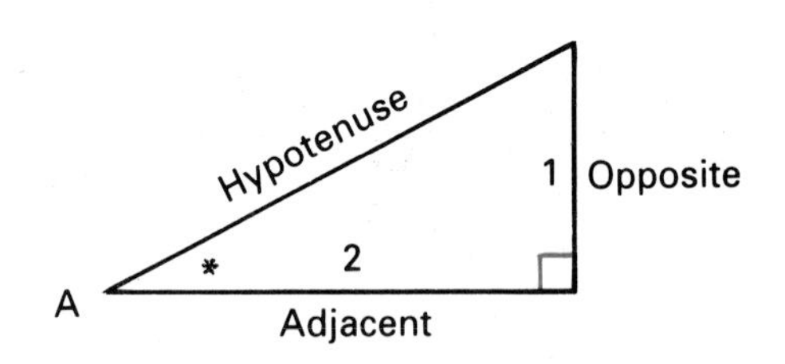

(i) Copy each triangle opposite, and mark the hypotenuse, then the opposite and adjacent sides for the starred angle.

(ii) Write down the ratio for tan A in each triangle.

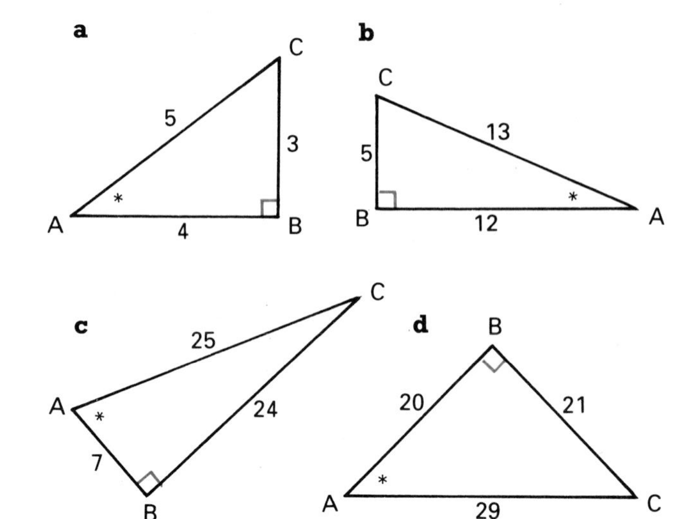

2 Write down the ratio for tan A in each right-angled triangle.

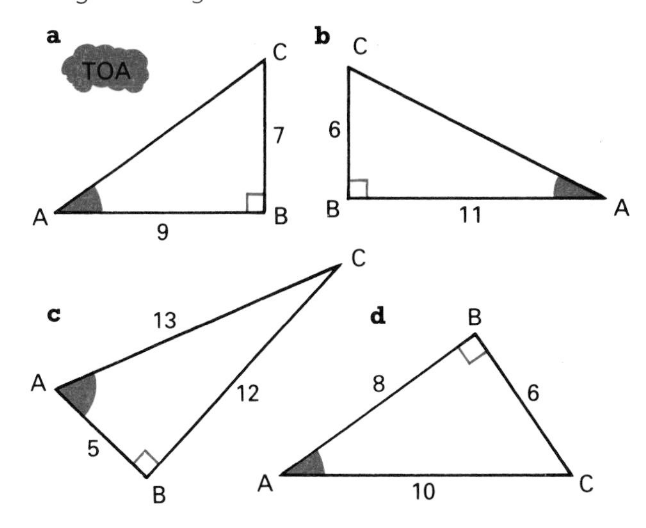

3 Calculate the value of tan A, correct to 2 decimal places, for each of these right-angled triangles.

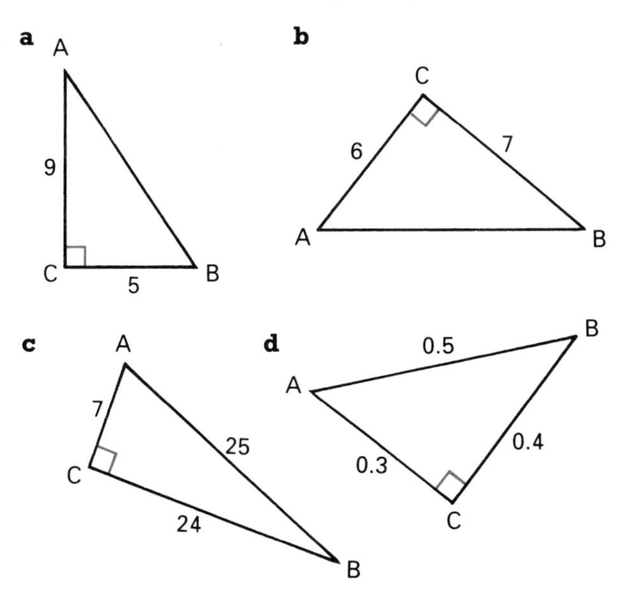

Your calculator has a list of angles and their tangents built into it. Check how yours operates.

Key tan 45, or 45 tan.

Which gives 1, the correct value for tan 45°? Use it in this way in future.

4 Use your calculator to list the tangents of these angles, correct to 2 decimal places.

Angle A	10°	20°	30°	40°	50°	60°	70°	80°
Tan A	0.18							

5 Find the tangents of these angles, correct to 2 decimal places:
a 34° **b** 47° **c** 61.5° **d** 77.7° **e** 2.7°

6 Find the values of these tangents, correct to 2 decimal places:
a tan 27° **b** tan 80.4° **c** tan 1.8°

Given tan A = 0.15, you can find acute angle A like this:

Key 2ndF tan .15, or .15 2ndF tan,

to get 8.53°, to 2 decimal places.

7 Use your calculator to find the acute angles, correct to 0.1°, for which:

a tan A = 0.7 **b** tan B = 0.46
c tan C = 1 **d** tan D = 1.8
e tan E = 2.95 **f** tan F = 573

8 For each triangle in question **3**:
(i) write down the ratio for tan B
(ii) calculate angle B, correct to 1 decimal place.

For example:

a (i) $\frac{9}{5}$ (ii) Key 2ndF tan (9 ÷ 5 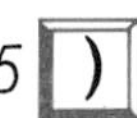

to get 60.9°.

A SURPRISE—THE CALCULATOR CAN'T COPE!

Try to find the value of tan 90°.
Find tangents of 89°, 89.9°, 89.99°, 89.999°, . . . how far can you go?
Explain what is happening in terms of the definition of the tangent in a right-angled triangle.

THE TANGENT RATIO IN ACTION

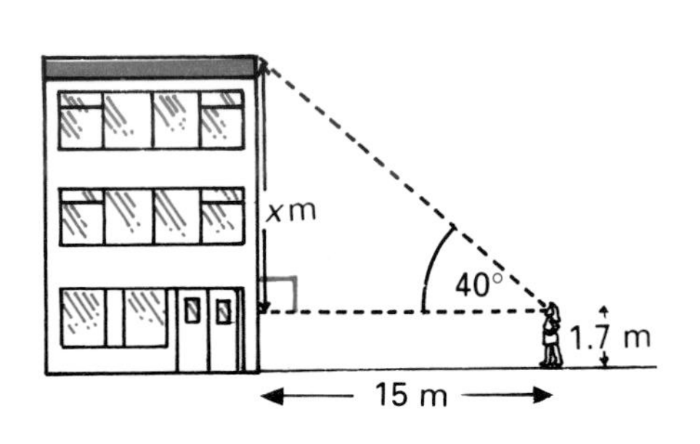

In question **8** of Looking Back you made a scale drawing to find the height of a school.
We can now use trigonometry instead.

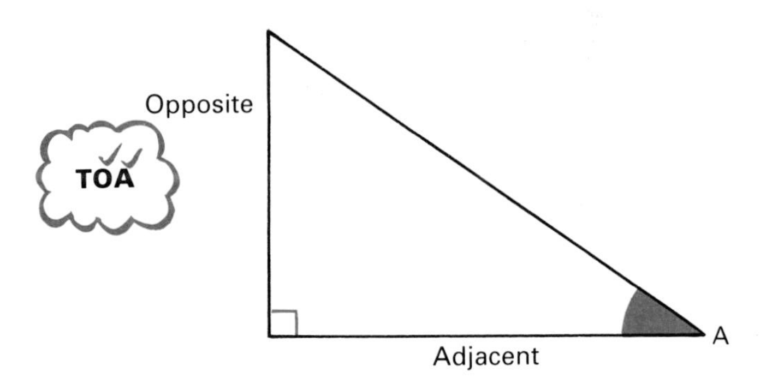

$$\text{Tan A} = \frac{\text{Opposite}}{\text{Adjacent}}$$

So $\tan 40° = \dfrac{x}{15}$

and $x = 15 \times \tan 40°$
$= 12.6$, correct to 1 decimal place.
The height of the school is $12.6 + 1.7\text{ m} = 14.3\text{ m}$.

Throughout this chapter, give lengths correct to 1 decimal place and angles correct to 0.1°, unless there are other instructions.

EXERCISE 3A

1 The angle of elevation of the top of the flagpole is 58°. Copy and complete the following to find the height of the pole to the nearest metre.

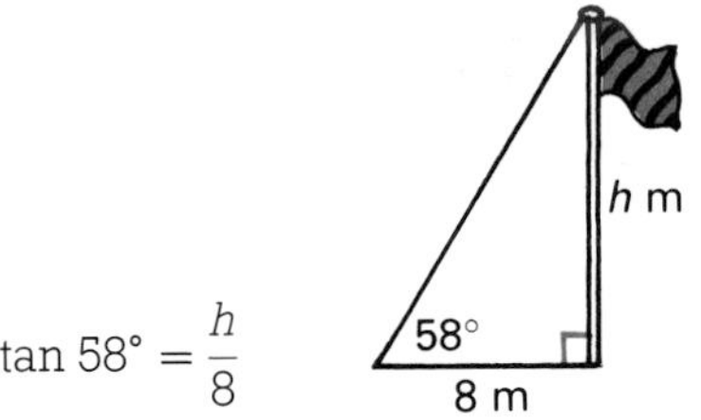

$\tan 58° = \dfrac{h}{8}$

So $h = \ldots \times \tan 58° = \ldots$, to 1 decimal place.
The pole is . . . m high.

2 Use the tangent of an angle to calculate x in each triangle below.

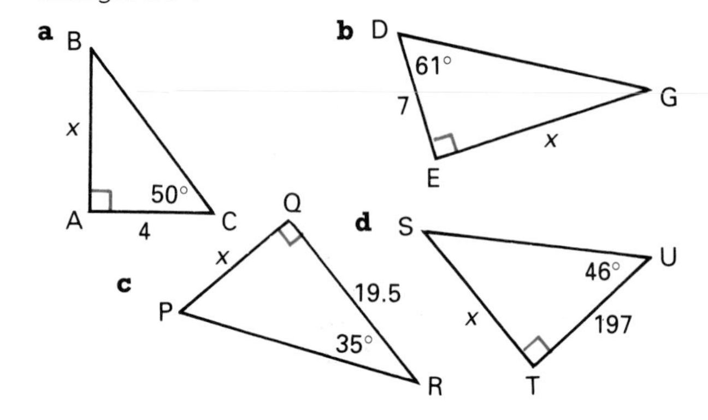

3 Calculate x in each diagram.

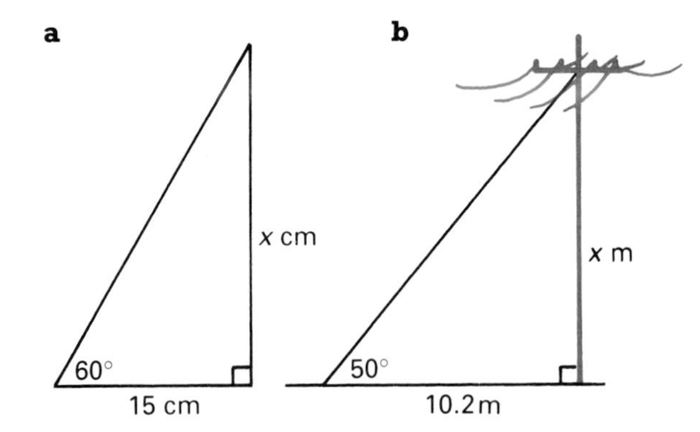

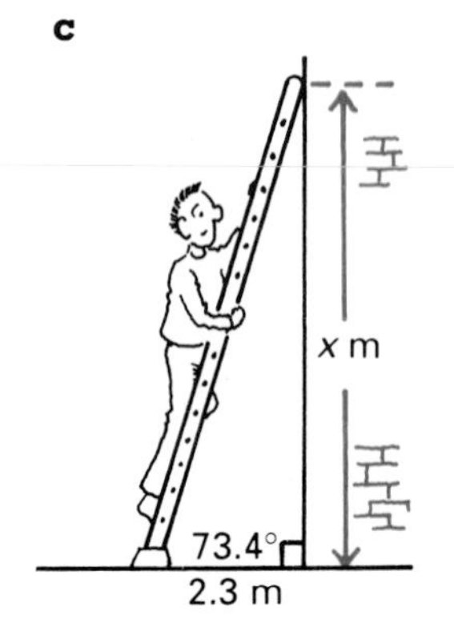

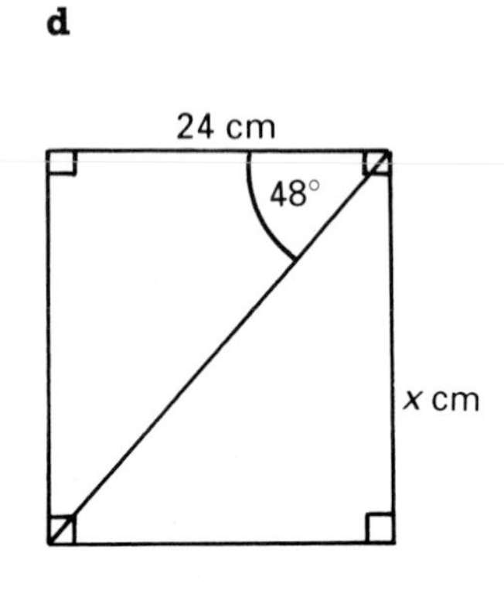

4 At 250 m from the base of the building the angle of elevation of its top is 32°. Calculate the height of the building, to the nearest metre.

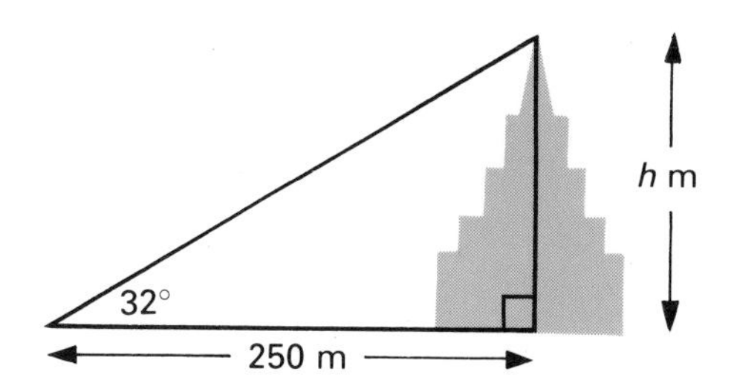

5 Michael measures the angle of elevation of this tree from a point 95 m from its foot. Calculate the height of the tree, to the nearest metre.

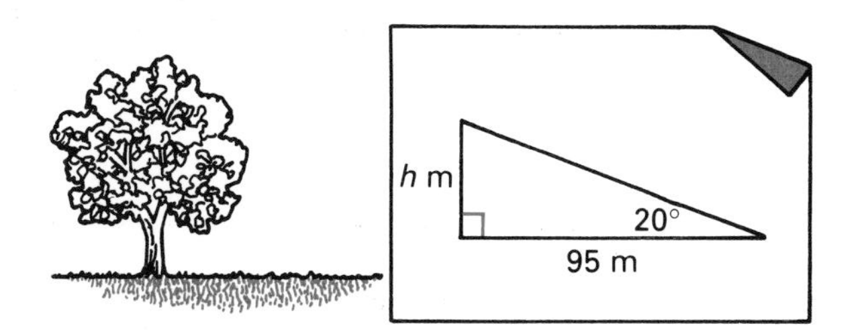

6 Here are some famous structures. The angle of elevation of their tops, measured from a distance of 250 m is circled. Use the tangent to calculate the height of each structure, to the nearest metre.

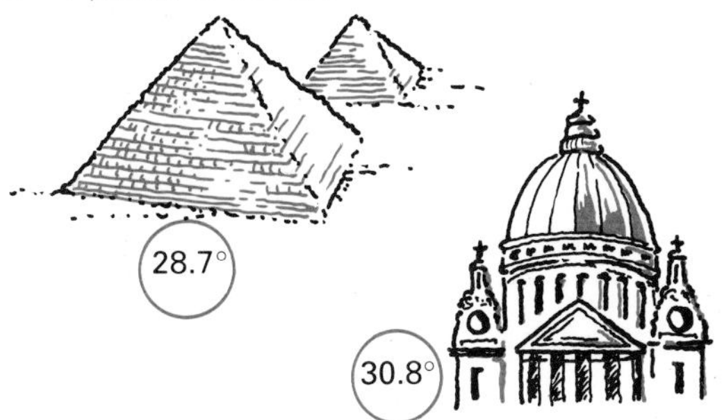

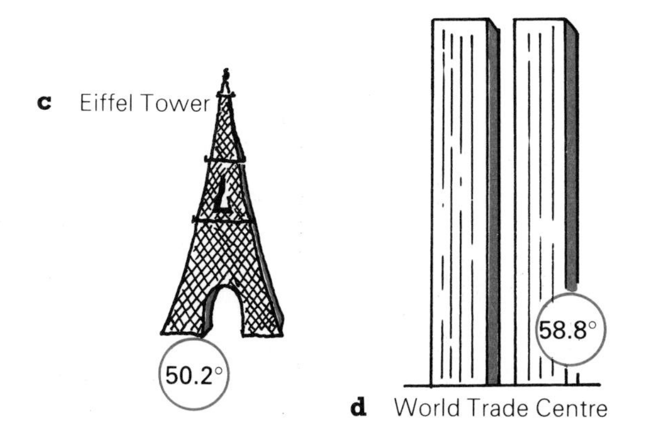

7 George has challenged Jack to find the width of the river. Jack marks point B directly across from a tree at C on the opposite bank. He measures 100 m to A and finds ∠BAC = 40°. Help him to calculate the river's width, *w* metres, to the nearest metre.

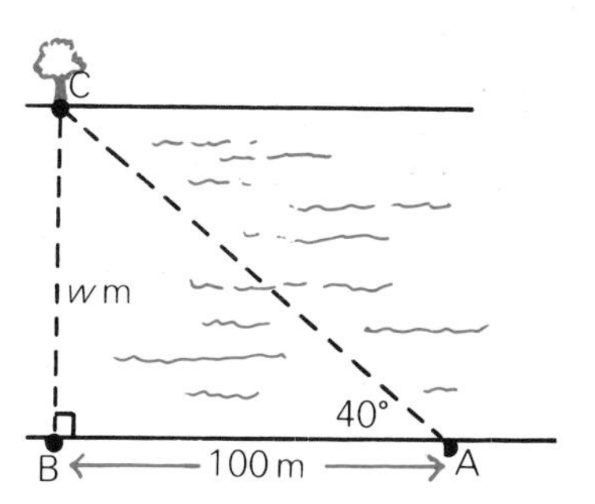

8 A youth group is trying to calculate the height of the church spire to find the winner of a 'Guess the height' competition at the fete. Calculate the height to the nearest 10 cm.

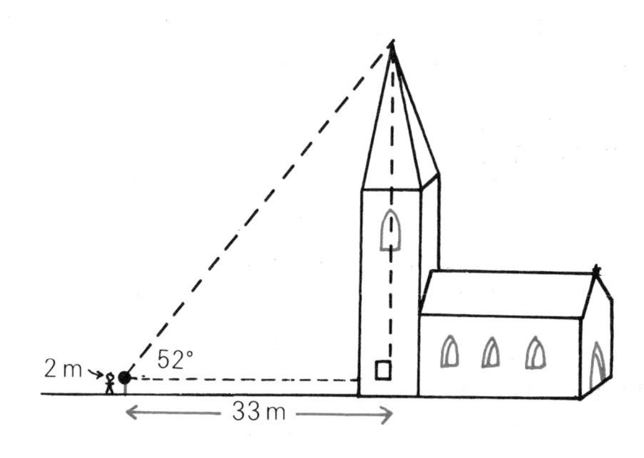

9 A buoy is anchored to a rock 100 m from the foot of a cliff. From the top of the cliff the angle of depression of the buoy is 18°. Calculate the height of the cliff, to the nearest metre.

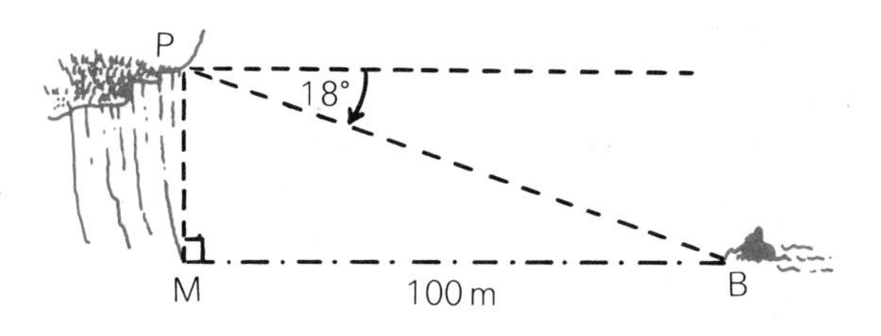

10 In △ABC, base AB = 25 cm, ∠ABC = 90° and ∠BAC = 45°. Calculate the length of BC, using:

a tan A

b the special kind of triangle that ABC is.

CALCULATING THE ADJACENT SIDE

A plank of wood 18 cm across has a corner sawn off at an angle of 42° to the side. Calculate d, correct to the nearest whole number.

Method 1

$\tan 42° = \frac{18}{d}$

$d \tan 42° = 18$ (multiplying each side by d)

$d = \frac{18}{\tan 42°}$ (dividing each side by $\tan 42°$)

$= 20$

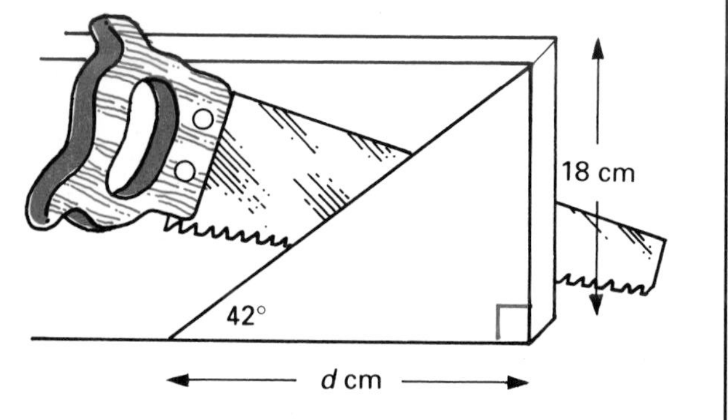

Method 2

Use $\tan 48° = \frac{d}{18}$, then $d = 18 \times \tan 48° = 20$

EXERCISE 3B

1 Calculate x in each triangle.

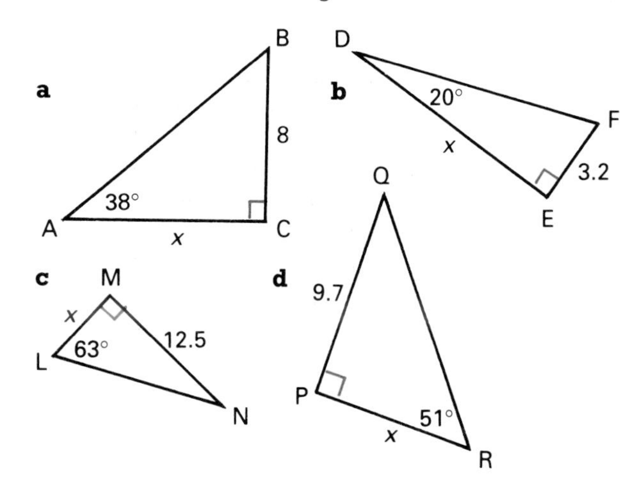

2 Calculate x in each diagram.

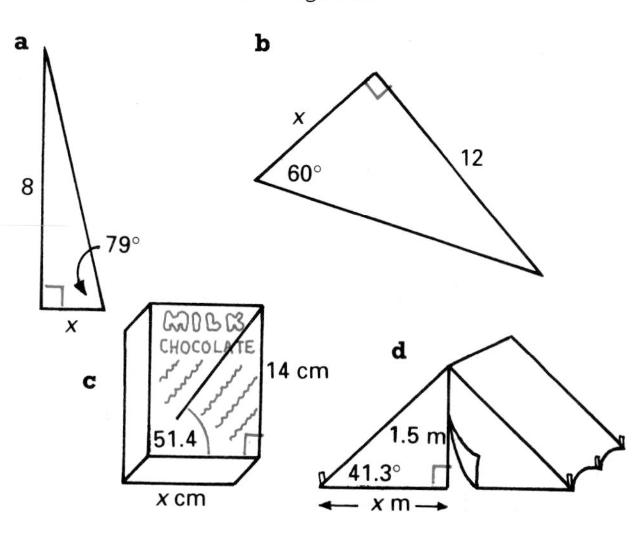

3 A rescue helicopter at a height of 150 m spots a dinghy at an angle of depression of 26°. Calculate the horizontal distance between the dinghy and the helicopter, to the nearest metre.

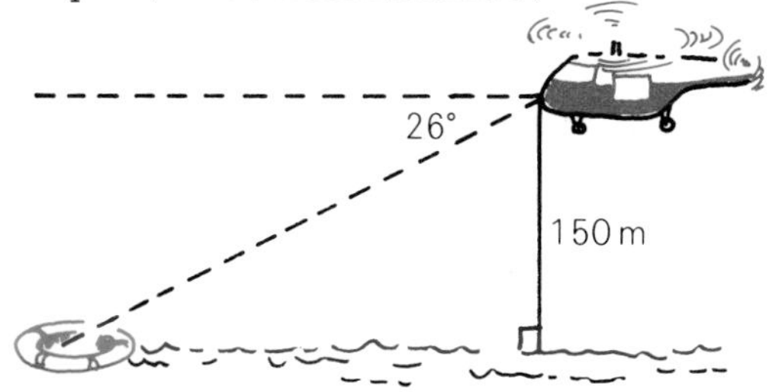

4 Mary's garden path is in the shape of a parallelogram. One of its diagonals is at right-angles to a side.

a Calculate the distance d metres.

b What area of crazy paving does Mary need to cover the whole path?

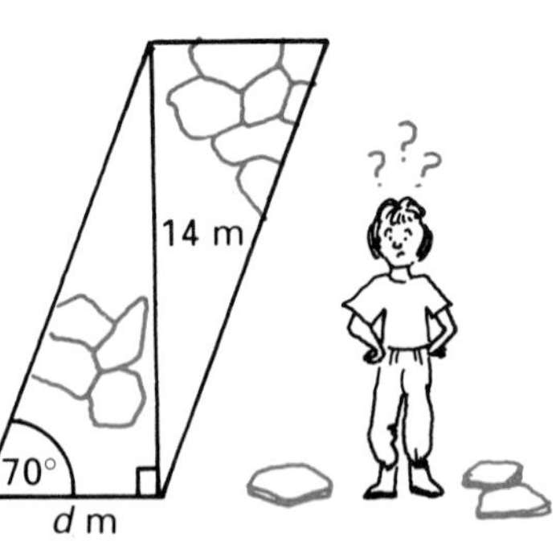

5 Eastport is due east of Westport. Midway bears 022° from Westport. Calculate the distance from Westport to Midway, to the nearest km.

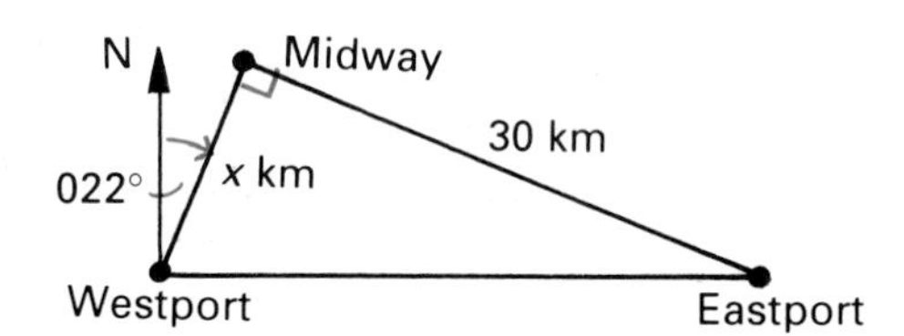

CALCULATING AN ANGLE

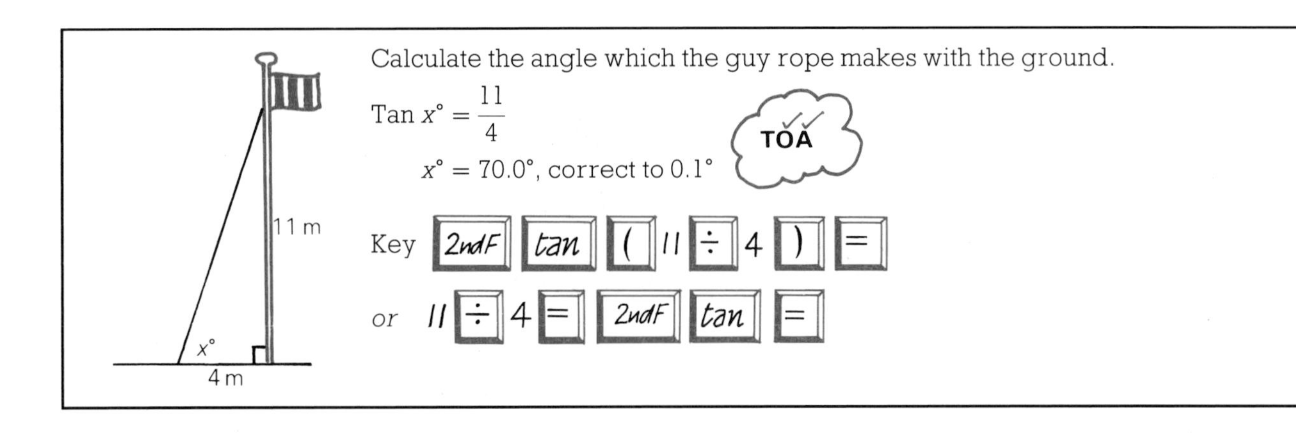

Calculate the angle which the guy rope makes with the ground.

$\text{Tan } x° = \frac{11}{4}$

$x° = 70.0°$, correct to 0.1°

EXERCISE 4

1 Calculate $x°$ in these triangles.

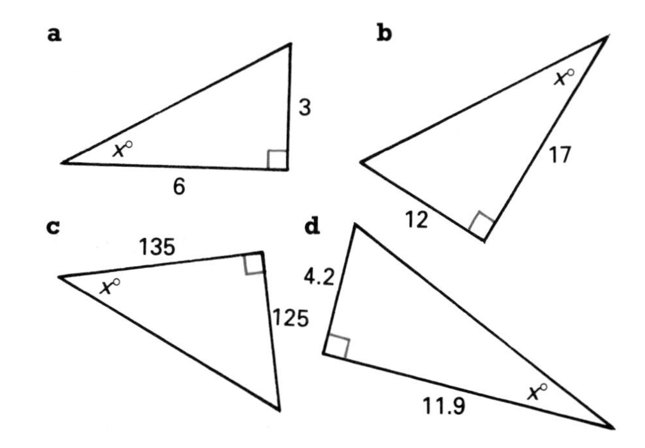

2 Calculate $x°$, to the nearest degree.

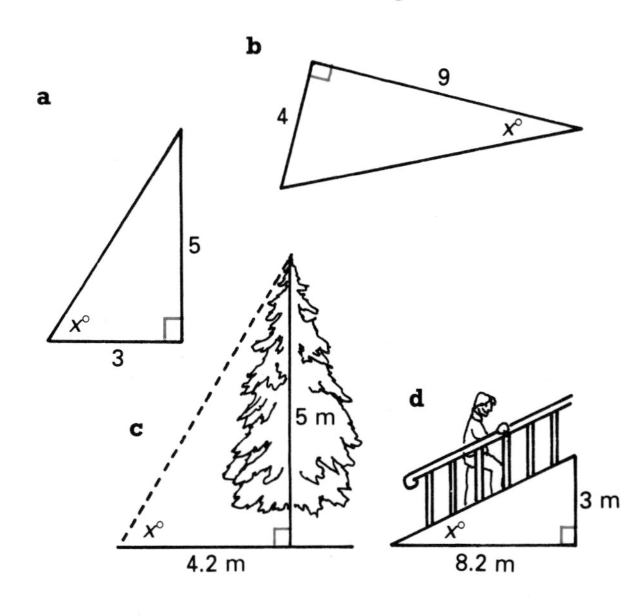

3 Ruth is 175 cm tall. In sunshine, the shadow she casts on the ground is 350 cm long. What is the angle of elevation of the sun?

4 Dave is making a garden gate.
Calculate: **a** $x°$ **b** $y°$.

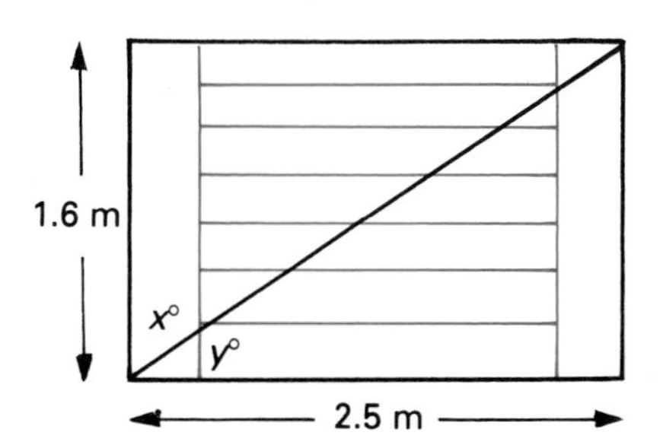

5 Harry is making a kite.
Calculate its angles.

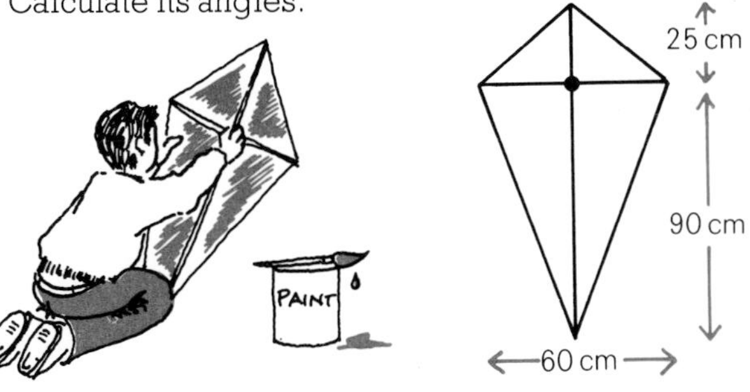

CHALLENGE

You now have all the mathematical tools required to make a cardboard scale model of your own school. Make all the necessary measurements, using an anglemeter and a plumb line, to find the angles of elevation. Work in groups, with each group building a different part of the school.

THE SINE OF AN ANGLE

The ladder will slip if the angle it makes with the ground is less than 40°. Can we calculate the angle? Using only the given lengths the tangent ratio won't work (TOA) ... we want to link the 3 m long side *opposite* the angle and the 4 m long *hypotenuse*.

We define a new ratio, $\dfrac{\text{Opposite side}}{\text{Hypotenuse}}$, called the **sine** of the angle.

$$\textbf{S}\text{in A} = \frac{\textbf{O}\text{pposite side}}{\textbf{H}\text{ypotenuse}}$$

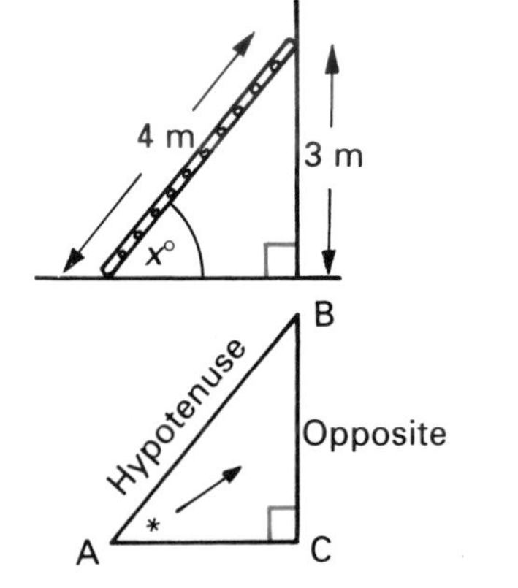

EXERCISE 5

1 Again tables of sines are built into your calculator. Use the [sin] key to list the sines of these angles, correct to 2 decimal places.

Angle A	10°	20°	30°	40°	50°	60°	70°	80°	90°
Sin A									

2 Use [2ndF] [sin] to find the acute angles in this table, correct to 0.1°.

Angle A							
Sin A	0.5	0.9	0.1	0.25	0.77	0.469	0.956

3 Will the ladder in the box above this exercise slip?

4 In this right-angled triangle, $\sin A = \dfrac{\text{Opp}}{\text{Hyp}} = \dfrac{4}{5}$.

(i) Copy each triangle below, and mark the hypotenuse and the side opposite A.

(ii) Then write down the ratio for sin A in each triangle.

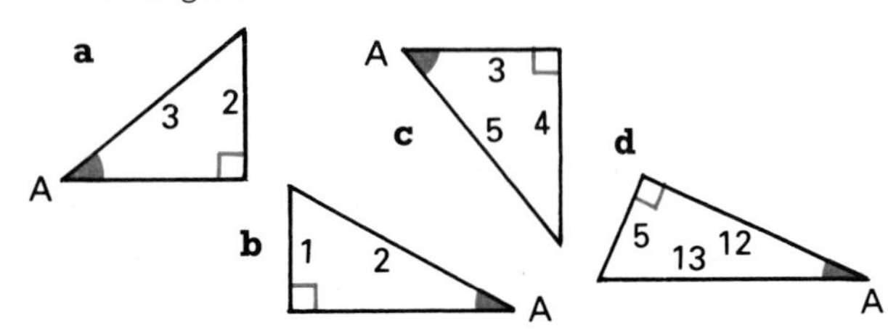

5 For each triangle below:

(i) write down the ratio for sin A

(ii) calculate the size of angle A.

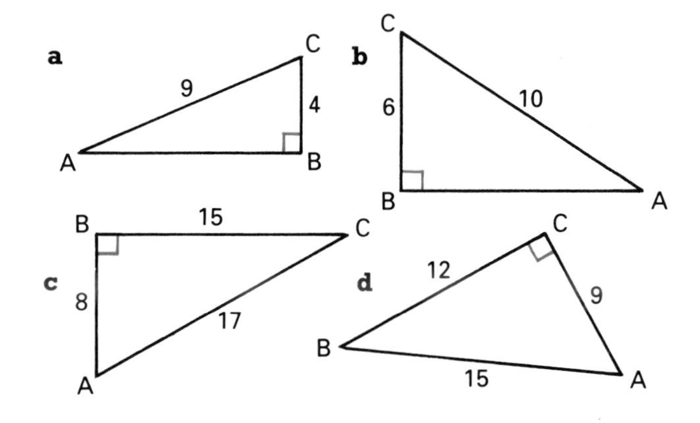

6 Calculate *x* in each right-angled triangle shown here.

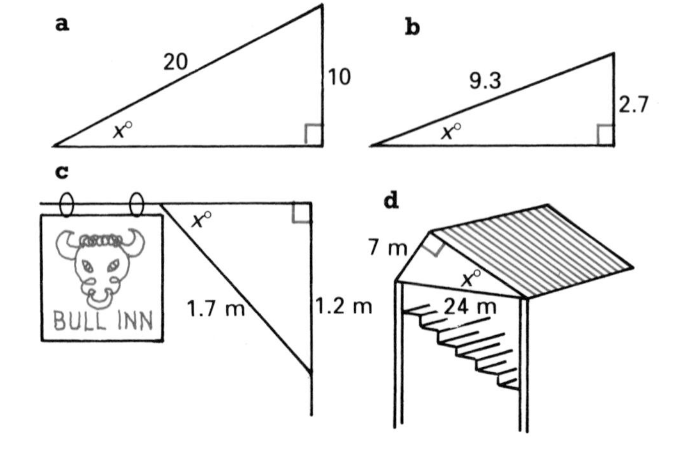

THE SINE RATIO IN ACTION

Charlie runs up the 4 metre-long plank to the top of the wall. Calculate his height above the ground when he reaches the top.

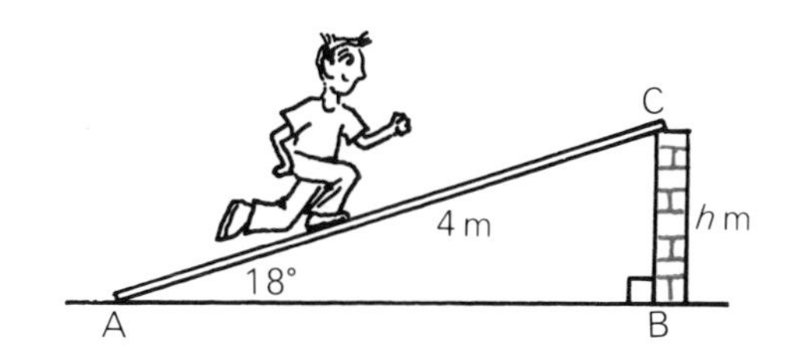

In $\triangle ABC$, $\angle B = 90°$.

$$\text{Sin A} = \frac{\text{Opposite}}{\text{Hypotenuse}}$$

$$\sin 18° = \frac{h}{4}$$

$$h = 4 \times \sin 18°$$

$$= 1.24\text{, correct to 2 decimal places}$$

His height is 1.24 m.

EXERCISE 6A

1 Calculate x in each triangle.

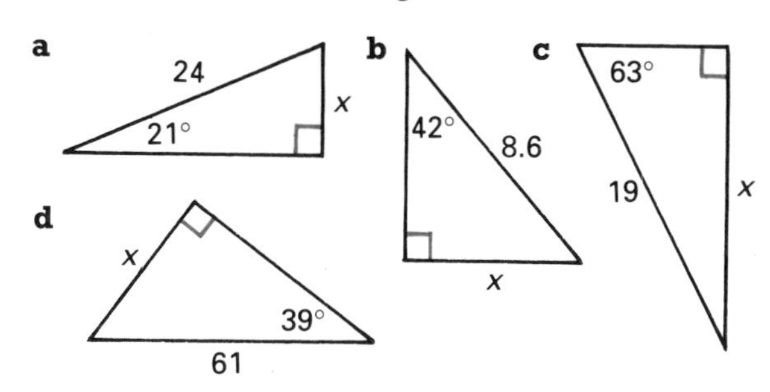

2 Calculate d in each diagram.

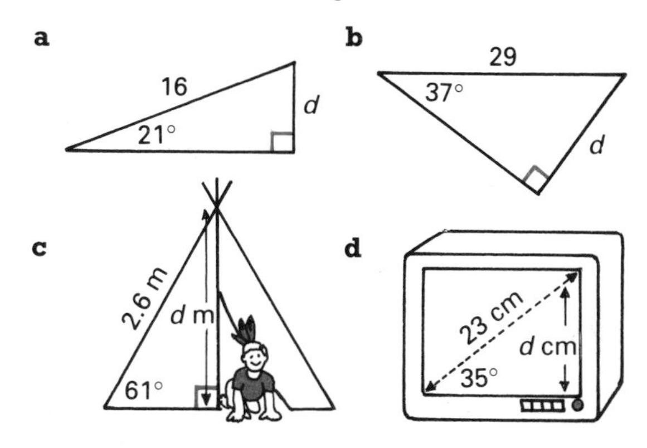

3 The cable car climbs at 40° to the horizontal.
AB = 60 m and BC = 40 m.
Calculate the vertical height the car has gained from:
a A to B **b** B to C.

4 

This sign shows that the road rises 1 m for every 9 m of road. Calculate the angle between the road and the horizontal.

5 It's a fine breezy day, and Yoko is flying her kite.

a She lets out 30 m of string, pulled taut, at 40° to the ground. Calculate the height of the kite.

b The wind drops, and the kite falls to a point 15 m above the ground, with the string still taut. What angle does the string now make with the ground?

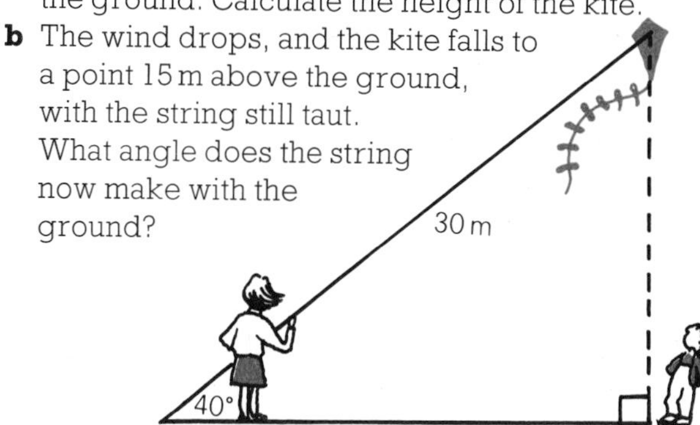

6 A ship is 8 km south of an oil rig, and is following a course 038°.

a What is the size of $\angle$ABC when the ship is closest to the oil-rig?

b How close to the oil-rig does it pass?

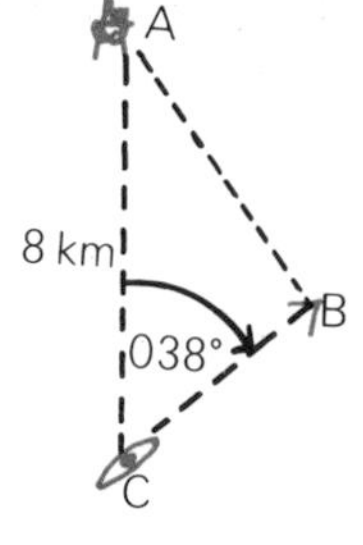

7 Paul is playing with a toy boomerang. Calculate:

a $\angle$ACB **b** $\angle$MCB

c $\angle$CBM **d** CM

e AM

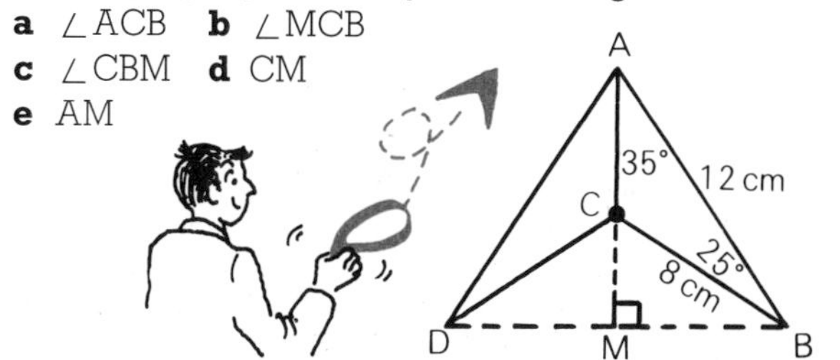

CALCULATING THE HYPOTENUSE

Calculate the length of the sloping edge of this wall-support, to the nearest cm.

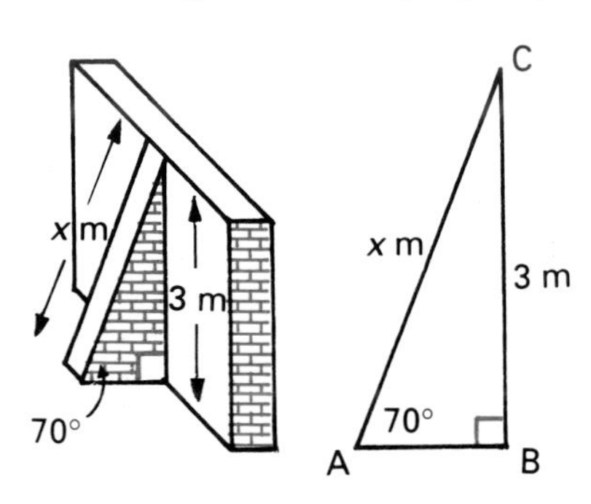

$$\text{Sin A} = \frac{\text{Opposite}}{\text{Hypotenuse}}$$

$$\sin 70° = \frac{3}{x}$$

$$x \sin 70° = 3$$

$$x = \frac{3}{\sin 70°} = 3.19, \text{ correct to 2 decimal places}$$

The edge is 3.19 m long.

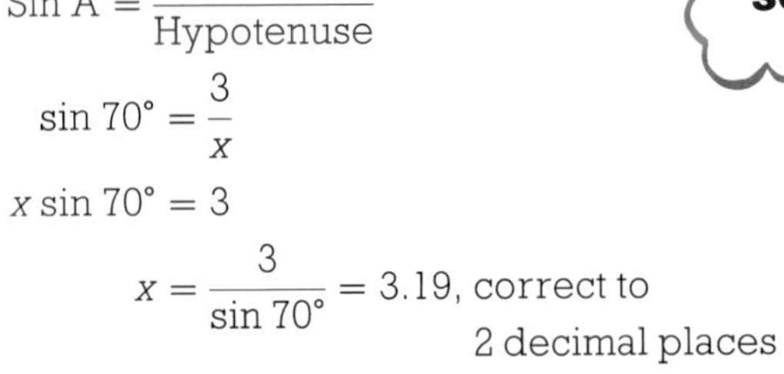

EXERCISE 6B

1 Calculate h, correct to 1 decimal place, in each triangle.

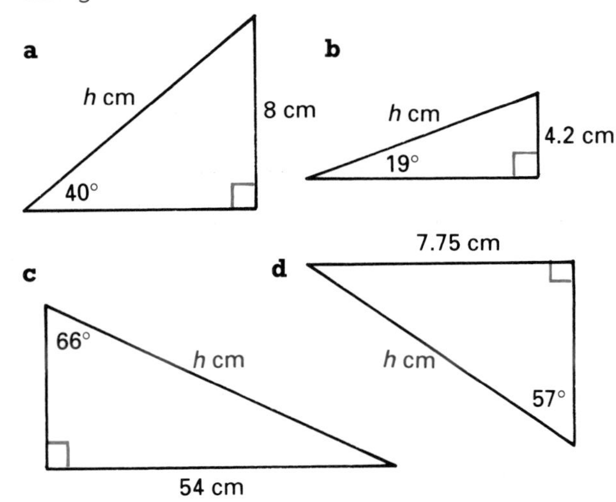

2 Calculate the length of the hypotenuse of each right-angled triangle.

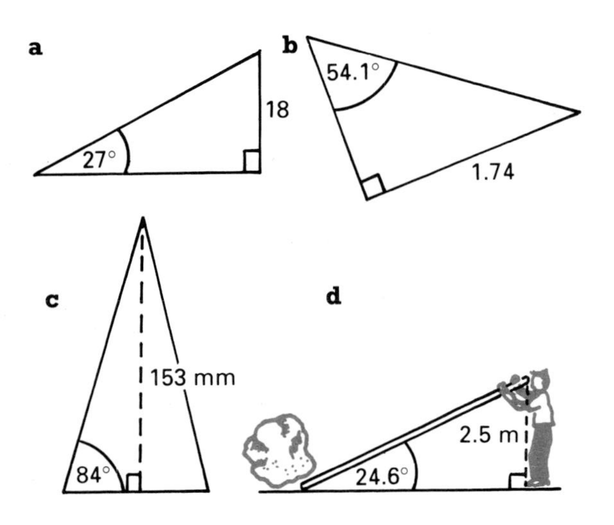

3 Number 9 Tay View is a small bungalow.

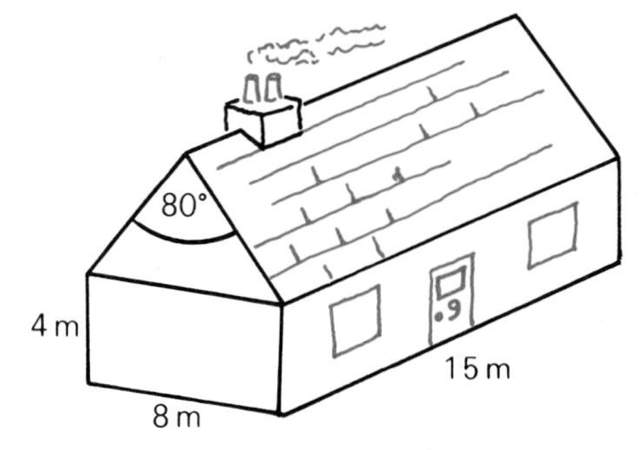

Calculate, to the nearest centimetre:

a the length of the sloping edge of the roof

b the height of the roof ridge above the ground.

4 The *Mary Anne* sets course from her harbour, O, on a bearing of 040°, and sails for 80 km. She then changes course to 140° and sails until she is due east of the harbour.

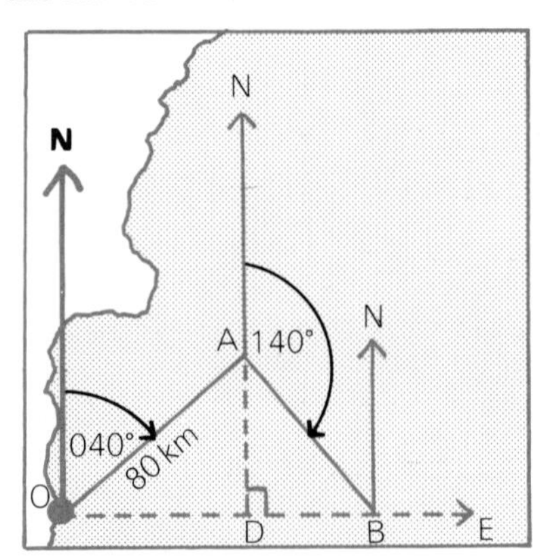

a What is the furthest north the *Mary Anne* goes from the harbour?

b What is the length of her journey from A to B?

c How far is she from the harbour at B?

THE COSINE OF AN ANGLE

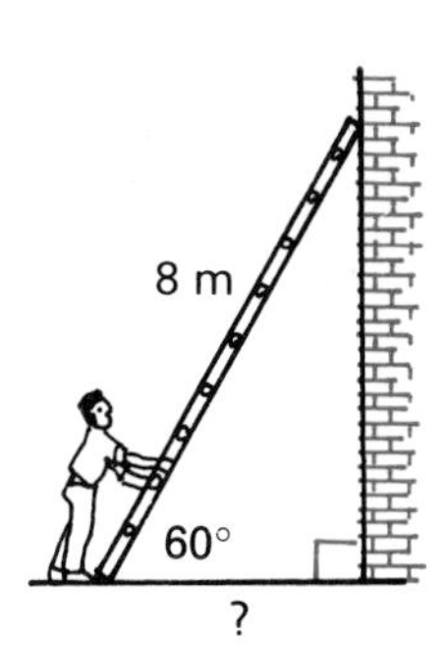

For safety, Sean aims to set the ladder at 60° to the ground. How far out from the wall should the foot be?

This time we have to link the hypotenuse and the other side, the one adjacent to A (next to A).

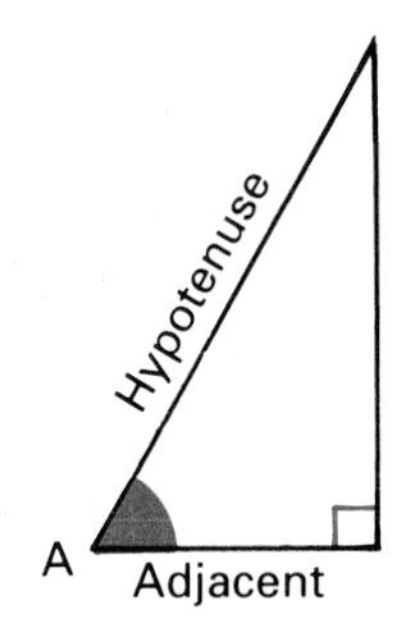

So we define another new ratio, $\frac{\text{Adjacent side}}{\text{Hypotenuse}}$, called the **cosine** of the angle.

$$\textbf{C}\text{os A} = \frac{\textbf{A}\text{djacent side}}{\textbf{H}\text{ypotenuse}}$$

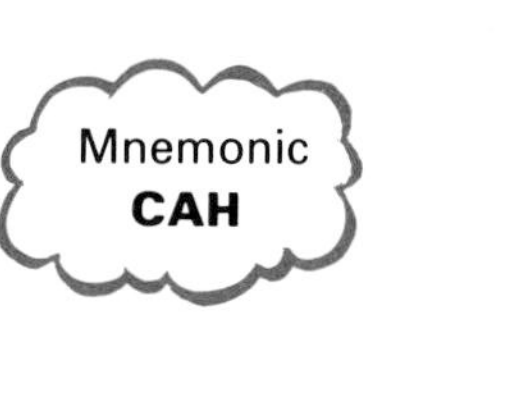

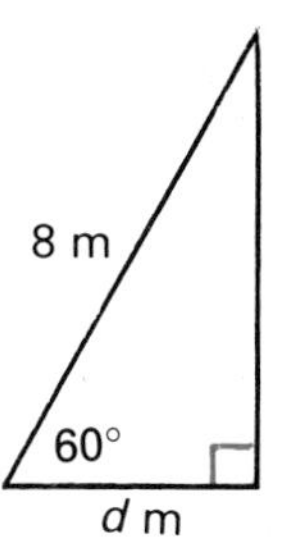

For the ladder, $\cos 60° = \frac{d}{8}$

$d = 8 \cos 60° = 4$

The foot of the ladder should be 4 m from the wall.

EXERCISE 7A

1 Use the [cos] key on your calculator to list these cosines, correct to 2 decimal places.

Angle A	10°	20°	30°	40°	50°	60°	70°	80°	90°
Cos A									

2 Use [2ndF] [cos], or [$\cos^{-1}$], to find the acute angles in this table, correct to 0.1°.

Angle A							
Cos A	0.5	0.9	0.1	0.25	0.77	0.469	0.956

3 (i) Write down the ratio for cos A in each triangle.
(ii) Calculate the size of angle A.

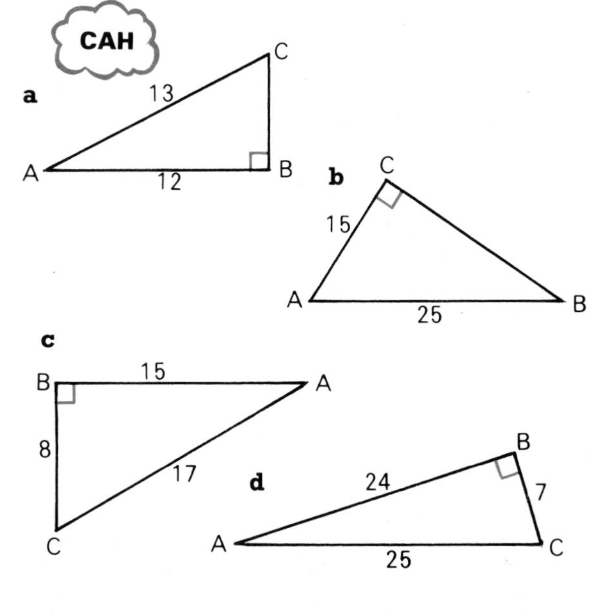

4 Use the cosine ratio to calculate x in each right-angled triangle.

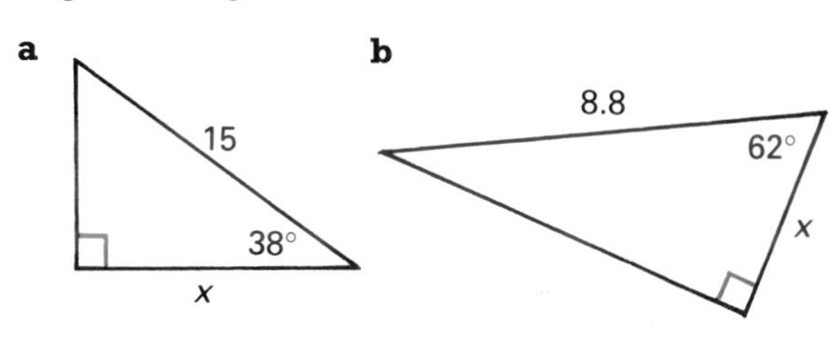

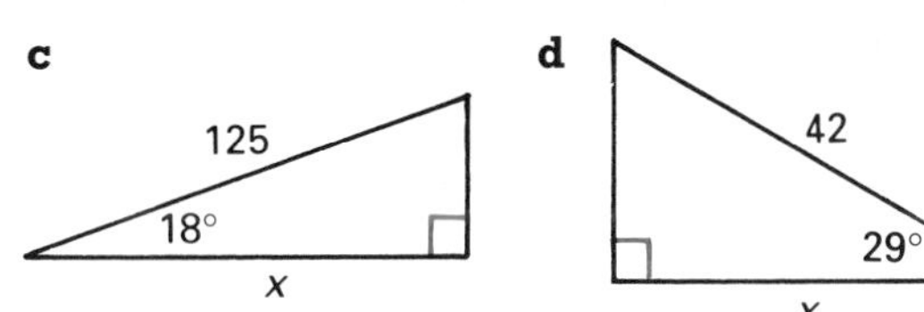

5 Calculate the submarine's angle of dive.

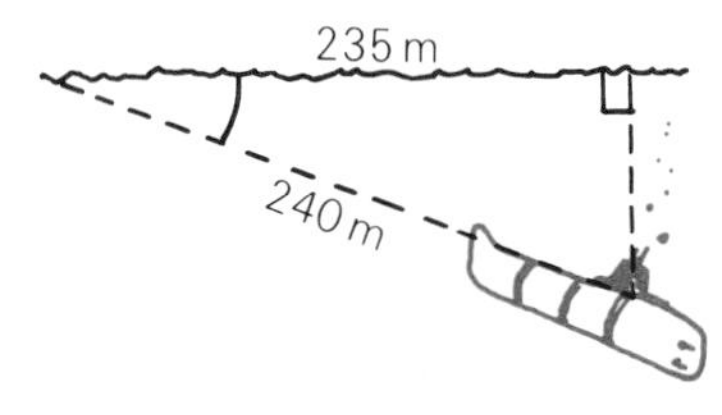

6 Calculate the lorry's tipping angle.

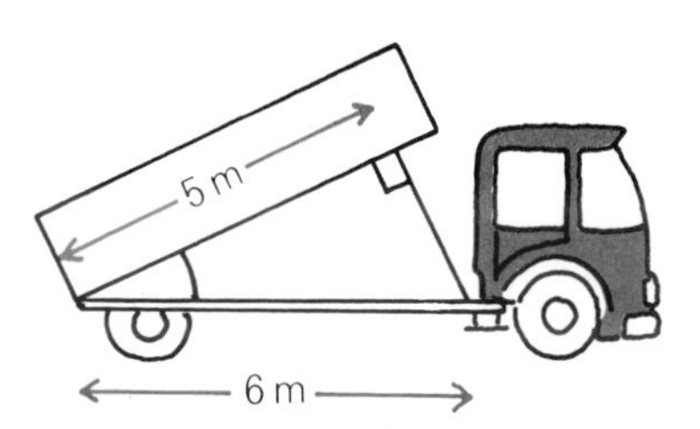

7 Calculate the angle of the ramp.

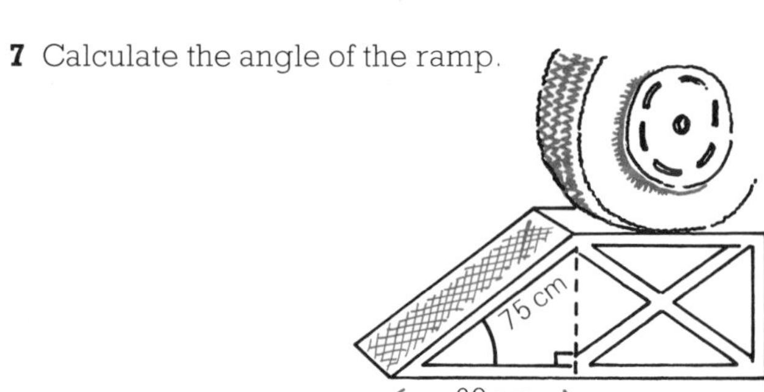

8 A trawler sets a course of 060° from the harbour. It keeps to this course for 12 km. How far:
a east **b** north, is it from the harbour then?

EXERCISE 7B

1 Calculate QR, the length of the horizontal part of the crane.

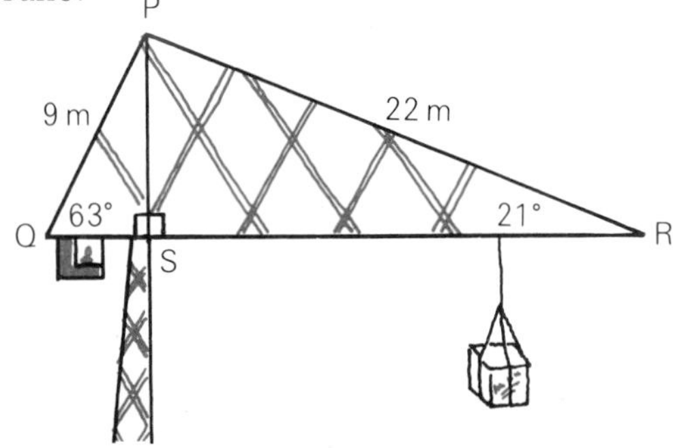

2 Calculate d.

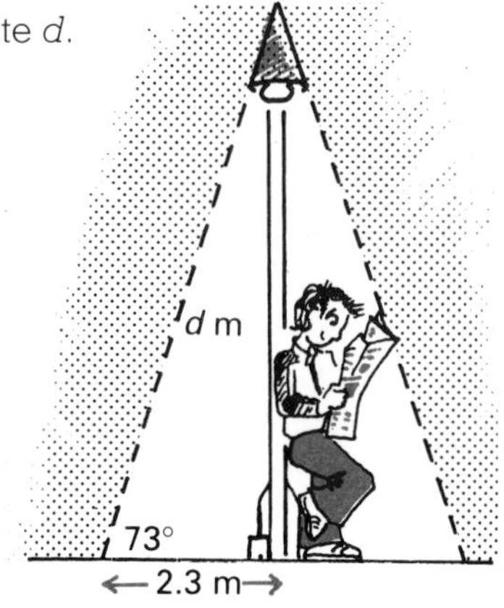

3 Calculate:
a the focal length AF of the lens
b the distance BF for a lens of focal length 90 mm and $\angle$AFB = 24.3°

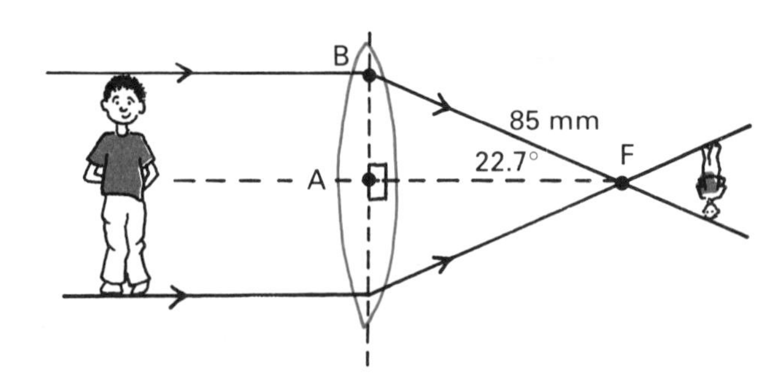

4 Calculate x and y.

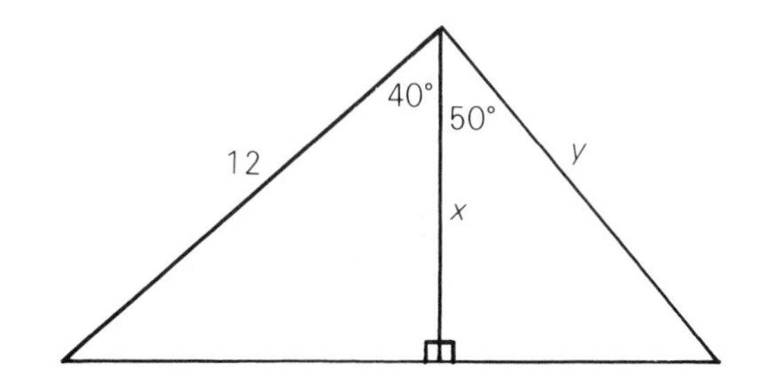

WHICH RATIO?

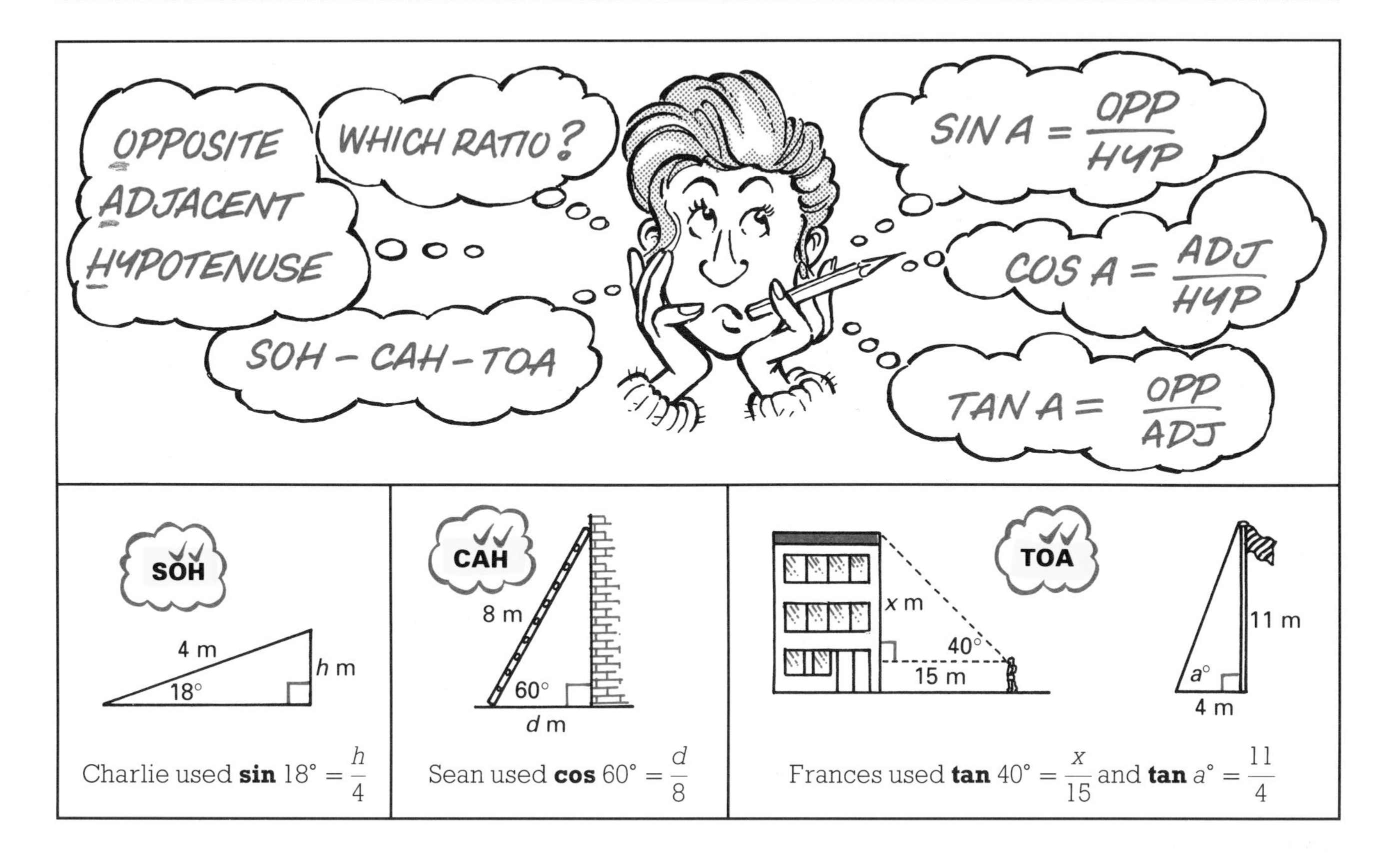

Charlie used **sin** $18° = \frac{h}{4}$

Sean used **cos** $60° = \frac{d}{8}$

Frances used **tan** $40° = \frac{x}{15}$ and **tan** $a° = \frac{11}{4}$

EXERCISE 8A

1 Choose your ratio and calculate d.

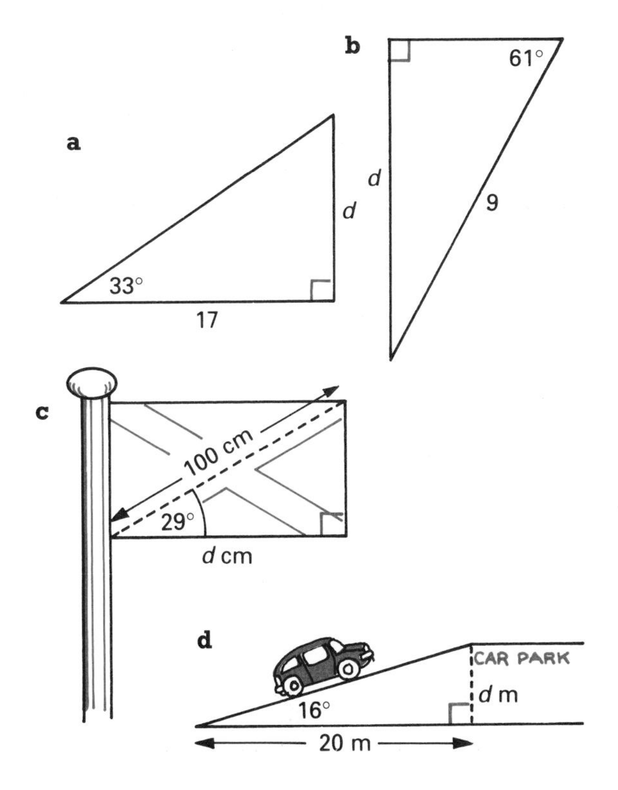

2 Calculate $x°$.

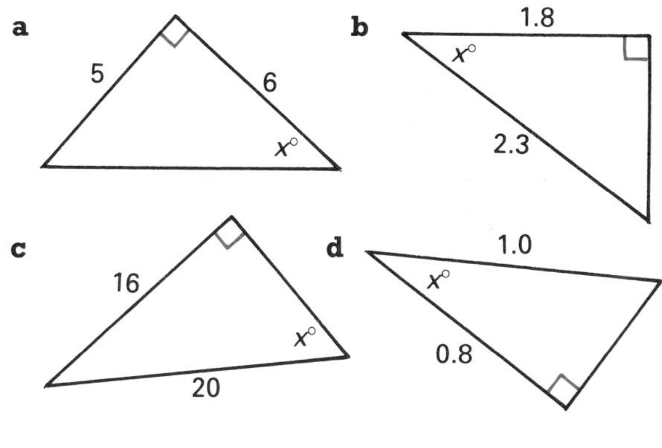

3 The angle of elevation of the top of this building is 56.7° from a point 250 m from its base. Show this in a right-angled triangle, and calculate the height of the building to the nearest metre.

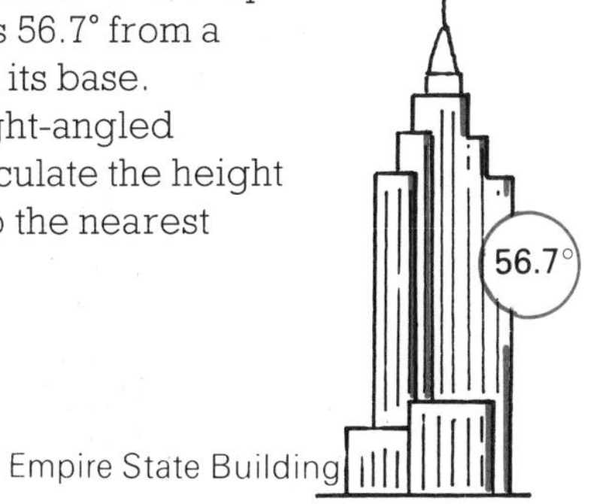

4 The supporting wire to the pole is 12 m long, and is at 72° to the ground. How far up the pole is it fixed, in m and cm, correct to the nearest cm?

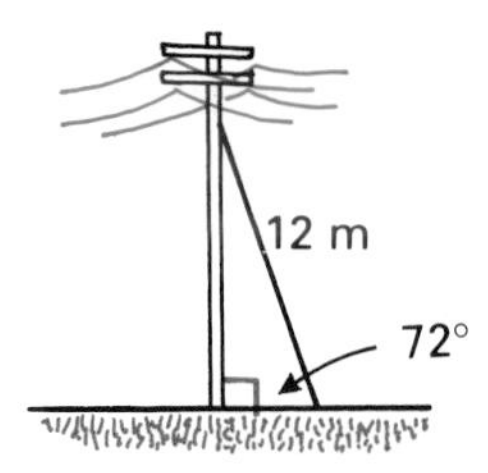

5 Which of the edges marked t cm and k cm is longer? By how many mm?

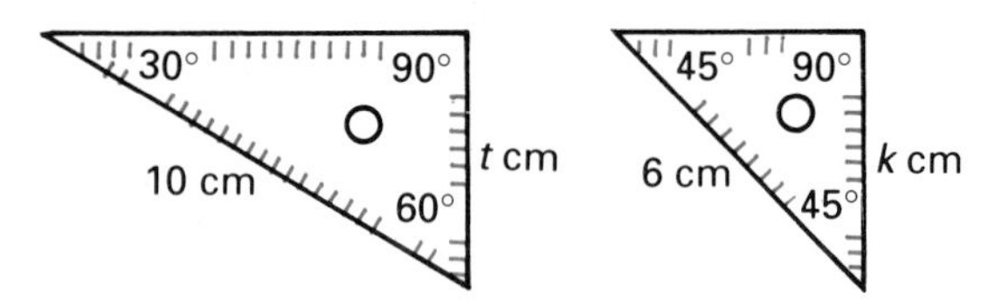

6 A mast 125 m tall casts a shadow 75 m long. Calculate the angle of elevation of the sun.

7 Is it easier to score from the penalty spot at football (A) or hockey (B)? Calculate the angles at A and B to compare the shooting angles. What do you find?
Are there other things to think about?

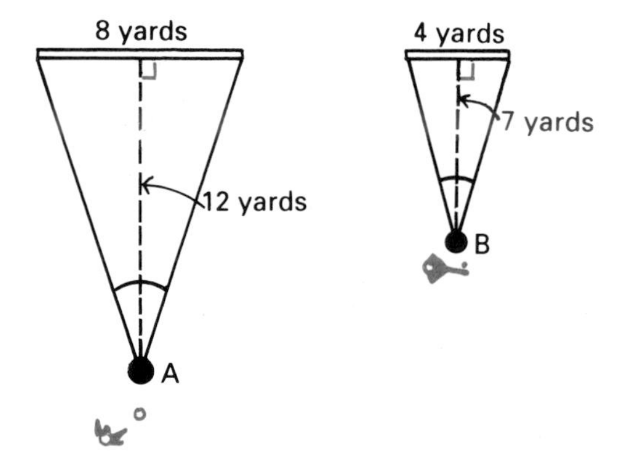

8 From a hot-air balloon vertically above A, the angle of depression of a boat B is 22°. Calculate the height of the balloon, to the nearest metre.

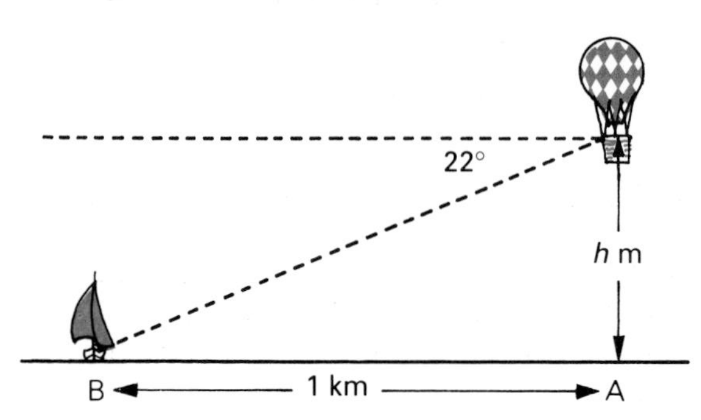

9 A trawler sets a course of 290° from the harbour. It keeps to this course for 15 km. How far, to the nearest 0.1 km:
a west **b** north, is it from the harbour then?

10 Frank makes a diagram of a rectangular garden gate. Calculate:
a x **b** y, using trigonometry
c y, using Pythagoras' Theorem
d the area of the gate.

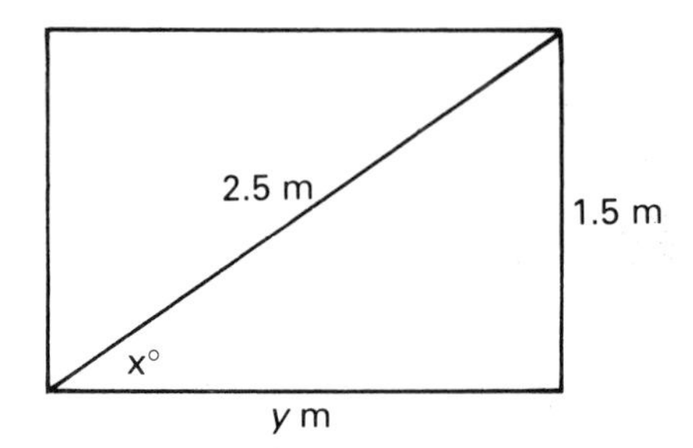

11 A ladder 5 m long is placed against the wall of a house, with its foot 2 m from the wall. Calculate:
a the angle between the ladder and the ground
b the height up the wall the ladder reaches, using:
(i) trigonometry
(ii) Pythagoras' Theorem.

12

The wire connectors on this electric vehicle are rhombus-shaped. Calculate:
a AE **b** BE **c** ∠BAE **d** AB

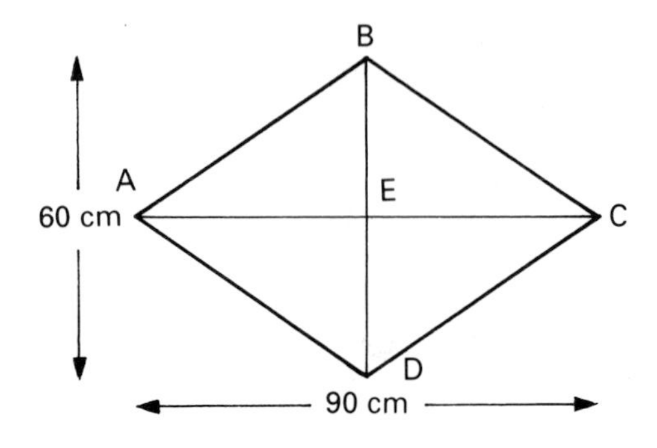

EXERCISE 8B

1 From the window of his house, 150 m from a factory chimney, Nick estimates that the angle of elevation of the top is 36°, and the angle of depression of the foot is 6°. Calculate the height of the chimney.

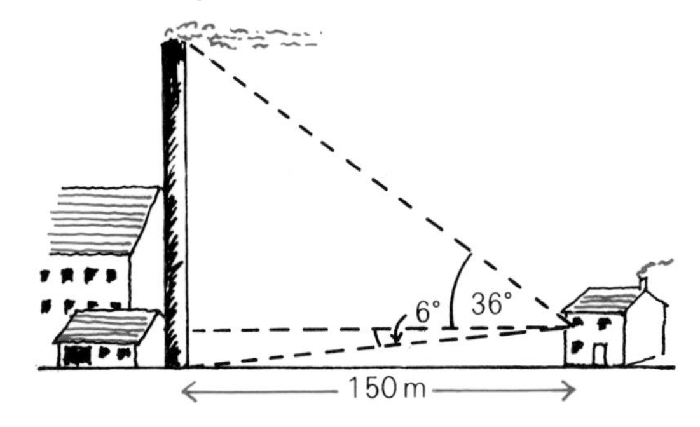

2 a Show that the angle between the roof of the car-port and the horizontal is 27°, to the nearest degree.

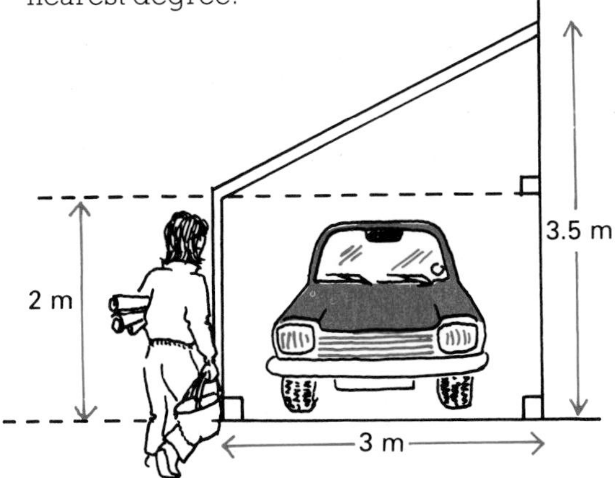

b Calculate the length of the sloping roof by:
(i) trigonometry (ii) Pythagoras' Theorem.

3 a Copy the isosceles triangle ABC, and draw its axis of symmetry, meeting BC at D.

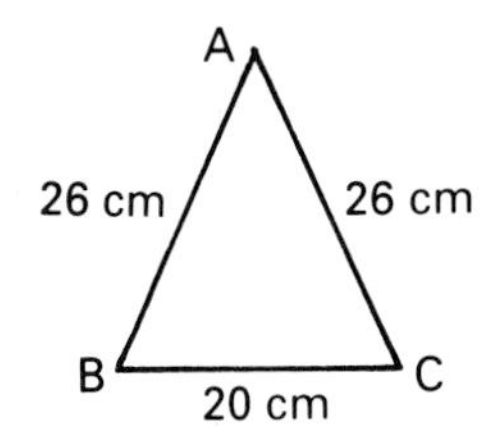

b Calculate:
(i) BD (ii) AD (iii) the area of △ABC
(iv) the angles of △ABC.

4 A rectangle is 15 cm long and 12 cm broad. Calculate the angles between:
a the diagonals and the sides
b the diagonals.

5 The roof space in a house has a wall along the centre. Calculate:
a AD **b** BD **c** the area of △ABC.

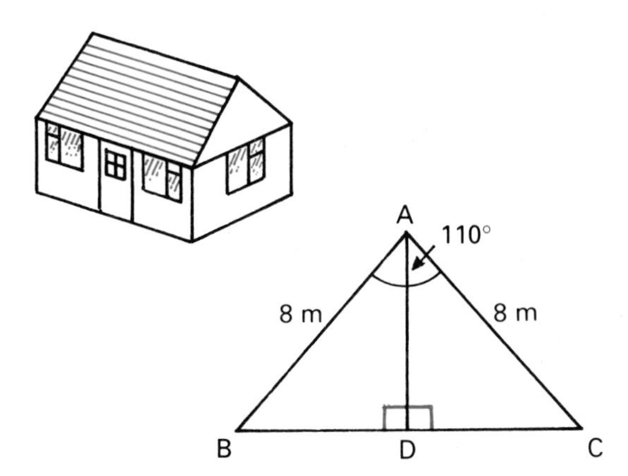

6 The spotlight on the ballet dancer has a 10° wide beam. Calculate the width of the beam:
a on the stage (AD)
b on the stage when the dancer and the centre of the beam are directly below C.

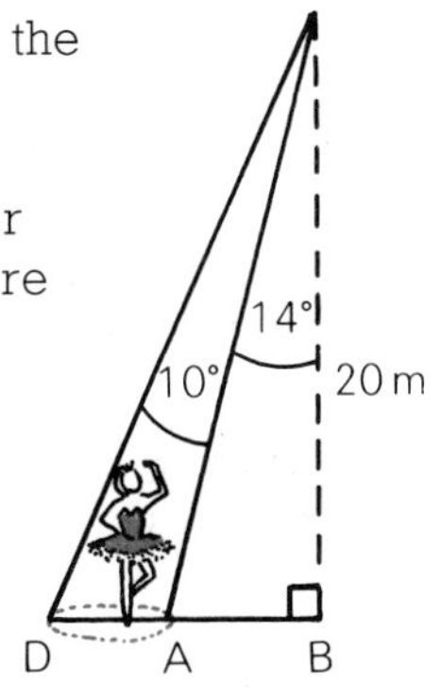

7 The slope, or gradient, of AB is measured by the tangent of the angle which AB makes with the horizontal line AC.

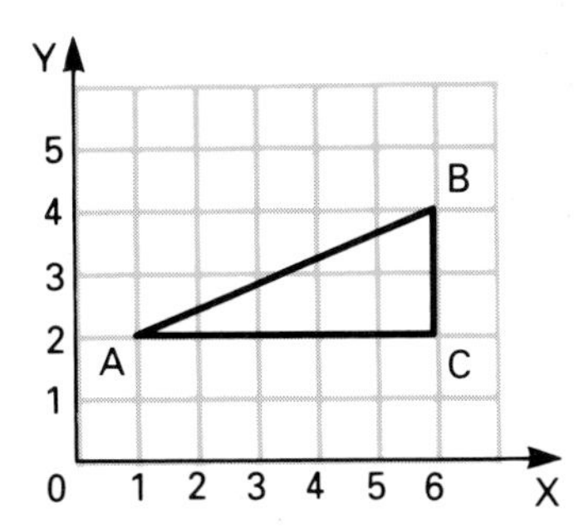

a Calculate the gradient of AB, and the size of ∠BAC.
b Calculate the gradients of the lines joining:
(i) P(2 1) and Q(7, 2) (ii) R(0, 3) and S(2, 6), and the angles these lines make with the horizontal.

8

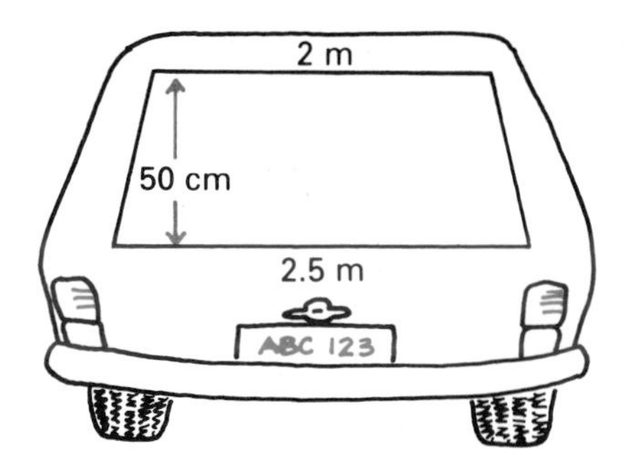

The rear windscreen of the car is in the shape of a symmetrical trapezium. Calculate:

a the angle between the side and the foot of the window

b the length of the rubber seal right around the window.

9 a Prove that the formula for the height of the building in the diagram below is $H = h + d \tan x°$.

b Use the formula to calculate H, given $h = 1.5$, $d = 100$ and $x = 22$.

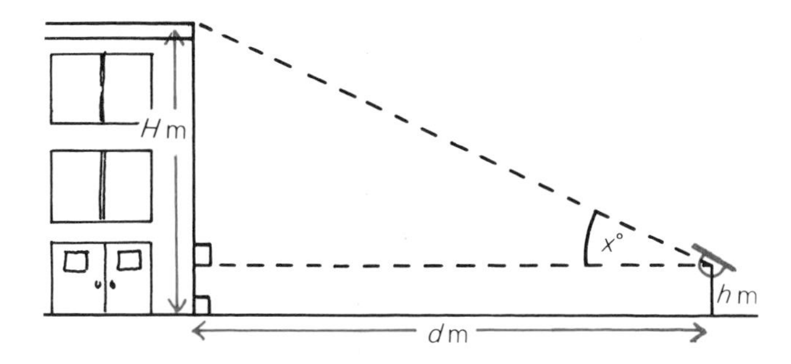

10 a Prove that AD = $b \sin$ C.

b Find a formula for the area of △ABC which involves a, b and sin C.

c Use your formula to calculate the area of △ABC, given $a = 8.5$, $b = 7.5$ and ∠ACB = 36°.

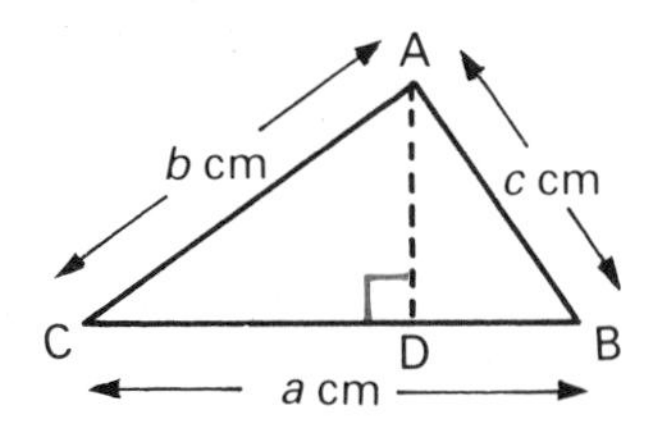

11 The area of this triangle is 80 cm^2.
Calculate: **a** h **b** x

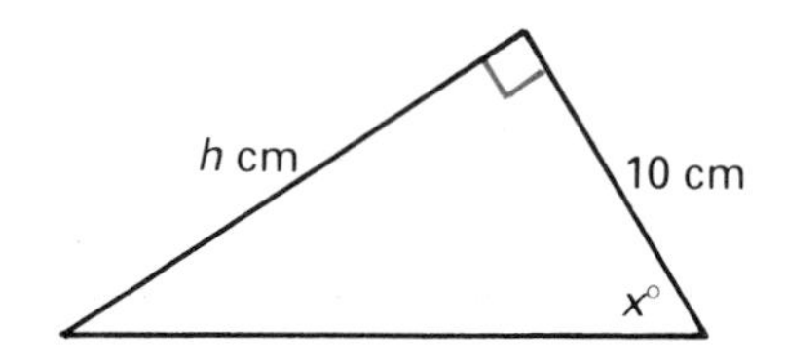

INVESTIGATIONS

***1* a** *As angle A increases from 0° to 90°, what happens to sin A, cos A and tan A?*

b *Is there any connection between the sine of an angle and the cosine of its complement? Why is this?*

c *Why are sin A and cos A never greater than 1?*

d *Why is sin 10° nearly the same as tan 10°?*

e *Find angle A when sin A = cos A, and 0° < A < 90°.*

***2* a** *Make a table of values for angle A and sin A every 10° from 0° to 90°.*
Draw a graph of A against sin A to get part of the 'sine wave'. Would you be prepared to continue it in a sketch to 180°, or even 360°? Your calculator could help!

b *Do this again for angle A against cos A.*

CHALLENGE

The detective's dilemma

Monsieur Poirpont knew that the fatal bullet came from the High Rise flats. It entered the cottage across the road through window A, killing the unfortunate victim before embedding itself in the wall at B.

Not for nothing was Monsieur Poirpont a graduate in mathematics at the Sorbonne. He sketched the front view of the flats, and the side view and plan view of the flats and cottage. Then he made some measurements and, after a few calculations, commanded 'Arrest the person in flat number X'. Can you find X?

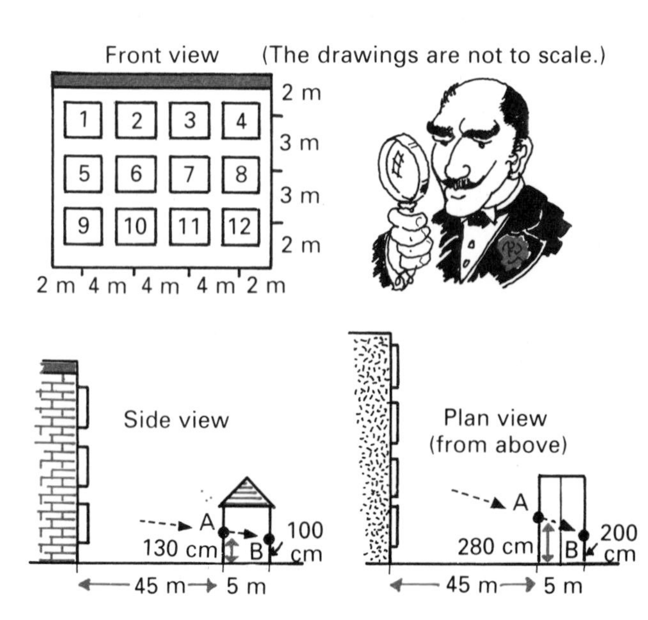

CHECK-UP ON TRIGONOMETRY

1 Write down ratios for these sines, cosines and tangents:

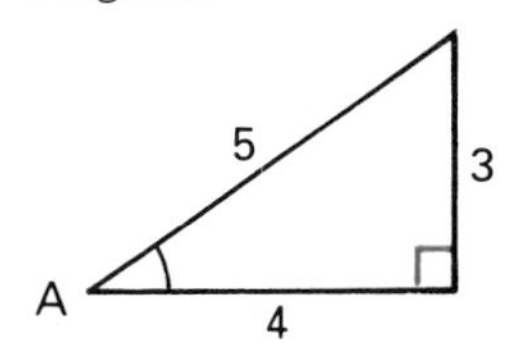

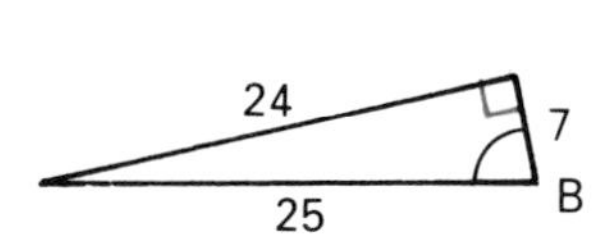

a sin A and cos A **b** sin B and tan B

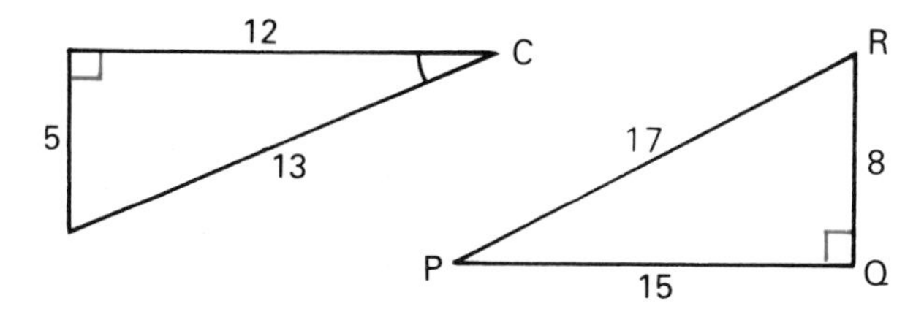

c cos C and tan C **d** sin P and cos R

2 Calculate x in each triangle.

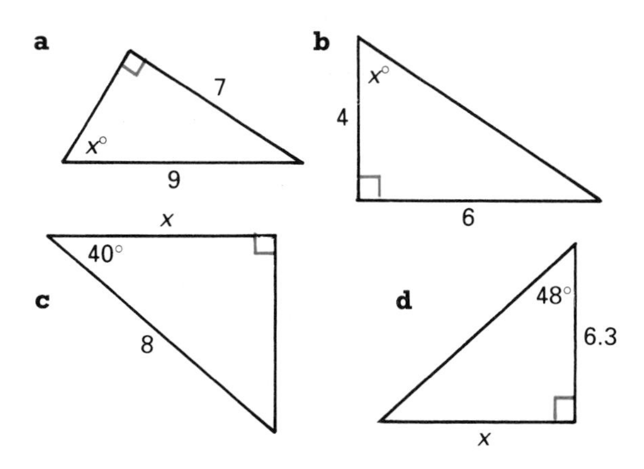

3 $\triangle$ABC is isosceles, with $\angle$BAC = 160°. Calculate:
a the length of altitude AD
b the length of the base BC
c the area of $\triangle$ABC.

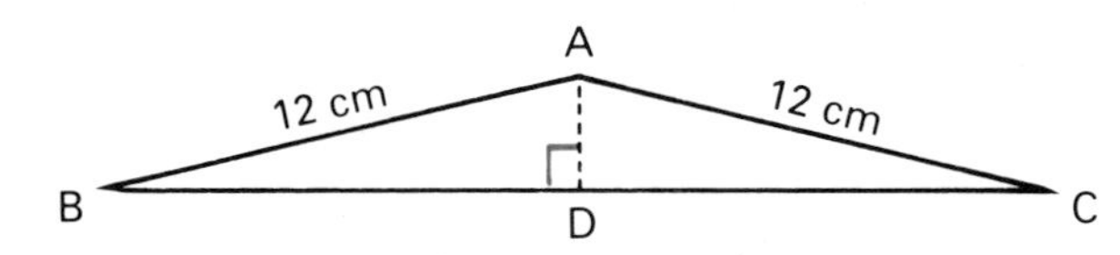

4 ABCD is a parallelogram. Calculate:
a its altitude **b** its area, to the nearest cm^2.

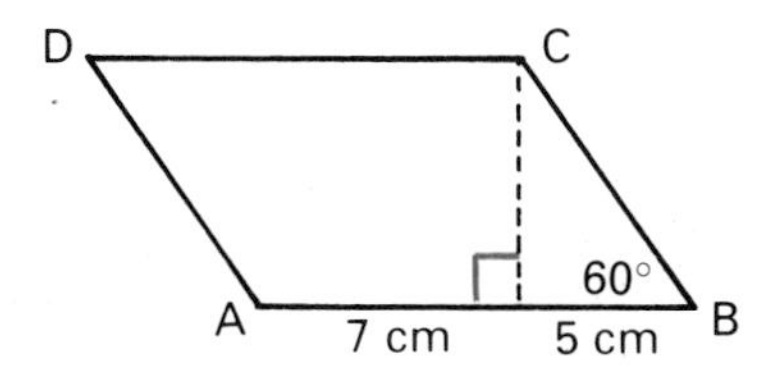

5 The bookshelf is supported by bracket AB. Calculate:
a BC **b** AB

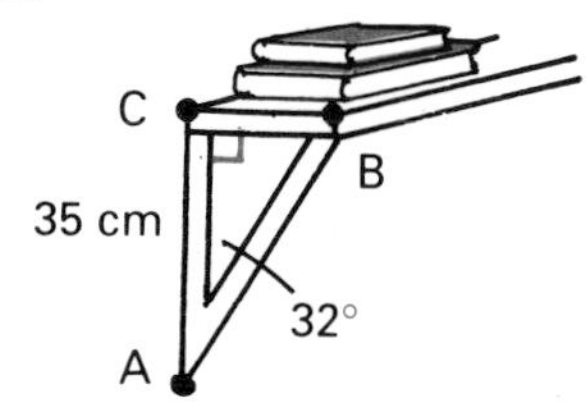

6 Tom is shining his torch straight at the wall.

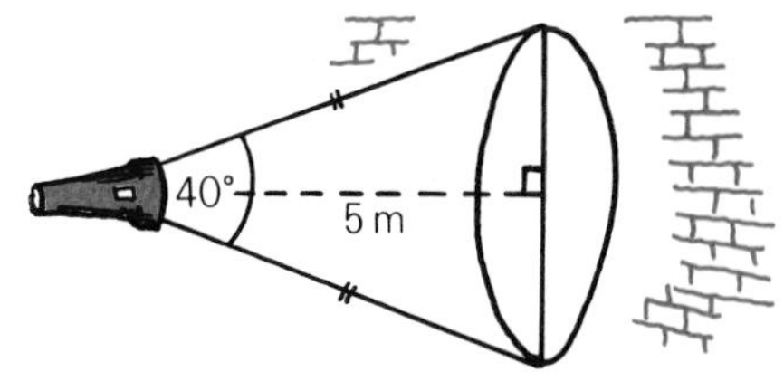

Calculate:
a the radius of the circle of light on the wall at a distance of (i) 5 m (ii) 10 m
b the area of the circle in each case.

7 Calculate the sizes of the angles, perimeter and area of this wall support.

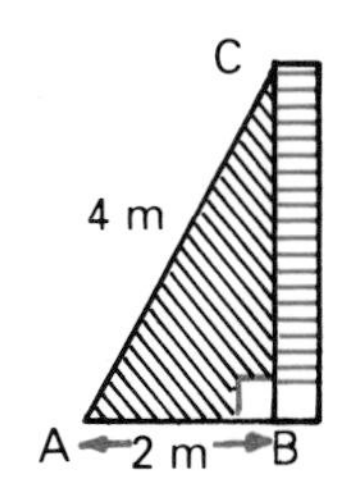

8 In the stair diagram below calculate:
a x
b y, using:
(i) trigonometry
(ii) Pythagoras' Theorem.

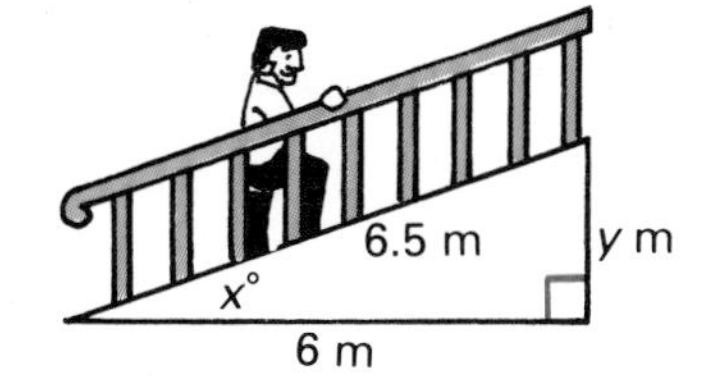

9 This castle stands on a hilltop. Calculate the distance from A to B, to the nearest metre.

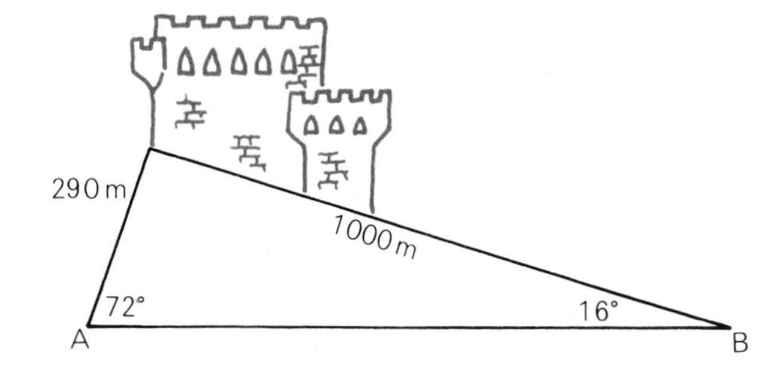

10 SIMULTANEOUS EQUATIONS

LOOKING BACK

1 Copy and complete this table:

x	-2	-1	0	1	2
$2x$					
$x+1$					
$4-x$					

2 Add these pairs of terms:

a $6x$, $5x$ **b** $2x$, $-2x$ **c** $-x$, $-3x$ **d** $-2x$, x

3 Subtract these pairs of terms:

a $7y$, $4y$ **b** $7y$, $-4y$ **c** $-2y$, y **d** $-3y$, $-2y$

4 Write down the coordinates of A, B, C and D.

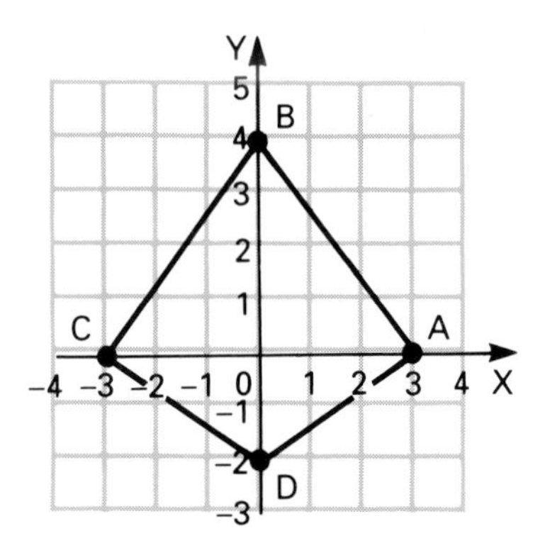

5 Given $x = 2$, find the value of y if:

a $y = 3x+1$ **b** $y = 2x-5$ **c** $y = -x+2$

6 Given $y = -3$, find the value of x if:

a $x = y+4$ **b** $x = 2y+1$ **c** $x = 1-y$

7 Solve these equations:

a $3x = 12$ **b** $2y = -6$ **c** $3t+1 = 13$

d $2x = x+5$ **e** $3x-7 = 5-x$ **f** $3y-2 = 4y-5$

8 $y = 2x-4$. Find:

a y when $x = 0$ **b** x when $y = 0$.

9 Name the points on the diagram at which:

a $x = 2$ **b** $y = 1$ **c** $y = -4$

d $y = x$ **e** $y = 2x$ **f** $x = 4$ and $y = -3$

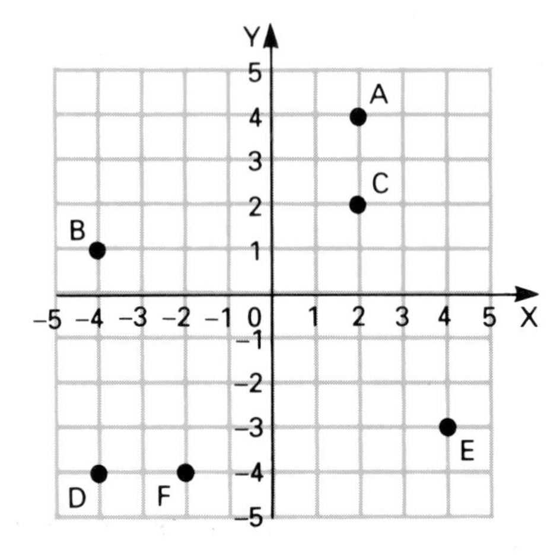

10 Write down expressions for the total cost of:

a 8 of these books

b x of these books

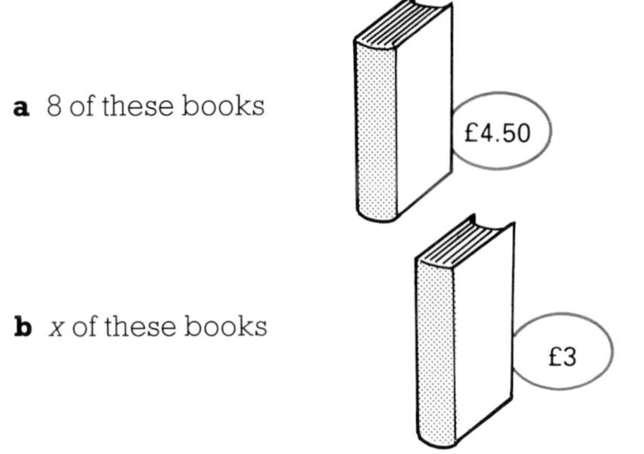

c 5 of these books and 3 of the pens.

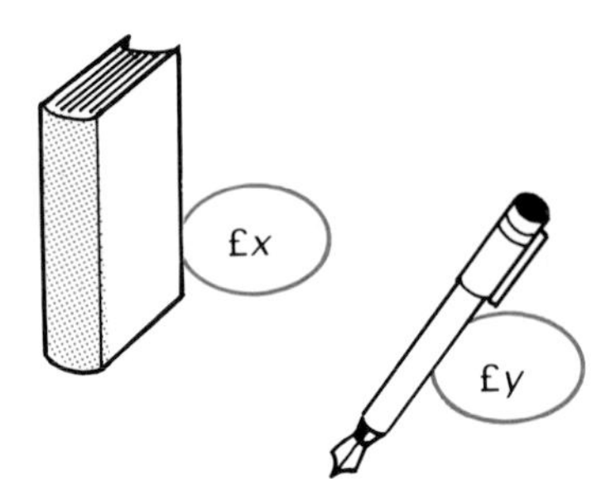

If x and y are whole numbers, and the total cost is £32, find the possible cost of a book and of a pen.

CLASS DISCUSSION/INVESTIGATION

Newtown Youth Group are having a Treasure Hunt. Part of the football field is marked off in 10 metre squares.
The treasure (a voucher for sports goods) is buried at one of the crossing points. The competitors are given these two clues to help them find the treasure:

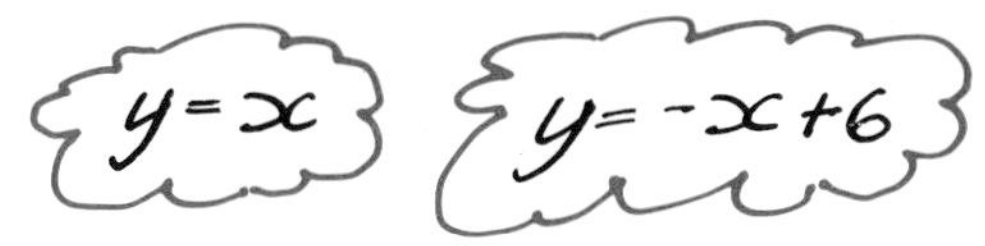

1 How many crossing points are there?

2 Copy the grid on squared paper. Imagine you are taking part in the Treasure Hunt. Can you find the treasure, using the two clues?

3 If you find a method that works, try it for these pairs of clues:

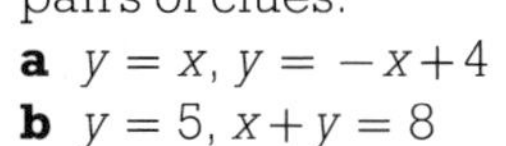

a $y = x$, $y = -x+4$
b $y = 5$, $x+y = 8$

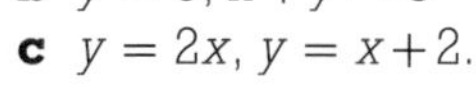

c $y = 2x$, $y = x+2$.

4 Can you think of another way to find the treasure?

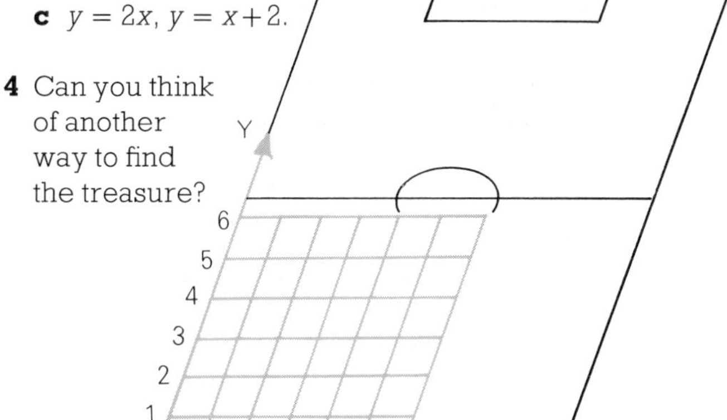

STRAIGHT LINE GRAPHS

(i) The first clue is $y = x$. Possible locations of the treasure at the crossing points are:

x	0	1	2	3	4	5	6
y	0	1	2	3	4	5	6

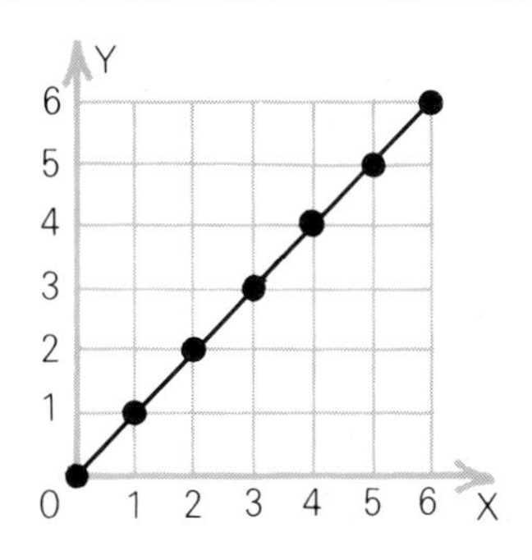

The locations are marked by the seven dots. If the treasure had been hidden *anywhere* on the straight line through the points, the competitors would have to search the whole of the line from $x = 0$ to $x = 6$.

(ii) The second clue is $y = -x+6$. Possible locations for this are:

x	0	1	2	3	4	5	6
y	6	5	4	3	2	1	0

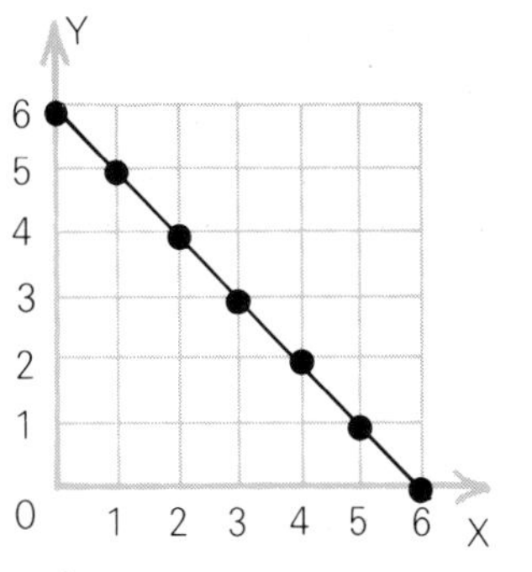

All crossing points on the line are shown on this graph.
Taken together, the clues lead to the actual position of the treasure at the point (3, 3).

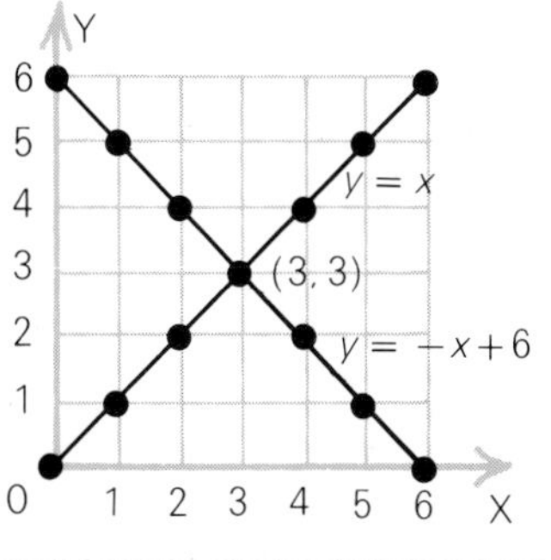

EXERCISE 1

For each pair of clues in this exercise:
a make a table of values
b plot the points on squared paper
c draw the straight lines through the points
d find the point where the treasure is hidden.

1 $y = x$ and $y = 4-x$, for x from 0 to 4.

2 $y = 2x$ and $y = 6-x$, for x from 0 to 6.

3 $y = x+1$ and $y = 5-x$, for x from 0 to 6.

4 $y = x+2$ and $y = 2-x$, for x from -2 to 2.

5 $y = -x$ and $y = x+4$, for x from -4 to 0.

6 $y = -2x$ and $y = -2$, for x from -2 to 2.

7 $x+y = 6$ and $x-y = 4$, for x from 0 to 6.

8 $2x+y = 6$ and $x-2y = 8$, for x from 0 to 8.

BRAINSTORMER

What can you say about the person who buried treasure and gave the clues $y = 2x$ and $y = 2x+3$?

SKETCHING STRAIGHT LINES

Graphs with equations of the form $y = ax+b$ are always straight lines, so you only need to plot two points to draw them. Choose the points where the line crosses the x and y axes.

Example
Sketch the graph of $y = 2x-6$.
When $x = 0$, $y = -6$, so $(0, -6)$ is on the line.
When $y = 0$, $2x-6 = 0$, giving $x = 3$, so $(3, 0)$ is also on the line.

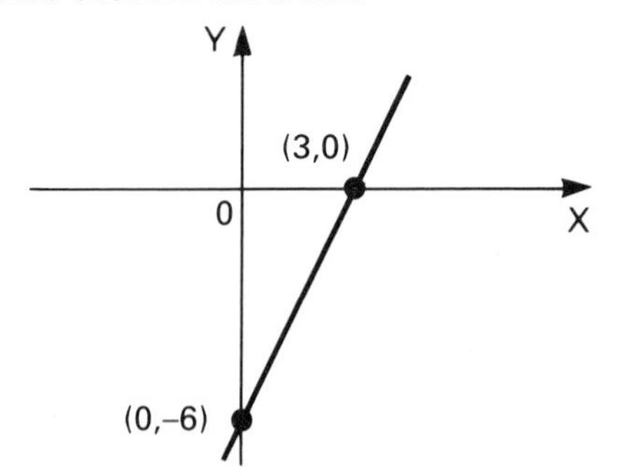

EXERCISE 2

Find the points where these lines cross the axes, then sketch the lines on plain paper.

1 $y = x-4$ **2** $y = 3x-6$ **3** $y = x+3$

4 $x+y = 5$ **5** $x-y = 8$ **6** $x-y = -4$

7 $2x+3y = 6$ **8** $5x-2y = 10$ **9** $3x-4y+12 = 0$

10 $y=8x$ (Take $x = 1$ for one point)

11 $2x+y = 0$

(i) Solving simultaneous equations by straight line graphs

Examples
By drawing graphs of the equations,
solve the pairs of equations $\left.\begin{aligned}3x+2y &= 12\\ 3x-2y &= 0\end{aligned}\right\}$

a $3x+2y = 12$:
when $x = 0$, $2y = 12$, so $y = 6$, giving $(0, 6)$
when $y = 0$, $3x = 12$, so $x = 4$, giving $(4, 0)$

b $3x-2y = 0$:
when $x = 0$, $y = 0$, giving $(0, 0)$
when $x = 6$, $18-2y = 0$, so $y = 9$, giving $(6, 9)$.

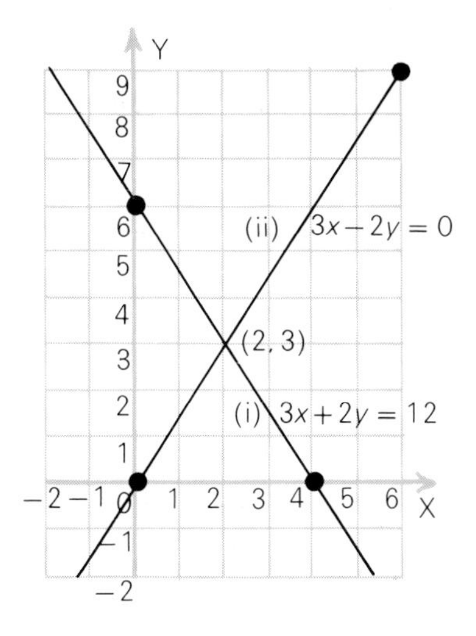

At $(2, 3)$, where the graphs cross, $3x+2y = 12$ **and** $3x-2y = 0$.
So the solution of the *pair of equations* is $(2, 3)$, or $x = 2$, $y = 3$.

Note Since we are looking for values of (x, y) that satisfy the two equations at the same time, equations like these are often called **simultaneous equations**.

EXERCISE 3A

By drawing graphs of the equations on squared paper, solve each pair of simultaneous equations in this exercise.

1 $x+y=6$, $x+2y=8$
2 $x+3y=6$, $x+y=4$
3 $x-y=4$, $x-2y=2$
4 $x+y=6$, $x-y=4$
5 $x+2y=6$, $x-y=3$
6 $y=x+3$, $y=-x-3$
7 $x+3y=3$, $x-3y=9$
8 $y=2x+4$, $y=-x-2$
9 $3x-y=7$, $y=5$

EXERCISE 3B

Solve these pairs of equations by drawing their graphs.

1 $2x-y=0$, $x+y=-3$
2 $x+y=0$, $x-y=0$
3 $y=x$, $y=8-x$
4 $2x-y=7$, $x+y=5$
5 $y=2x+6$, $y=6-2x$
6 $2x+y+4=0$, $x-2y+7=0$

Use 2 mm squared paper for questions **7**, **8** and **9**.

7 $x+y=2$, $2x-4y=1$
8 $2x+3=0$, $3x-y+1=0$
9 $x+y=6$, $5y-5x=6$

(ii) Solving simultaneous equations by substitution

Richard is twice as old as his sister Laura. Also, he is 5 years older than her. He challenges Peter to find their ages.
Peter takes Richard's age as y years and Laura's as x years, and makes this pair of simultaneous equations: $y = 2x$
$y = x+5$

The trouble is that when he draws these lines they don't meet on the graph paper. He realises that where they *do* meet, their y-values must be the same.

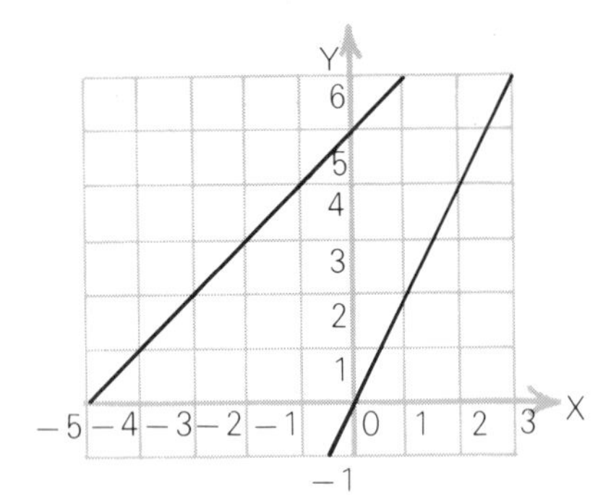

$\mathbf{y} = 2x$
$\mathbf{y} = x+5$
So $2x = x+5$ .... (i)
giving $x = 5$
$y = 2x = 10$

So Richard is 10 years old and Laura is 5.

Note In (i), $2x$ is substituted for y so the method is called **substitution**.
The solution of the simultaneous equations is (5, 10).

EXERCISE 4

Solve these simultaneous equations by substitution.

1 $y = 2x$, $y = x+10$
2 $y = 3x$, $y = 20-x$
3 $y = 2x-8$, $y = x+1$
4 $y = 4x+12$, $y = x+12$
5 $y = 1-2x$, $y = x+10$
6 $y = 3x+1$, $y = 6-2x$
7 $y = 3x$, $y = x+1$
8 $y = -2x$, $y = -x+1$
9 $y = -x$, $y = 3x+4$
10 $y = 2x$, $3x+y = 5$
11 $y = 2x-3$, $x+y = 3$
12 $2x+y = 8$, $y = x-1$

CLASS DISCUSSION/INVESTIGATION

Use these two bits of information to find the cost of a child's ticket and the cost of an adult's ticket.
Explain your method and check your answer!

EXERCISE 5

Think out the cost of an adult's ticket and a child's ticket for each of these—trial and improvement could help!

1

2

3

4

5

6

7

8

(iii) Solving simultaneous equations by elimination

Question **8** of Exercise 5 asked you to find the cost of an adult's ticket and a child's ticket, given:

Here is Alison's solution.
If an adult's ticket costs £x, and a child's £y, then:

$$2x+6y = 24$$
$$2x+4y = 19$$

Subtract $2y = 5 \ldots\ldots$ (i)

$$y = 2\tfrac{1}{2}$$

Put $y = 2\frac{1}{2}$ in the first equation.

$$2x+15 = 24$$
$$x = 4\tfrac{1}{2}$$

So the tickets cost £4.50 and £2.50 each.

Note When the two equations are subtracted, x is eliminated in (i), so this method is called **elimination**. The solution of the simultaneous equations is $(4\frac{1}{2}, 2\frac{1}{2})$.

EXERCISE 6

Solve these simultaneous equations by adding or subtracting them to eliminate x or y, etc.

1 $x+y = 12$, $x-y = 6$

2 $x+y = 15$, $x-y = 5$

3 $x+y = 10$, $x-y = 3$

4 $a+3b = 3$, $a-3b = 9$

5 $c+4d = 22$, $c-4d = -18$

6 $2e+3f = 9$, $2e-3f = 15$

7 $3p-q = 8$, $3p+q = 4$

8 $2p+q = -2$, $4p-q = -1$

9 $u+2v = 2$, $u-2v = -8$

10 $2x+y = 17$, $x+y = 9$

11 $5x+y = 4$, $2x+y = 1$

12 $3x+3y = 3$, $x+3y = 3$

13 $2a-b = 11$, $a-b = 5$

14 $5m-2n = 13$, $m-2n = 1$

15 $7u-3v = 10$, $2u-3v = 5$

16 $5x-2y = 11$, $-x+2y = 1$

17 $3p-5q = -1$, $p-5q = 3$

18 $4s-3t = 15$, $2s+3t = 3$

CHALLENGE

When do you add the equations, and when do you subtract them? Can you find a 'rule'?

A slight difficulty

John always looks for difficulties. 'What about these?' he asks Alison. 'Your method won't work for them.'

$$\left.\begin{aligned} x+2y &= -3 \\ 3x+y &= 1 \end{aligned}\right\}$$

Add: $4x+3y=-2$ *Subtract:* $-2x+y=-4$

John is right. Neither x nor y is eliminated.

Alison's solution

$$\begin{aligned} x+2y &= -3 \quad \times\mathbf{1} \quad & x+\mathbf{2y} &= -3 \\ 3x+y &= 1 \quad \times\mathbf{2} \quad & 6x+\mathbf{2y} &= 2 \\ & \textit{Subtract} & -5x \quad &= -5 \\ & & x &= 1 \end{aligned}$$

or

$$\begin{aligned} x+2y &= -3 \quad \times\mathbf{1} \quad & x+\mathbf{2y} &= -3 \\ 3x+y &= 1 \quad \times\mathbf{-2} \quad & -6x-\mathbf{2y} &= -2 \\ & \textit{Add} & -5x \quad &= -5 \\ & & x &= 1 \end{aligned}$$

Put $x=1$ in the first equation:

$$\begin{aligned} 1+2y &= -3 \\ 2y &= -4 \\ y &= -2 \end{aligned}$$

The solution is $(1, -2)$

Remember:

An equation is like a balance. You must do the same to **each side**—add, subtract, multiply or divide.

EXERCISE 7A

Solve these simultaneous equations by first multiplying both sides of the equations by suitable numbers.

1 $x+2y=8$
$2x-y=1$

2 $3x+y=6$
$x-2y=2$

3 $4x-y=7$
$x+2y=4$

4 $x+2y=6$
$2x-y=7$

5 $x+3y=1$
$2x-y=2$

6 $5x+y=4$
$x-2y=3$

7 $3a+2b=4$
$a+b=1$

8 $4c+2d=2$
$c+d=0$

9 $5e+f=1$
$e+2f=2$

10 $3x-4y=14$
$x+y=0$

11 $5x-2y=7$
$x-3y=4$

12 $x-y=1$
$3x-2y=4$

13 $2x+y=1$
$x+2y=-1$

14 $x-3y=-1$
$2x-y=-2$

EXERCISE 7B

Solve:

1 $3x+2y=15$ $(\times\mathbf{3})$
$5x-3y=25$ $(\times\mathbf{2})$

2 $2x-3y=7$
$7x+2y=12$

3 $4x+2y=-6$
$x-4y=3$

4 $4a-3b=0$
$5a-4b=-1$

5 $2c-3d=12$
$3c-2d=13$

6 $2e+3f=-5$
$3e+2f=0$

7 $3x+4y=20$
$4x-3y=10$

8 $11x+3y+7=0$
$2x+5y-21=0$

9 $3x+5y-23=0$
$5x+2y-13=0$

10 $6p+7q-5=0$
$7p+8q-6=0$

EXERCISE 8

Solve these simultaneous equations by the most suitable method.

1 $3x+y=11$
$2x-y=4$

2 $5x-3y=14$
$2x-3y=2$

3 $y=2x+1$
$x+y+1=11$

4 $4x-y=5$
$x+3y=-2$

5 $6x+5y=16$
$x+y=3$

6 $y=4x$
$2x-3y=20$

7 $2x+3y=1$
$5x+2y=-3$

8 $x+y=4$
$x-y=7$

9 $3x-y=3$
$5x-y=6$

10 $x+y=10$
$y-x=-1$

11 $7x-5y=-1$
$3y=4x$

12 $2x+3y+4=0$
$3x+4y+5=0$

CHALLENGES

1 *Choose any two values for x and y, and make up a pair of simultaneous equations which can be solved to give your chosen values of x and y. Check the solution.*
Then make up another pair of equations with the same solution, and check again.

2 *Solve the simultaneous equations $y=5-x$ and $y=2x+14$ by all three methods—graphs, substitution and elimination.*

USING SIMULTANEOUS EQUATIONS TO FIND EQUATIONS OF STRAIGHT LINES

A free-fall parachutist calls out the distance *d* metres he falls from a marked position after time *t* seconds.

x	3	6	9	12
y	25	40	55	70

The graph of y against x is a straight line. So they are connected by an equation of the form $y=ax+b$.

When $x=3$, $y=25$, so $25=3a+b$
When $x=6$, $y=40$, so $40=6a+b$

Subtract $-15=-3a$
$a=5$

Put $a=5$ in the first equation: $25=15+b$
$b=10$

So the equation of the graph is $y=5x+10$.

Check When $x=12$, $y=70$.

Note Using the graph, when $x=7$, $y=45$.

Using the equation, when $x=20$, $y=110$.

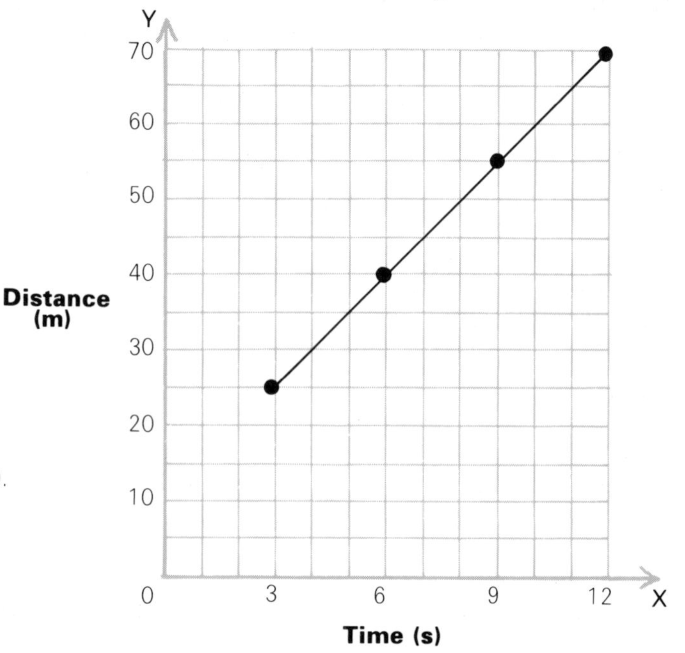

EXERCISE 9

1 A straight line with equation $y = ax+b$ passes through the points P(1, 6) and Q(−1, 2). Make two simultaneous equations in a and b, solve them and write down the equation of the line $y = \ldots x + \ldots$.

2 Repeat question **1** for the pairs of points below:
a (2, 5) and (1, 4) **b** (3, 2) and (1, −2)

3 In an experiment, measurements of x and y give a straight line graph of y against x. So $y = ax+b$. When $x = 11$, $y = 9$ and when $x = 3$, $y = 1$. Calculate a and b, and write down the equation of the graph.

4 In another experiment, the graph of y against x is a straight line, so $y = ax+b$. When $x = 10$, $y = 32$ and when $x = 20$, $y = 62$. Calculate a and b, and write down the equation of the line.

5 The pressure P at depth x in the ocean is of the form $P = ax+b$. At the surface (depth = 0) $P = 15$, and when $x = 60$, $P = 45$.
a Calculate a and b, and write down the equation.
b Draw a graph for $0 \leqslant x \leqslant 100$.
c Estimate the pressure when $x = 80$, using:
(i) the graph (ii) the formula.
d Why is $P = 15$ at the surface?

6 The school's Young Enterprise Group decide to produce scribble pads. Before they start production there is a basic cost of £a. Thereafter the cost of producing each pad is b pence.
a Prove that the formula for the total cost C pence for n pads is $C = 100a+nb$.
b For 100 pads, $C = 1400$ and for 50 pads $C = 900$. Find the group's working formula for C.
c Calculate the cost of producing 250 pads.

INVESTIGATIONS

1 a *Investigate the solution of this system of equations.*
$x+2y = 8$, $3x+y = 9$, $2x-y = 1$ *and* $x-y = -1$.
Include graphs of the equations on the same diagram in your answer.
b *Investigate the possible number of different points in which four straight lines can meet. Illustrate in diagrams.*

2 *Investigate the solutions of these pairs of equations. Try different methods, for example 'trial and error', tables of values, graphs, substitution.*

a $y = x$, $y = x^2$

b $y = x$, $y = x^3$

c $y = x$, $y = \frac{16}{x}$

d $y = x$, $x^2+y^2 = 8$

CHALLENGE

Can you find and use a formula?
On its first day of flowering a Christmas cactus plant had three flowers. Two more flowers opened the next day. The total number of flowers, f, after n days is given by the formula f = an+b, where a and b are numbers.

a *(i) Check that for n = 1 and f = 3, a+b = 3.*
(ii) Make another equation involving a and b.
(iii) Solve the two equations for a and b.
(iv) Complete the formula f = . . . n + ...
b *If the plant first flowered on 23rd December, how many flowers did it have on 1st January (assuming that none had withered)?*

USING SIMULTANEOUS EQUATIONS TO SOLVE PROBLEMS

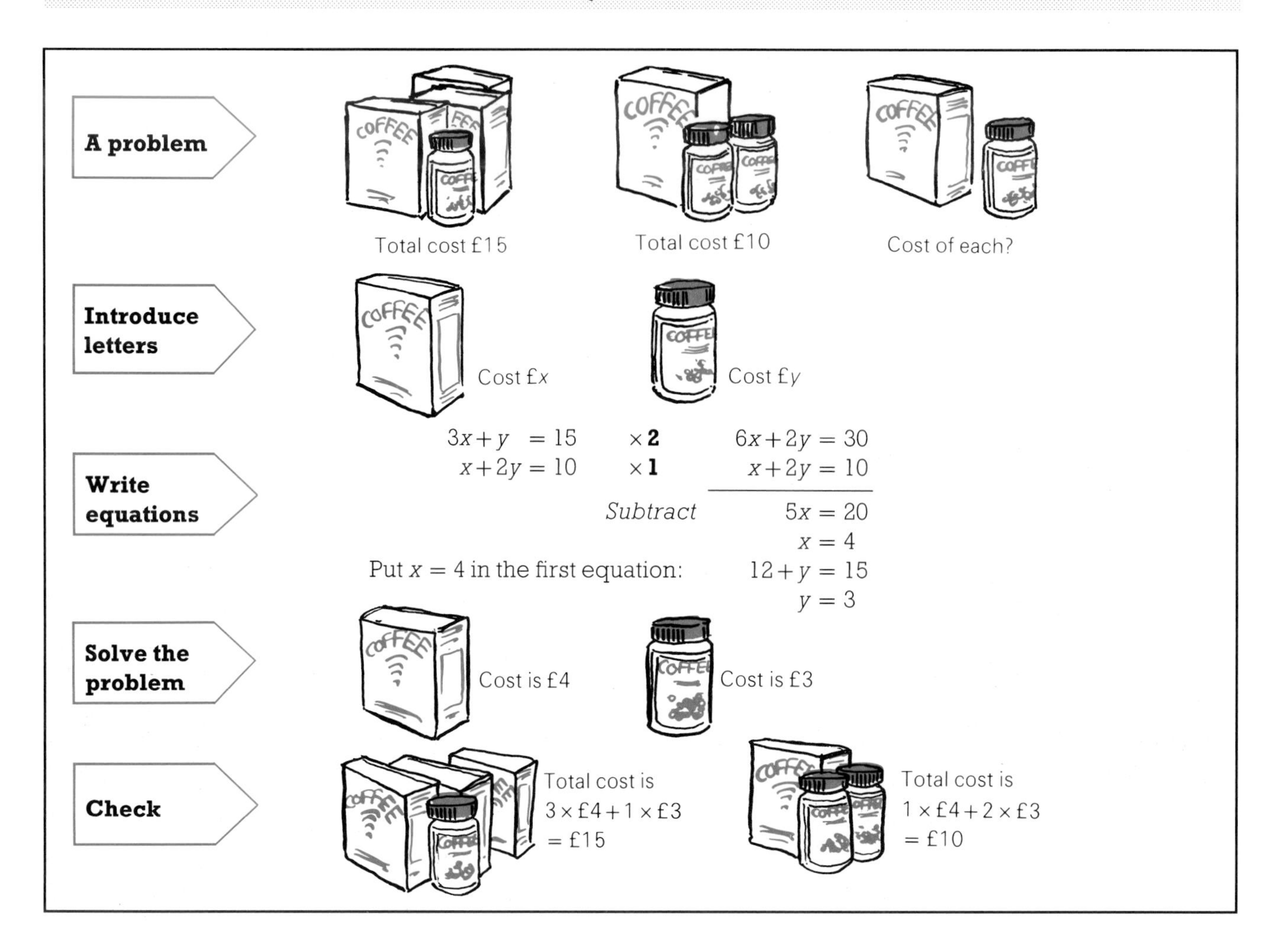

EXERCISE 10A

Write down a pair of simultaneous equations for each picture, then solve them to answer the question.

1 Cost of: **a** a book **b** a pen?

Total cost £9

Total cost £12

2 Cost of: **a** foil **b** glue?

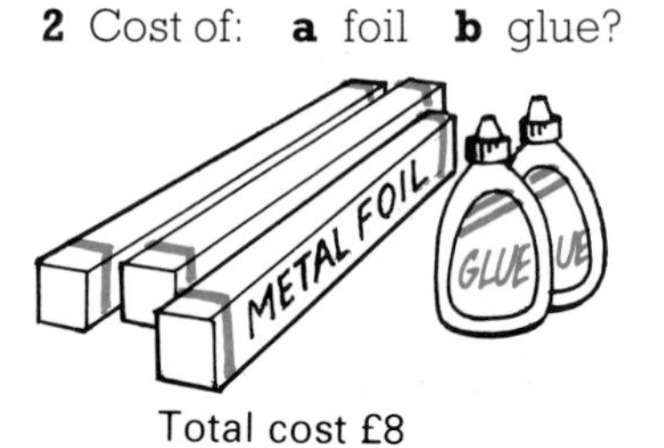

Total cost £8

Total cost £7

3 Length of: **a** skein of wool **b** ball of wool?

Total length 30 m

Total length 28 m

4 Playing time of: **a** Video 1 **b** Video 2?

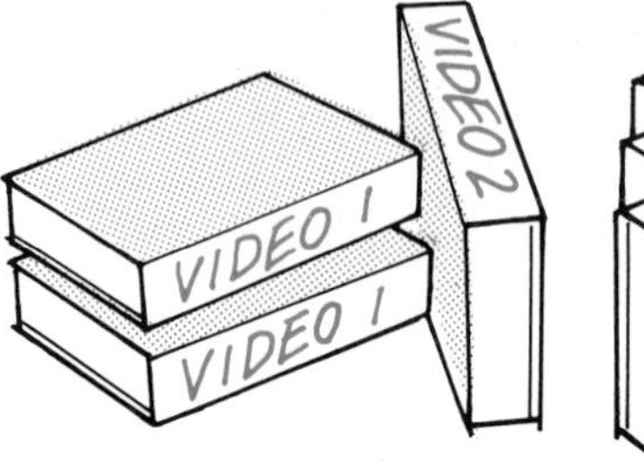

Total playing time 13 hours

Total playing time 22 hours

5 Cost of hiring a video?

Hiring cost £12 (Subtract!)

Hiring cost £3 (Subtract!)

6 Cost of: **a** milk **b** yoghurt?

Cost £2.85

Cost £3.05

EXERCISE 10B

1 For a concert, 2 adults' tickets and 5 children's tickets cost £27. 1 adult's and 3 children's cost £15. Find the cost of each kind of ticket.

2 2 CDs and 3 cassettes cost £48. 1 CD and 4 cassettes cost £44. What is the cost of each?

3 An adult's train fare is £2 more than a child's. The adult's fare is twice the child's. Find the cost of each fare.

4 Two posts laid end to end measure 12 metres. Side by side, one is 6 metres longer than the other. Find the length of each post.

5 The difference between the length and breadth of a rectangular park is 50 m. The perimeter of the park is 200 m. Calculate the length and breadth.

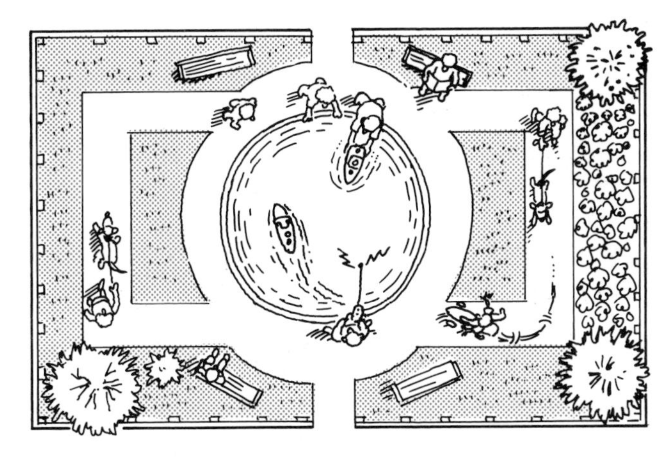

6 In an Australian Rules football game, West scored 2 goals below the bar and 5 above. East scored 3 below and 2 above. The final score was 'West 17, East 20'.

a How many points were awarded for goals below the bar, and how many for goals above the bar?

b The return game was drawn, 7 points each. How might this number of points have been scored?

CHECK-UP ON SIMULTANEOUS EQUATIONS

1 Can you find the treasure?

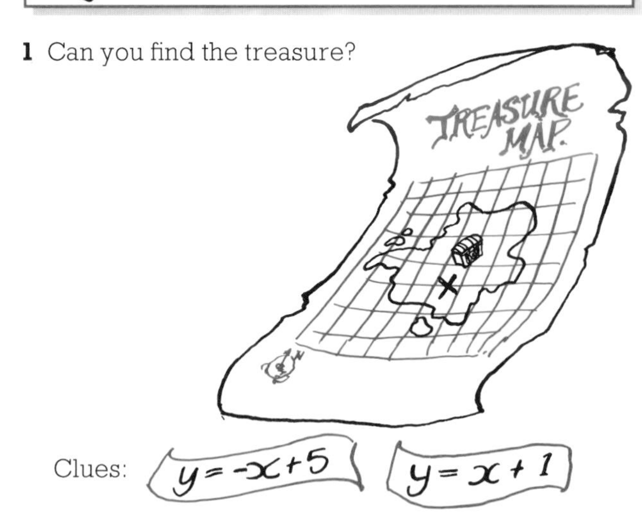

Clues: $y = -x + 5$ $y = x + 1$

a Make a table for each clue, taking $x = -2, -1, 0, 1, 2, 3, 4, 5$.
b Plot the two sets of points.
c Draw straight lines through the points.
d Find where the treasure is hidden.

2 On plain paper *sketch* the straight lines with equations:
a $y = x - 5$ **b** $x + y = 4$ **c** $3x - 5y = 15$

3 Solve these simultaneous equations by drawing their graphs on squared paper:
a $y = x + 1$, $x + 2y = 8$ **b** $y = 2x$, $x - 2y = -6$

4 Look at the diagram, then write down:
a the coordinates of P
b the equation of each straight line.

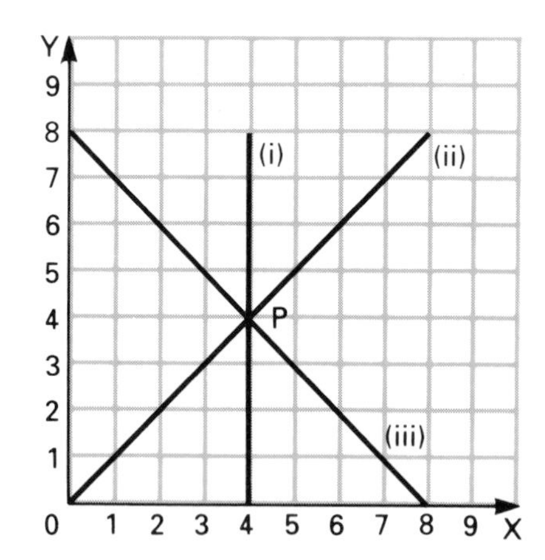

5 Solve these pairs of equations by substitution:
a $y = 5x$, $y = 3x - 6$ **b** $y = 3x$, $x + y + 8 = 0$

6 Solve these equations by first eliminating x or y:
a $3x + 2y = 11$, $x - 2y = 9$ **b** $2x - 4y = -3$, $6x - 4y = 3$

7 A straight line with equation $y = ax + b$ passes through the points (2, 3) and (−1, 0). Find the equation of the line in terms of x and y.

8 Choose any method to solve each pair of simultaneous equations.
a $3x + 4y = 24$, $x - y = 1$ **b** $y = 10x$, $y = 2x + 4$
c $3x + 5y = 11$, $2x + 4y = 9$ **d** $x + y + 1 = 0$, $4x + 2y - 1 = 0$

9 On a ferry crossing, 3 caravans and 2 cars cost £205, and 2 caravans and 3 cars cost £195. Find the cost for a car and for a caravan.

AREAS AND VOLUMES

LOOKING BACK

1

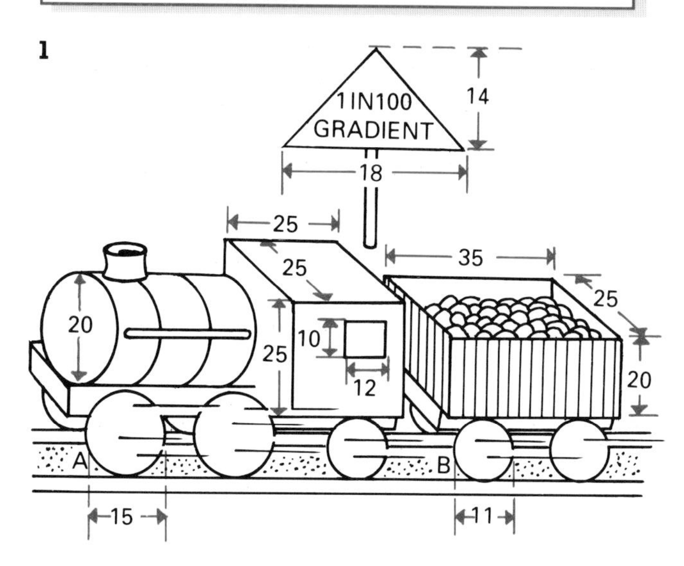

Ian has made a model engine and tender for his technical project. All the measurements are in centimetres. Calculate, correct to the nearest unit:

a the circumference of:
   (i) wheel A
   (ii) wheel B

b the area of:
   (i) the cab window
   (ii) the triangular sign
   (iii) the front of the engine's boiler

c the volume of: (i) the cab (ii) the tender.

2 $\pi = 3.14159\ldots$ Write down its value, correct to:

a the nearest whole number

b 1 decimal place

c 3 significant figures.

3 Calculate the area of each triangle below.

a

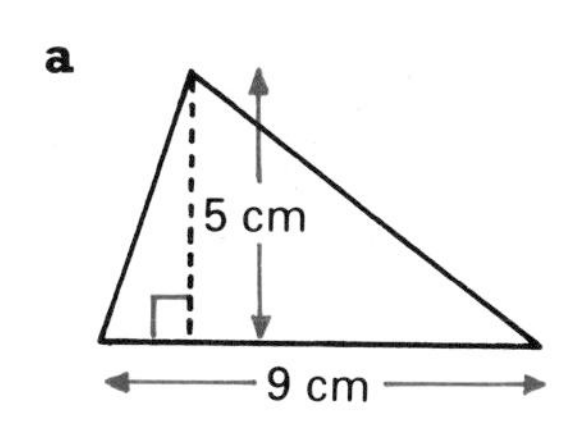

b

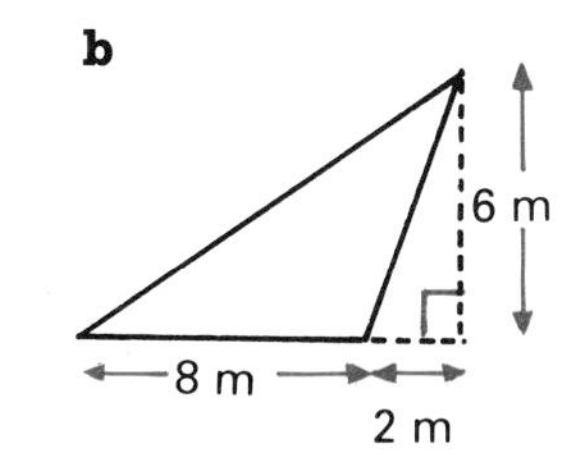

4 Use Pythagoras' Theorem to calculate x.

a **b**

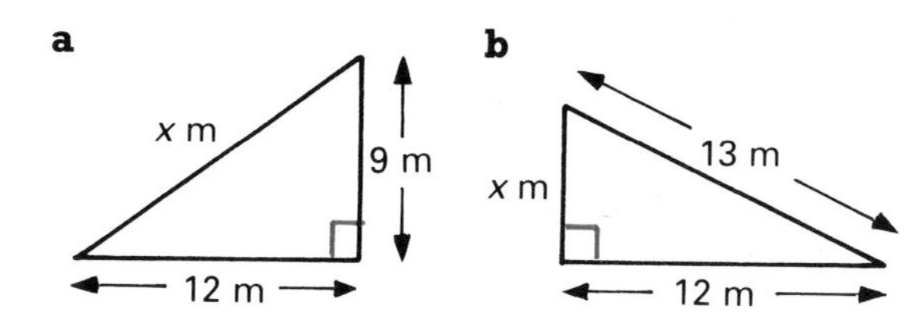

5 Use trigonometry to calculate x, correct to 1 decimal place.

a **b**

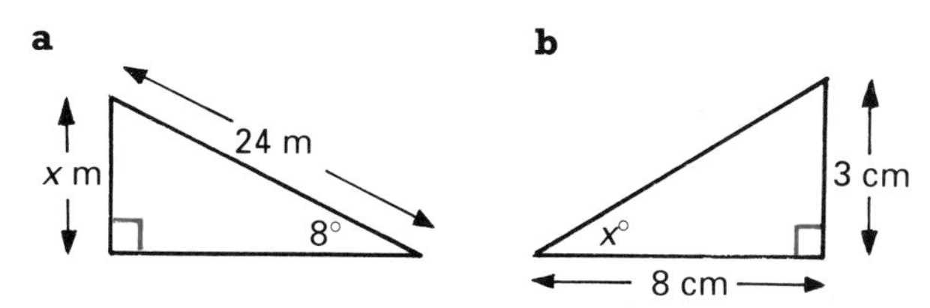

6 Calculate, to the nearest unit, on the gradient sign in question **1**:

a the length of the sloping edges

b the angle between the sloping edges and the horizontal.

RECTANGLES, TRIANGLES AND CIRCLES

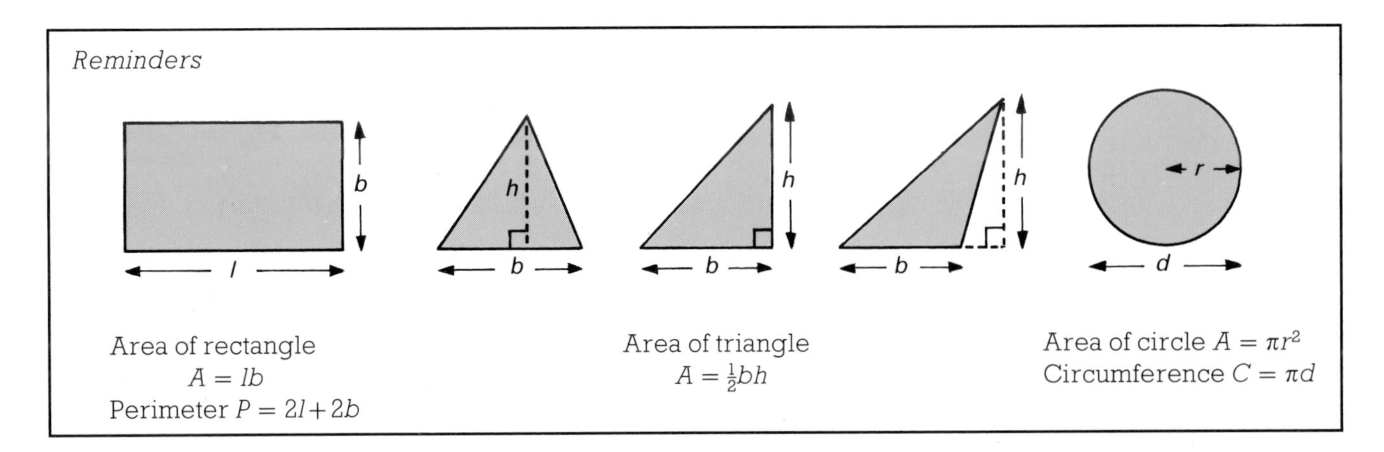

In this chapter use the π key on your calculator and give answers correct to 3 significant figures, unless there are other instructions.

EXERCISE 1A

1 Calculate the perimeter and area of each shape.

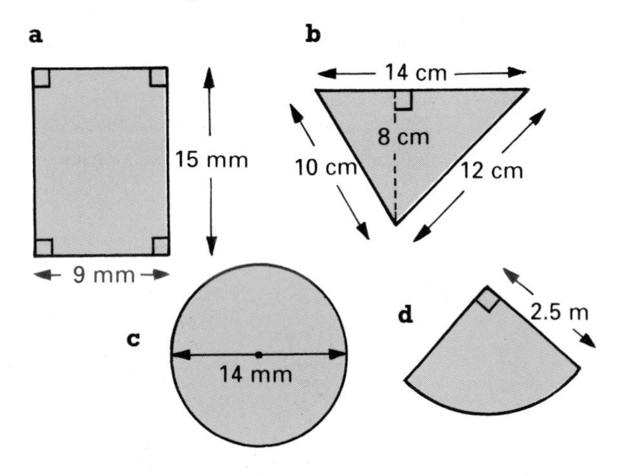

2 Calculate the area of red gravel in each garden. All shapes are rectangular or circular.

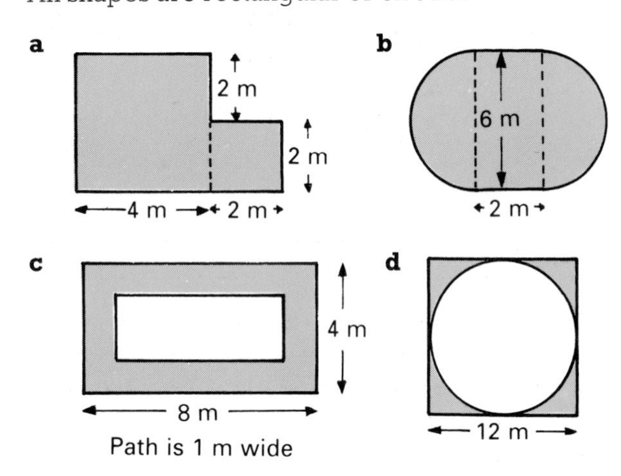

3 The shape of this old castle window is a square and a semi-circle combined. Calculate:
a its perimeter
b its area.

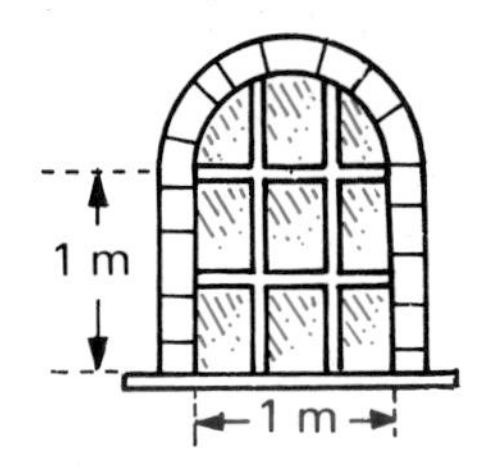

4 Tom is laying these slabs on a rectangular patio 420 cm long and 360 cm broad.
a How many slabs will he need?
b Draw a diagram to show how he should lay them to avoid waste.

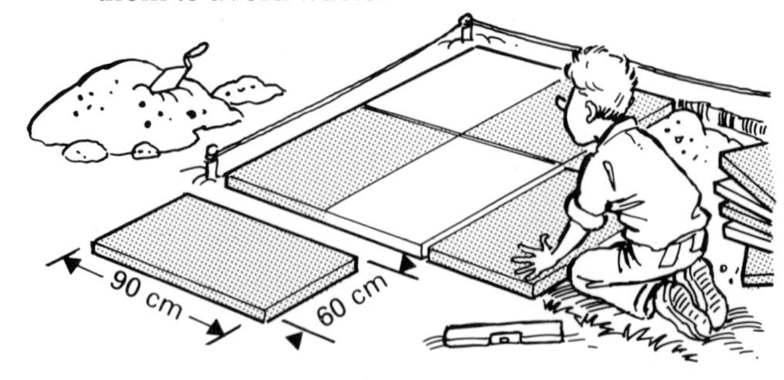

5 In the school shop Mr Smart sells rubbers which measure 5 cm by 3 cm by 2 cm. They come tightly packed in boxes, 12 cm by 10 cm by 9 cm deep.
a Draw a diagram to show the bottom layer.
b How many layers are there?
c How many rubbers are in a box?

EXERCISE 1B

1 Mr Green's garden is square, with equal quarter circle flowerbeds at the corners. Calculate:
a the perimeter of the lawn
b the area of the lawn.

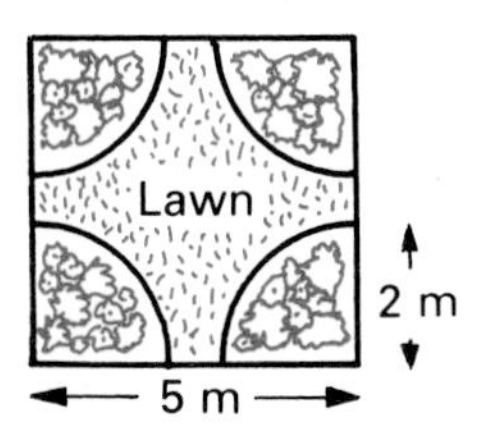

2 a Sunnyside's end walls have to be roughcast. The cost depends on the area. Calculate the area of each end wall.

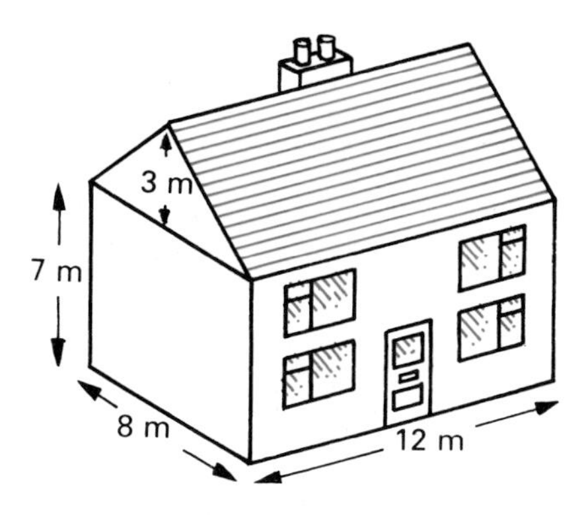

b The roof also needs to be reslated. Calculate:
(i) the length of a sloping edge
(ii) the cost for the whole roof, at £60 per m^2.

3 Caroline measures the circumference of a circular tree trunk, 245 cm. Calculate the trunk's:
a diameter **b** area of cross-section.

4 a With the help of diagrams, calculate the area of:
(i) the largest circle that can be drawn in a square of side 10 cm
(ii) the largest square that can be drawn in a circle of diameter 10 cm.
b What percentage of the outer shape does each fill?

5 Find the areas of these red semi-circular metal supports.

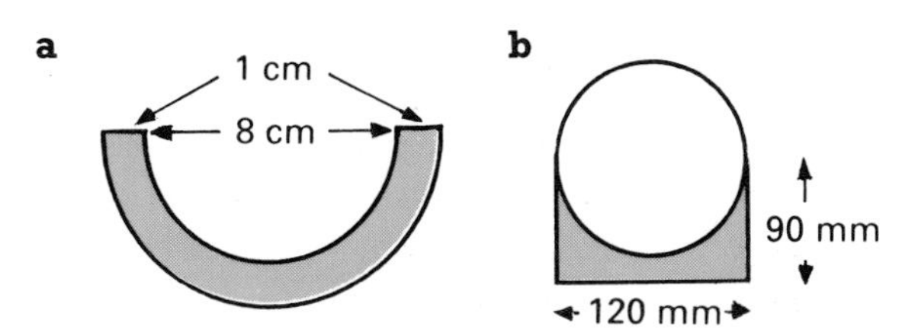

BRAINSTORMER

Rachel cuts out a sector of a circle in paper, and folds it to make a party hat for her young brother Andrew. Calculate the circumference of the base and the sloping area of the hat.

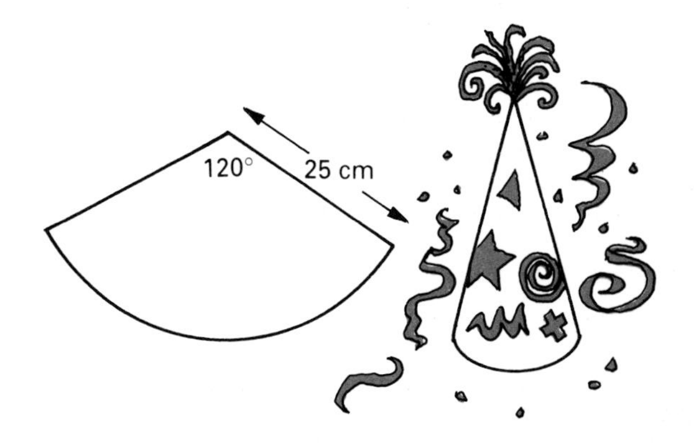

CHALLENGE

Mr Habib has 20 m of fencing. He plans to fence three sides of a rectangular plot, using the wall for the fourth side. What length and breadth would give him the greatest area?

INVESTIGATION

In some rectangles with sides a whole number of units long, the number of units in their perimeters is equal to the number of units in their areas. How many such rectangles can you find? Try to work methodically.

AREAS OF QUADRILATERALS

EXERCISE 2A

(i) Rhombus and kite

For a rhombus or kite, the area of each small triangle is half the area of its surrounding rectangle.
So the area of the whole rhombus or kite is half the area of its surrounding rectangle.

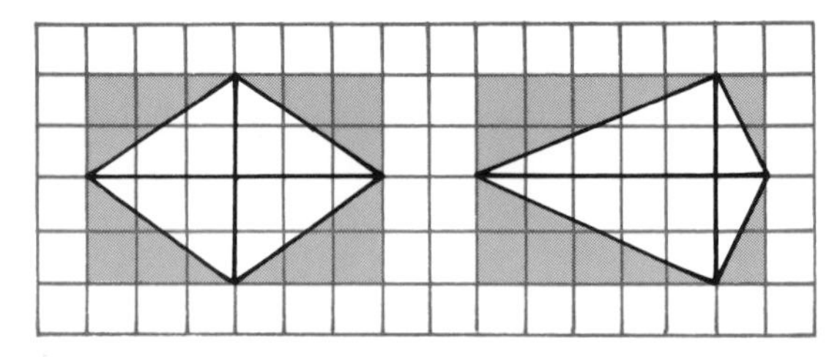

Area of rhombus or kite
$= \frac{1}{2}$ area of surrounding rectangle
$= \frac{1}{2}$ diagonal × diagonal

1 Calculate the areas of these kites and rhombuses.

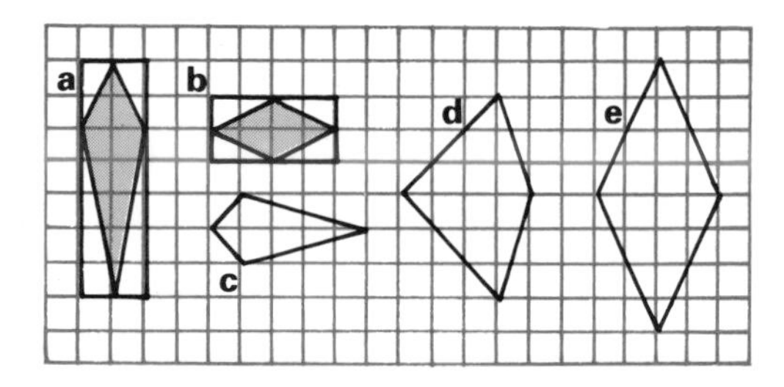

2 Calculate the areas of these kites and rhombuses.

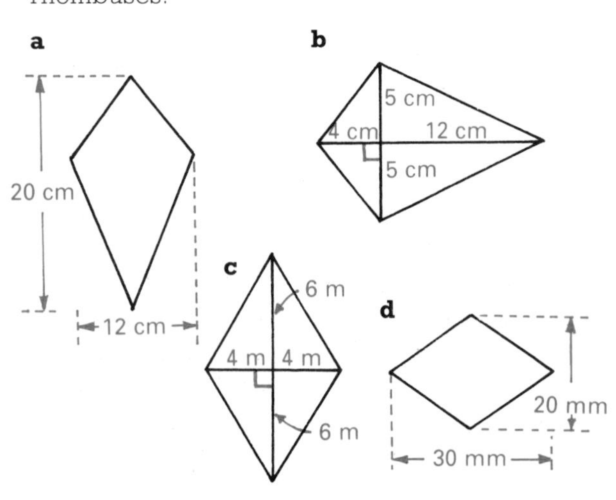

3 The upwards 'lift' for a kite depends on its area. Calculate the areas of these kites, and list them in order of lift. All measurements are in cm.

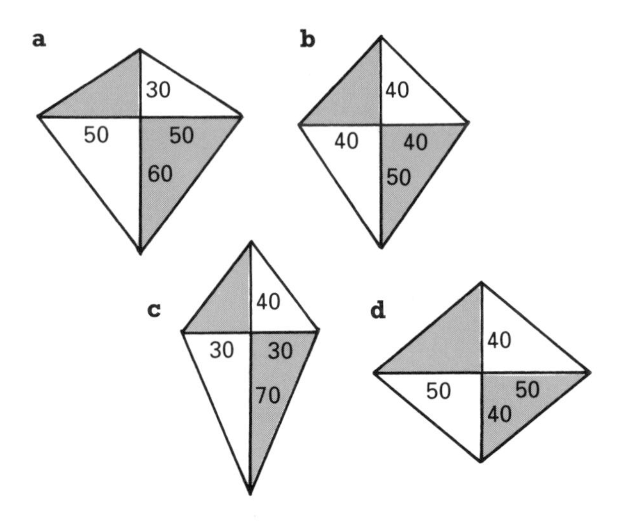

(ii) Parallelogram

The area of the parallelogram
= the area of the rectangle. Why?

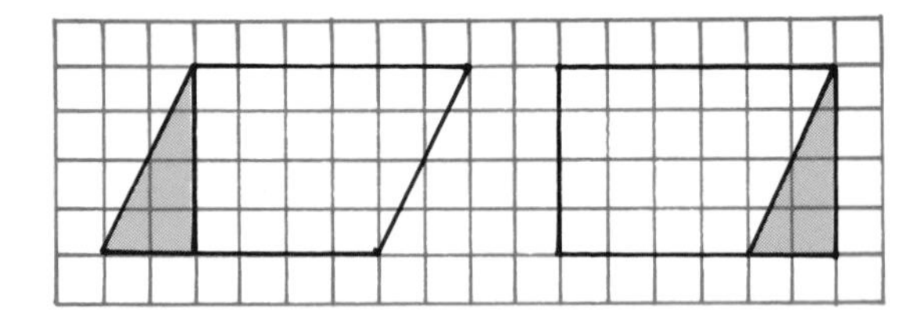

So area of parallelogram
= area of rectangle of same base and height
= base × height

4 Calculate the areas of these parallelograms.

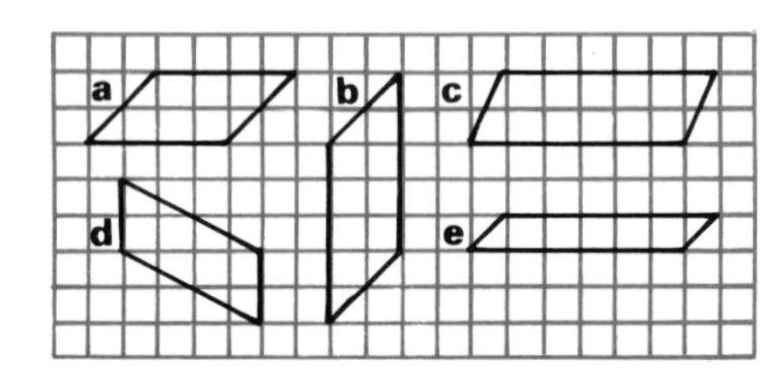

5 Calculate the areas of these parallelograms.

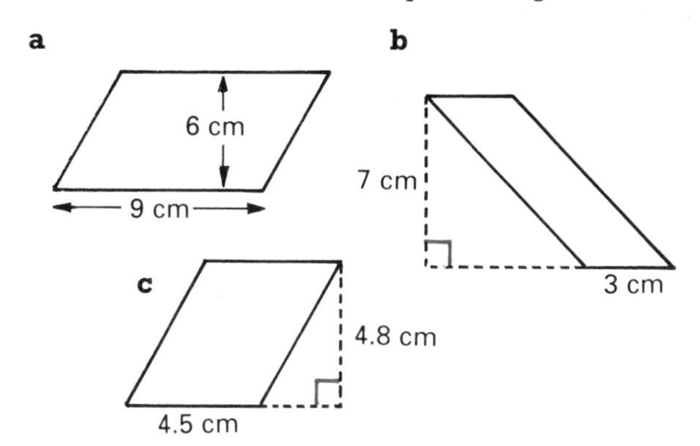

6 Use Pythagoras' Theorem to calculate the height of each parallelogram below; then calculate its area.

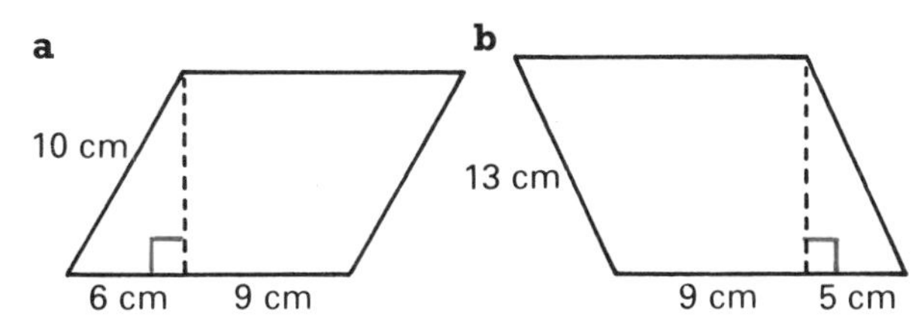

7 Calculate the area of each parallelogram side of this step ladder, which is 75 cm high.

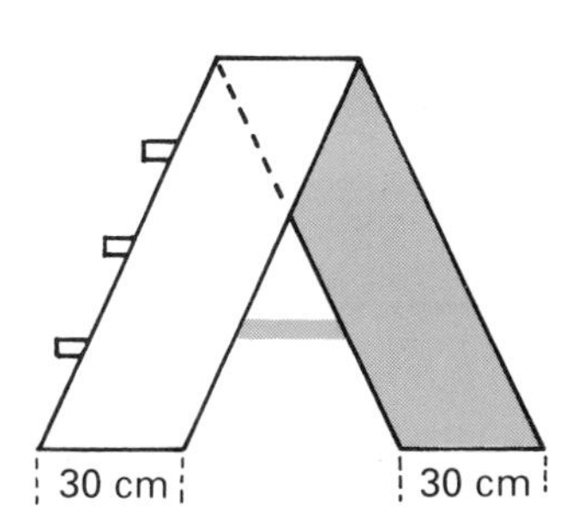

(iii) Trapezium

Area of trapezium ABCD

= area of $\triangle$ABC + area of $\triangle$ACD

$= \frac{1}{2}ah + \frac{1}{2}bh$

$= \frac{1}{2}(a+b)h$

Area of trapezium

= $\frac{1}{2}$ sum of parallel sides × height

8 ABCD is a trapezium. Calculate the areas of:

a $\triangle$s ABC and ACD

b the trapezium.

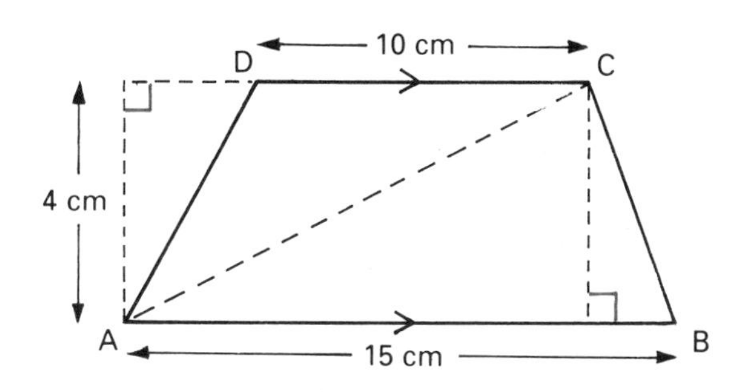

9 Calculate the area of each trapezium.

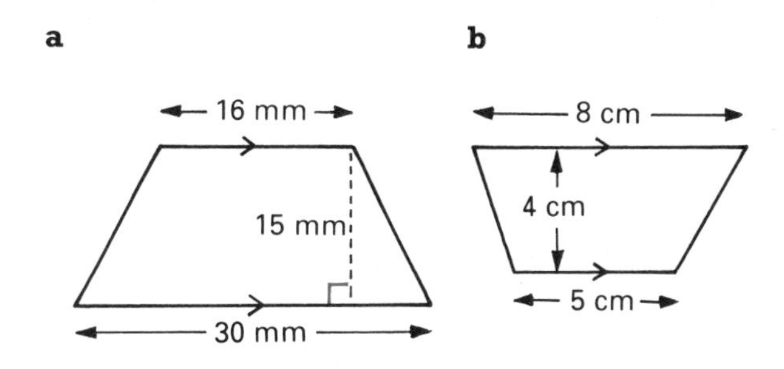

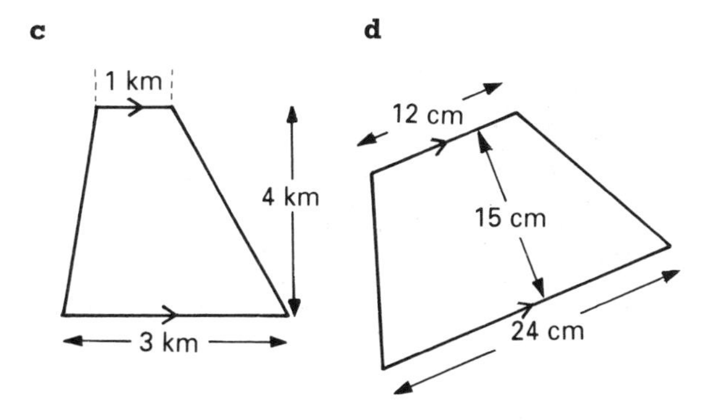

10 Calculate the areas of the end wall of the shed (including the window), and the side of the lampshade.

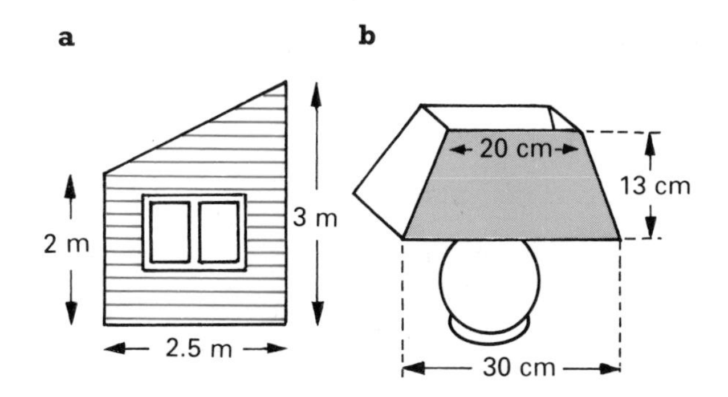

EXERCISE 2B

1 Calculate the areas of these quadrilaterals. Make sure that lengths are in the same units.

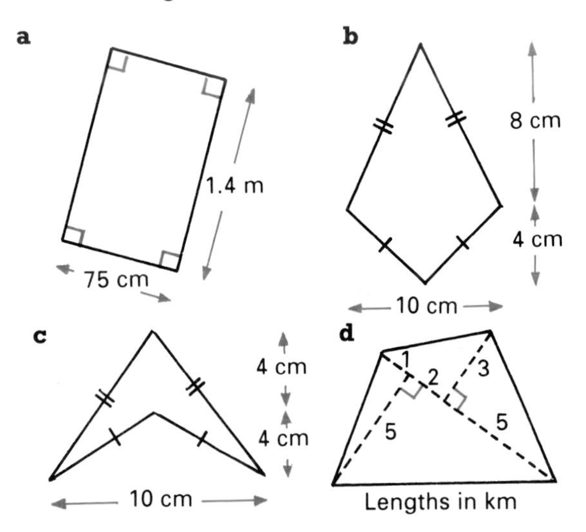

2 Use Pythagoras' Theorem to calculate h. Then calculate the areas of the shapes.

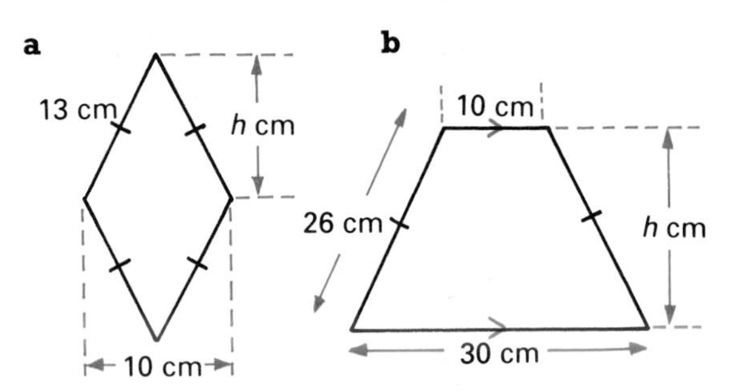

3 Use trigonometry to help you calculate the areas of the parallelogram and kite, correct to 1 decimal place.

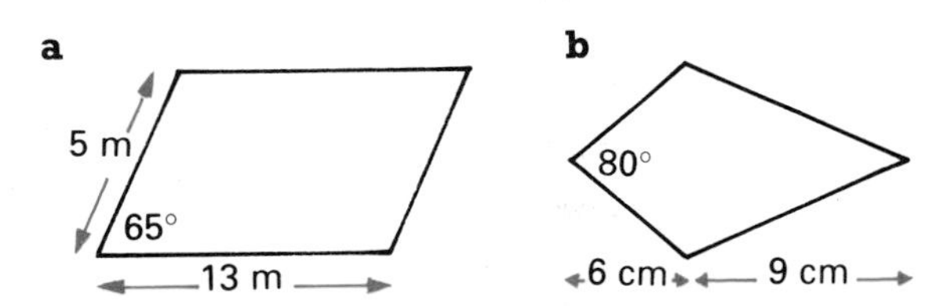

4 **a** Calculate the total area of the five rhombus-shaped stained glass panels.
b What percentage of the window is clear glass?

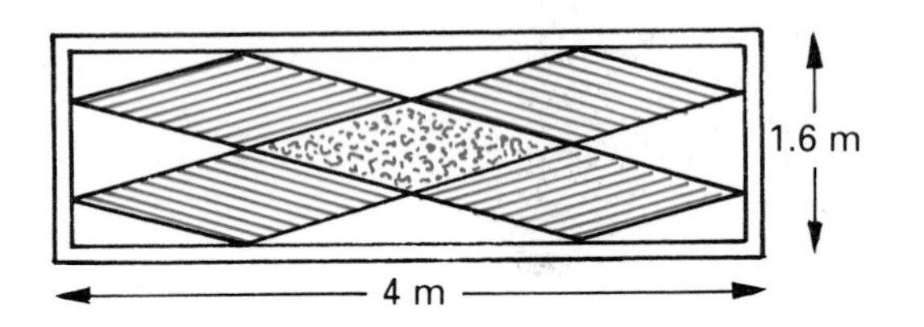

5 Write down formulae for the areas of these quadrilaterals:

a a parallelogram

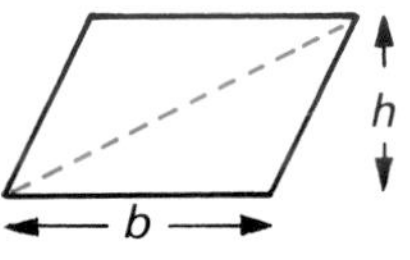

b a rhombus

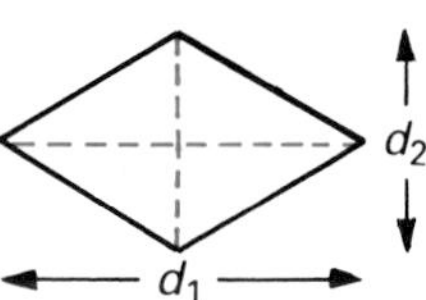

c a trapezium.

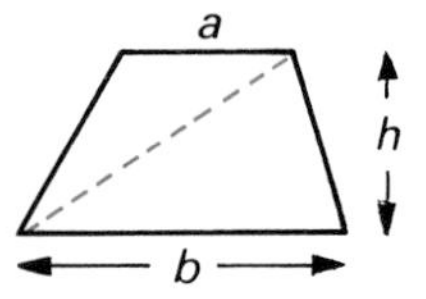

6 Find the areas of these shapes, but be careful—there is more data than you need.

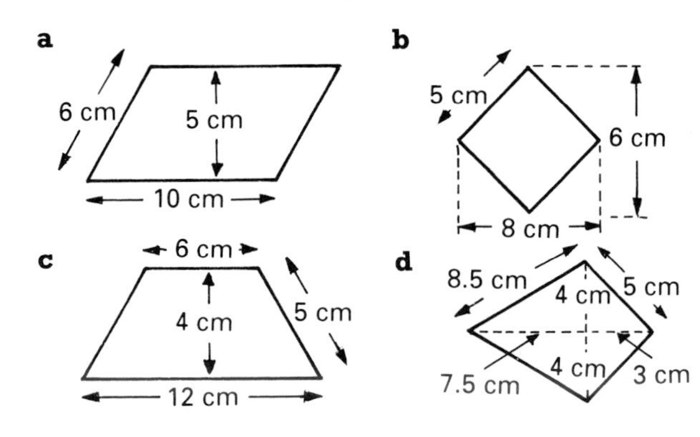

7 **a** As you approach the motorway roundabout the stripes get narrower. Why?
b Calculate the total painted (shaded) area. (Can you see a quick way?)

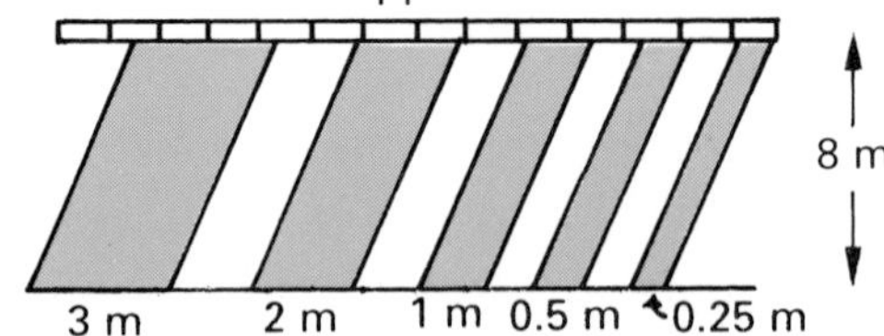

CHALLENGE

Sketch each of the following shapes with an area of 16 cm², and whole numbers of cm in the lengths of the sides:

a *a square* **b** *a rectangle* **c** *a rhombus*
d *a kite* **e** *a parallelogram*

PRISMS

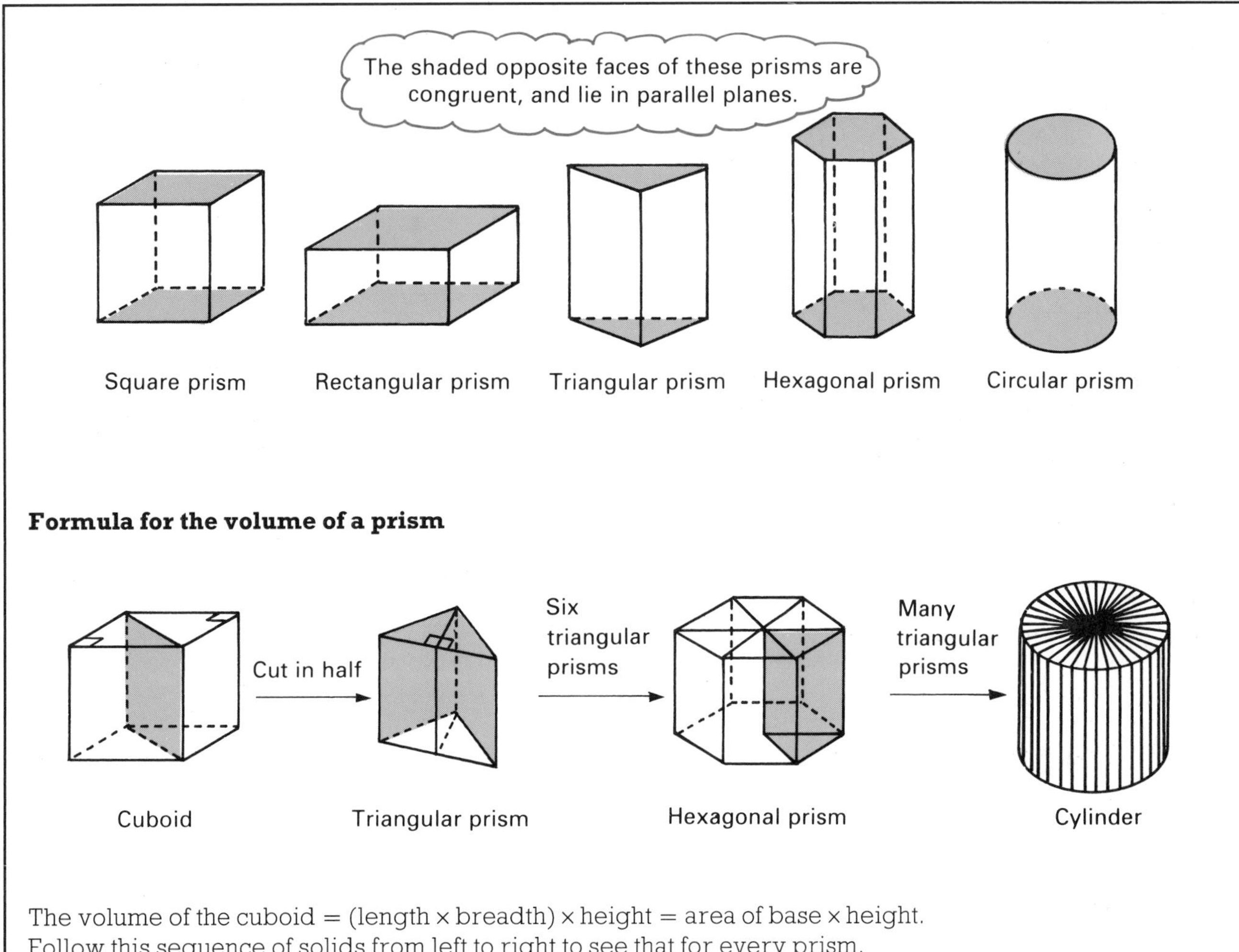

Formula for the volume of a prism

The volume of the cuboid = (length × breadth) × height = area of base × height.
Follow this sequence of solids from left to right to see that for every prism,

Volume = area of cross-section × height, or $V = Ah$

EXERCISE 3A

1 What is the usual name for a:
- **a** square prism
- **b** rectangular prism
- **c** circular prism?

2 Each small block is a cube or half-cube of side 1 cm. Calculate the volume of each prism structure **a–d** by:
(i) counting the blocks in it
(ii) using the formula $V = Ah$.

a

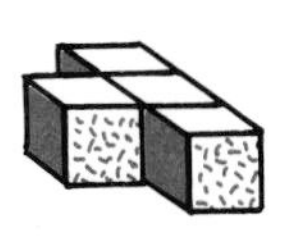

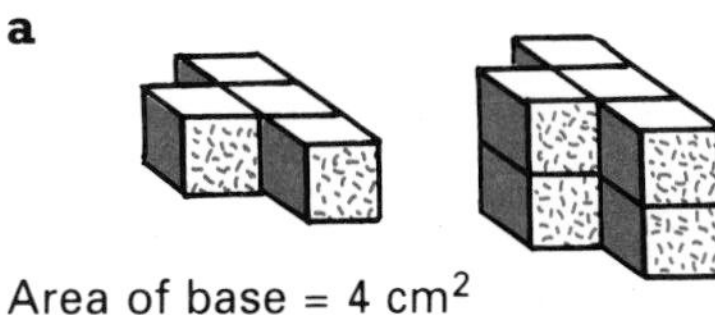

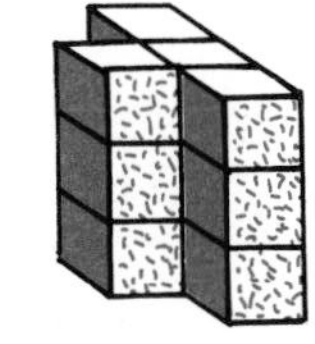

Area of base = 4 cm^2

b

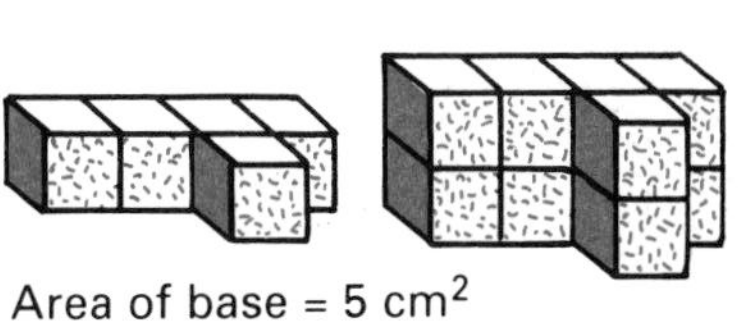

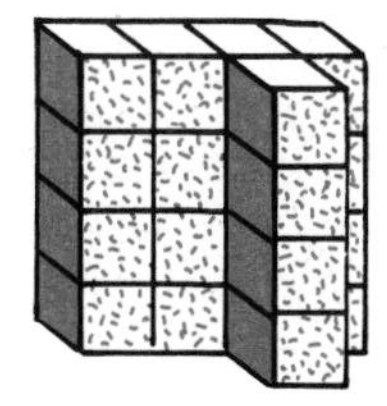

Area of base = 5 cm^2

c

Area of base = 0.5 cm^2

d

Area of base = 3.5 cm^2

3 Which of these are prisms (with congruent cross-sections)?

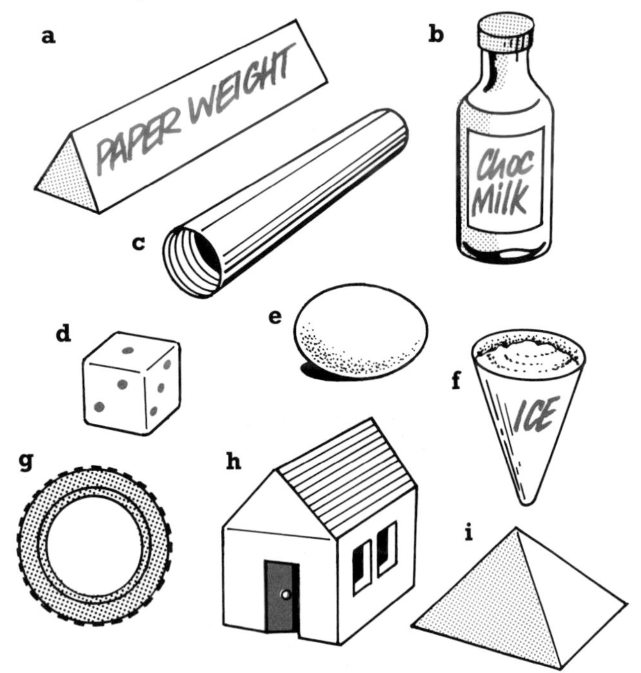

4 Calculate the volumes of these boxes and tins.

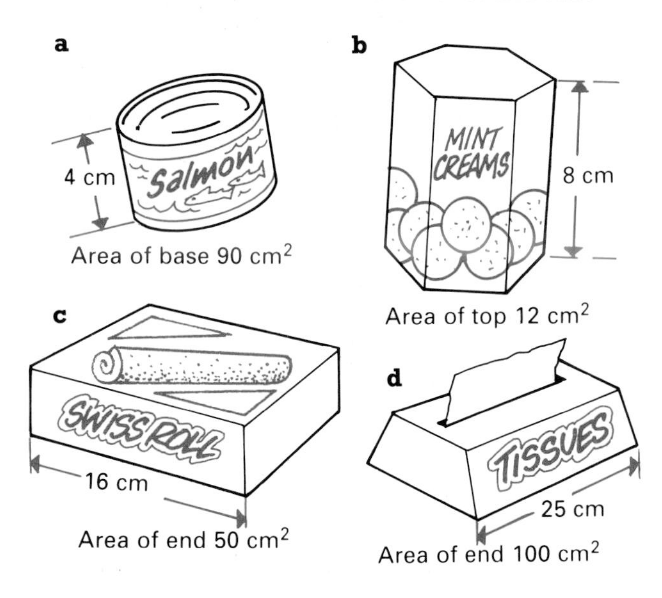

5 These wooden mouldings are 5 cm long, and are made from cuboids with square 1 cm by 1 cm cross-sections. Calculate:

(i) the area of the end facing you

(ii) the volume of each block.

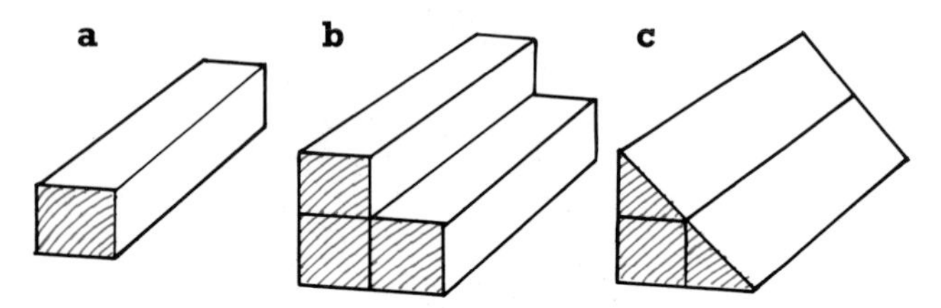

6 Sunshine Soaps sell bars of soap in various shapes. Calculate the volumes, to the nearest cm³.

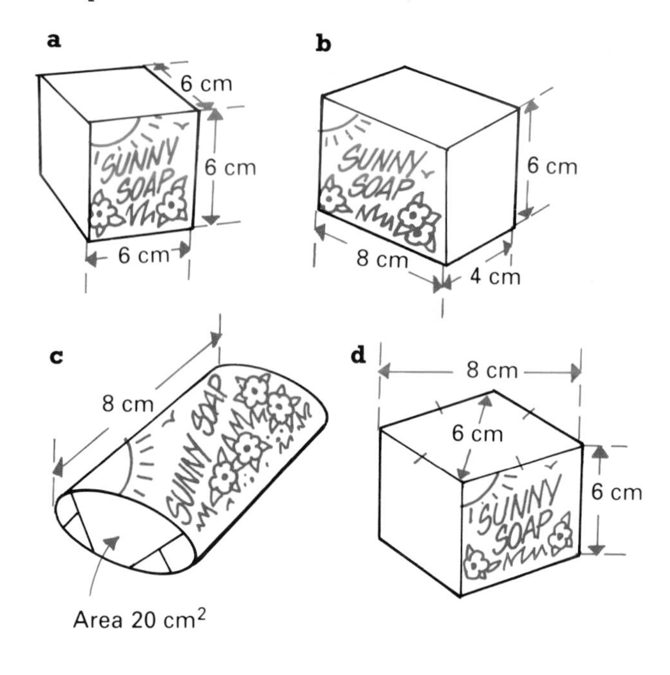

7 Calculate:

(i) the area of the top or bottom of each triangular prism

(ii) the volume of each prism.

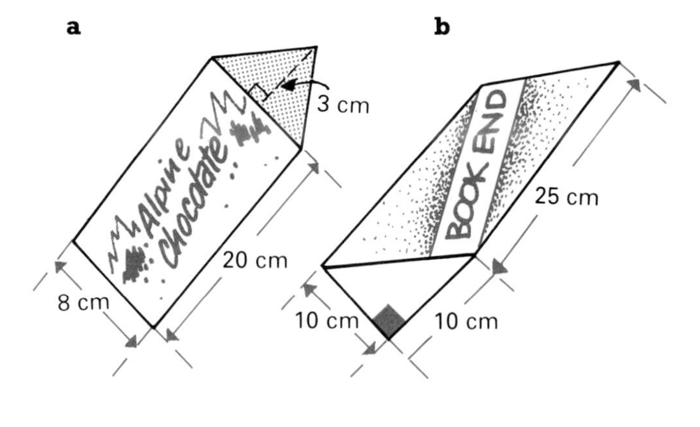

8 Calculate the volume of this garage.

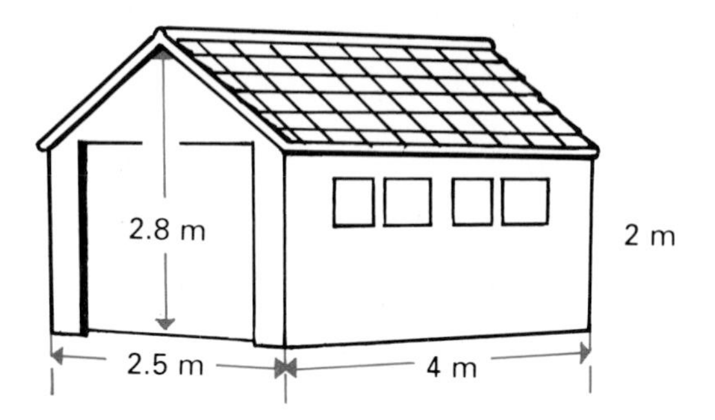

EXERCISE 3B

1 Calculate the volume of each prism. Notice that in $V = Ah$, 'A' means 'area of congruent cross-sections' and 'h' can be the 'length' of the prism. Units are metres.

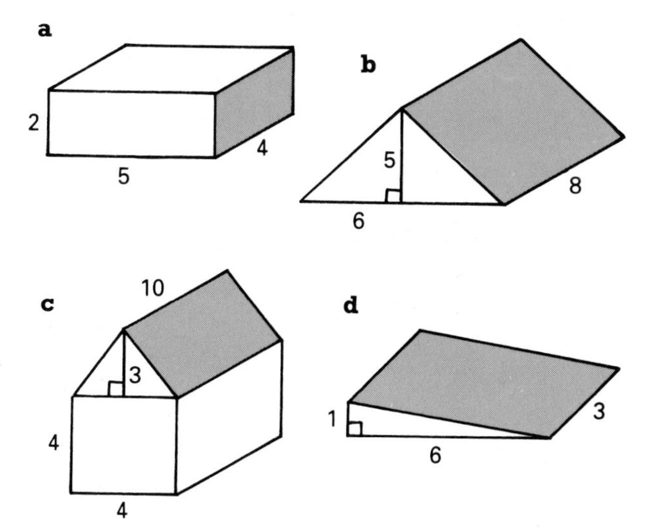

2 **a** Which solid can be made from this net?
b Calculate:
(i) its total surface area
(ii) its volume.

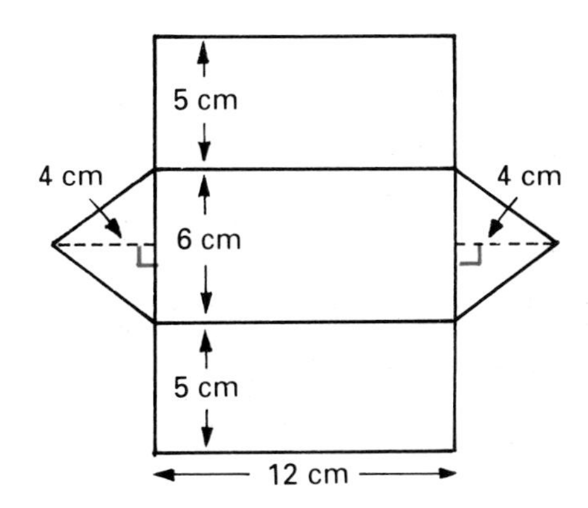

3 Calculate:
a the volume of the tent
b the length of a sloping edge
c the total area of canvas used in the tent.

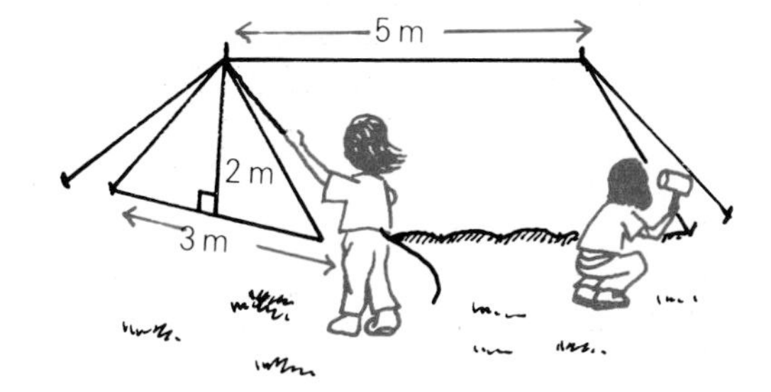

4 This concrete buttress supports the wall. Calculate the volume of concrete needed to make it.

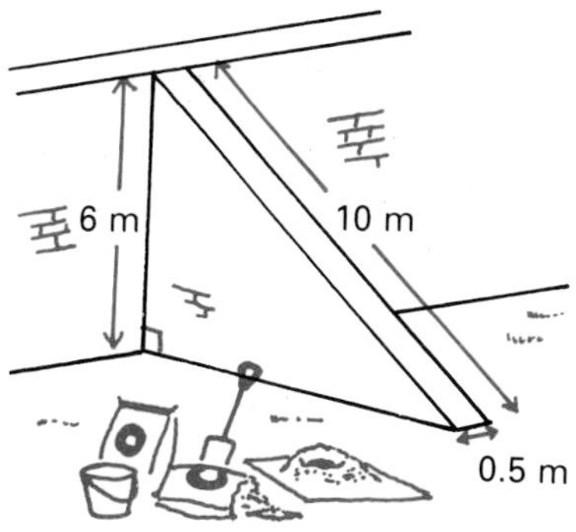

5 The cross-section of a car's petrol tank is a trapezium. Calculate, in litres, the volume of petrol in a full tank.

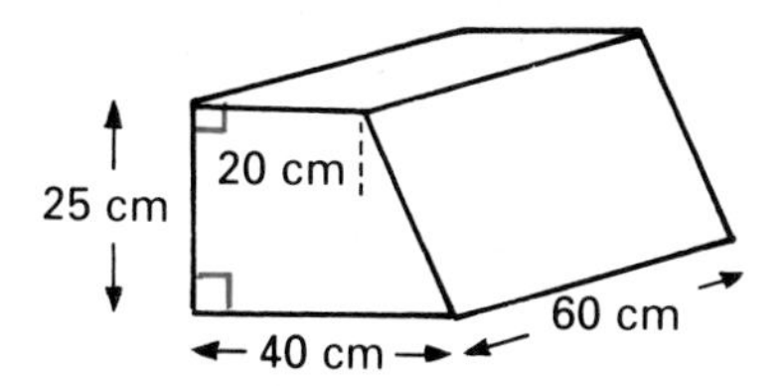

6 **a** List the dimensions of all the pieces of wood needed for this corner cupboard.

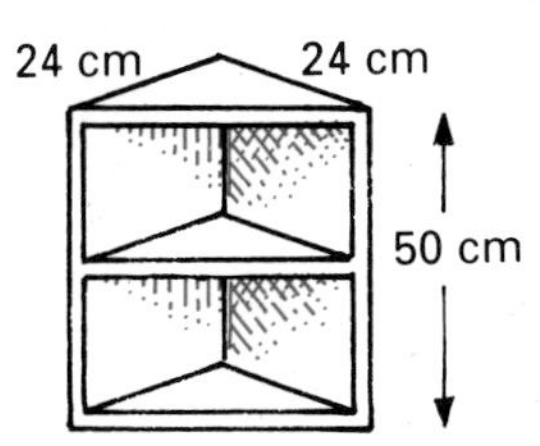

b Calculate:
(i) the total area of wood required
(ii) the volume of the cupboard
(iii) the width of the front, to the nearest cm
(iv) the area of glass needed for a front door.

7 The roof of this lean-to shed slopes up at an angle of 17° to the horizontal. Calculate:
a the height of the back wall, to the nearest cm
b the volume of the shed.

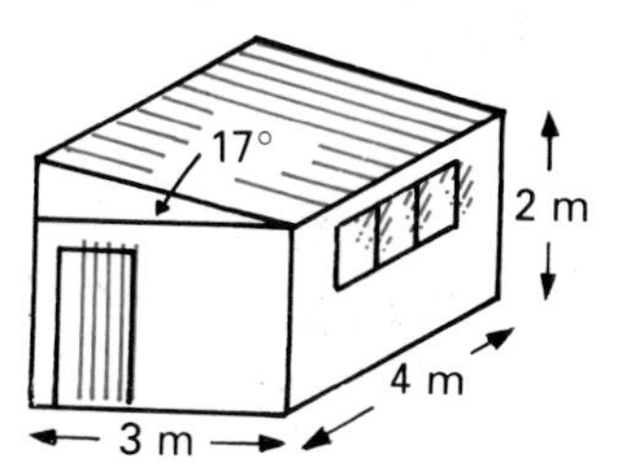

INVESTIGATION

1 *A design problem. First of all calculate the volumes of these solids made from cuboids.*

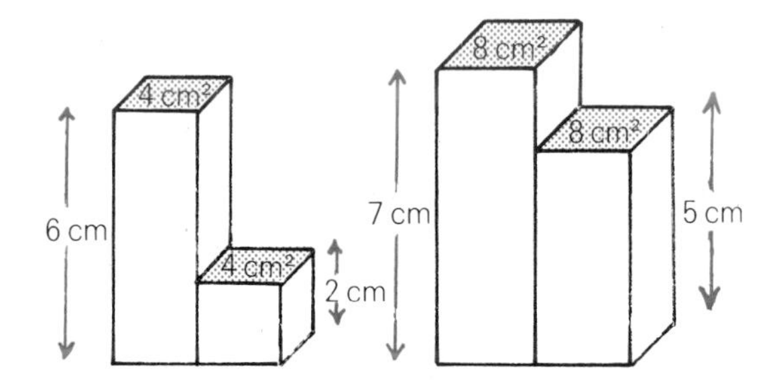

2 *A skin-care company has developed a sun-tan lotion in two strengths. The company decides to sell them in two-part containers, one holding twice as much as the other. Their designer Joan was asked to design a suitable container.*

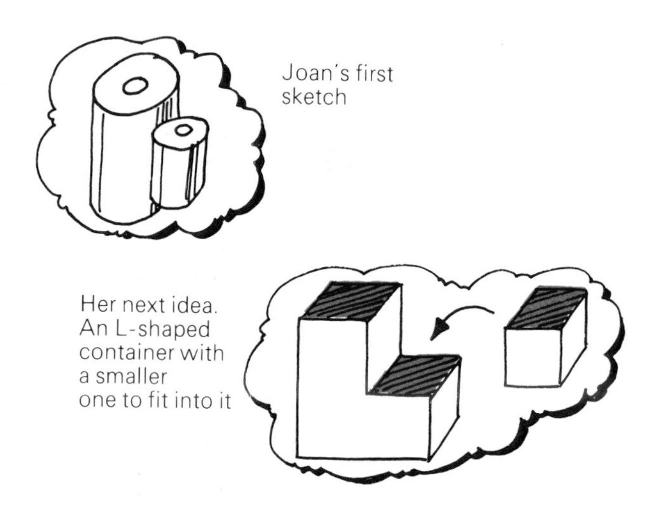

Joan's problem is now mathematical. The larger container has to hold twice as much as the smaller one. She decided that the cross-section of each part should be square, with area 6 cm^2.

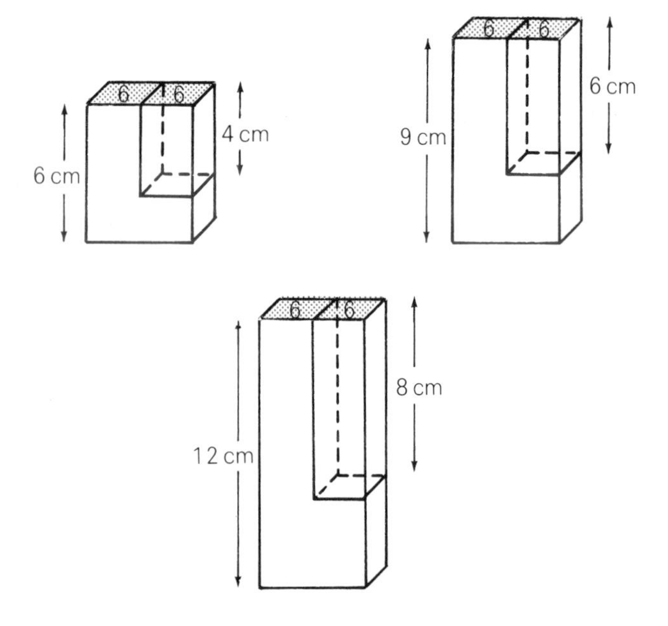

a *Calculate the volumes of the two bottles in each model.*
b *Do you see the connection between the heights of the bottles in each pair?*
c *If the height of the L-shaped bottle is x cm, and the height of the other bottle is y cm, make an equation connecting x and y.*
Joan streamlined her design, with the two bottles fixed together.

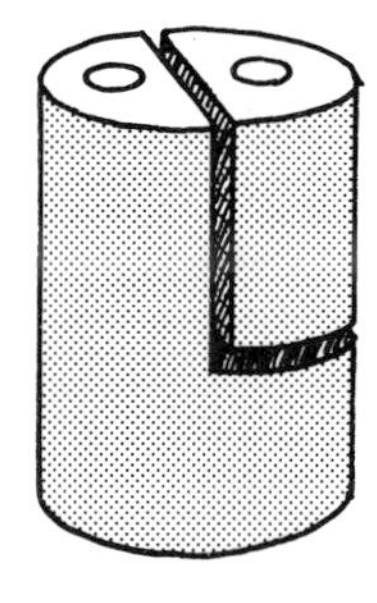

3 *Try designing pairs in which one volume is:*
a *three times* **b** *four times, the other.*

CYLINDERS ARE SPECIAL

(i) Volume of a cylinder

Cylinders have their own formulae.

For every prism, $V = Ah$

So for a cylinder, $V = \pi r^2 h$

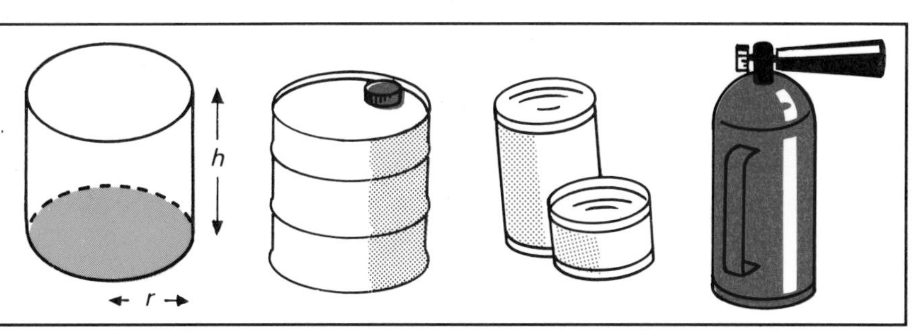

EXERCISE 4A

1 Calculate the volume of each cylinder.

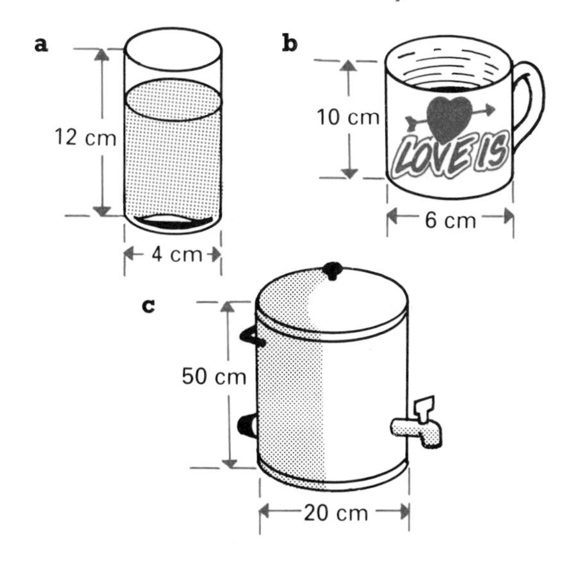

2 Which jar holds more cream? How much more, to the nearest cm³?

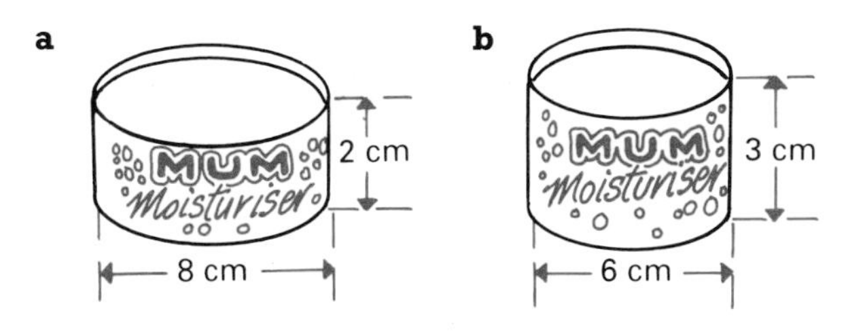

3 Calculate (in litres) the volume of the oil tank for the Jeffreys' central heating. (1 m³ = 1000 litres)

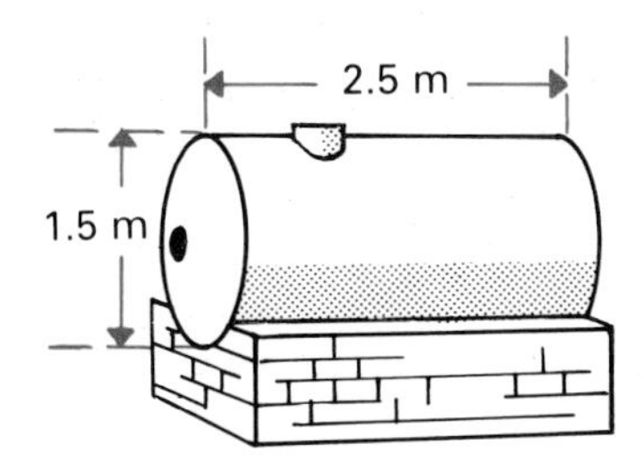

4 The end of the storage barn consists of a rectangle and a semi-circle. Calculate:

a the area of each end

b the volume of the barn.

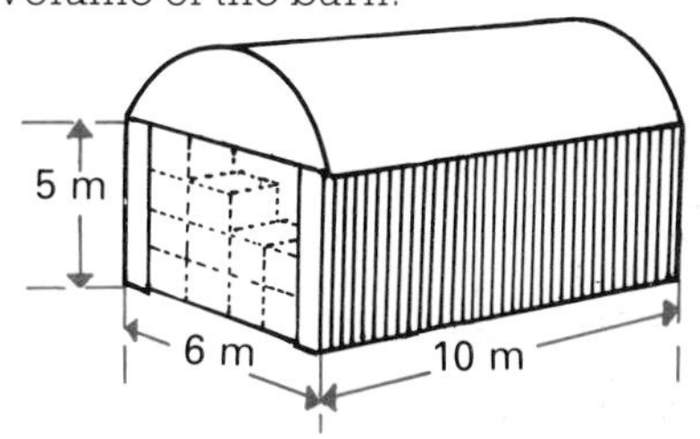

5 The cylindrical jug is full of raspberry juice which is poured into rectangular moulds to make raspberry ice bars. How many bars can be made?

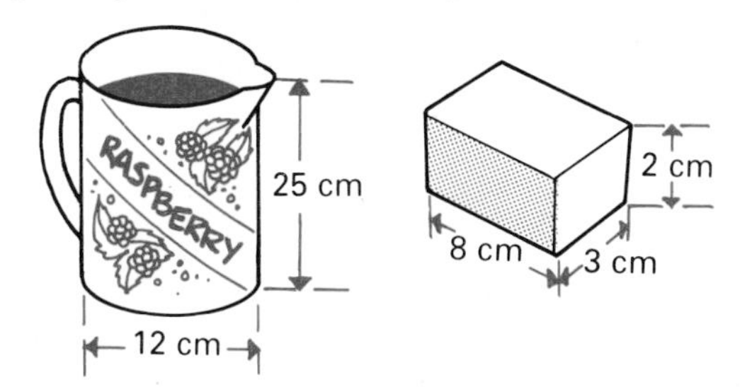

6 These grain storage silos are 4.5 m high, and 3.2 m in diameter. The lorry's load space is 4 m by 2.5 m by 1 m.

a What total volume of grain can the three silos hold?

b How many journeys must the lorry make to remove all the grain?

EXERCISE 4B

1 Tins of peaches are packed in boxes like this.

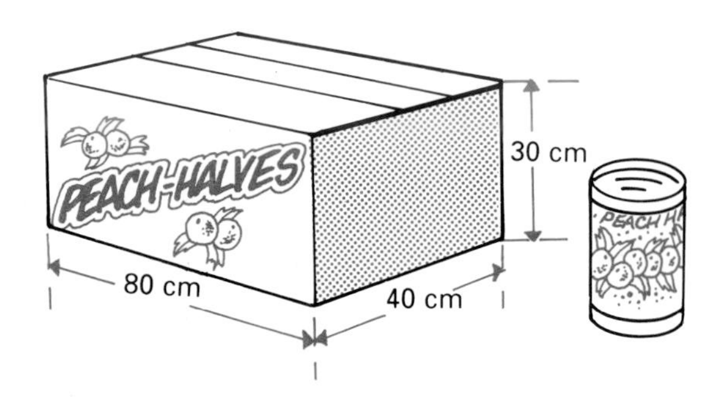

Each tin has diameter 8 cm and height 10 cm.

a How many tins can be packed into the box?

b What percentage of space in the box is not used (to the nearest 1%)?

2 20 m of lead pipe has to be removed from Newland Nursery School. The cross-section of the pipe is bounded by two concentric circles of radii 2.5 cm and 2.7 cm.

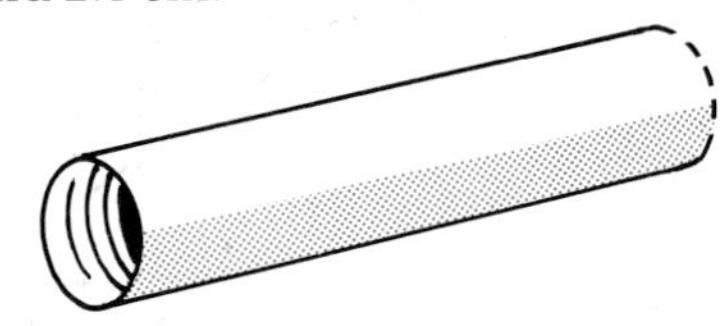

Calculate:

a the area of cross-section of the pipe

b the volume of lead to be removed

c the weight of the lead, to the nearest kg (1 cm^3 weighs 11.6 g).

3 **a** Leaving π in your answers, calculate the volume of:

(i) the cylinder shown

(ii) a cylinder with twice the height

(iii) a cylinder with twice the radius.

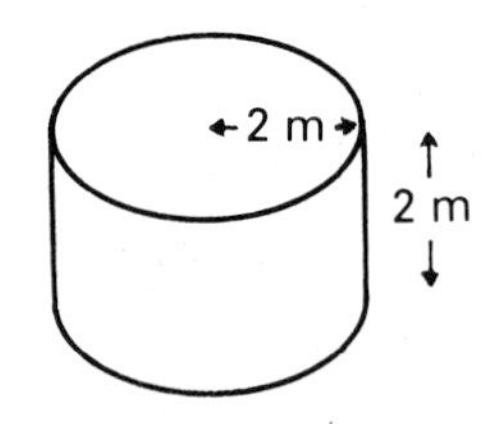

b Find the ratios of volume (i) to volume (ii), and volume (i) to volume (iii).

4 Two cylinders each have a volume of 500 cm^3.

a The height of one is 20 cm. Calculate its diameter.

b The diameter of the other is 20 cm. Calculate its height.

5 The manufacturer of Fruity Juice decides to increase the volume of the can by 25%. How can he do this by:

a increasing the height only

b increasing the radius only?

PRACTICAL PROJECTS

1 *Make the necessary measurements, and calculate the volume of a pencil, a coin, a piece of chalk or a CD. Then find its density (the weight of 1 cm^3 of the material).*

2 *Calculate the volumes of some cylindrical cans, and compare your answers with the values given on the labels. (Remember that 1 ml = 1 cm^3.)*

BRAINSTORMER

You can make two different open-ended cylinders with a rectangular sheet of paper.

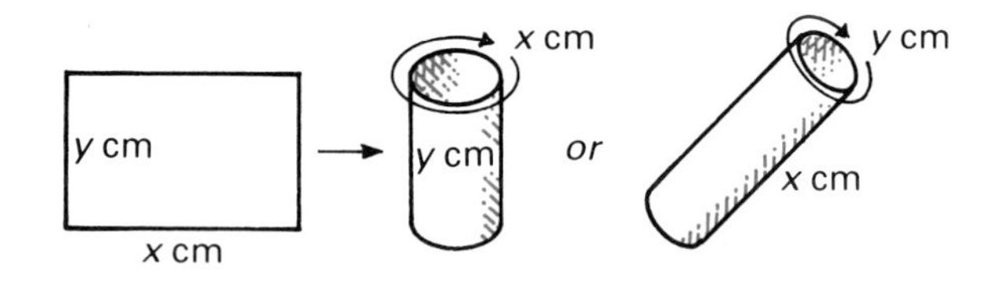

a *Do you think that the volumes are equal?*

b *Find a formula for each volume.*

c *Find the ratio of the volumes when:*

(i) $x = y$ (ii) $x = 2y$ (iii) $x = 3y$ (iv) $x = ny$.

CHALLENGE

Raincatcher Company use rectangular sheets of metal which are made into rain gutters with square or semi-circular ends. Which shape would catch more rain? How much more?

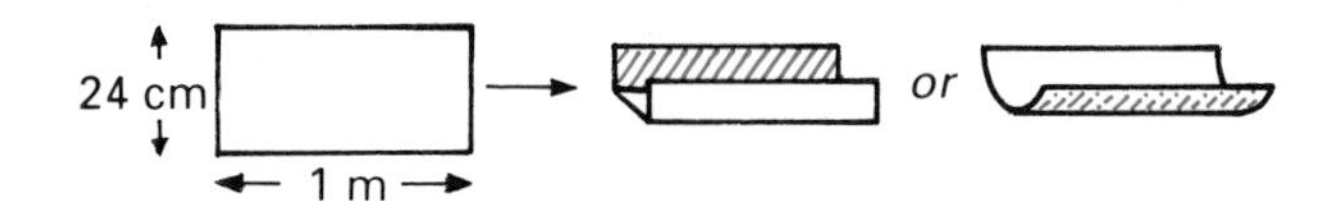

(ii) Surface area of a cylinder

The label round the side of the tin is rectangular, so the curved surface area of the cylindrical tin is
$\pi d \times h = \pi dh$

Curved surface area $A = \pi dh$

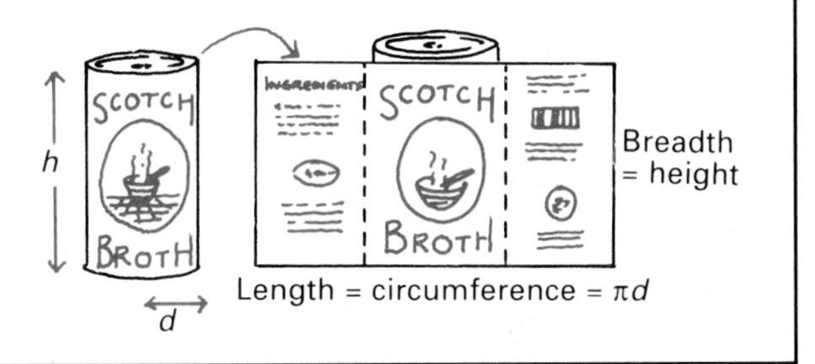

EXERCISE 5

1 Calculate the area of each label, given its length and breadth.

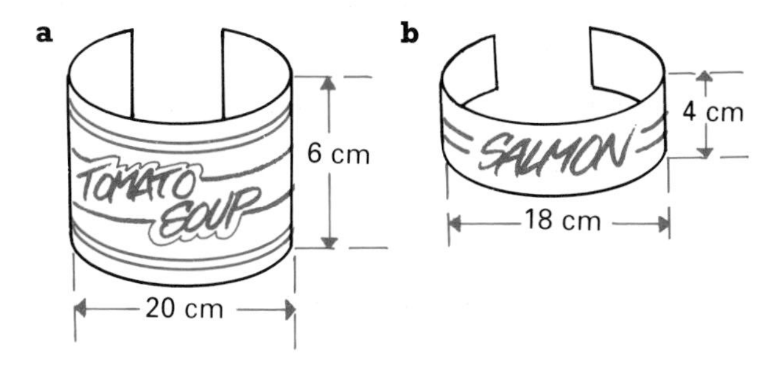

2 Calculate the area of each label, given the diameter of the base of the tin.

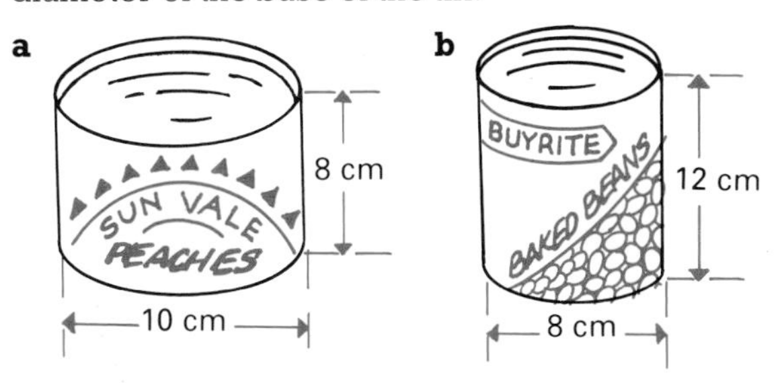

3 Calculate the total surface area (top, bottom and curved surface) of this tin.

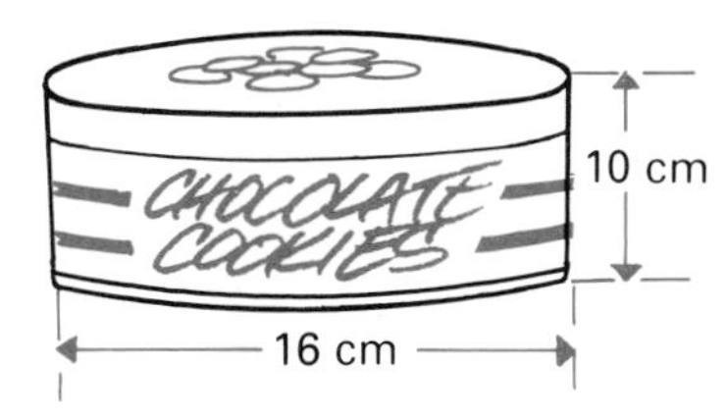

4 The tin of soup is unopened, and the pan is open at the top. Calculate each of their:
(i) outside surface areas
(ii) volumes.

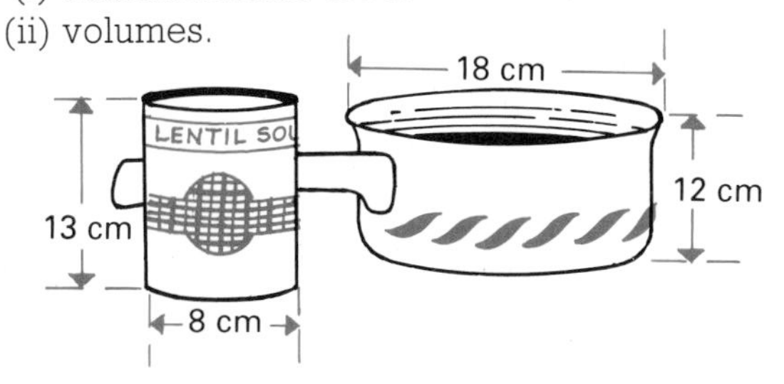

5 Write down a formula for:
a the area of the curved surface of the can
b the area of the top of the can plus the area of the bottom of the can
c the total surface area of the can—top, bottom and side.

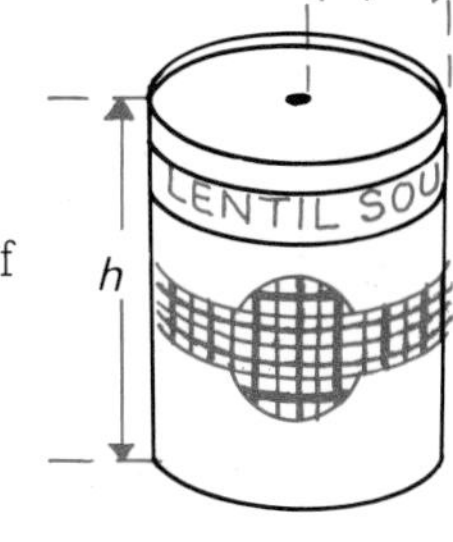

6 Three formulae for the surface areas of open and closed cylinders:
a $A = \pi dh + 2\pi r^2$
b $A = \pi dh + \pi r^2$
c $A = \pi dh$
Which formula would you use for each object?

7 Tom Brown is rolling the local cricket pitch.

a How many turns does the roller make if he rolls a length of 20 m?
b What area (in m^2) has he rolled after 10 turns of the roller?
c The roller is full of sand to give it weight. What volume of sand (in m^3) is in the roller?

CHALLENGE

Calculate the total surface area of metal in this washer.

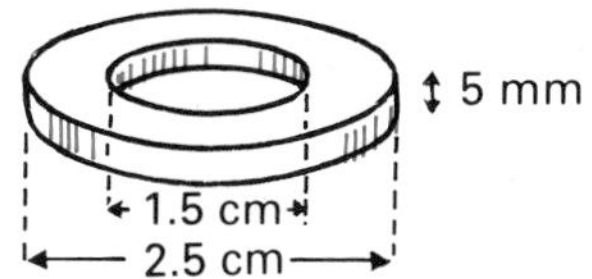

DIMENSIONS: LENGTH, AREA AND VOLUME

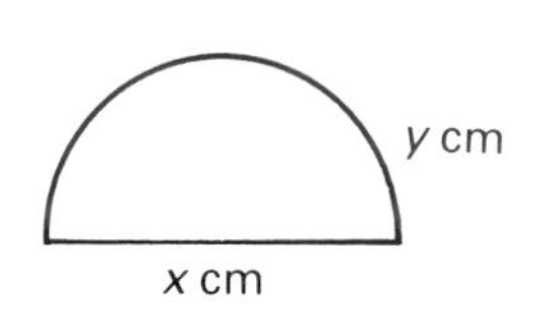

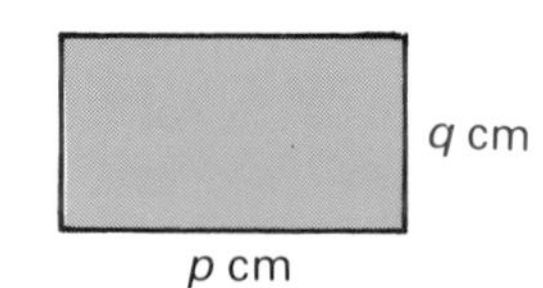

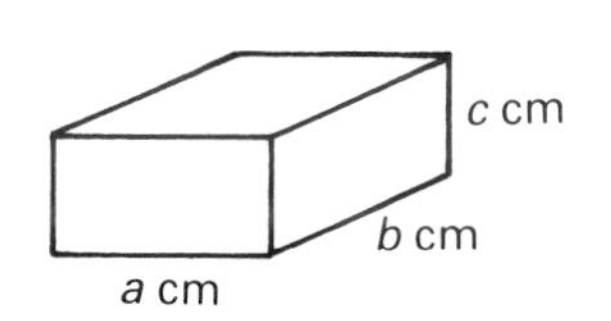

Perimeter P cm (1 dimension)
$P = x + y$ (sum of lengths)

Area A cm^2 (2 dimensions)
$A = pq$ (product of two lengths)

Volume V cm^3 (3 dimensions)
$V = abc$ (product of three lengths)

EXERCISE 6

In this exercise, $a, b, c, \ldots, u, v, w$, etc., represent measurements.

1 Write down formulae for the perimeters (P cm), surface areas (A cm^2) and volumes (V cm^3) of the cube and cuboid.

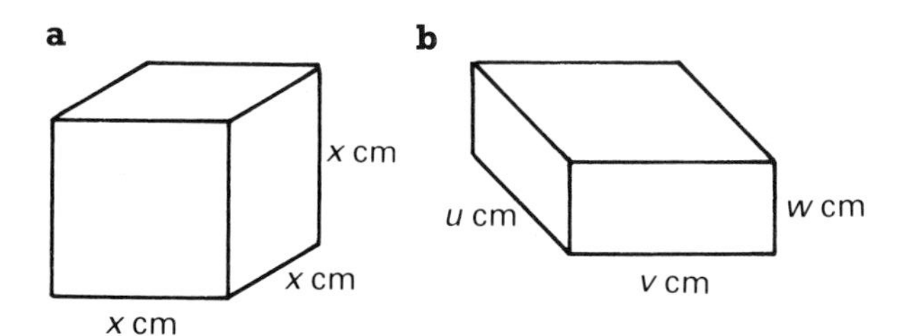

2 These are formulae for circles. How can you tell at a glance which are for areas and which are for perimeters?
a $M = \pi r^2$ **b** $N = 2\pi r$ **c** $W = \pi d$ **d** $X = \frac{1}{4}\pi d^2$

3 Which formulae below represent perimeters, areas or volumes?
a $P = a + b + c$ **b** $Q = a^3$
c $R = b + c$ **d** $S = bc$

4 These are formulae for the volume, area and circumference of this sphere. Which is which?
a $B = \frac{4}{3}\pi r^3$ **b** $K = 2\pi r$ **c** $S = 4\pi r^2$

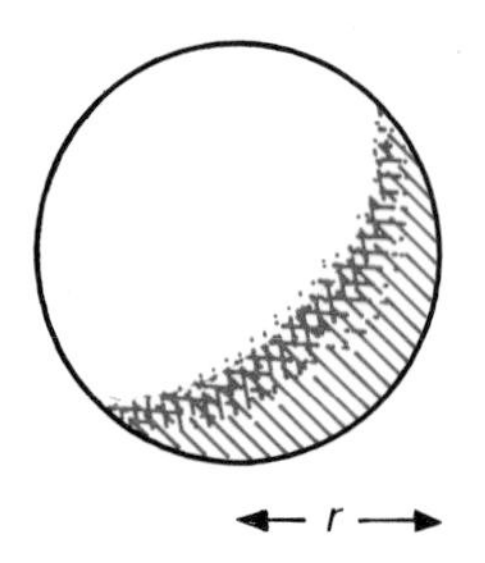

5 These are formulae for the volume, area and distance from apex to base, round base and back to apex, of this cone. Which is which?
a $R = \pi rs + \pi r^2$ **b** $S = \frac{1}{3}\pi r^2 h$ **c** $T = 2s + 2\pi r$

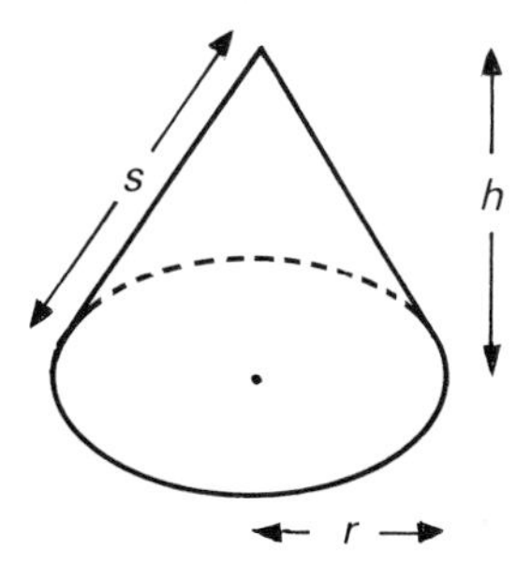

6 List these formulae under the headings perimeter, area, volume.
$G = a^2$, $H = b^3$, $J = pqr$, $K = f + g$, $L = \pi x^2$, $M = \pi y$, $N = p - q$, $P = x^2 y$, $R = mn + uv$

7 a Write down formulae for the perimeter (P cm), surface area (A cm^2) and volume (V cm^3) of the prism below. The sides are rectangles and right-angled triangles. Lengths are in cm.
b Check that the dimensions of your formulae (1, 2 and 3 respectively) are correct.

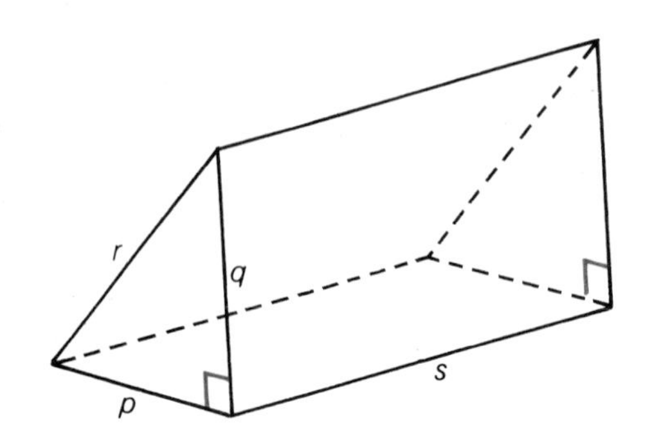

CHECK-UP ON AREAS AND VOLUMES

1 Calculate the area of each shape.

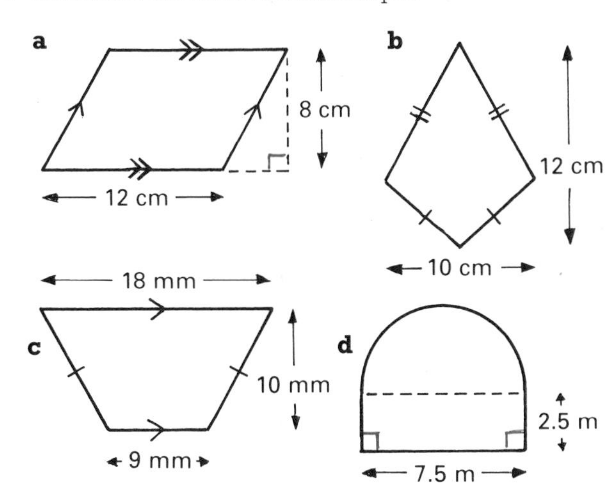

2 Calculate the volume of each prism.

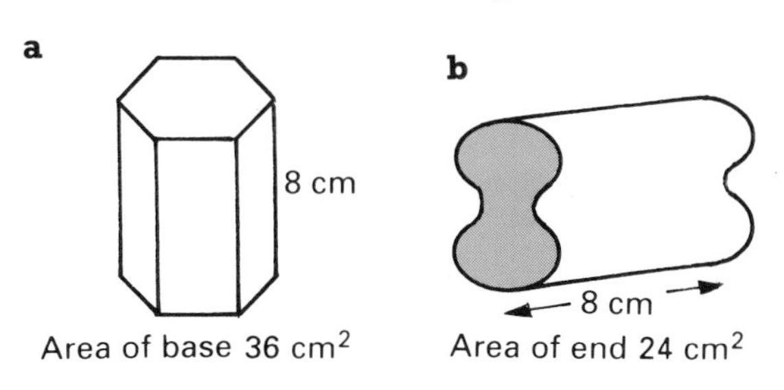

3 Calculate:

a the height of the slope, using Pythagoras' Theorem

b the area of the largest triangle.

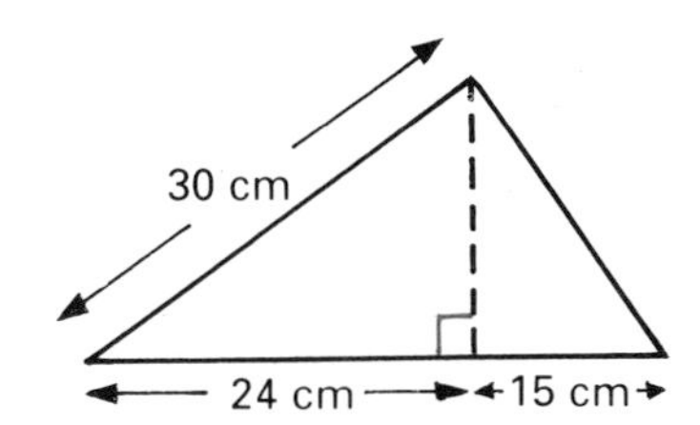

4 The ramp slopes up at 10° to the horizontal, leading to a horizontal car park. Calculate:

a the height of the car park

b the area of the wall supporting the ramp and the car park.

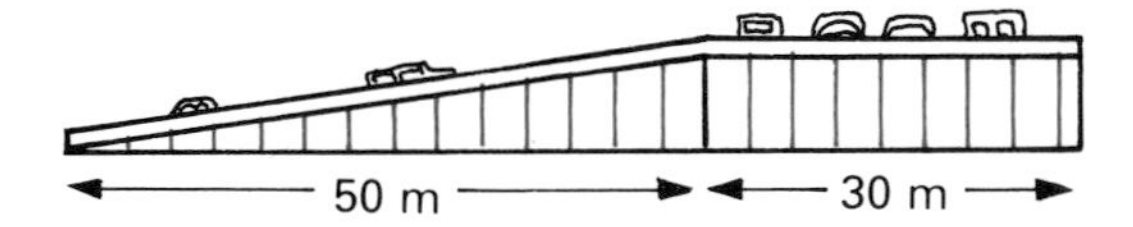

5 Calculate the volume of this wedge of cheese before the mouse started eating it.

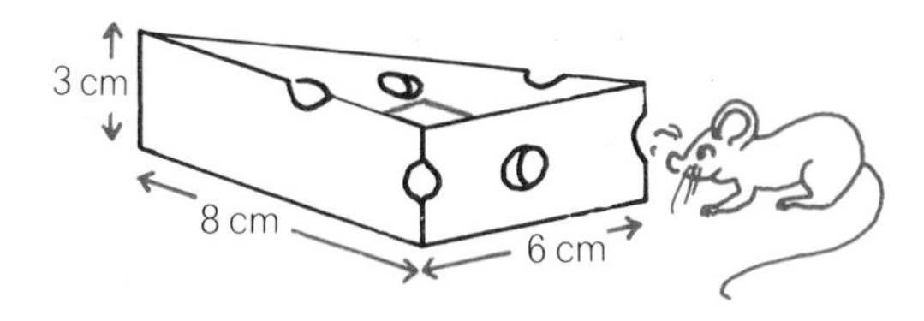

6 **a** Draw a net for this triangular prism.

b Calculate the surface area and volume of the prism.

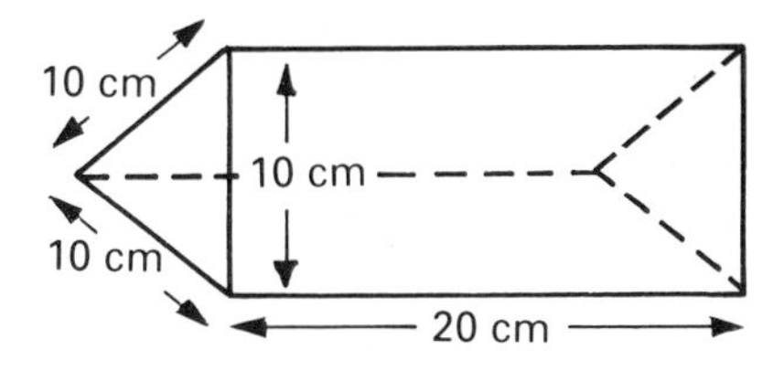

7 In a geometry set the protractor is a semi-circle with diameter 10 cm, and the set-square has sides about the right angle 7.5 cm and 12.6 cm long. Both are 1 mm thick. Calculate the volume of plastic in each, in cm³.

8 Calculate the volume and surface area of each solid.

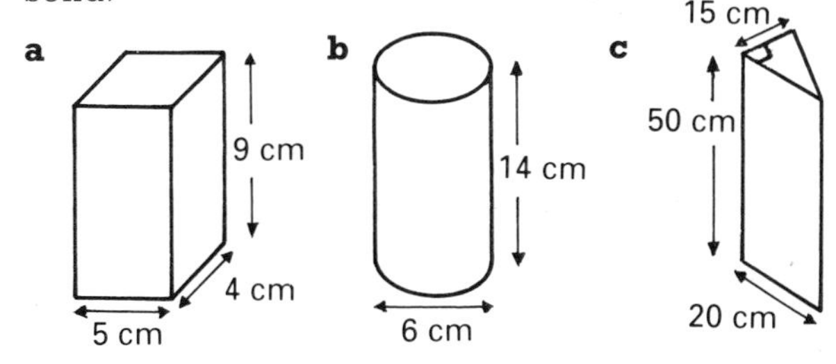

9 **a** Show that 1 m³ = 1000 litres.

b Calculate the weight of water (in tonnes) in the swimming pool, given 1 litre weighs 1 kg and 1 tonne = 1000 kg.

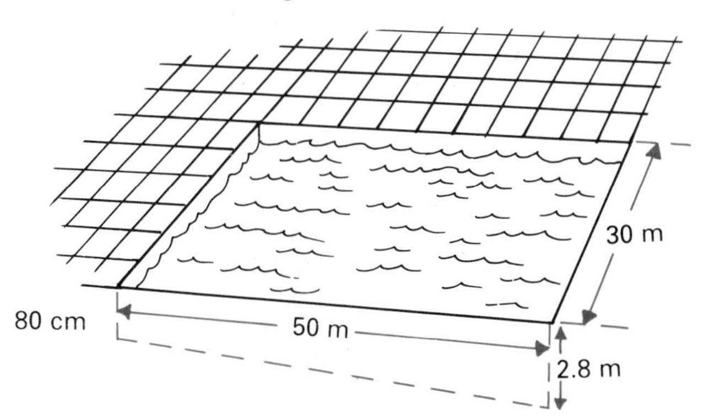

FACTORS

LOOKING BACK

1 $6 = 6 \times 1 = 3 \times 2$
In the same way, write down *all* the pairs of factors of:
a 8 **b** 9 **c** 10 **d** 11 **e** 12

2 Which number, apart from 1, is a factor of:
a 4, 6 *and* 8 **b** 10, 15 *and* 20?

3 $2t = 2t \times 1 = 2 \times t$
Write down all the pairs of factors of:
a $3m$ **b** x^2

4 Calculate:
a $36 \div 9$ **b** $y^2 \div y$ **c** $2p \div p$ **d** $4x^2 \div 2x$

5 Write down the first four terms of sequences with nth terms:
a $2n+4$ **b** $3n-3$

6 In these gardens, write down the area (in m²) of:
(i) the grass
(ii) the flowers
(iii) the whole garden.

a

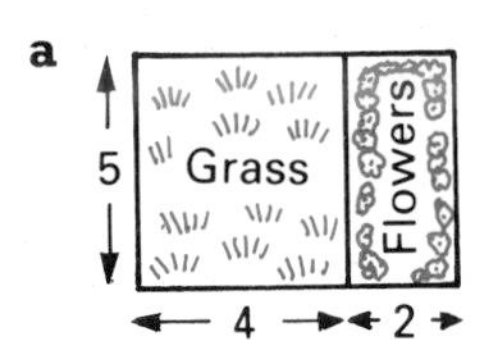

b

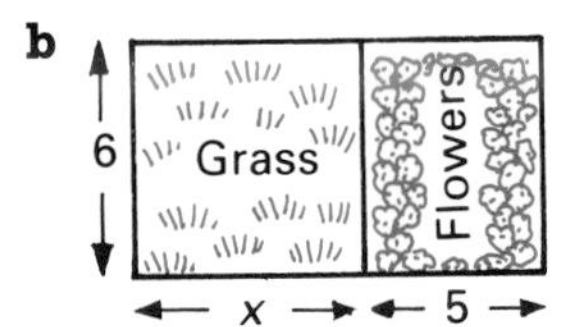

7 Multiply out:
a $3(x+1)$ **b** $4(y-2)$ **c** $5(2n+1)$
d $6(1-m)$ **e** $a(b+c)$ **f** $x(x-y)$

8 Multiply out, then simplify:
a $2(x+1)+3(x-1)$ **b** $5(y-2)-2(y-5)$

9 Write down two expressions for the area of each garden, one with brackets and the other without brackets.

a

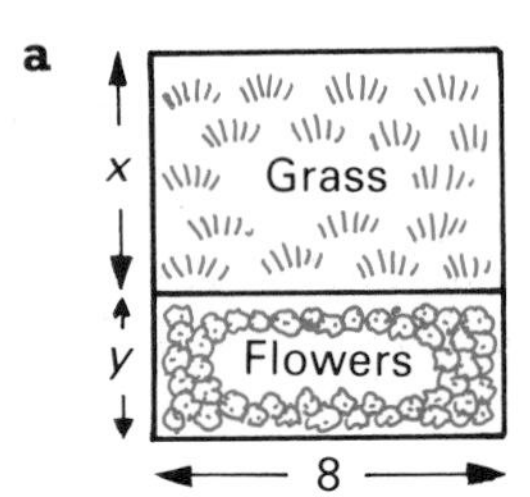

b

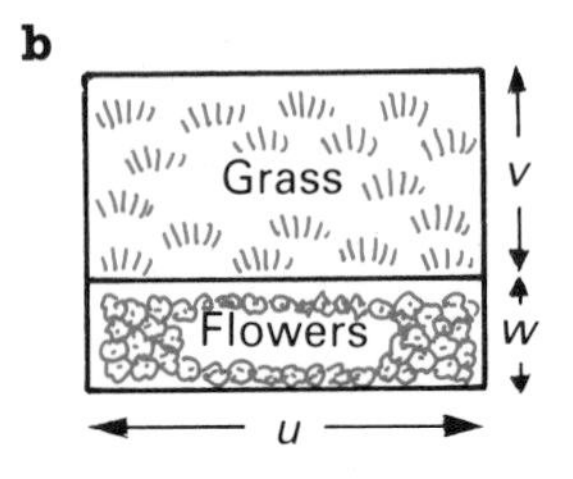

10 Multiply:
a $(x+2)(x+3)$ **b** $(y-1)(y-4)$
c $(3z+1)(2z-1)$ **d** $(5u-2)(5u+2)$
e $(v+5)^2$ **f** $(2w-1)^2$

11 a Find a formula for the nth term of each sequence:
(i) 1, 3, 5, 7, . . . (ii) 4, 7, 10, 13, . . .
b Use the formula to find the twentieth term. Then check by continuing each sequence.

CHALLENGE

The dominoes can be arranged in this square pattern:

Domino 1 is in the left-hand corner

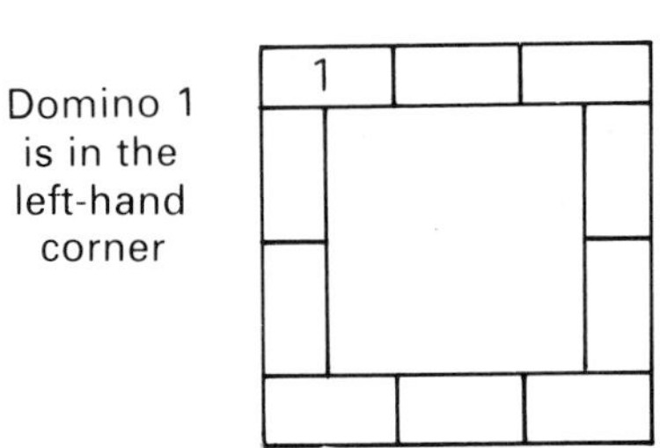

Where the dominoes touch, the expressions are equal

Domino 1

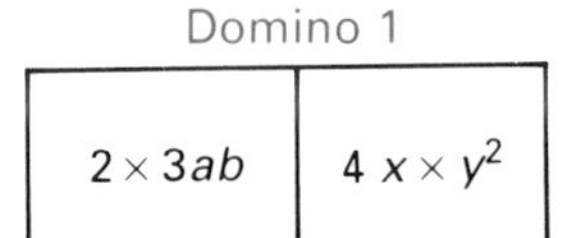

Domino 2

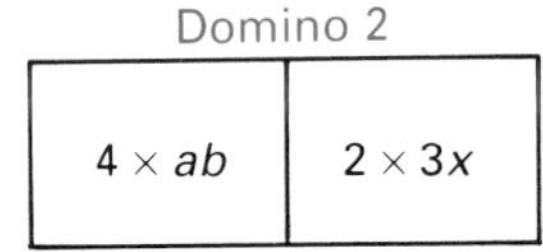

Domino 3

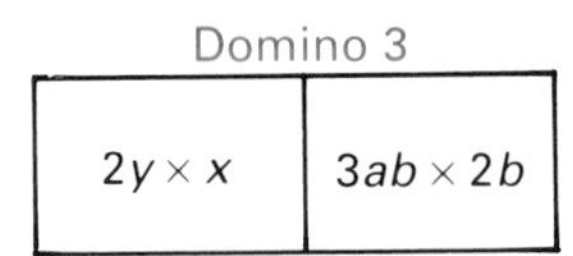

Domino 4

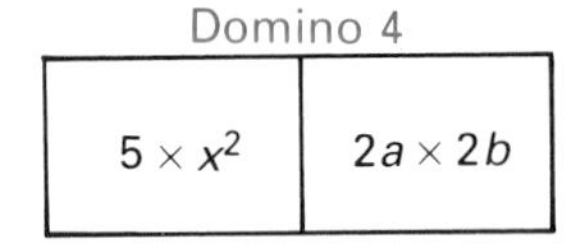

Domino 5

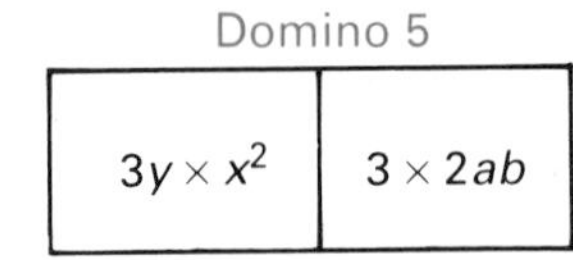

Domino 6

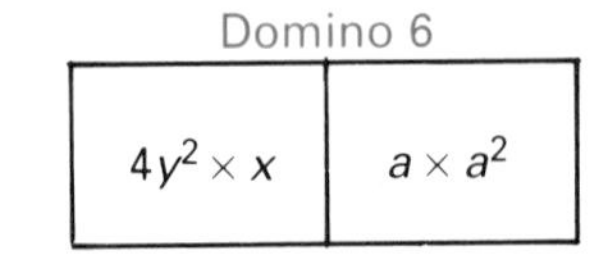

Domino 7

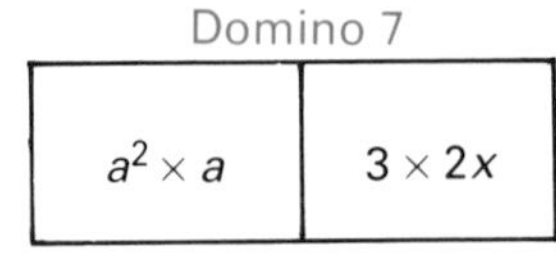

Domino 8

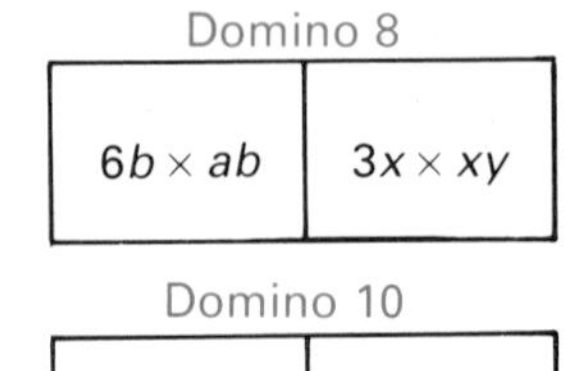

Domino 9

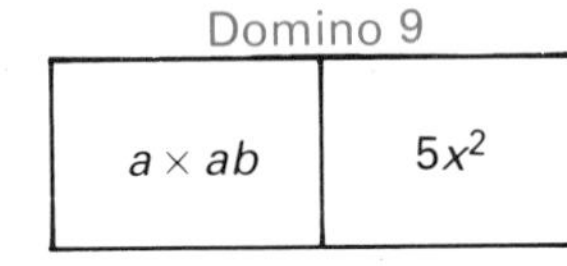

Domino 10

$2x \times y$	a^2b

Copy the square pattern, and fill in the rest of the domino numbers.

FACTORS AND HIGHEST COMMON FACTORS

All these cards can go into the envelope, as the numbers are **factors** of 6.
They divide into 6 exactly.

1 2 3 6
THE FACTORS OF 6

Every factor has a partner, like this:

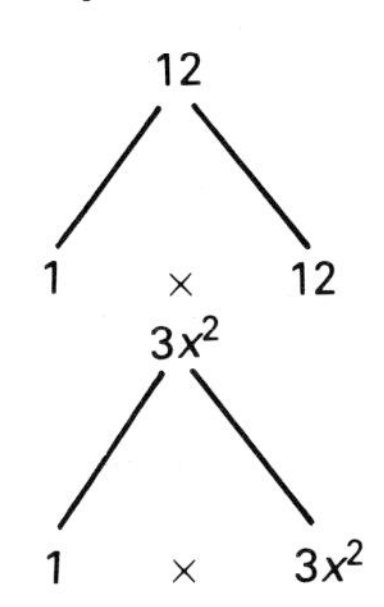

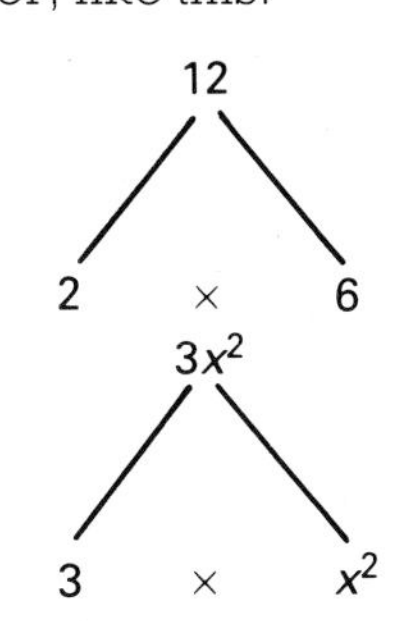

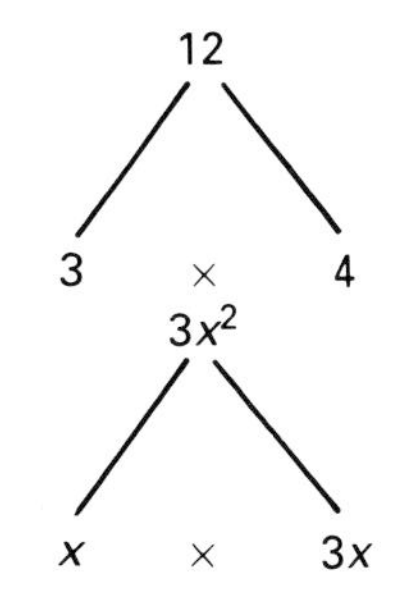

The factors of 12 are 1, 2, 3, 4, 6 and 12

The factors of $3x^2$ are 1, 3, x, $3x$, x^2 and $3x^2$

EXERCISE 1

1 Using the method shown above, find all the factors of:
a 6 **b** 16 **c** xy **d** b^2 **e** $3a$

2 Write down all the factors of:
a 4 **b** 8 **c** 13 **d** 14 **e** 20

3 You are given one factor. Write down its partner.
a 2 is a factor of 18 **b** 7 is a factor of 7
c 3 is a factor of 6 **d** 3 is a factor of $6x$
e x is a factor of $6x$ **f** x is a factor of x^2

4 Write down the partner of the factor of each term in the table.

Term	35	$8x$	$12y$	$8u$	$20x$	$6a^2$	$12b^2$	y^3
Factor	5	2	6	$2u$	$4x$	$2a$	$3b$	y

Highest common factor (hcf)
(i) 2 is a common factor of 8 and 12, because 2 divides into 8 and 12 exactly.
4 is also a common factor of 8 and 12. Why?
4 is the **highest common factor** of 8 and 12. Why?
(ii) 2, 3 and 6 are common factors of $30x$ and 12. The highest common factor is 6.
(iii) 3, x and $3x$ are common factors of $3x$ and $6x^2$. The highest common factor is $3x$.

5 Write down the highest common factor (hcf) of:
a 6 and 9 **b** 12 and 16 **c** 4 and 8
d 20 and 30 **e** 5 and $10x$ **f** $8y$ and 4
g $15k$ and 10 **h** $6m$ and 12 **i** n and n^2
j $4x$ and x^2 **k** $4x$ and $2x^2$ **l** $7y$ and $21y^2$

6 Write down the hcf of each of these pairs of cards:
a 1 and 2, 1 and 3, 1 and 4, 1 and 5
b 2 and 3, 2 and 4, 2 and 5
c 3 and 4, 3 and 5
d 4 and 5

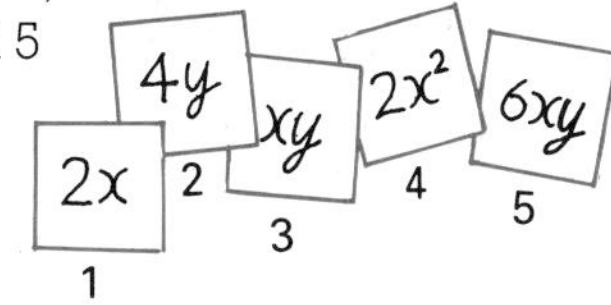

7 Write down the hcf of:
a 6, 9, 12 **b** 5, 10, 15 **c** $2x$, $3x$, $4x$
d $2x$, $4x$, $6x$ **e** 3, $6x$, $9x^2$ **f** ab, ac, ad

INVESTIGATION

Investigate the number of (different) factors that the expressions on the nth cards have. The results for the first few cards should give you a sequence to follow.

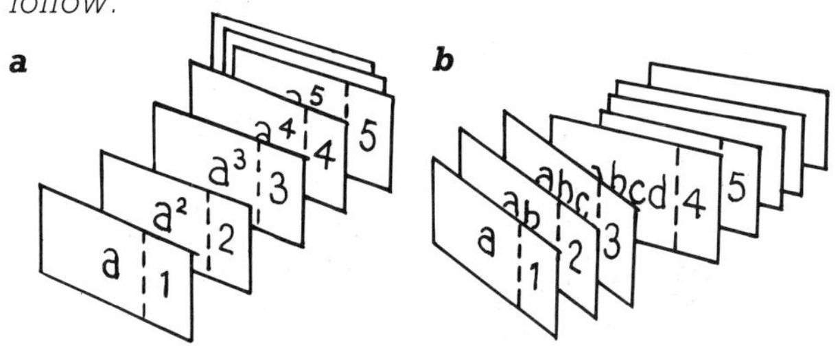

(i) Common factors

a Removing brackets . . . $3(x+5) = 3x+15$
Bringing back brackets . . . $3x+15 = 3(x+5)$

3 is the hcf of $3x$ and 15 — $x+5$ comes from the factor partners

b Removing brackets . . . $2x(x-1) = 2x^2-2x$
Bringing back brackets . . . $2x^2-2x = 2x(x-1)$

hcf of $2x^2$ and $2x$ — Factor partners

Examples

1 Factorise $6x+15$. Think it out like this:

Pick out a factor common to $6x$ and 15. → Is it the highest common factor? → If it is, put its partners in brackets. → $6x+15 = 3(2x+5)$

2 Factorise $9y^2-6y$.

3 is a common factor. → y is a common factor. → $3y$ is the highest common factor. → $9y^2-6y = 3y(3y-2)$

Good advice: Check your answer by multiplying out.

EXERCISE 2A

1 In each part:
(i) write down the highest common factor of the two terms
(ii) factorise the terms.

a $4x+12$ **b** $9x+15$ **c** $24a+8$
d $21y-14$ **e** $5x-10x^2$ **f** x^2+x^3

2 Factorise, then check by multiplying out mentally:

a $6x+9$ **b** $4y+6$ **c** $10y+15$
d $5a-10$ **e** $7m-14$ **f** $6b+12$
g $3+9p$ **h** $4-6q$ **i** $18-9r$
j $20s+10$ **k** $18t-10$ **l** $8u-12$
m $36v-45$ **n** $20w+12$ **o** $24-38t$

3 Factorise, then check:

a x^2+x **b** y^2-y **c** n^2+3n
d m^2-2m **e** $2k^2+4k$ **f** $9t^2-3t$
g $4a^2-6a$ **h** $5b^2-3b$ **i** $4d+9d^2$
j $12c^2+30c$ **k** $6d^2-8d$ **l** $10e+50e^2$
m $24f-40f^2$ **n** $18g^2-45g$ **o** $42w+18w^2$

4 Factorise, then check:

a $2x+4y+6$ **b** $3m+9n+6$
c $3a-12b+9$ **d** $5x^2+10x+20$
e $4y^2-4y+4$ **f** $8a^2+8a+20$
g $2p^2-8p+10$ **h** x^3+x^2+x

5 Factorise these to find quick mental calculations:

a $6\times3+6\times7$ **b** $4\times85+4\times15$
c $9\times999+9\times1$ **d** $7\times18-7\times8$
e $8\times111-8\times11$ **f** $7.5\times12+7.5\times8$
g $\frac{1}{4}\times97+\frac{1}{4}\times3$ **h** $897\times64-897\times63$

6 Sarah is buying stamps at the Post Office. Use a common factor for a quick calculation of the cost. Check by calculator.

7 Mr Jack buys 12 cans of dogfood at 65p each and 12 cans of catfood at 35p each. What is the total cost of his shopping? Use a common factor in your calculation, then check by calculator.

8 At the supermarket John had to arrange one long shelf of bottles in batches of three dozen; Kola at 77p per bottle, Orange at 59p, Lemon at 47p, Lime at 65p and Pineapple at 52p. Calculate the total value of the bottles on the shelf. Use a common factor in your calculation, then check by calculator.

EXERCISE 2B

1 By first writing down formulae for the areas of triangles ABC and ADC, show that the area of the trapezium is $\frac{1}{2}h(a+b)$.

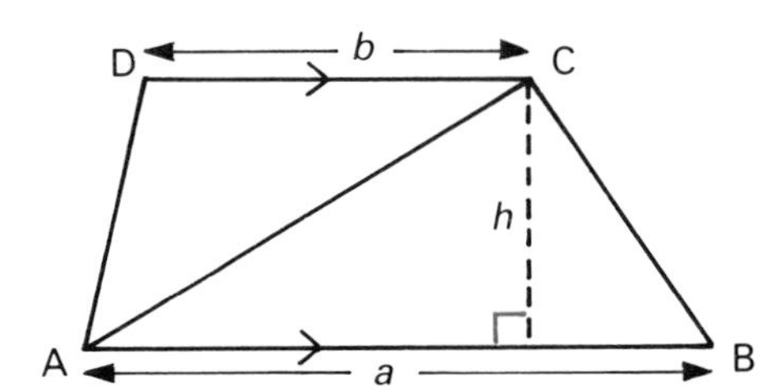

2 Show how to find these formulae for the areas of the running track and the end view of the house:

a

$2r$

$2r$

$A = r^2(\pi + 4)$

b

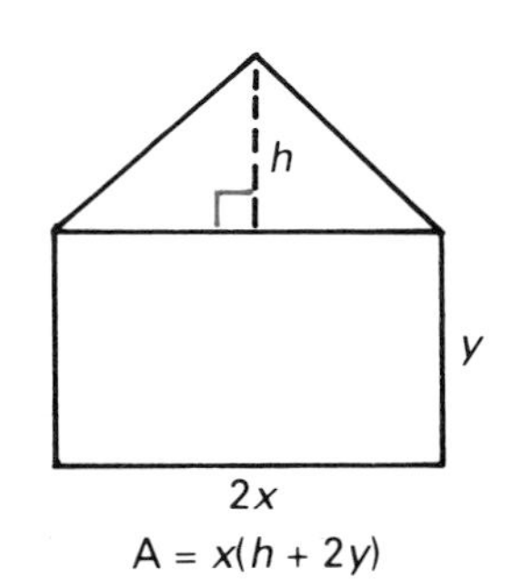

$A = x(h + 2y)$

3 Jason, the junior in John Allison's, Engineers, was asked to calculate the area of the cross-section of a circular hollow shaft. His Rule Book stated: 'Square each radius, subtract, and multiply by 3.14.'

a Prove that the rule gives a good approximation for the area.

b Calculate the area, to 3 significant figures.

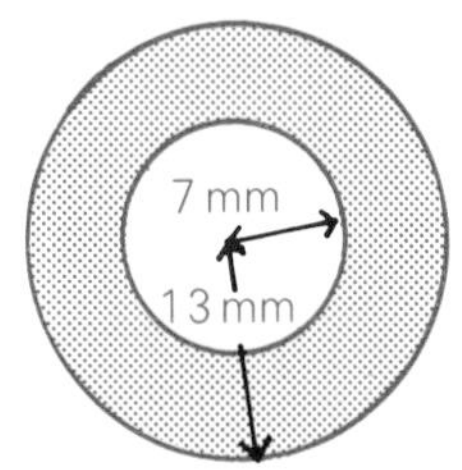

4 Take pairs of cards, like this:
(i) 1 and 2, 1 and 3, 1 and 4
(ii) 2 and 3, 2 and 4 (iii) 3 and 4
In **a**, add the terms and factorise fully.
In **b**, subtract the second from the first, and factorise fully.

a **b**

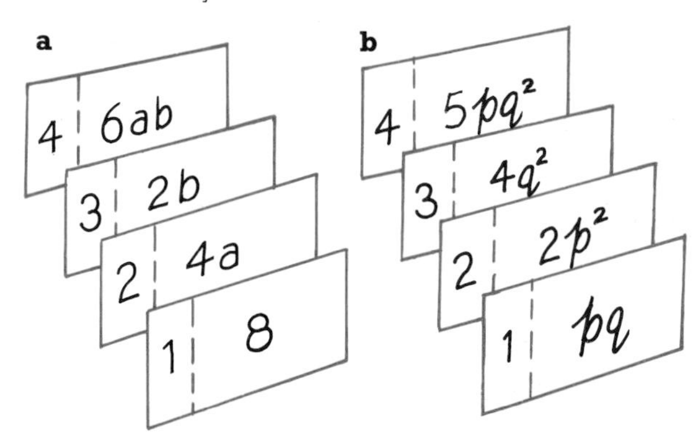

(ii) Factors of a difference of two squares

CLASS DISCUSSION/INVESTIGATION

a Complete the contents of the torn table.
b What pattern links the first and last column?
c What are the factors of $x^2 - y^2$?

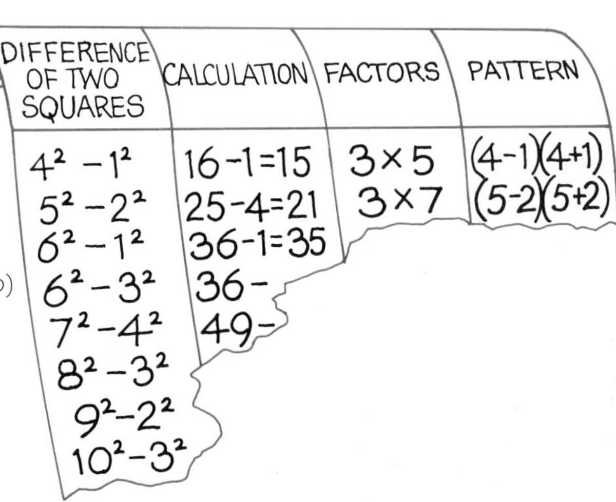

DIFFERENCE OF TWO SQUARES	CALCULATION	FACTORS	PATTERN
4^2-1^2	$16-1=15$	3×5	$(4-1)(4+1)$
5^2-2^2	$25-4=21$	3×7	$(5-2)(5+2)$
6^2-1^2	$36-1=35$		
6^2-3^2	$36-$		
7^2-4^2	$49-$		
8^2-3^2			
9^2-2^2			
10^2-3^2			

Removing brackets . . . $(a-b)(a+b) = a(a+b) - b(a+b)$
$= a^2 + ab - ba - b^2$
$= a^2 - b^2$

Bringing back brackets . . . $a^2 - b^2 = (a-b)(a+b)$

This factorises a difference of two squares, $a^2 - b^2$.

EXERCISE 3A

$$a^2 - b^2 = (a-b)(a+b)$$

1 Factorise these. For example:

$25 - x^2 = 5^2 - x^2 = (5-x)(5+x)$

a $a^2 - b^2$ **b** $c^2 - d^2$ **c** $p^2 - q^2$ **d** $x^2 - 2^2$
e $y^2 - 3^2$ **f** $z^2 - 4^2$ **g** $5^2 - u^2$ **h** $7^2 - v^2$
i $1 - w^2$ **j** $t^2 - 9$ **k** $k^2 - 4$ **l** $m^2 - 36$
m $d^2 - 100$ **n** $100 - d^2$ **o** $m^2 - n^2$ **p** $a^2 - 1$
q $64 - n^2$ **r** $y^2 - 81$

2 Factorise these by first writing them as a difference of two squares. For example:

$4p^2 - 25 = (2p)^2 - 5^2 = (2p-5)(2p+5)$

a $4x^2 - 9$ **b** $4y^2 - 49$ **c** $9a^2 - 1$
d $4a^2 - 25$ **e** $9b^2 - 4$ **f** $16c^2 - 9$
g $4x^2 - y^2$ **h** $p^2 - 9r^2$ **i** $100a^2 - b^2$
j $64c^2 - d^2$ **k** $x^2 - 16y^2$ **l** $t^2 - 36u^2$
m $4a^2 - 9b^2$ **n** $9c^2 - 16d^2$ **o** $16e^2 - 25f^2$
p $1 - 100k^2$ **q** $81n^2 - 1$ **r** $49x^2 - 81y^2$

3 Factorise these to find a quick method of calculation.

a $99^2 - 1^2$ **b** $67^2 - 33^2$ **c** $18.5^2 - 8.5^2$
d $111^2 - 110^2$ **e** $(\frac{3}{4})^2 - (\frac{1}{4})^2$ **f** $9.95^2 - 0.05^2$

4 Take pairs of cards in set **a**, subtract the terms as shown, and factorise each pair of terms:
Card 1–2, 1–3, 1–4; Card 2–3, 2–4; Card 3–4.
Repeat for set **b**.

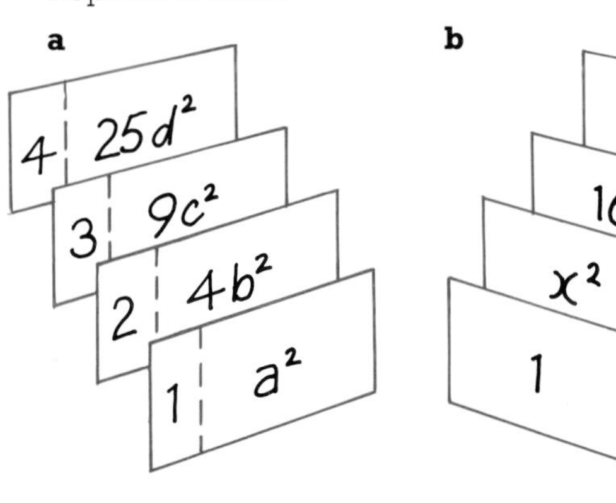

EXERCISE 3B

1 Factorise, by taking out the common factor first. For example,

$3a^2 - 12b^2 = 3(a^2 - 4b^2) = 3(a-2b)(a+2b)$.

a $2x^2 - 8$ **b** $3y^2 - 3$ **c** $5a^2 - 45$
d $4b^2 - 16$ **e** $2a^2 - 2b^2$ **f** $9c^2 - 36d^2$
g $10e^2 - 90f^2$ **h** $11g^2 - 11h^2$ **i** $ax^2 - ay^2$
j $\pi R^2 - \pi r^2$ **k** $ka^2 - 4kb^2$ **l** $np^2 - 9nq^2$
m $x^3 - 4x$ **n** $25b - b^3$ **o** $2y^3 - 32y$
p $27a^3 - 48a$

2 Use Pythagoras' Theorem to calculate x, with the help of the factors of a difference of two squares:

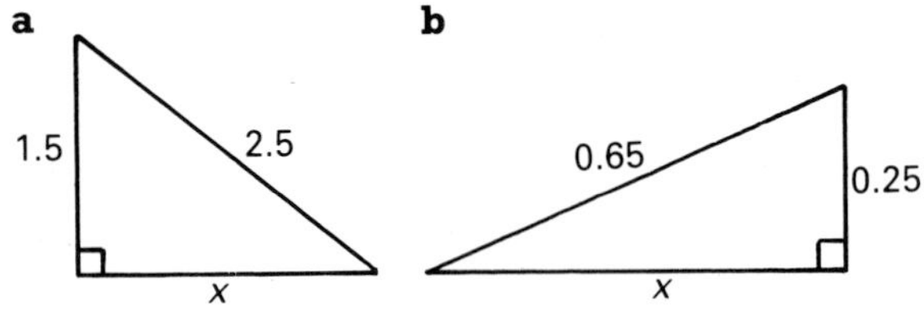

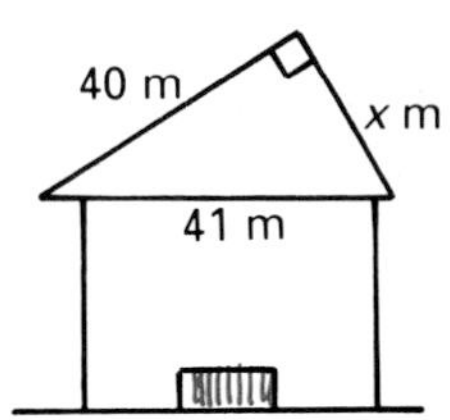

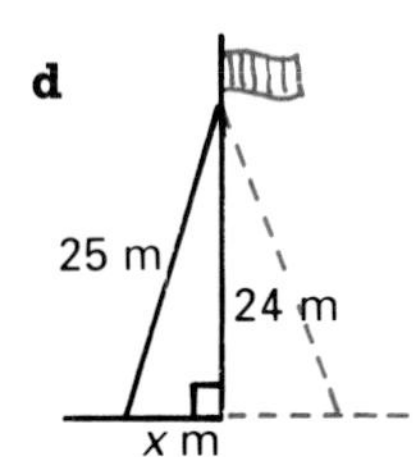

3 A concrete shaft is bounded by a square of side 4 m, inside a square surround of side d m.

a Prove that the cross-sectional area of the concrete is $(d-4)(d+4)\,\text{m}^2$.

b Calculate the area if $d = 5.5$ m.

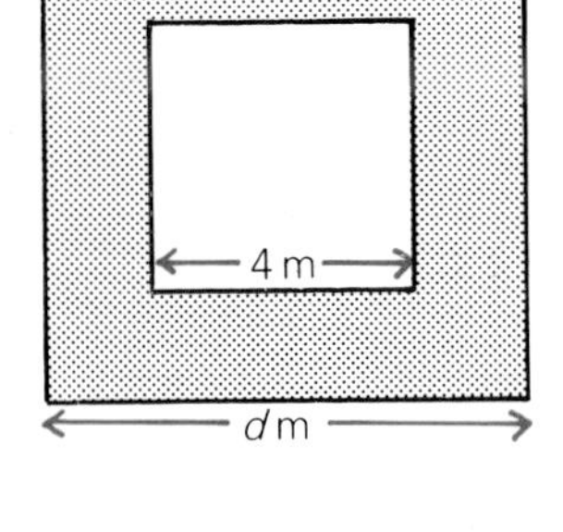

4 Jason the junior engineer found a better way to calculate the area of the circular hollow pipe. 'Add and subtract the radii, multiply the answers, then multiply by 3.14.'

a Can you justify this rule?

b (i) Use the rule when $R = 55$ and $r = 45$.
(ii) Check by calculator.

5 Pair the expressions: Card 1–2, . . . Card 1–4, Card 2–3, etc., then factorise each difference fully.

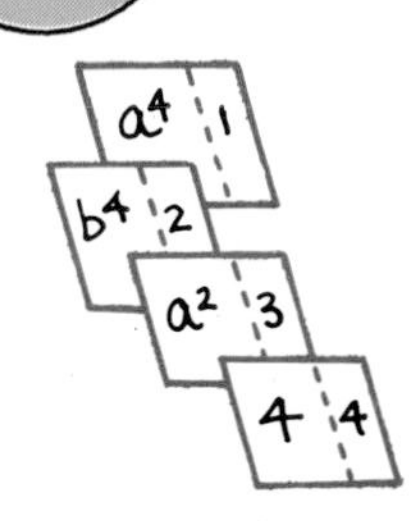

(iii) Factors of trinomials, or quadratic expressions

$(x-4)(x+1)$
$= x^2+x-4x-4$
$= x^2-3x-4$

How can we reverse the process, and factorise x^2-3x-4?
Here is one way:

First terms	*Last terms*	*Inner and outer terms* (Sum $-3x$)
x^2-3x-4	x^2-3x-4	
$=(x\quad)(x\quad)$	$=(x\quad 4)(x\quad 1)$	$4x$, x; and $-4x+x=-3x$
	or $(x\quad 2)(x\quad 2)$	$2x$, $2x$

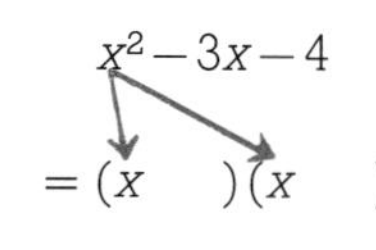

So $x^2-3x-4=(x-4)(x+1)$. *Check last terms* $-4\times 1=-4$✓

Examples

a Factorise $x^2-7x+12$

	Inners and outers (Sum $-7x$)
$x^2-7x+12$	
$=(x\quad 12)(x\quad 1)$	$12x$, x
or $(x\quad 6)(x\quad 2)$	$6x$, $2x$
or $(x\quad 4)(x\quad 3)$	$4x$, $3x$; and $-4x-3x=-7x$

So $x^2-7x+12=(x-4)(x-3)$. *Check last terms* $-4\times -3=12$✓

b Factorise x^2+5x-6

	Inners and outers (Sum $5x$)		
x^2+5x-6			
$=(x\quad 6)(x\quad 1)$	$6x$, x; and $6x-x=5x$	$(x+6)(x-1)$.	*Check* $6\times -1=-6$✓
or $(x\quad 3)(x\quad 2)$	$3x$, $2x$; and $3x+2x=5x$	$(x+3)(x+2)$.	*Check* $3\times 2=6$ (should be -6)

So $x^2+5x-6=(x+6)(x-1)$

Note An expression of the form ax^2+bx+c is often called a **trinomial** (with three terms).

EXERCISE 4A

Factorise the expressions in questions **1–8**.

1 a x^2+3x+2 **b** x^2+6x+5 **c** x^2+2x+1
d $x^2+7x+10$ **e** $x^2+11x+10$ **f** x^2+5x+6
g $y^2+8y+15$ **h** t^2+6t+8 **i** t^2+9t+8

2 a n^2-2n+1 **b** k^2-6k+9 **c** p^2-3p+2
d q^2-4q+4 **e** $r^2-10r+21$ **f** s^2-7s+6
g $u^2-12u+35$ **h** $v^2-10v+16$ **i** w^2-6w+5

3 a $a^2+10a+25$ **b** $a^2-10a+25$ **c** $t^2+8t+16$
d $t^2-8t+16$ **e** $t^2+10t+16$ **f** $t^2-10t+16$
g $t^2+17t+16$ **h** $t^2-17t+16$ **i** $p^2-20p+100$

4 a x^2+6x+9 **b** x^2-6x-7 **c** x^2+4x+4
d $y^2+9y+20$ **e** y^2+y-6 **f** y^2-y-2
g z^2+4z+4 **h** $z^2-10z+16$ **i** $z^2-3z-10$

5 a m^2-2m-3 **b** n^2+4n-5 **c** p^2+p-20
d $q^2+12q+36$ **e** $r^2-4r-21$ **f** s^2+6s+8
g t^2-t-56 **h** u^2+u-72 **i** $v^2+16v+64$

6 a x^2-x-6 **b** x^2-x-20 **c** y^2+y-2
d m^2+m-30 **e** $p^2+3p-10$ **f** $q^2+2q-15$
g t^2-t-20 **h** $u^2-2u-24$ **i** $v^2-2v-15$

7 a $x^2+7x+12$ **b** $x^2-7x+12$ **c** y^2+y-12
d y^2-y-12 **e** $z^2+4z-12$ **f** $z^2-4z-12$
g $a^2-8a+12$ **h** $b^2+11b-12$ **i** $c^2-13c+12$

8 a $v^2+8v-20$ **b** $w^2-5w-36$ **c** $x^2-10x-11$
d $y^2+9y+18$ **e** $z^2-11z+18$ **f** $x^2+14x+49$
g $m^2-12m+36$ **h** $n^2-13n-30$ **i** $p^2+21p-72$

EXERCISE 4B

1 The area of a square is $y^2-8y+16\,m^2$.

a Find the length of its side, in terms of y.

b What values can y not have?

2 Repeat question **1** for a square of area $y^2-22y+121\,cm^2$.

3 The area of a rectangle is $k^2-3k-4\,cm^2$. Find the expressions for possible lengths and breadths of the rectangle, assuming that $k > 4$.

4 Posts are placed 1 metre apart to link the fence round a rectangular field. There are x posts along the breadth of the field. The area of the field is $x^2+4x-5\,m^2$.

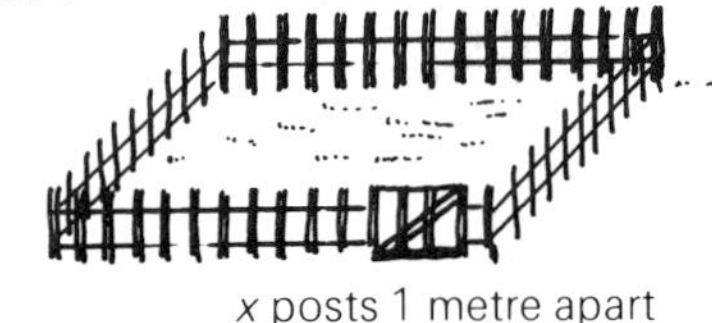

x posts 1 metre apart

a Write down an expression for the breadth of the field.

b Find expressions for the length and perimeter of the field.

c Find expressions for the length and area of a square field which has the same perimeter.

INVESTIGATION

a *For what values of c can x^2+x-c be factorised in the form $(x+a)(x+b)$, where a and b are integers (positive and negative whole numbers, and zero)?*
Hint: Find c as a product of two factors differing by 1. Can you find a formula?

b *Investigate x^2+2x-c and x^2+3x-c.*

c *Compare your three formulae. What do you notice?*

EXERCISE 5

Example Factorise $3x^2+2x-8$.

	Inners and outers (Sum $2x$)
$3x^2+2x-8$	
$=(3x\quad 8)(x\quad 1)$	$8x$, $3x$
or $(3x\quad 1)(x\quad 8)$	x, $24x$
or $(3x\quad 4)(x\quad 2)$	$4x$, $6x$; and $-4x+6x=2x$
or $(3x\quad 2)(x\quad 4)$	$2x$, $12x$

$3x^2+2x-8=(3x-4)(x+2)$. *Check* $-4\times 2=-8$

Summary

1 Fix the first terms.

2 List the last terms.

3 Combine the inner and outer terms.

4 Check the last terms.

Factorise the expressions in questions **1–4**.

1 a $2x^2+3x+1$ **b** $3y^2+4y+1$
c $2x^2+5x+2$ **d** $2x^2+9x+7$
e $5y^2+11y+2$ **f** $3a^2+5a+2$
g $3b^2+7b+2$ **h** $4c^2+5c+1$
i $4d^2+4d+1$ **j** $6t^2+5t+1$
k $5x^2+7x+2$ **l** $11y^2+13y+2$
m $6a^2+7a+1$ **n** $12m^2+7m+1$

2 a $2x^2-3x+1$ **b** $3x^2-4x+1$
c $5a^2-6a+1$ **d** $6y^2-5y+1$
e $12b^2-8b+1$ **f** $24x^2-11x+1$
g $2m^2-5m+2$ **h** $3t^2-5t+2$
i $2y^2-7y+3$ **j** $4k^2-7k+3$
k $9e^2-6e+1$ **l** $9f^2-10f+1$
m $5g^2-16g+3$ **n** $4x^2-8x+3$

3 a $2n^2+n-1$ **b** $3a^2-2a-1$
c $5b^2+4b-1$ **d** $3x^2+x-2$
e $3x^2-x-2$ **f** $5y^2+3y-2$
g $24d^2+2d-1$ **h** $24m^2+5m-1$
i $8a^2+10a-3$ **j** $8b^2-2b-3$
k $6c^2+17c-3$ **l** $6d^2+7d-3$
m $9x^2-3x-2$ **n** $12y^2+5y-2$

4 a $25x^2-10x+1$ **b** $1+4x+3x^2$
c $1-x-2x^2$ **d** $9a^2-19a+2$
e $4y^2-11y+6$ **f** $6y^2-5y-6$
g $15-7t-2t^2$ **h** $1-3u-18u^2$
i $5+11v-12v^2$ **j** $6-5d-6d^2$
k $1-8x+16x^2$ **l** $10y^2+3y-4$
m $18b^2+b-4$ **n** $12y^2-7y-12$

MAKING SURE

EXERCISE 6A

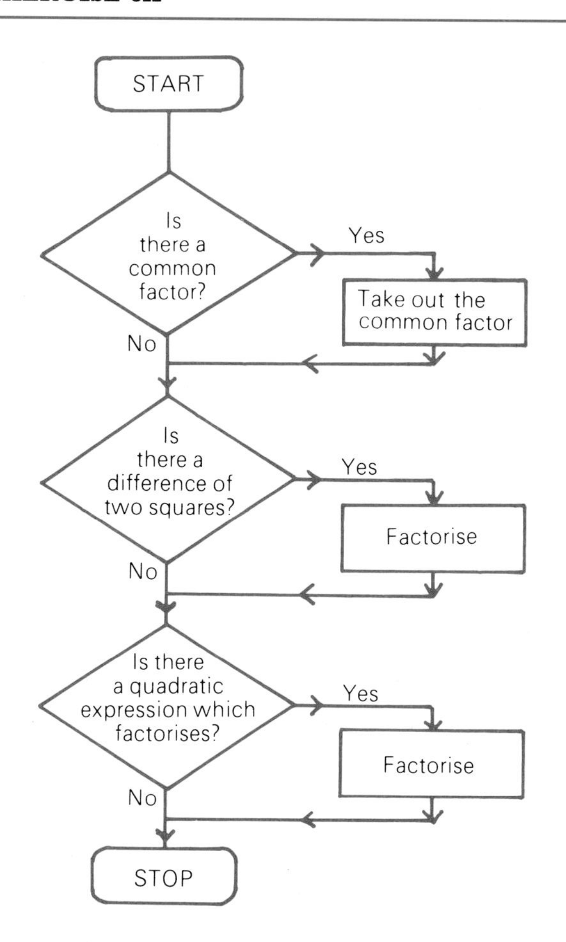

Examples Factorise fully:

a $2x^3-8x$
$= 2x(x^2-4)$
$= 2x(x-2)(x+2)$

b $2-4x+2x^2$
$= 2(1-2x+x^2)$
$= 2(1-x)(1-x)$
$= 2(1-x)^2$

Factorise fully:

1 $3x+6$
2 $8-2x$
3 x^2+x
4 $y-y^2$
5 $4a-2a^2$
6 c^2-d^2
7 p^2-4
8 $9-y^2$
9 $9x^2+6x$
10 x^2-xy
11 $36-k^2$
12 $ab+ac$
13 $6a^2-3b^2$
14 $4a^2+2ab$
15 $2a^2-2b^2$
16 p^2+6p+9
17 t^2-3t+2
18 u^2-u-6
19 v^2+2v-3
20 $x^2+2x-15$
21 a^2-100
22 $2y^2-18$
23 x^3-x
24 x^3-2x^2
25 x^3-4x
26 $3xy+6ay$
27 $4mn-6np$
28 $5p^2-45$
29 $2n^2-2n-144$
30 $2m+2m^2+2m^3$
31 $3x^2-12x$
32 $2x^3-18x$
33 $3p^2-3p$
34 $4y^2+6y-10$
35 $3t^3-27t$
36 $6m^2n-8mn^2$

EXERCISE 6B

Factorise fully:

1 $36-y^2$
2 $2abc-8abd$
3 $1-2x+x^2$
4 $6y+16y^3$
5 $14x^2+20x+6$
6 $2x^2+2x-12$
7 $4x^2-9y^2$
8 $2-50d^2$
9 $4s^2-2s-2$
10 $4k^2-11k+6$
11 m^2-9n^2
12 xyz^2+x^2yz
13 $a^2b^2-4b^2$
14 $6x^2-8x-8$
15 $6x^2-17x+12$
16 $y-y^3$
17 $4+8a-12a^2$
18 $x^2y^4-x^4y^2$
19 $200a^2-2$
20 $2-4b+2b^2$
21 $x^2+2xy+y^2$
22 $2a^2+3ab-2b^2$
23 u^2-u^6
24 $k^2-k^3+k^4$
25 $6p^2-p-15$
26 $x^2+xy-6y^2$
27 $(a+b)^2-c^2$
28 $(p-q)^2-1$
29 $a(x+y)+b(x+y)$
30 $3p+3q+ap+aq$
31 $\pi^2R^2+3\pi Rh+2h^2$
32 $(6x)^2-(5y)^2$
33 $(a+b)^2-(b+c)^2$
34 $36m^2-59m+24$
35 x^4+2x^2+1
36 x^4-x^2-6

BRAINSTORMER

A side road w metres wide, part of which is circular, joins the motorway. Prove that the distance across the entrance, d metres, is given by:

$$d^2 = w(R_1+R_2)$$

Calculate d, given $w = 8$ and $R_1 = 60$.

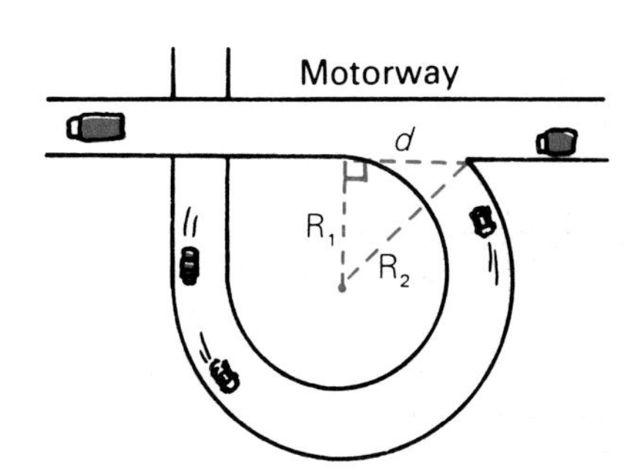

FACTORS AND SEQUENCES

EXERCISE 7

Example
Find a formula for the nth term of the sequence 4, 10, 16, 22, . . . and use it to *prove* that every term in the sequence is a multiple of 2.

4 10 16 22 . . . (differences 6 6 6) The differences are all 6, so the nth term is $6n + \ldots$

The first term is $6 \times 1 + \ldots = 4$, so the missing number is -2, and the nth term is $6n - 2$.

Factorising, $6n - 2 = 2(3n - 1)$, which is a multiple of 2.

1 Find a formula for the nth term of each sequence below, and use it to prove that the terms of the sequence are multiples of the given number.
- **a** 8, 14, 20, 26, . . . multiples of 2
- **b** 9, 15, 21, 27, . . . multiples of 3
- **c** 8, 20, 32, 44, . . . multiples of 4
- **d** 6, 24, 42, 60, . . . multiples of 6

2 A formula for the nth term of each sequence is given.
- (i) Use it to write down the first four terms of the sequence.
- (ii) Factorise the formula, and use this to write down the first four terms as products of factors, with the smaller factor first.
- (iii) Check that corresponding terms in the sequences are equal.

a $n^2 - 1$ **b** $n^2 + n$ **c** $n^2 + 3n + 2$
d $n^2 - 2n + 1$ **e** $\frac{1}{2}n^2 + \frac{1}{2}n$

3 Find formulae for the nth terms of the sequences marked in this number spiral. Use them to predict the 100th number in each sequence as a product of factors. For example:

sequence 6, 20, 42, . . . nth term is $2n(2n + 1)$
2×3 4×5 6×7 100th term is 40 200

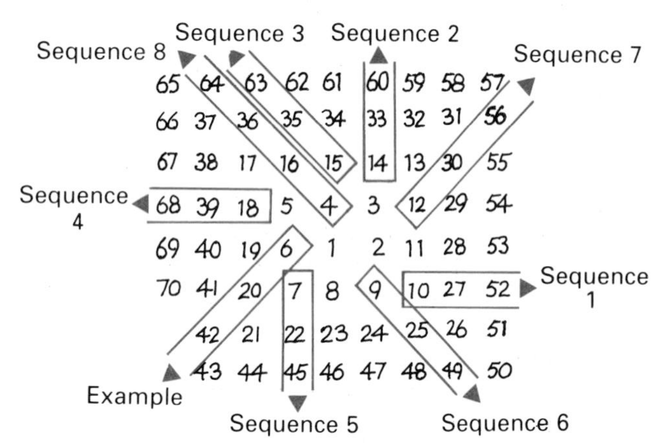

CHALLENGE

(i) Find a formula for the nth term of each sequence.
(ii) Factorise the formula, and name the largest number that every term in the sequence is a multiple of.

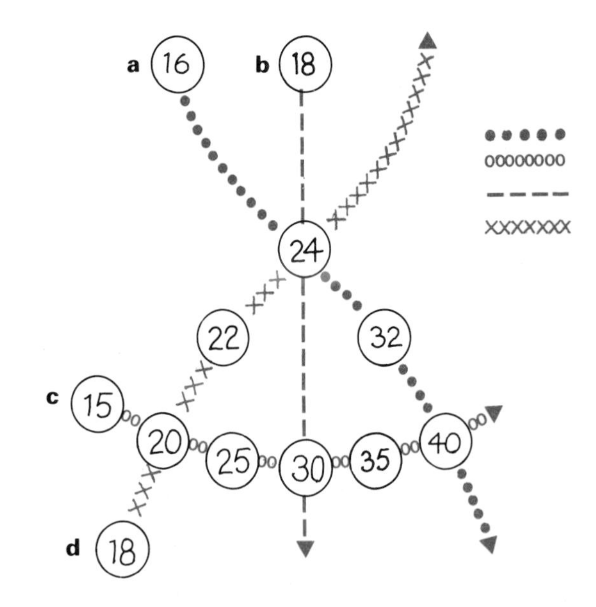

CLASS DISCUSSION/INVESTIGATION

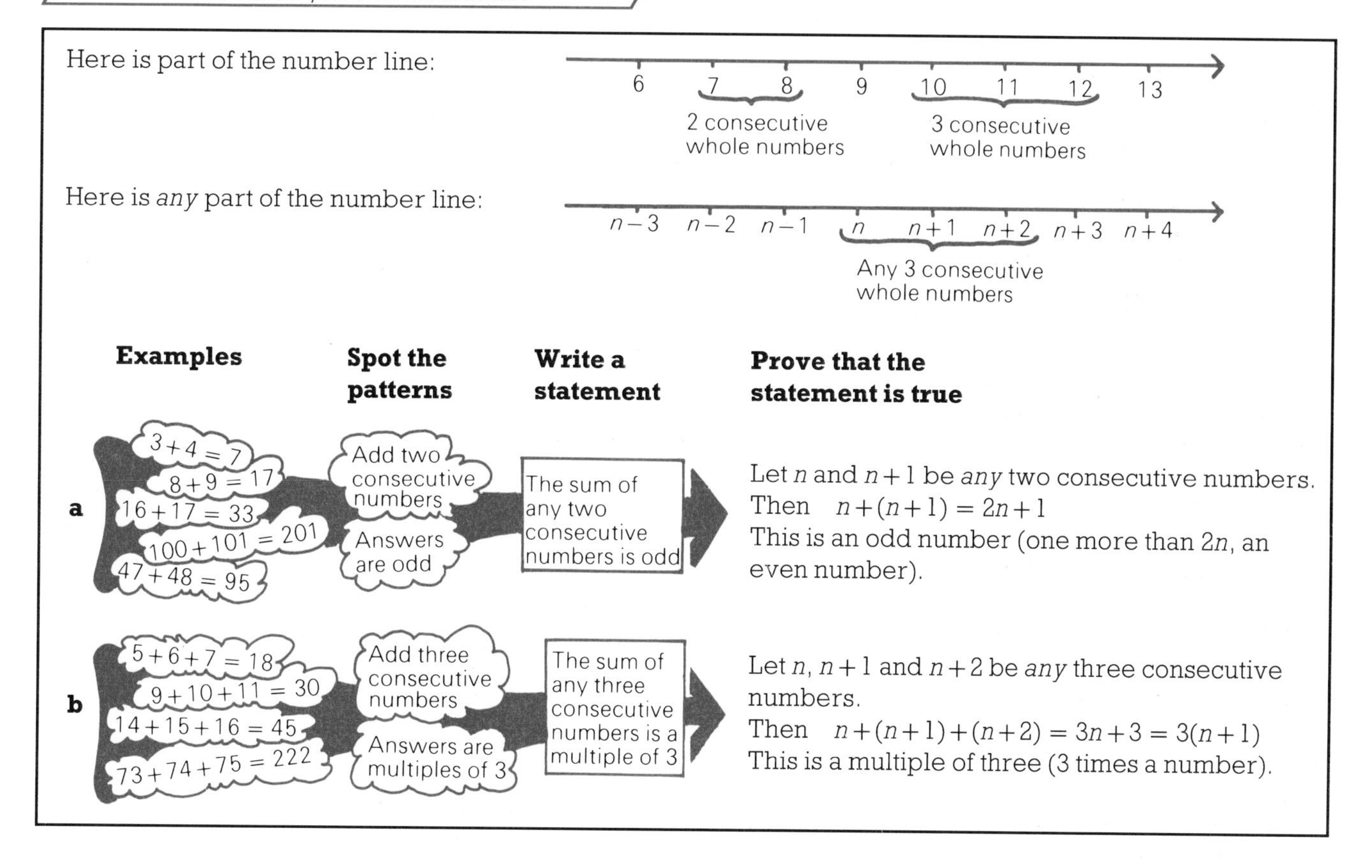

1 Work out each sum. Spot any patterns, and then write a statement. Can you prove that your statement is true?

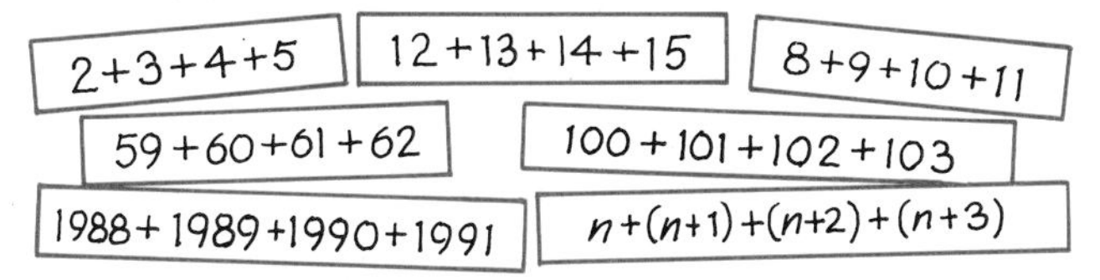

2 Use these examples to write a statement.
Work out each sum first.
Can you prove that your statement is true?

6+7+8+9+10
10+11+12+13+14
1+2+3+4+5
50+51+52+53+54
$n+(n+1)+(n+2)+(n+3)+(n+4)$

3 Try question **2** again, taking $n-2$, $n-1$, n, $n+1$ and $n+2$ for the five consecutive numbers.

4 Extend this table as far as you can.
Describe any patterns in the results that you get.

Consecutive numbers added	3	4	5	6	7	8	
Sum is a multiple of . . .	3						

5 Even more odd number factors

Jane: 'If you square an odd number, you always get an odd number.'
Janet: '$1^2 = 1$, $3^2 = 9$, $5^2 = 25$, $9^2 = 81$. It seems to work, but . . . '
Jane: 'I'll prove it. $2n$ must be an even number, so $2n+1$ must be odd, $(2n+1)^2 = 4n^2+4n+1 = 4n(n+1)+1$'
Janet: 'Right. $4n(n+1)$ must be even, so adding 1 gives an odd number.'

Try to prove as many of these statements as you can.

a The sum of two odd numbers is even.
b The sum of two even numbers is even.
c The sum of an even and an odd number is odd.
d The product of two odd numbers is odd.
e The product of an even and odd number is even.
f The product of two even numbers is even.
g The sum of three consecutive numbers starting with an odd number is a multiple of 6.
h The sum of five consecutive numbers starting with an even number is a multiple of 10.
i One less than the square of an odd number is a multiple of 4.
j The difference of the squares of two odd numbers is a multiple of 4.

CHECK-UP ON FACTORS

1 Write down all the factors of:
a 24 **b** $10x$ **c** $2x^2$

2 Write down the partner for each of these factors:
a 9 is a factor of 54 **b** $2t$ is a factor of $8t^2$

3 Find the highest common factor of:
a 8 and 20 **b** $6x$ and 18 **c** $4y$ and $12y^2$

4 Factorise, where possible:
a $6p-24$ **b** $5-5q$ **c** x^2+2x
d $3a-6b+9c$ **e** $2\pi R-2\pi r$ **f** $2x-3y$

5 Factorise:
a m^2-n^2 **b** k^2-1 **c** p^2-16q^2
d $2a^2-18$ **e** $\pi R^2-\pi r^2$ **f** ax^2-4ay^2

6 Use factors to calculate:
a 86^2-14^2 **b** $0.876^2-0.124^2$

7 Factorise:
a x^2+6x+9 **b** $y^2-10y+25$ **c** z^2-z-30
d $u^2+3u-28$ **e** x^2-2x+1 **f** $2y^2+4y+2$

8 The area of a square is $t^2-12t+36\,\text{m}^2$. Find an expression for the length of its side. What values can t not have?

9 Factorise:
a $2a^2+5a+3$ **b** $2b^2-5b+3$ **c** $2c^2+5c-3$
d $2d^2-5d-3$ **e** y^2-y **f** y^2-4
g y^2-4y+4 **h** $2t-8t^3$ **i** $6a^2+7a-10$

10 Find formulae for the nth terms of these sequences, and use them to prove that the terms are multiples of the given numbers.
a 10, 15, 20, 25, . . . multiples of 5
b 6, 15, 24, 33, . . . multiples of 3

11 a Write down the first three terms of the sequences with nth term formulae:
(i) $4n^2-1$ (ii) n^2+5n+6
b Factorise each formula and use this to write down the first three terms as products. Check that they are the same as in **a**.

12 The diagram shows the cross-section of a hollow concrete support. The units are metres.

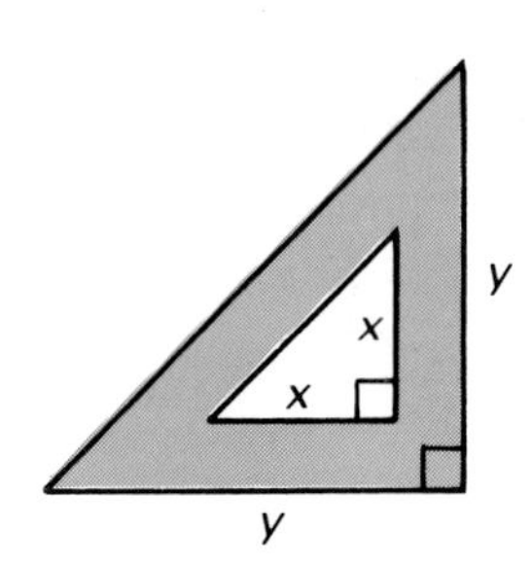

a Prove that the cross-sectional area is $\frac{1}{2}(y-x)(y+x)\,\text{m}^2$.
b If x and y differ by 2, and the area is $14\,\text{m}^2$, find the lengths of all three sides in each triangular part.

13 Prove that the sum of three consecutive:
a odd numbers is divisible by 3
b even numbers is divisible by 6.

13 MONEY MATTERS - PERSONAL FINANCE

LOOKING BACK

1 Pat went on a school trip to Paris. She changed £75 to francs at an exchange rate of 8.59 francs to the £.

a How many francs, to the nearest franc, did she receive?

b She spent 497 francs altogether. How many did she have left?

2 a What percentage of each price have you to pay at this sale?

b Express this percentage as:
(i) a decimal fraction
(ii) an ordinary fraction, in its simplest form.

3 Jessica Malcom's basic rate of pay is £6.90 an hour. Her normal working week is 37 hours, and she is paid 'time and a third' for overtime. Calculate:

a her normal weekly wage

b her earnings in a week when she worked 44 hours.

4 Tom Tracy sells computer discs. Each month his commission is 5% of sales over £3500. Calculate his commission in months when he had sales of:
a £8300 **b** £20 000.

5 Mr Speedie sent out this bill for replacing some old lead pipe. Calculate the total amount due.

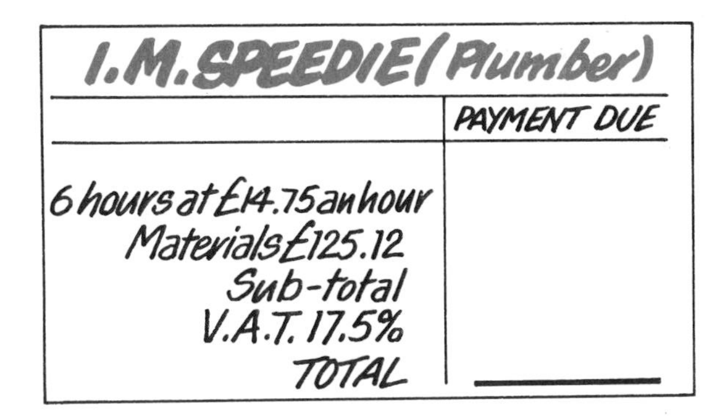

I.M. SPEEDIE (Plumber)

	PAYMENT DUE
6 hours at £14.75 an hour	
Materials £125.12	
Sub-total	
V.A.T. 17.5%	
TOTAL	

6 Jamal starts work with a salary of £10 000 a year, and annual increases of £250. Allison's starting salary is £9000 a year, but she will get annual increases of 6% of her previous year's salary. Calculate their salaries during their fourth year of work.

7 Jim Mason received interest of £900 from the Steady Growth Building Society in 1994.
How much was this:

a per month

b per week (to the nearest penny that would be paid)?

PAYSLIPS

Mr Dixon is a car mechanic.
Here is his payslip for a recent week.

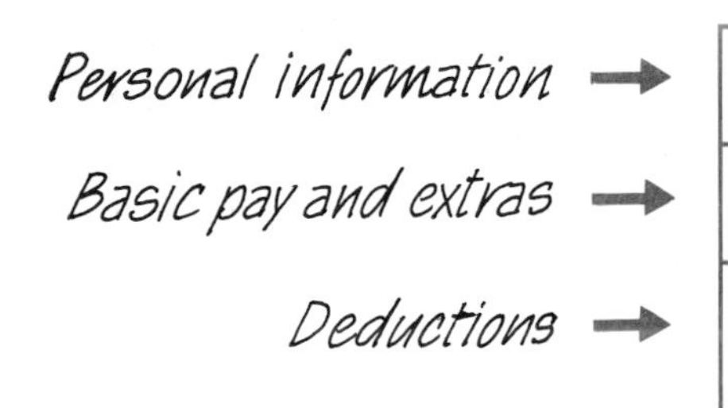

Employee 123	T. DIXON	Tax Code 516H	Week 52
Basic Pay 280.00	Overtime 40.00	Commission —	Gross Pay 320.00
Income Tax 48.76	Superannuation 16.00	National Insurance 24.88	Total Deductions 89.64
			Net Pay 230.36

EXERCISE 1

1 Look at the entries in Tom Dixon's payslip. Check the calculations for his:
a gross pay **b** total deductions **c** net pay.

2 Anita Morrison works in a computer office. Here is her payslip for the week ending 19th September. Calculate her:
a gross pay **b** total deductions **c** net pay.

Employee 386	ANITA MORRISON		Tax Code 330L	Week 24
Basic Pay 201.92	Overtime 15.94	Other Allowances —	Bonus —	Gross Pay
Income Tax 36.65	Superannuation —	National Insurance 15.69	Other Deductions —	Total Deductions
				Net Pay

3 Paul Williams works for an insurance company. Calculate his:
a gross pay **b** total deductions **c** net pay.

Employee 01010	PAUL WILLIAMS		Tax Code 510H	Month 8
Basic Pay 1237.50	Overtime —	Other Allowances 250.00	Bonus —	Gross Pay
Income Tax 178.25	Superannuation 74.25	National Insurance 116.89	Other Deductions 3.50	Total Deductions
				Net Pay

4

Susan Munn is employed in the office of the Fast Glaze window firm. Find the missing entries in her payslip.

Employee 87	SUSAN MUNN		Tax Code 360L	Month 11
Basic Pay	Overtime 28.72	Commission —	Bonus 16.84	Gross Pay 702.41
Income Tax	Superannuation 35.12	National Insurance 46.23	Other Deductions 3.79	Total Deductions
				Net Pay 523.91

5 Make out John Chang's month 10 payslip.
Employee number: 040506; Basic Pay: £887.92;
Commission: 4% of £1780; Income Tax: £127.33;
Superannuation: 6% of gross pay;
National Insurance: £69.33.

SUPERANNUATION (OR PENSION)

EXERCISE 2

1 Look at Tom Dixon's payslip again.

Employee 123	T. DIXON	Tax Code 516H	Week 52
Basic Pay 280.00	**Overtime** 40.00	**Commission** —	**Gross Pay** 320.00
Income Tax 48.76	**Superannuation** 16.00	**National Insurance** 24.88	**Total Deductions** 89.64
			Net Pay 230.36

He pays 5% of his gross pay for superannuation.

a Check the entry on his payslip for superannuation.

b How much does he pay for superannuation in a year, assuming that his overtime is the same each week?

2

Mrs Dixon works in a department store, and earns £660 a month. She pays 6% of this for superannuation. How much does she pay:

a per month **b** in a year?

3 The Dixons' daughter, Ann, works for a computer firm, and has just had her monthly salary increased to £1138. She now pays £62.59 superannuation monthly. What percentage is this of her monthly pay?

4 Employers also make contributions to their employees' superannuation schemes. Mr Dixon's employer pays an amount equal to 7% of Mr Dixon's gross pay. Including Mr Dixon's contributions (5% of his pay), calculate the total sum paid for his superannuation in:

a 1 year **b** 30 years (at the same rate of pay).

NATIONAL INSURANCE (NI)

GROSS INCOME		National Insurance Contribution
WEEKLY	ANNUALLY	
Less than £56	Less than £2912	0
£56–£420	£2912–£21 840	2% of the first £56 (or £2912) 9% of the rest
over £420	over £21 840	£33.88 (or £1761.76)

EXERCISE 3

1 Check the deduction of £24.88 on Tom Dixon's payslip on page 162.

2 How much NI would Mr Dixon have deducted on a gross pay of:
a £54 in week 1
b £282 in week 2
c £190 in week 3?

3 Calculate Ann's annual NI contribution for a monthly gross salary of:
a £1138 **b** £1580 **c** £1880.

4 Stephen Stacker earned £55.80 each week in a part-time job. He was given a small pay rise, increasing his earnings to £57. Why was he annoyed when he received his new payslip?

5 Meena Kapoor earns £12 000 per annum. What percentage of this, correct to 1 decimal place, does she pay in NI?

6 Teachers, and employees in the Health Service and in most large companies, pay different rates from the ones in the table. Mr Horner earns £300 a week. His NI contribution is 9% of the first £56 he earns, and 6.85% of the remainder.
a Calculate his weekly contribution.
b How does this compare with someone in the Government scheme who earns £300 a week?

INCOME TAX

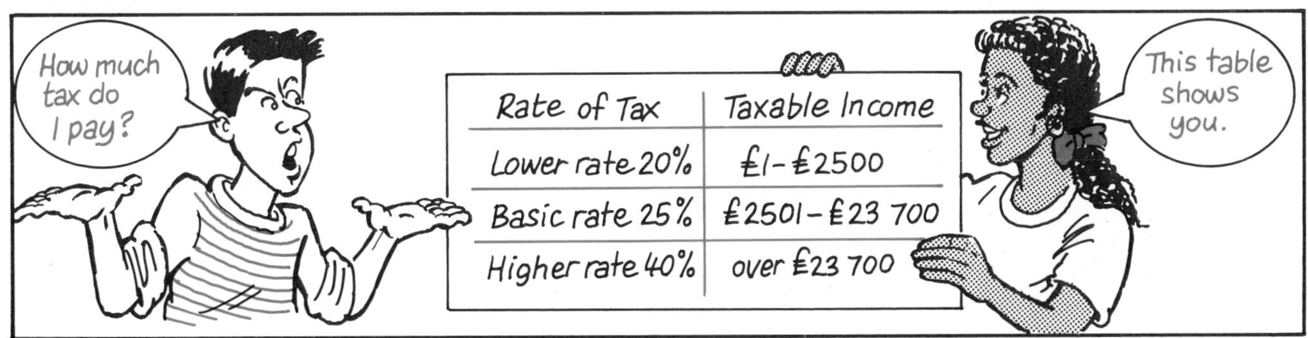

Rate of Tax	Taxable Income
Lower rate 20%	£1–£2500
Basic rate 25%	£2501–£23 700
Higher rate 40%	over £23 700

Example
Check the income tax deduction in Tom Dixon's payslip. Assume that his gross pay is £320 every week, and that tax is calculated on his annual pay, less superannuation and income tax allowances.

Annual pay: $52 \times £320 = £16\,640$
Superannuation: 5% of pay = £ 832
£15 808

Personal and married couple's allowances (£3445 + £1720): £ 5 165
Taxable income: £10 643

Tax due for the year:
20% of £2500 = £ 500
+25% of £8143 = £2035.75
£10 643 £2535.75

Weekly tax deduction: £2535.75 ÷ 52
= £48.76

EXERCISE 4A

1 Beryl earns £62 a week in a part-time job. Her tax allowance is £3445 per annum. Calculate her tax bill for the year.

2 Rob is an engineer. He is single, and earns £290 a week. Copy and complete:

Annual pay = $52 \times £290$ = £ . . .
Personal allowance = . . .
Taxable income = . . .
Income tax due = 20% of £ . . . = . . .
+25% of £ . . . = . . .
£ per year
= £ . . . . per week.

3 Shona Bates is a waitress. In her annual income tax return she enters her wage as £7540, and 'tips' (also taxable) as £430. She has a personal tax allowance of £3445. Calculate:
a her taxable income
b the amount of tax she has to pay
(i) annually (ii) weekly.

4 Brian earns £350 gross, weekly, as a garage foreman. Calculate:
a his gross pay for a year
b his taxable income (after deduction of his personal tax allowance)
c the tax he pays
(i) annually (ii) weekly
d his net weekly pay.

5 Calculate the amount of income tax paid annually by:
a Mr T. Jones, married, annual salary £19 105 (he has a personal and married couple's allowance)
b Miss M. Taylor, annual salary £26 980
c Mr B. Davidson, single, weekly pay £283.10
d Mrs A. Burns, weekly pay (part-time job) £49.50.

6 Find the entries for the empty boxes in Sally Sellar's monthly salary slip. Assume that her basic pay, commission and National Insurance are the same every month, and that she has an allowance of ten times her tax code number. What is her take-home pay for the month?

Employee 99999	SALLY SELLAR		**Tax Code** 516H	**Month** 4
Basic Pay 1325.25	**Overtime** —	**Commission** 170.75	**Bonus** —	**Gross Pay**
Income Tax	**Superannuation** —	**National Insurance** 117.66	**Other Deductions** —	**Total Deductions**
				Net Pay

EXERCISE 4B

1 Richard Richman is a head teacher and earns £38 000 per year. He is married and so claims the married couple's allowance. His superannuation payments are 6% of his gross salary. Copy and complete:

Allowances
Superannuation payments = £
Personal allowance = £
Married couple's allowance = £

Total allowances = £

Salary = £38 000
Total allowances = £. . . . . .

Taxable income = £

Income tax due (See table on page 165.)
20% of £ . . . . = ...
25% of £ . . . . = ...
40% of £ . . . . = ...

£

2 Action Rovers' top goal scorer Martin James earns £95 000 a year. Calculate his annual tax bill under PAYE. He is not married and is not in a pension scheme at present.

3 During one tax year Sam Swinger's earnings as a professional golfer were £126 540. His tax allowances totalled £18 500.

a How much income tax had he to pay?

b How much had he left after tax?

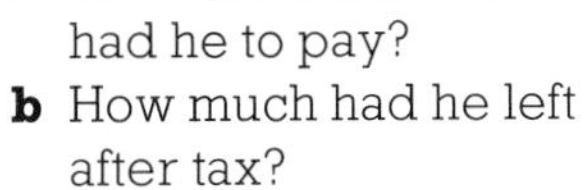

4

Bridget writes financial articles for newspapers. She predicts new tax rates, shown in the table.

Income (£)	Description	Rate of tax
0–10 000	Tax-free	0%
10 001–20 000	Basic rate	20%
>20 000	Higher rate	50%

She draws this graph to help her readers.

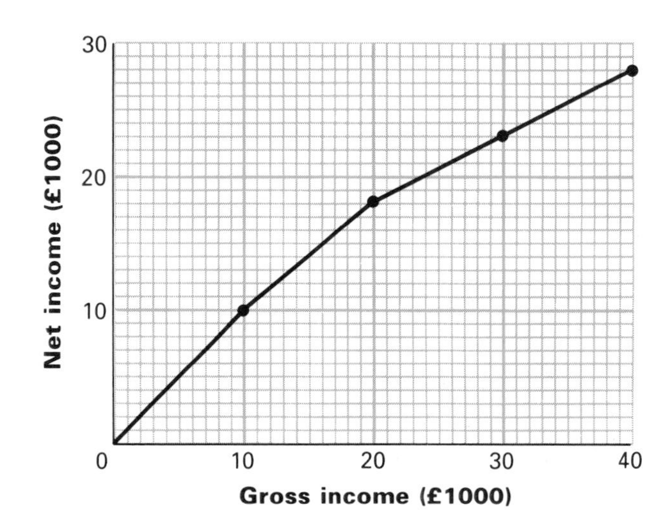

a Use the graph to write down the net incomes (after tax) for gross incomes (before tax) of:
(i) £10 000 (ii) £20 000
(iii) £30 000 (iv) £40 000.

b Check your answers to **a** by calculation.

c What percentage is each net income of its gross income in **a**?

INVESTIGATION

Find out:

a *when the financial year starts and ends*

b *who gives the 'Budget' speech*

c *when and where the speech is made*

d *what the present allowances and rates of income tax are. Why do these change nearly every year?*

LIFE INSURANCE

What are the advantages and disadvantages of each kind of policy?
(See also the tables of costs below.)
The Safe and Secure Insurance Company calculates the premiums you pay from these tables.

Monthly premiums for every £1000 insured

Whole-life (with profits)

Age		Non-smoker	Smoker
Male	Female		
16–25	16–32	1.90	2.30
26	33	1.90	2.35
27	34	1.95	2.45
28	35	1.95	2.55
29	36	2.00	2.65
30	37	2.05	2.70
31	38	2.10	2.80
32	39	2.15	2.95

Endowment (with profits)

Age		10 years		20 years	
Male	Female	Non-smoker	Smoker	Non-smoker	Smoker
16–25	16–32	9.65	11.29	4.58	5.36
26	33	9.66	11.30	4.60	5.39
27	34	9.67	11.31	4.61	5.40
28	35	9.67	11.31	4.62	5.40
29	36	9.68	11.33	4.63	5.41
30	37	9.69	11.34	4.64	5.42
31	38	9.70	11.35	4.65	5.44
32	39	9.72	11.38	4.67	5.46

EXERCISE 5

1 Why are the monthly premiums in the tables:
- **a** greater as the starting age increases
- **b** less for females than males
- **c** more for smokers than non-smokers?

2 Write down the premiums for a £1000:
- **a** whole-life policy for
  - (i) a male non-smoker aged 26
  - (ii) a female smoker aged 39
- **b** endowment policy for
  - (i) 10 years, for a male smoker aged 21
  - (ii) 20 years, for a female non-smoker aged 35.

3 

Michael and Nicola are 32-year-old twins. How much less does Nicola pay *in a year* than Michael for a whole-life insurance policy of £1000, if:
a neither smokes **b** both smoke?

4 Calculate the monthly premium for a 10 year endowment policy of £5000 for Elspeth Holmes, a 34-year-old smoker.

5 Andrea, 20 years old and a non-smoker, works in a library. She takes out a 10 year endowment policy worth £7500. Copy and complete this calculation of her monthly premium:

Insurance *Premium*

£1000 ⟷ £9.65

£1 ⟷ $\frac{£9.65}{1000}$

£7500 ⟷ $\frac{£9.65}{1000} \times 7500 = £\ldots$

6 Calculate the monthly premiums for these people:

	Name	Age	Policy	Term	Sum insured
a	Justin Case (male, non-smoker)	26	Whole-life	Life	£10 000
b	Fallon Downie (female, non-smoker)	22	Endowment	20 years	£6000
c	Al Dunn (male, smoker)	18	Endowment	10 years	£5000
d	J. McClean (female, non-smoker)	38	Whole-life	Life	£25 000

7 Jim Thomson, 31, a keen sportsman, doesn't smoke. He takes out a 20 year endowment policy, with profits, giving £8000 cover.
- **a** Calculate his annual premium.
- **b** How much less would he pay for a whole-life policy?

8 At the age of 25, Alan McKechnie, a smoker, took out a whole-life policy worth £15 000. Tragically he was killed in an accident 12 years later. Calculate the difference between the amount he had paid and the amount his dependants will receive (before any profits are included).

9 Christine, a 26-year-old dentist, finds she can save £80 a month. She decides to put this into life insurance. What choices has she? How much cover can she get (to the nearest £100)? What difference does it make whether she smokes or not?

10 The Safe and Secure Insurance Company realises that inflation will eat into the value of the money they pay out. So they offer a new 10 year endowment policy. For £1000 cover, the first year's premium for a 20-year-old non-smoker is £120. Both premiums and cover increase each year by 5% of their previous year's value. Calculate the premium and the cover for:
- **a** the second year
- **b** the third year
- **c** the fourth year

COMPOUND INTEREST

Sally Brown put £1000 in the North-South Savings Bank at an annual rate of interest of 10%. Each year the interest is calculated on the previous year's balance, that is 'principal + interest'. This is called *compound interest*. Some Building Societies calculate interest every six months; some lending companies calculate it monthly.

North-South Savings Bank				
Date	Code	In	Out	Balance
20.1.92		1000	—	1000
20.1.93	INT.	100	—	1100
20.1.94	INT.	110	—	1210
20.1.95	INT.	121	—	1331

EXERCISE 6

1 George calculates how much money he will have in his Investment Account after three years. He has deposited £800 at 5% per annum, and will leave it to grow with compound interest. Copy and complete his calculation.
Taking P for the Principal and I for interest,

1st year: $P = 800$
$I = \underline{40}$ (5% of 800)

2nd year: $P = 840$
$I = \underline{\ldots}$ (5% of 840)

3rd year: $P = 882$
$I = \underline{\ldots}$ (5% of 882)

$\underline{\ldots}$, so total amount = £ . . .

2 As in question **1**, calculate the total amount of:

a £500 for 2 years at 10% per annum compound interest

b £400 for 2 years at 5% per annum compound interest

c £1000 for 3 years at 4% per annum compound interest

d £120 for 2 years at 7.5% per annum compound interest (round down).

Calculator short-cuts

In question **1**, the principal is multiplied by 1.05 each year. You can use

(i) the 'constant facility', or memory, giving:

800 → 1.05 [X] 800 = 840

→ 1.05 [X] 840 = 882

→ 1.05 [X] 882 ie £926.10

or

(ii) the [y^x] key, giving:

800 [X] 1.05 [y^x]
= 926.1

3 Use one or more of the short-cuts to answer questions **1** and **2** again.

4 Calculate the compound interest on £3456 for 10 years at 7.5% p.a.

5 How many years would it take for £100 to double at 10% per annum compound interest?

6 Erica deposits £850 with the Handshake Building Society, which pays 4% compound interest half-yearly.

a How much will she have altogether at the end of one year?

b Compare this with an *annual* 8% compound interest rate.

7 David Barr deposited £250 in a bank for three years, leaving the interest to be added to his account each year. The annual rate of interest dropped from 10% in the first year to 8% in the second year and 6% in the third year. How much money was in his account after three years?

BRAINSTORMER

Wayne discovered that his grandfather had put £10 000 in the Van Winkle Bank 100 years ago at 7.25% per annum compound interest. When Wayne went to take the money home in a case from the bank he had a problem. How big was his problem?

APPRECIATION AND DEPRECIATION

Houses tend to **appreciate**, or increase in value, year by year; cars usually **depreciate**, or decrease in value year by year. Why?

Example
Jim and Jenny bought a house for £80 000. In the following three years its value appreciated by 10%, 5% and 2%. Calculate its value each year.

First year Value = £80 000
Appreciation = 10% of £80 000 = £8000
Second year Value = £88 000
Appreciation = 5% of £88 000 = £4400
Third year Value = £92 400
Appreciation = 2% of £92 400 = £1848
Value after 3 years = £94 248

Note Another method: the value after one year is 110% of the original value = 1.1 × £80 000 = £88 000, and so on.

EXERCISE 7

1 Mr and Mrs Jackson bought a flat for £75 000. In each of the following two years its value appreciated by 8%. How much was it worth at the end of two years?

2 Jock's car depreciated rapidly. It cost him £4000.
a The first year it lost 25% of its value. What was it worth then?
b The second year it lost 20% of this value. What was it worth then?
c The third year it lost 15% of this value. What was it worth at the end of the three years?

3 Calculate the percentage appreciation of the value of this house:
a over the first year
b over the three year period.

4 Calculate the percentage depreciation in this car's value:
a over the first year
b over the three year period.

5 The value of the machinery in a factory depreciates by 7% annually. The machinery cost £960 000 new. Calculate its value after two years.

6 Mr Richman valued his shares at the beginning of each year. Use the graph to calculate the percentage change in their value each year, compared to the previous year's value.

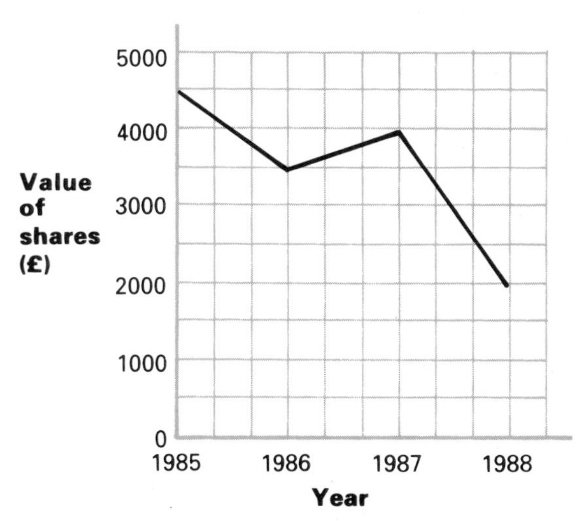

7 The Khans' house cost £175 000 in 1989. By 1992 its value had depreciated by 20%. By 1995, however, it was worth 12% more than in 1992. Calculate:
a its value in 1995
b the percentage change in value from 1989 to 1995.

8 To find the *real* appreciation in value of Jim and Jenny's house in the *Example* on page 171, the annual rate of inflation has to be taken into account. If the rate of inflation was 2% each year, the real appreciation in the first year was 10% − 2%; 8% of £80 000. Calculate the real appreciation each year in this way.

9 The value of Trisha's coin collection fell by 5%, to £430. Calculate its previous value. (Take this to be 100% of the value.)

APR (ANNUAL PERCENTAGE RATE)

In the advertisement the monthly rate of interest is 2%. But the **annual** rate is more than 12 × 2%, as it is calculated on the **loan + interest each month**. This means the interest is compounded **monthly**.

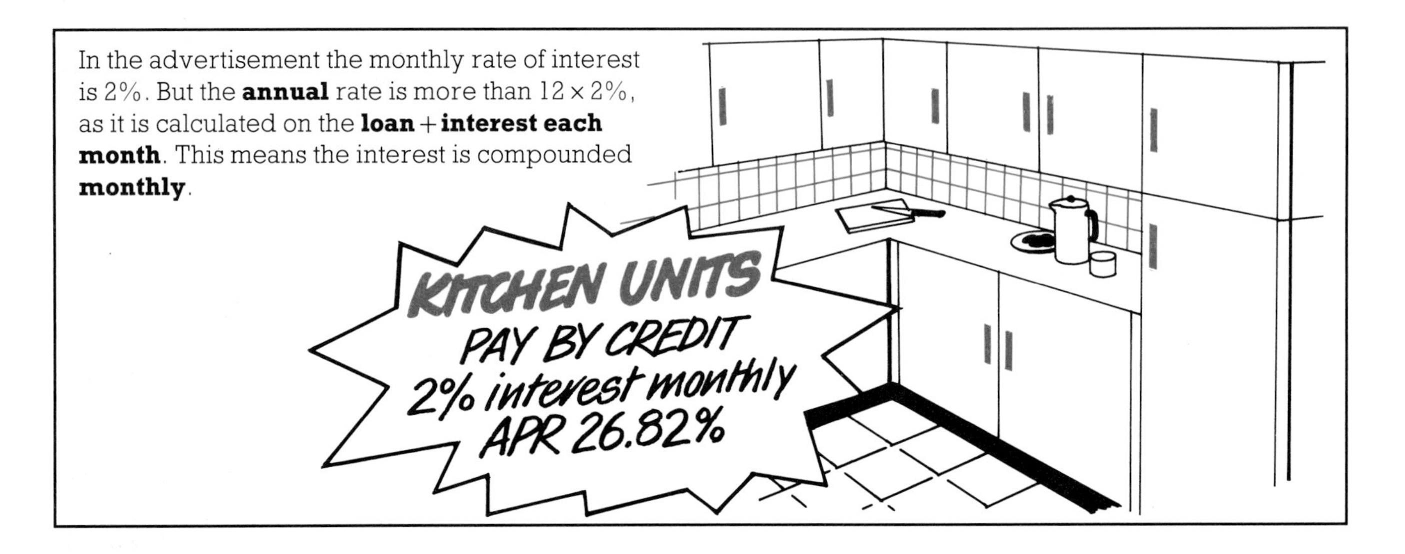

EXERCISE 8B

1 Use your calculator to check that a loan of £100 at 2% rate of interest per month becomes £126.82 at the end of 12 months. So the APR is 26.82%.

2 Some credit card companies charge 3% per month. Calculate this APR, correct to 2 decimal places.

3

Is Honest Joe really honest? He should say what the APR is. Can you work it out for him, to the nearest 1%?

4 Calculate the APR for a monthly interest rate of:
a 1% **b** 0%.

5 Cath borrowed £10 000 from the Bank of Brit at an APR of 19.9%. She agreed to pay back the loan, plus the interest, over 12 months. Calculate her equal monthly repayments.

6 By making an estimate, and improving it, find, correct to 2 decimal places, the monthly interest rate which produces an APR of 50%.

BRAINSTORMER

Mr Campbell takes out a loan for a year at a monthly rate of 3.5%. At the end of the year he clears his debt by paying £302.21. How much did he borrow?

INVESTIGATION

Investigate the APRs charged by different credit cards, shops, garages and finance companies. The rates are often given in newspaper advertisements. What are the highest and lowest APRs that you can find?

CHECK-UP ON MONEY MATTERS—PERSONAL FINANCE

1 Calculate Iain Fraser's gross pay and net pay.

Employee 321	I. FRASER	**Tax Code** 418H	**Week** 37
Basic Pay 285.50	**Overtime** 37.50	**Commission** —	**Gross Pay**
Income Tax 54.24	**Superannuation** 16.00	**National Insurance** 25.48	**Total Deductions**
			Net Pay

2 Hillary James earns £248 a week. She pays 6% superannuation, also 2% National Insurance on the first £56 and 9% on the rest.

a Calculate how much she pays weekly for superannuation and National Insurance.

b She also pays £39.31 tax and £3.40 union dues each week. What is her weekly take-home pay?

3 Tahir, a married man, earns £16 275 a year. He pays income tax at the rate of 20p in the £ on the first £2500 of taxable income and at 25p in the £ on the rest. His personal allowances total £5165. Calculate:

a how much income tax he has to pay

b how much more tax he will pay after a 5% rise in his salary.

4 Mike Milligan, a teacher of physical education, is 26 years old and a non-smoker. He can afford £55 a month for a life insurance policy. Use the tables on page 168 to find how much (to the nearest £100) he can insure his life for with:

a a whole-life policy

b a 20-year endowment policy.

5 Calculate the compound interest on:

a £400 for 2 years at 6% per annum

b £875 for 3 years at 4.5% per annum.

6 A local council takes out a loan of £1 400 000 to pay for a new swimming pool. The annual rate of compound interest is 13%. How much, to the nearest £, does the council owe at the end of four years if none of the loan has been repaid?

7 A company's computers depreciate by 12% annually; that is, at the end of each year their value is 12% less than at the beginning of the year. They cost £1 500 000 when they were new. Calculate their value after three years.

8 The monthly interest rate with Spenders Credit Card is 2.2%. Calculate the APR.

14 FORMULAE

LOOKING BACK

1 $x = 4$ and $y = -1$ Find the value of:

a $9x$ **b** $4y$ **c** $x-y$ **d** xy
e x^2 **f** $2y^2$ **g** $3x-4y$ **h** $(3x)^2$
i $\sqrt{x}$ **j** $\sqrt{(6x-y)}$ **k** $(x+y)^2$

2 Solve:

a $4t = 20$ **b** $3u-7 = 11$ **c** $2v = 3-v$
d $5(a-1) = 10$ **e** $\frac{x}{2} = 8$ **f** $\frac{y}{3} = -12$

3 Remove brackets:

a $6(y-2)$ **b** $x(a-b)$
c $-2(x+1)$ **d** $8x-3(2x+1)$

4 Write down formulae for:

a the perimeter P cm of this rectangle

b the area A cm^2 of this circle

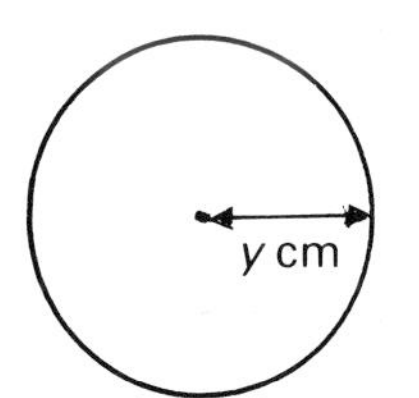

c the volume V cm^3 of this cuboid

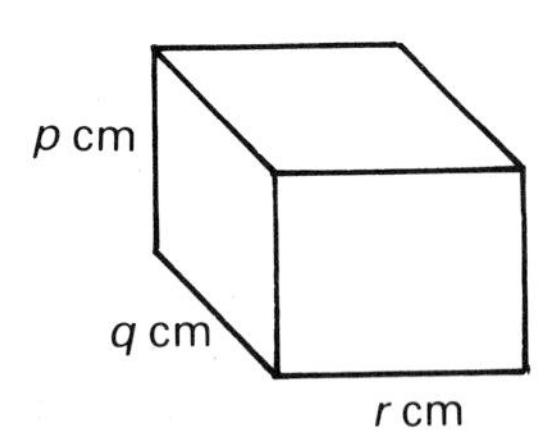

d the perimeter P cm of this 'half-moon' shape.

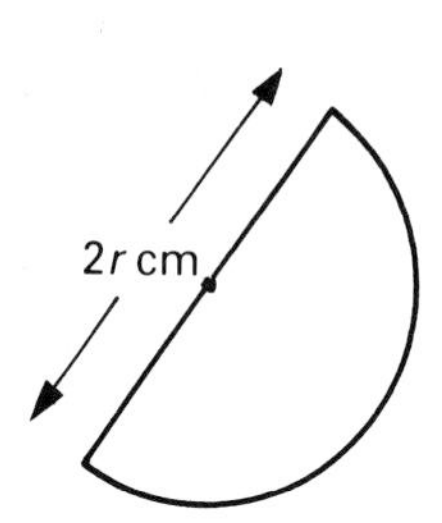

5 a Write down the nth term of the sequence 1, 4, 7, 10, . . .
b Calculate the 50*th* term.

6 Repeat question **5** for the sequence 6, 11, 16, 21, . . .

7 a An approximate formula which converts Celsius temperatures to Fahrenheit is $F = 2C+30$.
Calculate F when C is: (i) 15 (ii) 27.

b The other way round, $C = \frac{1}{2}(F-30)$.
Calculate C when F is: (i) 40 (ii) 100.

8 There are m girls and n boys on 3B's register.
a Make a formula for T, the total number of students present on a day when 3 are absent.
b There are twice as many girls as boys on the roll. Write down a formula for T in terms of n when everyone is present.

SUBSTITUTING NUMBERS IN FORMULAE

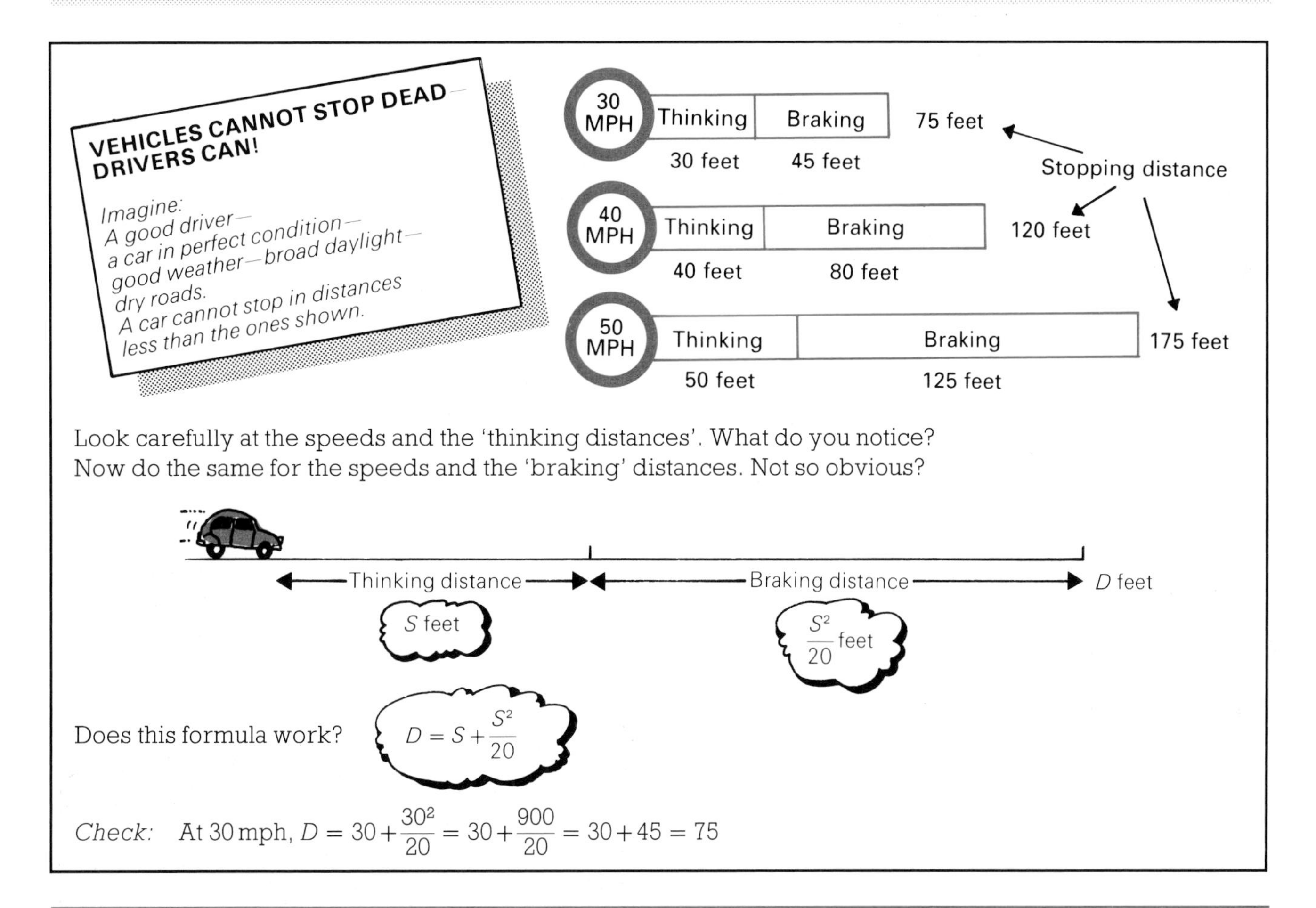

Look carefully at the speeds and the 'thinking distances'. What do you notice?
Now do the same for the speeds and the 'braking' distances. Not so obvious?

Does this formula work? $D = S + \frac{S^2}{20}$

Check: At 30 mph, $D = 30 + \frac{30^2}{20} = 30 + \frac{900}{20} = 30 + 45 = 75$

EXERCISE 1A

1 Use the formula $D = S + \frac{S^2}{20}$ to calculate the total stopping distances at speeds of:
a 10 mph **b** 70 mph.

2 In each part:
(i) write down the formula you choose
(ii) calculate the answer, correct to the nearest whole unit.

$A = lb$ $V = lbh$ $A = \pi r^2$ $C = \pi D$ $D = ST$

a A rectangle has length 6 cm and breadth 5 cm. What is its area?
b A circle has radius 3 m. What is its area?
c A circle has diameter 6 cm. Calculate its circumference.
d At 20 miles per hour for 6 hours, what distance is travelled?
e A cuboid has length 8 mm, breadth 4 mm and height 5 mm. What is its volume?

3 a A formula for the perimeter of the isosceles triangle is $P = 2a + b$. The lengths are in cm.
Calculate P when $a = 15$ and $b = 18$.

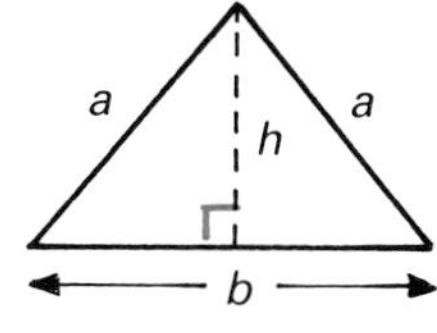

b The area of the triangle, A cm², is given by the formula $A = \frac{1}{2}bh$.
Calculate A when $b = 18$ and $h = 12$.

4 a The perimeter, P cm, of this trapezium is given by $P = a + b + 2c$.
Calculate P when $a = 6.5$, $b = 11.5$ and $c = 5$.
b The area, A cm², is given by $A = \frac{1}{2}h(a+b)$.
Calculate A when $a = 6.5$, $b = 11.5$ and $h = 4.5$.

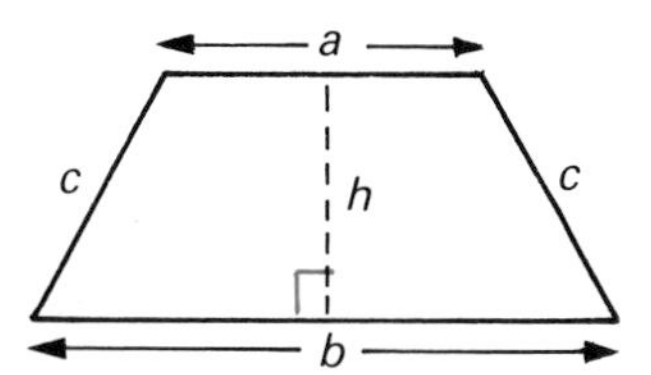

5 Which is warmer, Helsinki or Vienna?
Use this formula to find out.

6

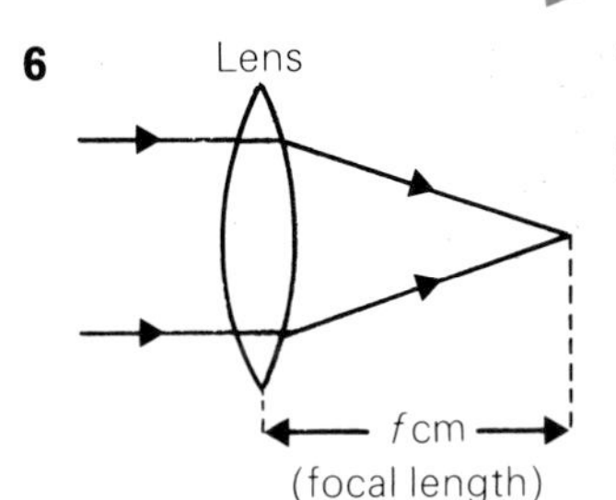

To calculate the power, p units, of a lens, use the formula:

a Which lens is more powerful?
b By how many units?

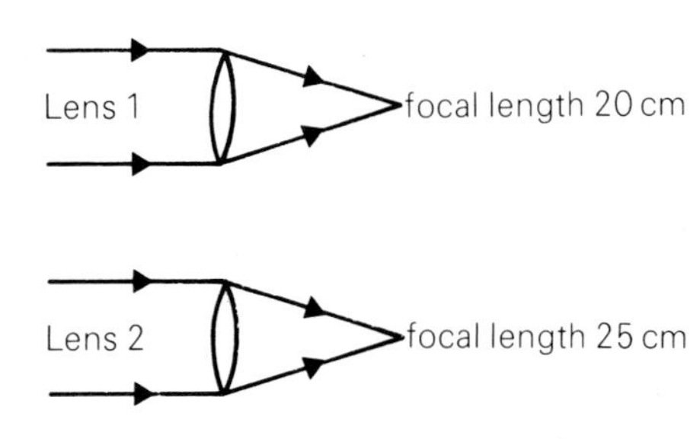

7

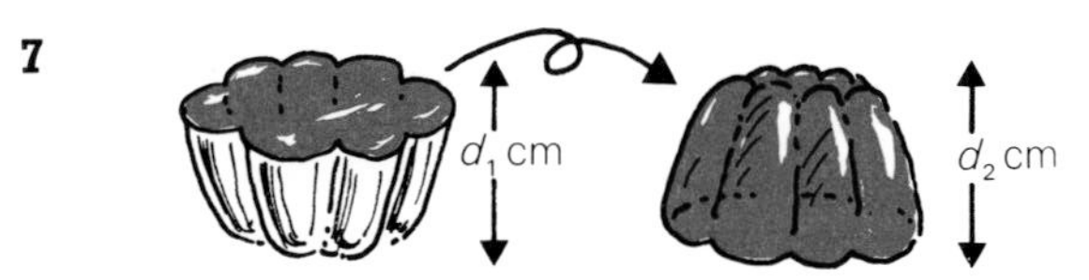

Jelly in mould　　Jelly turned out

The raspberry jelly is 20 cm deep. Turned out, it sags to 17 cm. A strawberry jelly starts at 25 cm, and sags to 21 cm. Use the formula to calculate the percentage sag for each jelly. Which jelly is firmer?

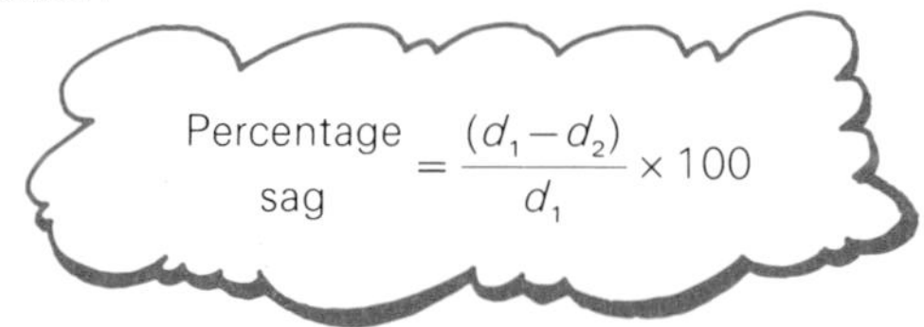

EXERCISE 1B

1 The vibration of the tuning fork, n cycles per second, is given by the formula $n = \frac{85\,000a}{d^2}$.

a Calculate n for forks (i) and (ii), to the nearest cycle per second.

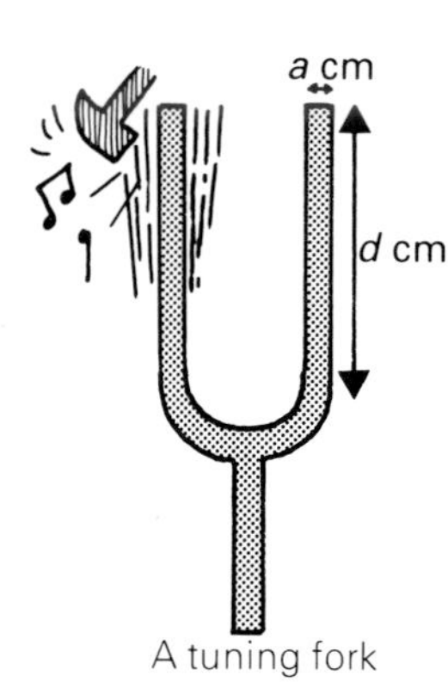

A tuning fork

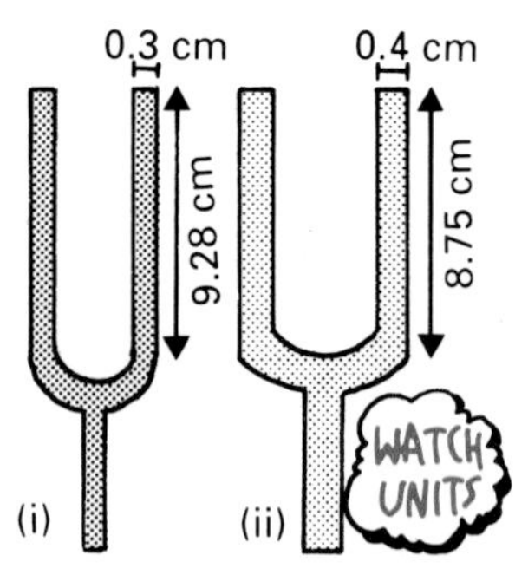

b Which note, D, F or A, does each fork play?

2 a Dropped from 1 metre onto a hard surface, a golf-ball bounced 49 cm. Calculate e, the coefficient of restitution, or 'bounciness', of the ball.

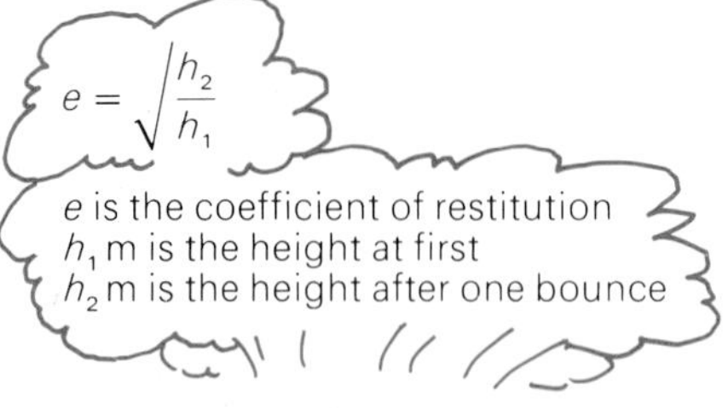

b The ball weighed 46 g, and was hit by a 200 g club-head travelling at 45 m/s. Calculate v m/s, the speed of the ball as it left the clubhead, to the nearest m/s.

$$v = \frac{Ms(1+e)}{M+m}$$

v m/s is the speed of the ball as it leaves the clubhead
M grams is the mass of the clubhead
m grams is the mass of the golf-ball
s m/s is the speed of the clubhead as it hits the ball

c Find the length L yards of the drive in **b**, to the nearest yard.

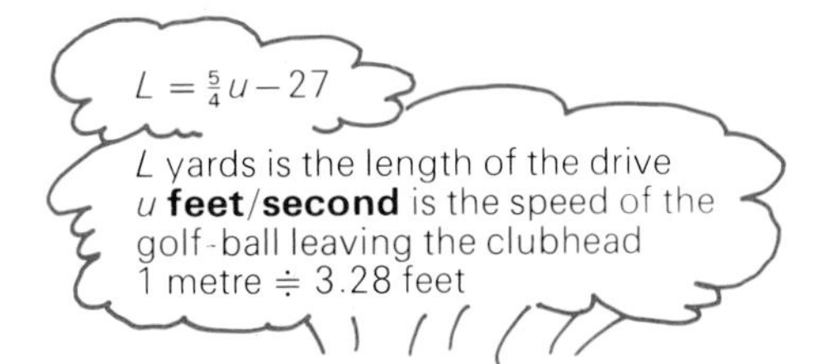

MAKE YOUR OWN FORMULAE

EXERCISE 2

1 Mr Jones runs a music shop. He buys CDs at £c, and sells them at £s. His profit is £p.
a Make a formula for p.
b Use it to find p when $c = 15$ and $s = 19$.

2 Nina has a starting salary of £9500 and annual increases of £250. Her salary after n years is £S.
a Make a formula for S.
b Use it to calculate S when $n = 6$.

3 Jim, Frank and Ali weigh x, y and z kg. Their average weight is A kg.
a Make a formula for A.
b Use it to find A when $x = 65$, $y = 77$ and $z = 54$.

4 Kim bought a personal stereo on hire purchase. The deposit was £D, and she has to make n payments of £p each. The total sum she will pay is £S.
a Make a formula for S.

b Use it to find S when $D = 5$, $n = 12$ and $p = 2$.

5 'The time T minutes' said Andrew 'to cook x kg of meat is 60 minutes plus 40 minutes per kg'.
a Make a formula for T.
b Use it to find T when $x = 2\frac{1}{2}$.

6 The Building Society manager offered Alan and Joy a loan, £L, made up of twice Alan's salary of £x, plus the amount of Joy's salary of £y.
a Make a formula for L.
b Use it to find L when $x = 15\,500$ and $y = 16\,000$.

7 A petrol tank with a capacity of L litres contains x litres. One litre costs p pence.
a Make a formula for the cost £C to fill up the tank.
b Use it to find C when $L = 40$, $x = 8$ and $p = 56$.

8 **a** Find a formula for T, the nth term of the sequence 2, 5, 8, 11, ...
b Use it to calculate T when $n = 50$.

9 Repeat question **8** for the sequence 12, 17, 22, 27, ...

10 A regular polygon with n sides is inscribed in a circle, centre O.

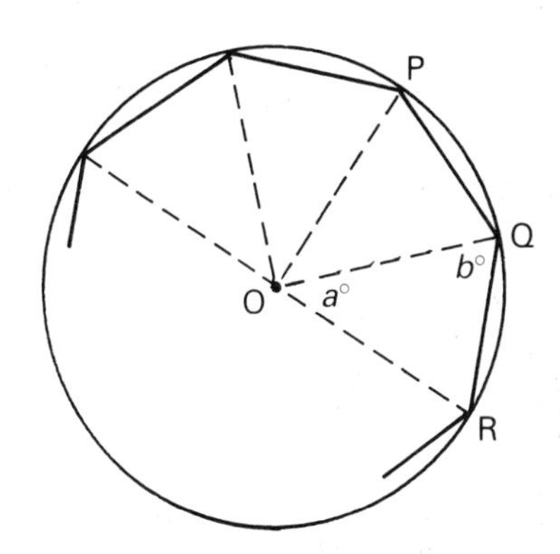

a Find a formula for a in terms of n.
b Use your answer in **a** to find a formula for b in terms of n.
c Write down a formula for the size of ∠PQR in the polygon in degrees, in terms of n.
d Use your three formulae to calculate the angles $a°$, $b°$ and ∠PQR for an octagon.

NEW FORMULAE FROM OLD: CHANGING THE SUBJECT 1

CLASS DISCUSSION

a Calculate the breadth.

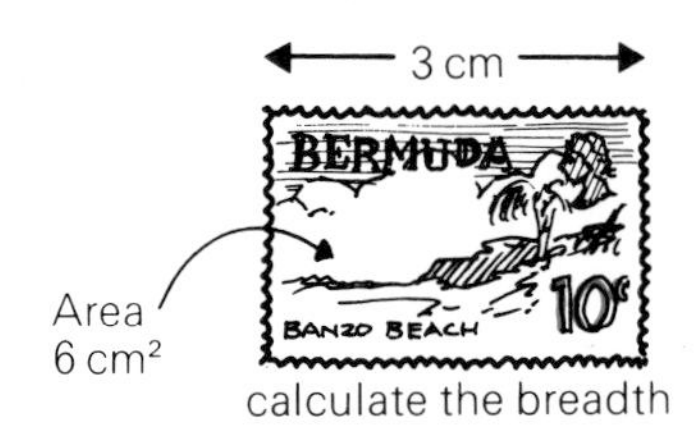

b Calculate the breadth.

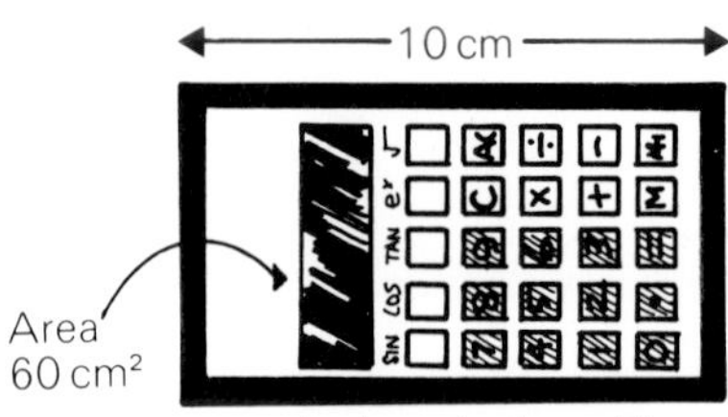

c How can the breadth (b cm) be calculated from the area (A cm^2) and the length (l cm)?

d Copy and complete the formula, $b = \ldots$

In the formula $A = lb$, A is the **subject** of the formula.

In the form $b = \frac{A}{l}$, we have **changed the subject** of the formula to b.

We can change the subject of a formula by using the rules for solving equations.

Examples
Change the subject of each formula to the letter in brackets.

a $a = b + c \quad (b)$
$a - c = b$ (subtracting c from each side)
So $b = a - c$

b $S = \frac{D}{T} \quad (D)$
$ST = D$ (multiplying by T)
So $D = ST$

c $V = IR \quad (R)$
$\frac{V}{I} = R$ (dividing by I)
So $R = \frac{V}{I}$

d $2x - u = v \quad (x)$
$2x = v + u$ (adding u)
$x = \frac{u+v}{2}$ (dividing by 2)

EXERCISE 3A

In questions **1–4**, change the subject to x. Use the rules for solving equations.

1 a $x + a = b$ (subtract a from each side)
b $x + 3 = b$ **c** $x + m = n$ **d** $x - p = q$

2 a $\frac{x}{a} = b$ (multiply each side by a)
b $\frac{x}{3} = b$ **c** $\frac{x}{4} = 9$ **d** $\frac{x}{s} = t$ **e** $\frac{x}{u} = v$

3 a $ax = b$ (divide each side by a)
b $5x = b$ **c** $cx = d$ **d** $mx = n$ **e** $rx = t$

4 a $2x + 3 = b$ **b** $2x + b = a$ **c** $3x - 1 = y$
d $3x - n = m$ **e** $px = q$ **f** $px = q + r$

5 A car travels D km in T hours at an average speed of S km/h.
Formula: $D = ST$
Change the subject to S.

6 The prism has volume V m^3, area of base A m^2 and height h m.
Formula: $V = Ah$
Change the subject to A.

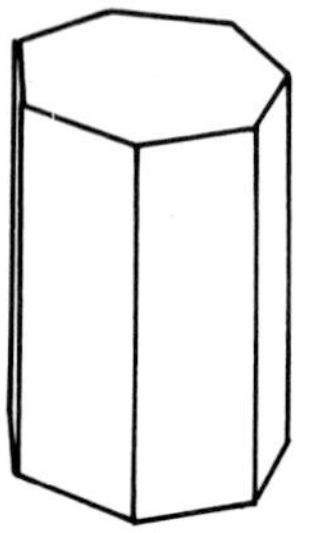

7 *Formula* for pyramid: $V = \frac{1}{3}Ah$
Change the subject to A.

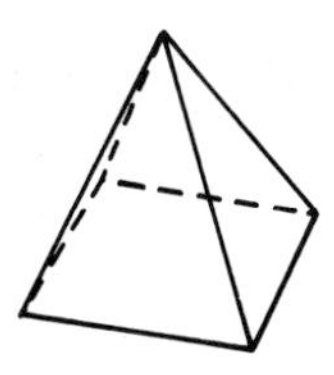

8 Area of rectangle is A km², perimeter is P km.
Formulae: $A = xy$, $P = 2x + 2y$
Change the subject to x in each formula.

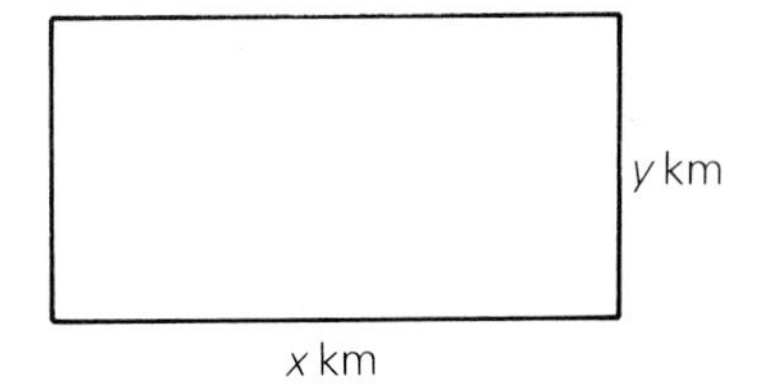

9 Circumference of circle is C cm, area is A cm².
Formulae: $C = 2\pi r$, $A = \pi r^2$
Change the subject to r in each formula.

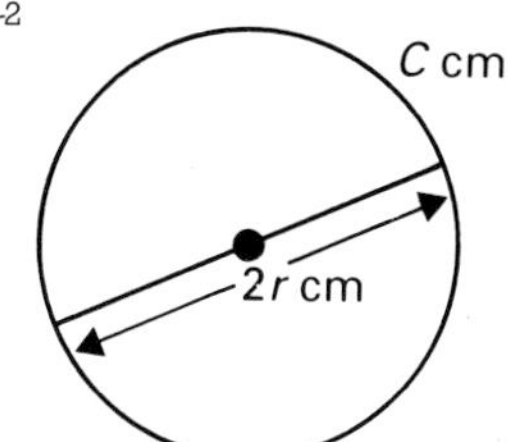

10 A formula for a distance in kilometres is $K = \frac{8}{5}M$, where M is in miles. Make M the subject.

11 The density D of a substance with mass M and volume V is $D = \dfrac{M}{V}$.
Change the subject of the formula to V.

12 A formula for calculating bank interest is
$I = \dfrac{PTR}{100}$. Make R the subject.

13 For velocity, a formula is $v = u + ft$.
Change the subject to: **a** u **b** f.

14 The volume of a cuboid with two square ends is $V = x^2h$.
Change the subject to: **a** h **b** x.

A 3-term triangle trick
When a formula is of the form $A = BC$, A, B or C can be made the subject by covering it up, as shown.

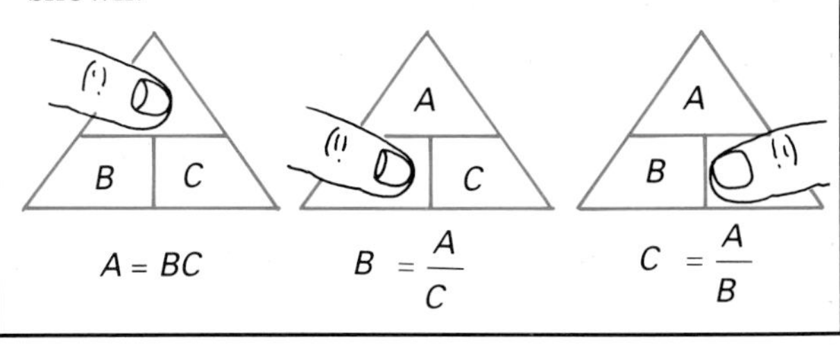

15 Sketch a triangle for each formula below, and cover up part to change the subject to the given letters:

a $D = ST$ (i) S (ii) T
b $A = LB$ (i) B (ii) L
c $x = \dfrac{y}{m}$ (i) y (ii) m
d $D = \dfrac{C}{\pi}$ (i) C (ii) π
e $A = \pi r^2$ (i) r^2 (ii) π
f $y = 2x$ (i) x
g $F = ma$ (i) m (ii) a
h $M = \dfrac{s}{d}$ (i) s (ii) d
i $e = ir$ (i) i (ii) r

EXERCISE 3B

Good advice: Think about changing the subject of a formula for quicker calculations.

1
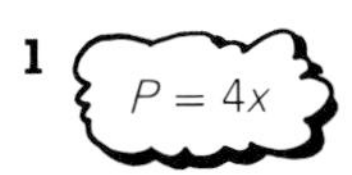

a Make x the subject of the formula.
b Calculate the length of the edge of a table with perimeter 220 cm.

2
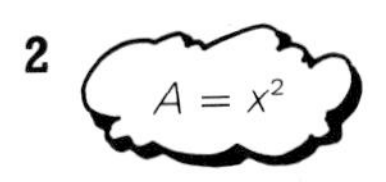

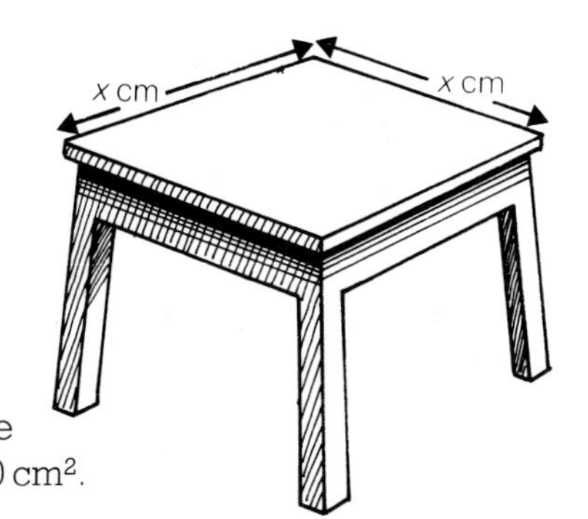

a Make x the subject of the formula.
b Calculate the length of the edge of a table with area 1600 cm².

3

M minutes, H hours. Change the subject of the formula to H, and calculate H when $M = 300$.

4

D days, W weeks.
a Write down the formula.
b Change the subject to W, and calculate W when $D = 91$.

5 FAST CAR HIRE
£50 + £5 a day

The cost £C for n days is given by the formula $C = 50 + 5n$.

a Calculate the cost for 12 days hire.

b Change the subject to n, and calculate n when $C = 90$.

6 a Make a formula for the surface area, A cm^2, of this cube.

b Change the subject of the formula to a.

c Calculate the length of side of a cube with surface area 34.56 cm^2.

7 The surface area S cm^2 of a sphere with radius r cm is $S = 4\pi r^2$.

a Change the subject to r.

b Calculate, correct to 3 significant figures, the radius of a sphere with surface area 100 cm^2.

8 a For the sword below, calculate l when $h = 17$ and $b = 93$.

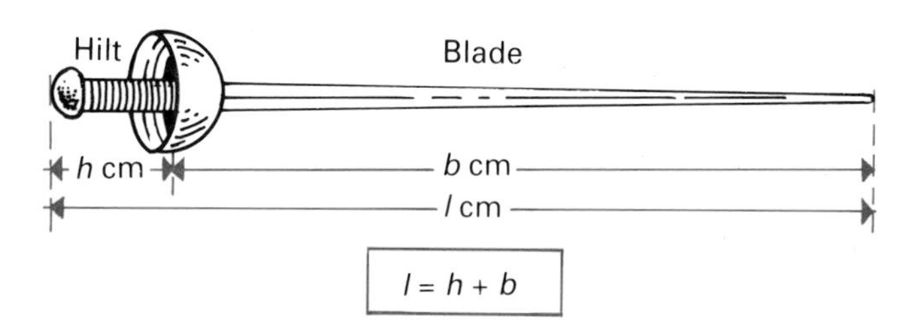

b Change the subject of the formula to b, and calculate b when $l = 130$ and $h = 20$.

9 The number of chirps per minute, c, made by a cricket is related to the Fahrenheit temperature, $F°$, by the formula $F = \frac{c}{4} + 37$.

a Make c the subject of the formula.

b (i) How many chirps does a cricket make at 50°F?

(ii) At what temperature do the chirps cease?

10 a Find a formula for T, the nth term of each of these sequences:

(i) 12, 14, 16, 18, . . .

(ii) $1^2+1, 2^2+1, 3^2+1, 4^2+1, \ldots$

b Change the subject of each to n.

c Calculate n when $T = 122$.

INVESTIGATION

V: Number of vertices
F: Number of faces
E: Number of edges

a *Copy and complete this table.*

Solid	V	F	E
Cuboid	8	6	12
Square pyramid			
Triangular prism			
Tetrahedron			
Pentagonal prism			
Pentagonal pyramid			

b *Find an equation connecting V, F and E for all the solids in the table.*

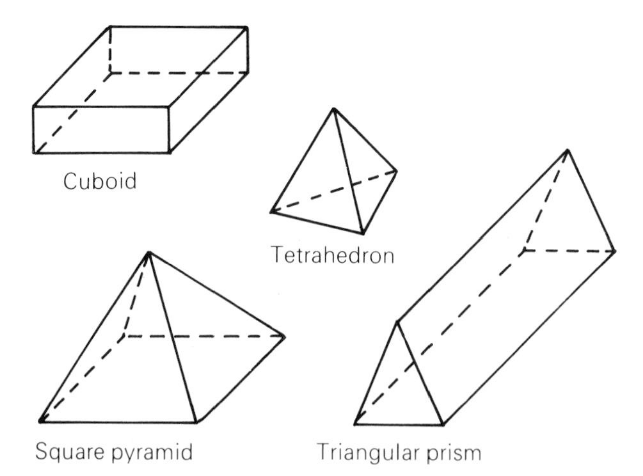

c *Use your equation to obtain a formula for:*
(i) V (ii) F (iii) E

d *An icosahedron has 30 edges and 20 faces. How many vertices does it have?*

e *Investigate whether your formula applies to other solids.*

CHANGING THE SUBJECT 2

More good advice:
1. Remove fractions.
2. Remove brackets.
3. Use rules for solving equations.

Examples

a Change the subject to x

$y = \frac{2}{3}(4x+5)$ . . . fraction
$3y = 2(4x+5)$ . . . brackets
$3y = 8x+10$ . . . . equation
$3y-10 = 8x$
$x = \frac{3y-10}{8}$

b Change the subject to r $(r > 0)$

$V = \frac{1}{3}\pi r^2 h$ . . . fraction
$3V = \pi r^2 h$ . . . . equation
$\frac{3V}{\pi h} = r^2$
$r = \sqrt{\frac{3V}{\pi h}}$

EXERCISE 4A

1 Change the subject to b for each card.

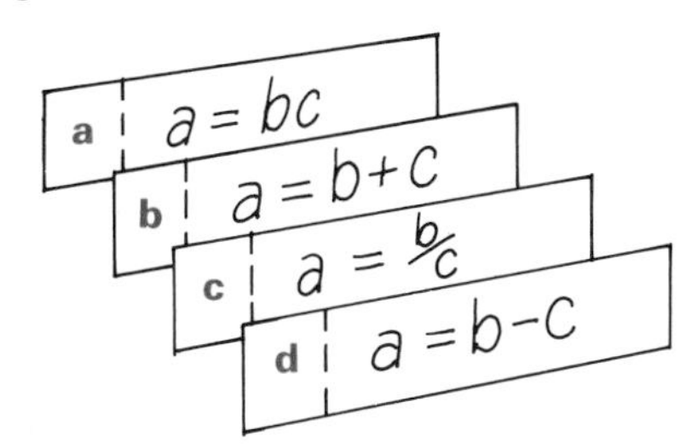

2 Change the subject of each formula to r.

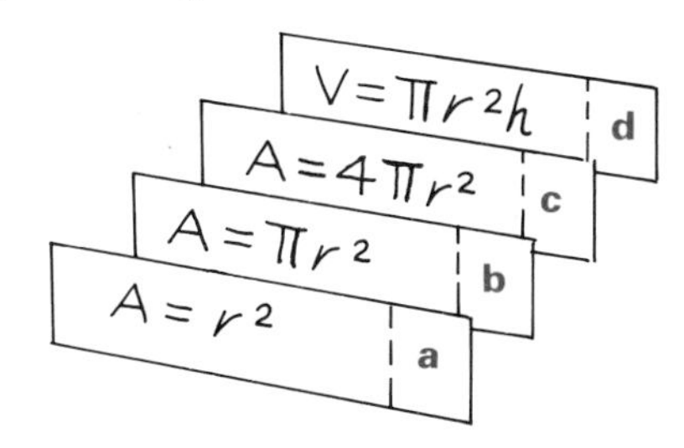

3 Change the subject of each formula to h.

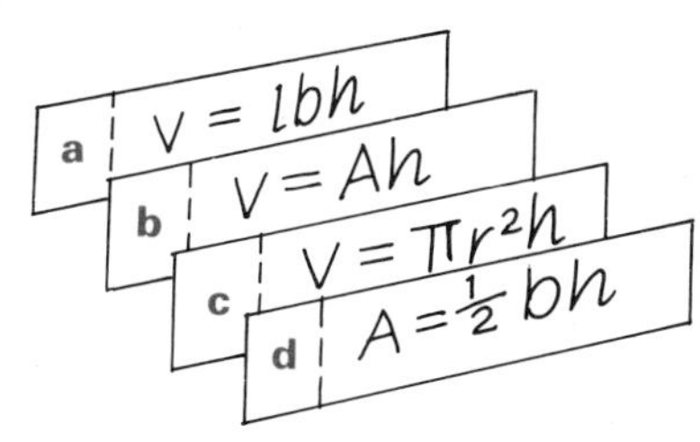

4 Change the subject of each formula to x.

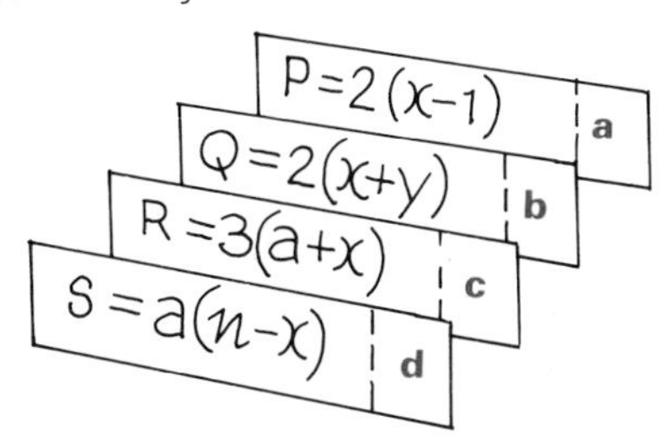

5 Change the subject to the letter in brackets.

a $\frac{V}{R} = I$ (R) **b** $a = \frac{1}{b}$ (b) **c** $I = \frac{C}{d^2}$ (d)

d $r = \sqrt{\frac{A}{\pi}}$ (A) **e** $a = \frac{b}{3c}$ (c) **f** $\frac{Q}{ab} = I$ (b)

g $M = \frac{3}{4d^2}$ (d) **h** $a^2 = b^2+c^2$ (c)

6 Change the subject to x:

a $b = a-x$ **b** $u = v+x$ **c** $t = s-2x$

d $y = 2x-1$ **e** $y = 3(x+1)$ **f** $y = \frac{1}{2}(x-3)$

g $y = \frac{3}{4}(x-1)$

7 The pressure, volume and temperature of a gas are connected by the formula $\frac{PV}{T} = K$. Change the subject of the formula to: **a** V **b** T.

8 In a loop-the-loop roller-coaster the minimum velocity allowed at the top of the loop is $v = \sqrt{gr}$, where r is the radius of the loop. Make r the subject.

9 The distance s travelled by a falling object in time t is given by the formula $s = \frac{1}{2}gt^2$. Change the subject to t.

10 a Change the subject of this formula to d.

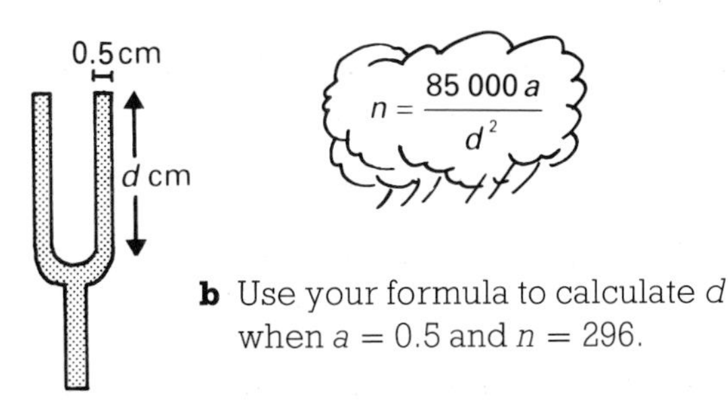

b Use your formula to calculate d when $a = 0.5$ and $n = 296$.

EXERCISE 4B

Examples

a Change the subject to F.

$C = \frac{5}{9}(F-32)$ . . . fraction

$9C = 5(F-32)$ . . . brackets

$9C = 5F-160$ . . . equation

$9C+160 = 5F$

$F = \frac{9C+160}{5}$

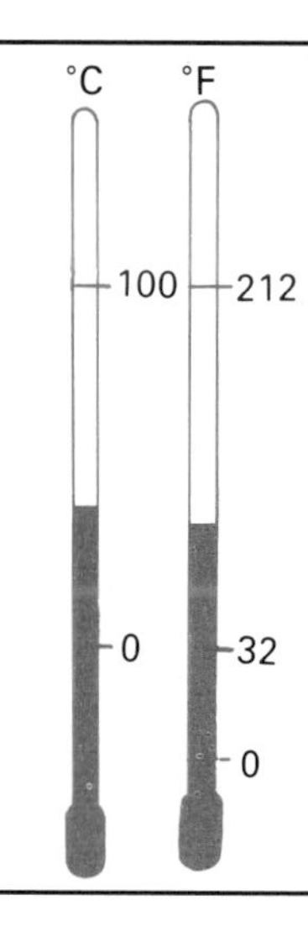

b Change the subject to x.

$y = \frac{x+1}{x-1}$ . . . fraction

$y(x-1) = x+1$ . . . brackets

$yx-y = x+1$ . . . equation, x-terms

$yx-x = y+1$

$x(y-1) = y+1$ . . . **x is a common factor**

$x = \frac{y+1}{y-1}$

1 The gravitational force between planets which are at distance d apart is given by the formula $F = \frac{GMm}{d^2}$.
Change the subject of the formula to d.

2 The formula for the volume of this sphere is $V = \frac{4}{3}\pi r^3$. Make r the subject of the formula.

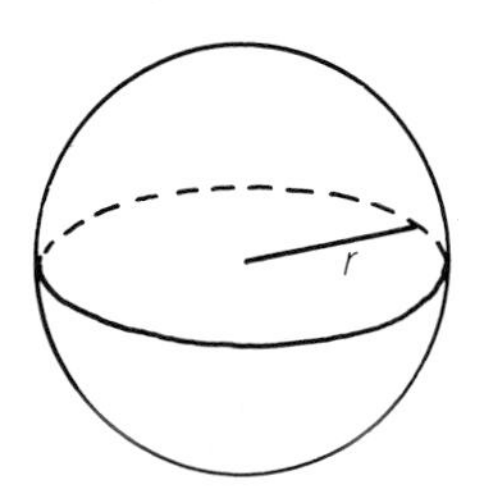

3 The time of swing of a pendulum of length l is given by $T = 2\pi\sqrt{\frac{l}{g}}$. Change the subject to l.

4 This running track with semi-circular ends has perimeter P m.

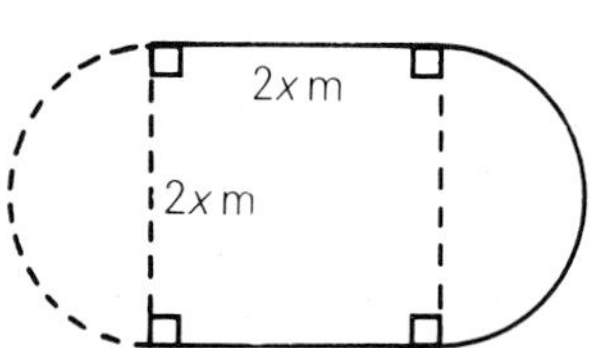

a Prove that $P = 2x(2+\pi)$.
b Make x the subject of the formula.

5 A formula for the area A m² of this trapezium is $A = \frac{1}{2}h(a+b)$. Make h the subject of the formula.

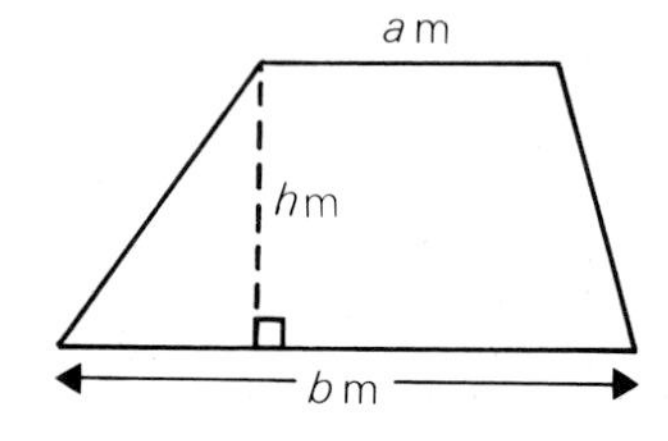

6 The golf ball leaves the club at speed u ft/s. The formula for the length of drive, L yards, is $L = \frac{5}{4}u-27$. Make u the subject of the formula.

7

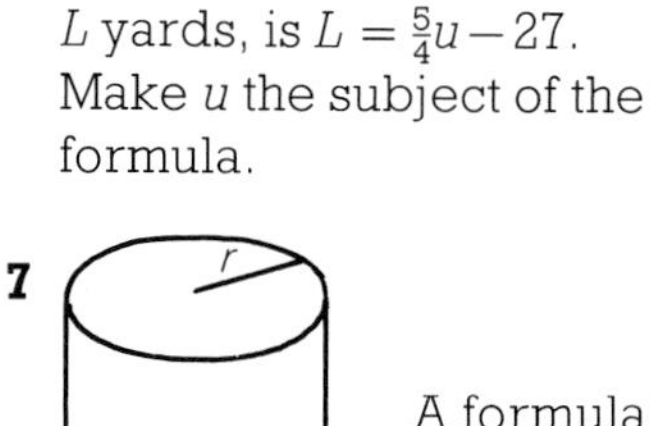

A formula for the total surface area of this cylinder is $A = 2\pi r(r+h)$. Change the subject of the formula to h.

8 Change the subject to x:
a $ax+bx = c$ (x is a common factor)
b $px = qx+r$
c $y = \frac{1}{2}(mx+n)$ **d** $y = \frac{a-x}{x}$ **e** $y = \frac{a-x}{a+x}$

9 The focal length, f cm, of this lens is given by the formula $\frac{1}{u}+\frac{1}{v} = \frac{2}{f}$.

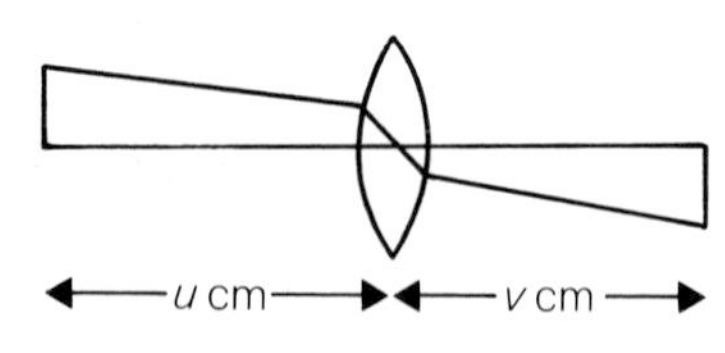

Make v the subject of the formula, by first multiplying each term by the lcm uvf.

10 The power, p units, of this lens pair is $p = \frac{f_1+f_2}{f_1f_2}$.

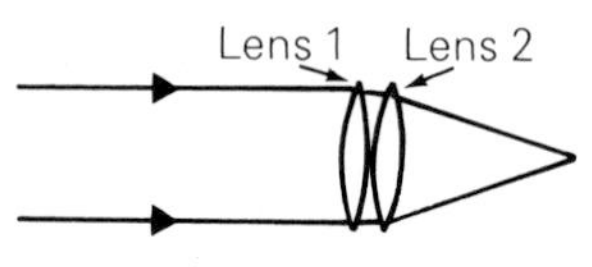

a Change the subject of the formula to f_1.
b Calculate f_1 when $p = 5$ and $f_2 = 0.3$.

THE UPS AND DOWNS OF FORMULAE

In a formula like $A = xy$ for the area of a rectangle, A, x and y are called **variables**, as they can take different values.
What happens to the value of A if y is fixed, and x:
a increases **b** decreases **c** is doubled?

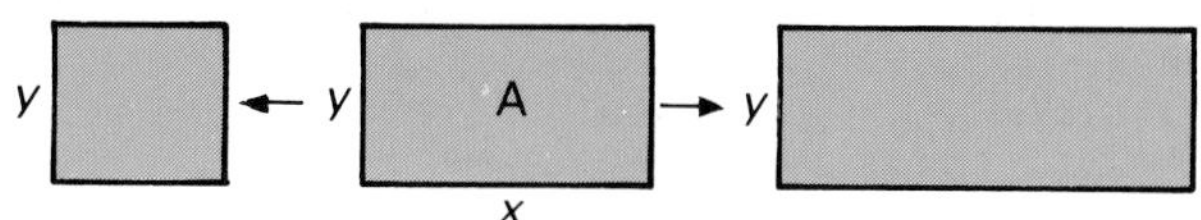

In **a**, the value of A increases, in **b** it decreases, and in **c** it is doubled.

EXERCISE 5

Any variables not mentioned in the changes are kept fixed.

1 In the distance-speed-time formula $D = ST$, what happens to D if T:
a increases **b** decreases
c is doubled **d** is halved?

2 In the electrical formula $I = \dfrac{240}{R}$, what happens to I if R:
a increases **b** decreases
c is doubled **d** is halved?

3 In the circle formulae $A = \pi r^2$ and $C = 2\pi r$ what happens to A and C if r:
a increases **b** decreases **c** is doubled?

4 At a fixed temperature, the pressure and volume of gas in a cylinder obey the 'law' $PV = K$, where K is constant. What is the effect on:
a P if V increases
b V if P decreases
c P if V is doubled?

5 A person's weight in space is given by
$W = w\left(\dfrac{r}{r+h}\right)^2$. What is the effect on W of:
a doubling w **b** increasing h?

6 The force of attraction between planets is
$F = \dfrac{Gm_1m_2}{d^2}$. What is the effect on F of:
a reducing m_1 and m_2
b halving d?

CHALLENGES

*1 Ohm (Ω) is the unit of resistance.
Volt (V) is the unit of electromotive force.
Ampere (A) is the unit of current.
Watt (W) is the unit of power.*

Look at these two formulae: $V = A\Omega$; $V = \dfrac{W}{A}$.

There is one more formula with V as subject. It involves Ω and W. Can you find it? Can you find more formulae for Ω?

2 An article is bought for £x and sold for £y. The profit is p % of the cost price. Make a formula with p as the subject.

INVESTIGATION

In 1772, the German astronomer Johann Bode gave the formula $D = \dfrac{3 \times 2^{n-2} + 4}{10}$ for the distance D, in astronomical units, of the nth planet out from the sun.
Use Bode's formula and the table of up-to-date data, to investigate the accuracy of the formula, and write a report. Note: $2^{-1} = \frac{1}{2}$; $2^0 = 1$. (Uranus, Neptune and Pluto had not even been discovered in 1772.)

Planet's number (n)	Planet's name	Actual distance, in Astronomical Units
1	Mercury	0.4
2	Venus	0.7
3	Earth	1
4	Mars	1.5
5?	—	—
6	Jupiter	5.2
7	Saturn	9.5
8	Uranus	19.2
9	Neptune	30.1
10	Pluto	39.5

CHECK-UP ON FORMULAE

1 For each question, choose the correct formula.

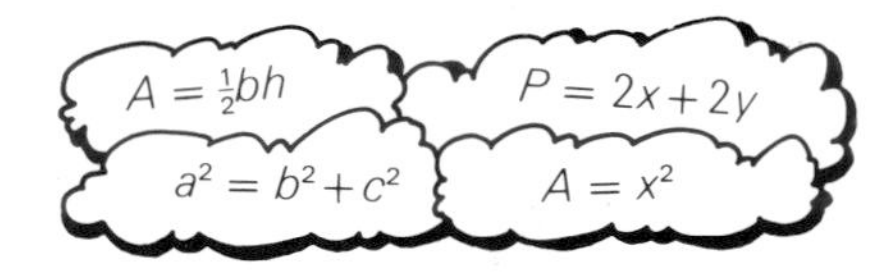

Use it to calculate:

a the area of a square of side 15 mm

b the length of the hypotenuse of a right-angled triangle with sides 7.5 cm and 10 cm long

c the perimeter of a rectangular field 125 m long and 88 m broad

d the area of a triangle with base 16 cm long and altitude 9 cm.

2 a Make a formula for the surface area A cm² of this cuboid.

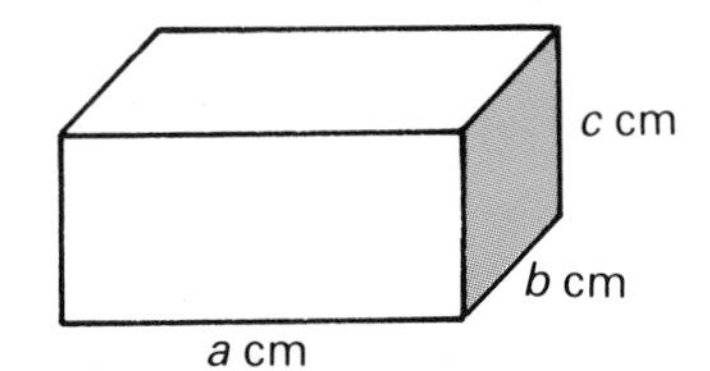

b Calculate A if $a = 8$, $b = 6$ and $c = 4$.

3 a Find a formula for the nth term, t, of the sequence 5, 12, 19, 26, . . .

b Change the subject to n, and calculate n when $t = 89$.

4 The square has side a cm long, and each triangle has altitude a cm.

a Explain why the area of the shape, A cm², is given by $A = 3a^2$.

b Calculate:

(i) A, given $a = 12$

(ii) a, given $A = 75$.

c Make a the subject of the formula.

5 The volume V of a cone with radius of base r and height r is $V = \frac{1}{3}\pi r^3$.
Change the subject of the formula to r.

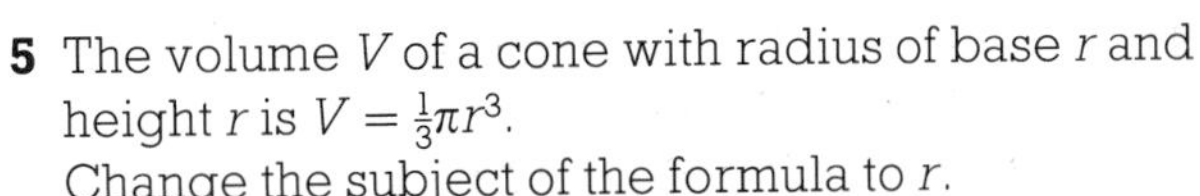

6 A snooker player pots 6 reds (1 point each), and another colour (x points each time) 5 times.

a Make a formula for the total points T scored.

b Make x the subject of the formula.

c If the total is 21, find the value of the colour.

7 A hollow bowl of depth h cm is cut from a sphere of radius R cm. The volume, V cm³, of the bowl is:

$$V = \frac{\pi h^2}{3}(3R - h).$$

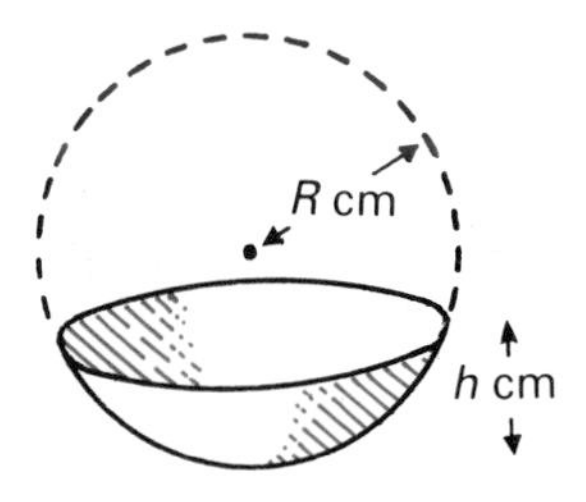

Make R the subject of the formula.

8 Make x the subject in each of these:

a $x + y + 1 = 0$ **b** $y = 5x - 2$ **c** $y = 3(x - 2)$

d $y = \frac{3}{4}(x + 1)$ **e** $x^2 - a^2 = b^2$ **f** $P = \sqrt{\frac{1}{x}}$

g $y = \frac{2x + 1}{x}$ **h** $y = \frac{2x + 1}{2x - 1}$

9 The straight part of this running track is $2r$ m long, and each semi-circle has radius r m.

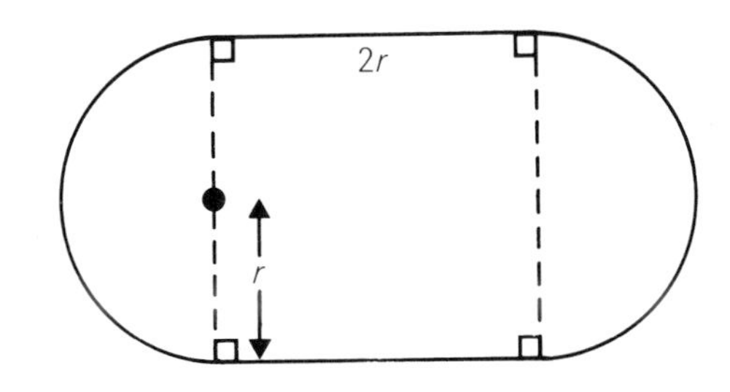

a (i) Make a formula for the area of the shape.
(ii) Calculate A, to the nearest m², given $r = 40$.

b (i) Make r the subject of your formula.
(ii) Calculate r, to the nearest metre, given $A = 17\,500$.

10 In the formula $h = \frac{A}{\pi r^2}$, what is the effect on h if:

a A increases **b** r increases

c A is halved **d** r is doubled?

11 $\frac{1}{R} = \frac{1}{R_1} + \frac{1}{R_2}$ is a formula for the resistance of an electrical circuit.
Change the subject of the formula to R_1.

15 PROBABILITY

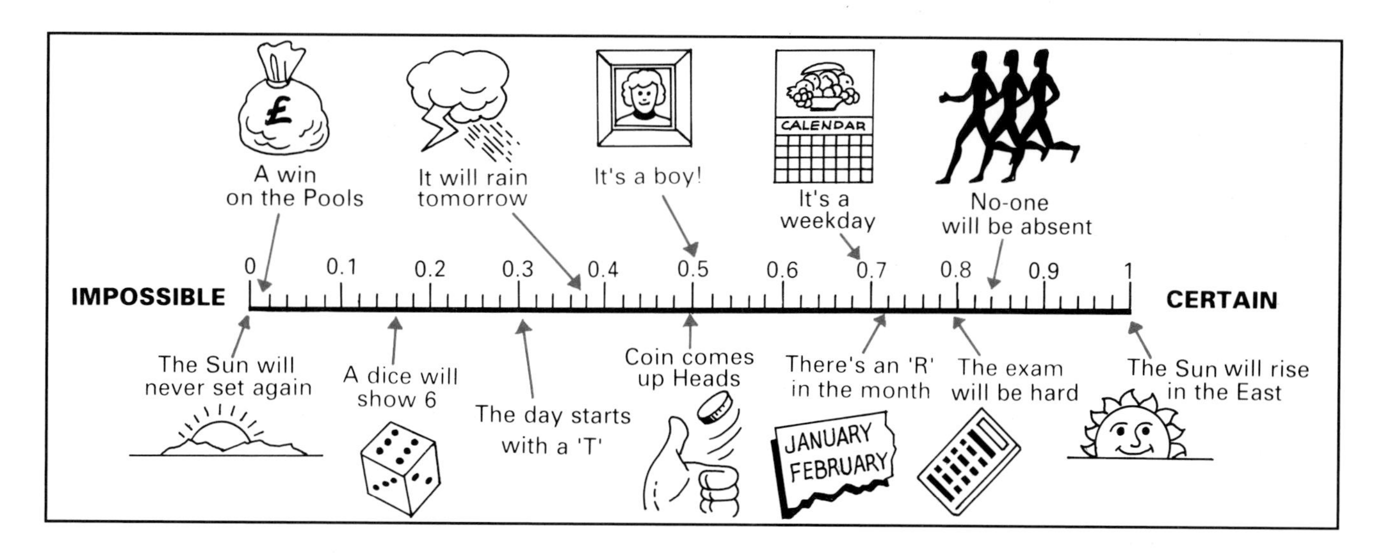

LOOKING BACK

1 One of the probabilities above is definitely wrong. Which one?

2

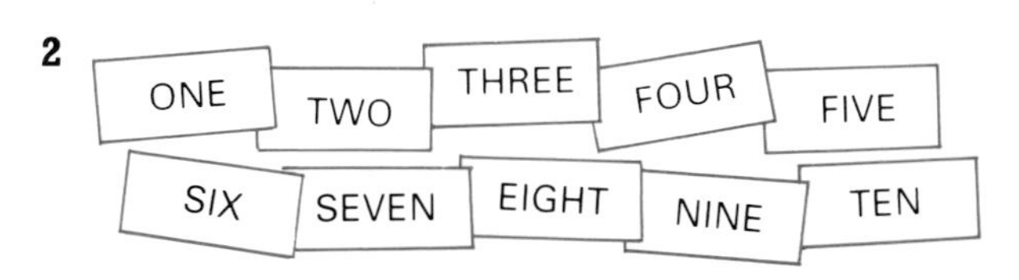

One of the above cards is chosen at random.

a Calculate the probability that it:
(i) has three letters
(ii) begins with 'F'
(iii) contains at least one 'E'
(iv) contains exactly one 'E'
(v) has six letters
(vi) has more than two letters.

b Copy this scale, and mark all the probabilities from **a** on it, using arrows:

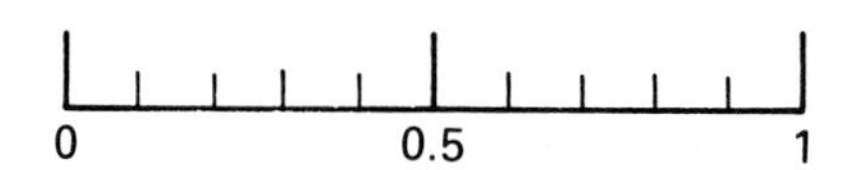

3 Red and white counters are mixed up in the bags as shown, and one is chosen at random. Copy and complete the tree diagrams.

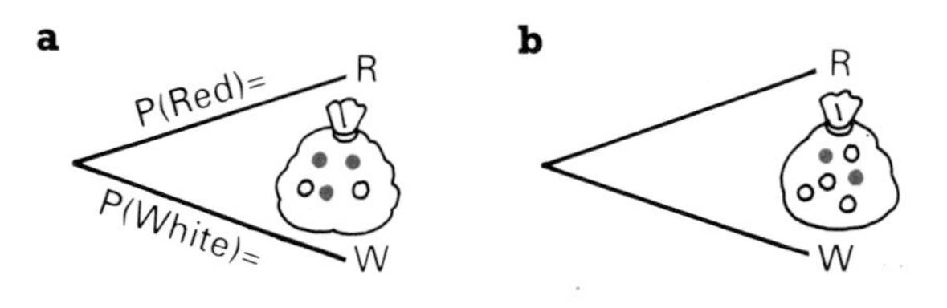

4 What is the probability of getting:

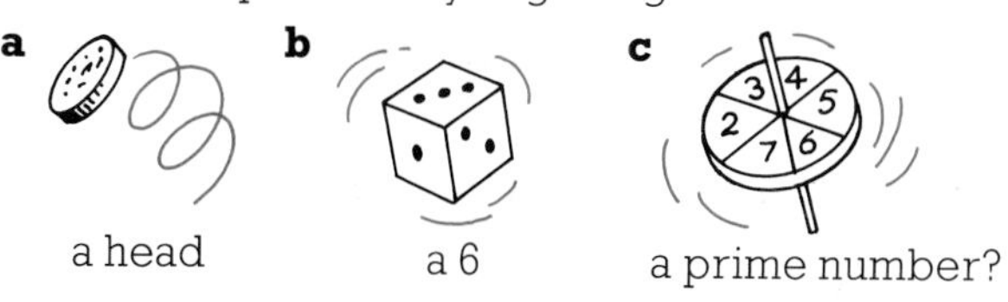

a a head **b** a 6 **c** a prime number?

5 **Survey—Experiment—Past data—Counting equally likely outcomes.**
Which of these methods would you choose to estimate or calculate the probability that:
a Galloping Fury will win the 3 o'clock race
b a Sparky matchbox will contain exactly the average number on the box
c Evelyn will pick a letter T at Scrabble
d a shoe, when dropped, will land on its sole?

6 Over the summer, Jason kept records of the arrival times of his morning train. He estimates that P(early) = 0.5 and P(on time) = 0.3.
a Estimate P(late).
b Out of 200 days, estimate the number of times Jason can expect the train to be:
(i) early (ii) on time (iii) late.

7 Peter kept a record of his scores at the short fourth hole on the Sealinks golf course. He calculated these probabilities.

Number of strokes	2	3	4	5	over 5
Probability	$\frac{1}{20}$	$\frac{1}{4}$	$\frac{1}{3}$	$\frac{1}{5}$	$\frac{1}{6}$

How many of each score should he expect in his next 20 rounds of golf?

COMBINING PROBABILITIES

When two coins are tossed we can show the possible outcomes:

(i) in a list (H, H), (H, T), (T, H), (T, T)

(ii) in a table of outcomes

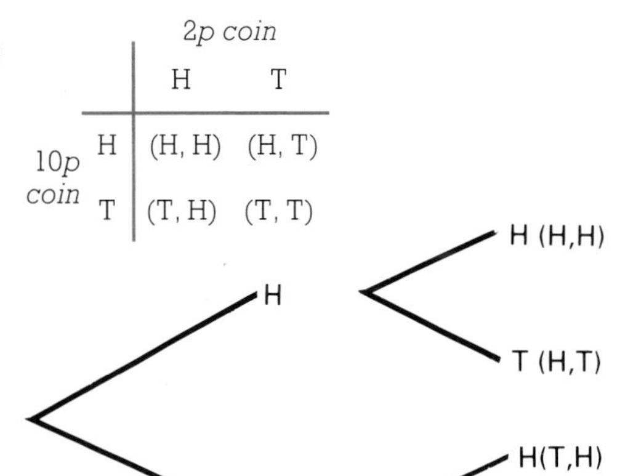

		2p coin	
		H	T
10p coin	H	(H, H)	(H, T)
	T	(T, H)	(T, T)

(iii) in a tree diagram.

P(both heads) = $\frac{1}{4}$, P(both tails) = $\frac{1}{4}$, P(one head and one tail, in any order) = $\frac{1}{2}$

EXERCISE 1A

1 Sally spins her racket to find out who serves first in badminton. The racket is equally likely to fall with its smooth side (S) or its rough side (R) up.

a Copy and complete this table of outcomes for two games.

		2nd game	
		S	R
1st game	S	(S, S)	
	R		

b Calculate: (i) P(both S) (ii) P(both R) (iii) P(one S and one R).

c Copy and complete the tree diagram, to check your answers to **b**.

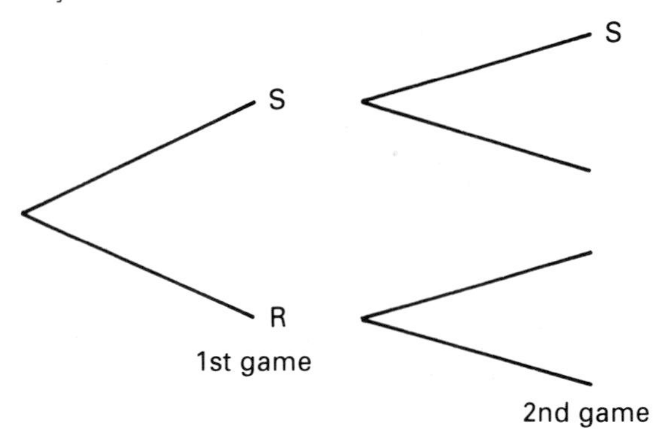

2 A game starts with these cards, which are turned over, face down.

Alison picks one at random from each pair.

a Copy and complete the table of outcomes.

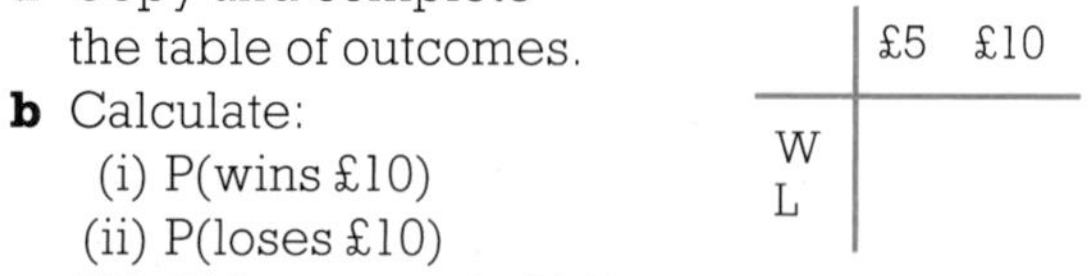

	£5	£10
W		
L		

b Calculate:

(i) P(wins £10)

(ii) P(loses £10)

(iii) P(does not win £10)

c Copy and complete the tree diagram.

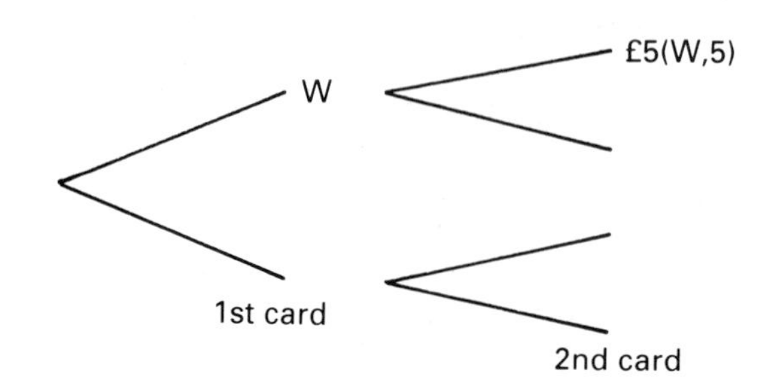

3 In a 'football forecast' game, a disc is marked Home (H) on one side and Away (A) on the other side.
There is also a spinner marked Win (W), Lose (L) and Draw (D).

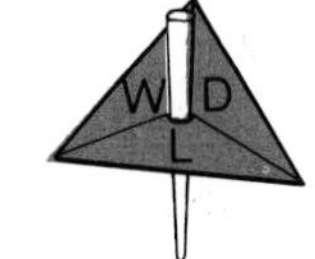

a Copy and complete the table of outcomes when the disc and the spinner are both used.

	W	D	L
H			
A			

b Calculate:
(i) P(a home win)
(ii) P(a win at home or away)

c 2 points are awarded for a win and 1 for a draw. What is the probability of scoring at least 1 point?

d Copy and complete the tree diagram.

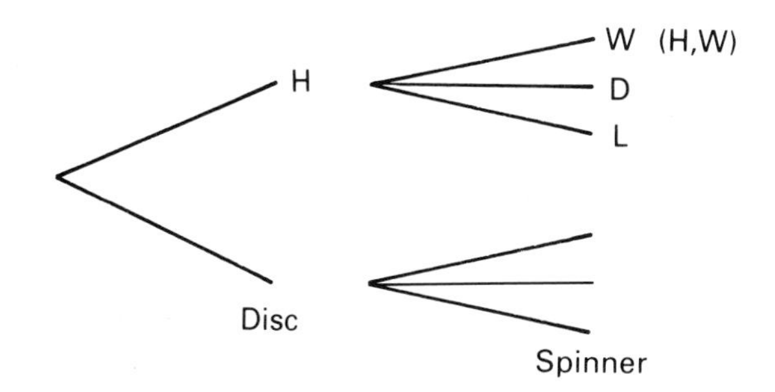

4 You get six colour-coordinated items when you buy this collection.

	Blouse (B)	T-shirt (T)	Jumper (J)
Shorts (Sh)	(Sh, B)		
Skirt (Sk)			
Slacks (Sl)			

Rita chooses a pair of items at random to see the effect, one from the shorts, skirt and slacks, and one 'top'.

a Copy and complete the table.

b Calculate the probability that her outfit:
(i) doesn't contain a skirt
(ii) contains a blouse, but not slacks
(iii) doesn't contain a T-shirt or shorts.

5 Maya had two spins.

a Copy and complete the tree diagram.

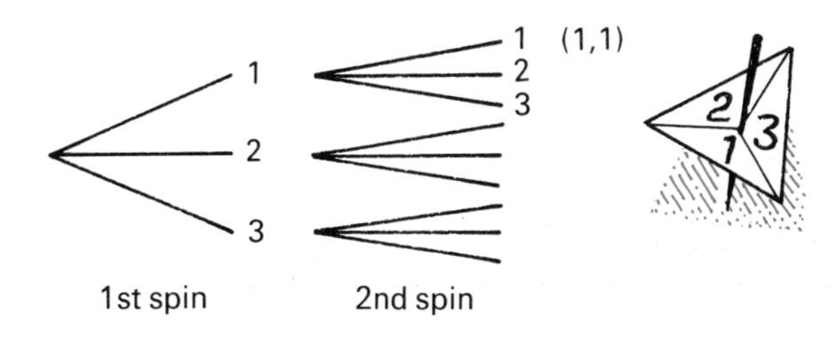

b Calculate:
(i) P(3, 3) (ii) P(2 and 3)
(iii) P(total of 4) (iv) P(total greater than 4).

EXERCISE 1B

1 Two drawers each contain a knife, fork and spoon. Nicole selects one at random from each drawer.

a List all the possible pairs in a table.

b Calculate:
(i) P(two knives) (ii) P(both items the same)
(iii) P(a knife and a fork).

c Illustrate by a tree diagram.

2 a Copy and complete this table for the toss of a red dice and a black dice.

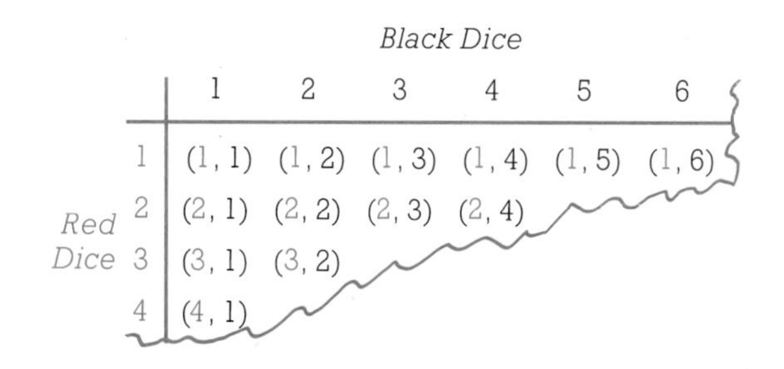

Black Dice

Red Dice	1	2	3	4	5	6
1	(1, 1)	(1, 2)	(1, 3)	(1, 4)	(1, 5)	(1, 6)
2	(2, 1)	(2, 2)	(2, 3)	(2, 4)		
3	(3, 1)	(3, 2)				
4	(4, 1)					

b How many possible outcomes are there?

c Calculate:
(i) P(6, 6) (ii) P(at least one dice shows 6)
(iii) P(both dice show the same score)
(iv) P(red score is greater than black score).

3 Alan's bike has three forward gears and six rear gears.

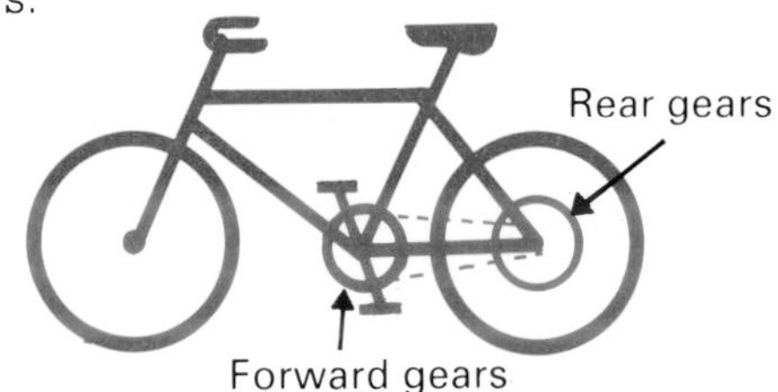

a Copy and complete this table of all gear pairings.

Rear gears

Forward gears	1	2	3	4	5	6
H	(H, 1)					
M						
L						

b If Alan selects gears at random, find the probability that he uses:
(i) a low forward gear
(ii) a high forward gear and rear gear 6
(iii) a middle forward gear and a rear gear of 4 or less.

4 Three coins are tossed.
a Copy and complete this tree diagram.

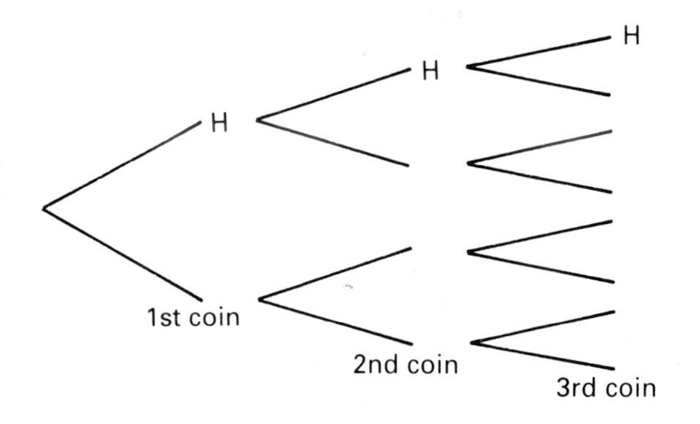

b How many equally likely outcomes are there?
c Calculate:
(i) P(H, H, H) (ii) P(T, T, T)
(iii) P(only one H) (iv) P(at least one H)

5 Lisa: 'I was born on a Tuesday'.
Ashmid: 'So was I'.
a Make a table of all possible pairs of days in the week when two people might have been born.
b Calculate:
(i) P(both were born on a Tuesday)
(ii) P(both were born on the same day of the week)
(iii) P(at least one was born at the weekend)
(iv) P(both were born on a weekday)

BRAINSTORMER

In a game, Claire uses a dice. Teresa uses two equilateral triangle spinners, adding the scores together.

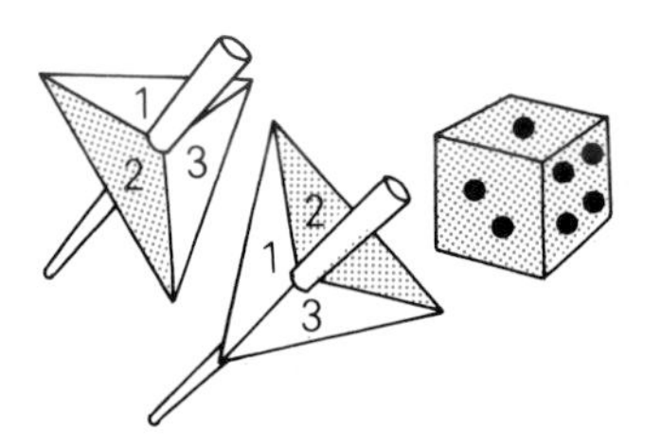

Claire rolls the dice, and Teresa spins the spinners. After each turn they compare scores; the higher score gets 1 point.
a *Who do you think is more likely to reach 20 points first?*
b *Show Teresa's possible scores for each pair of spins in a table. Use this to explain who is more likely to win, and why.*
c *Try the game with a partner.*

CHALLENGE

Draw a 3 × 3 grid of squares of side 4 cm. Throw a 1p coin (diameter 2 cm) 50 times onto the grid.

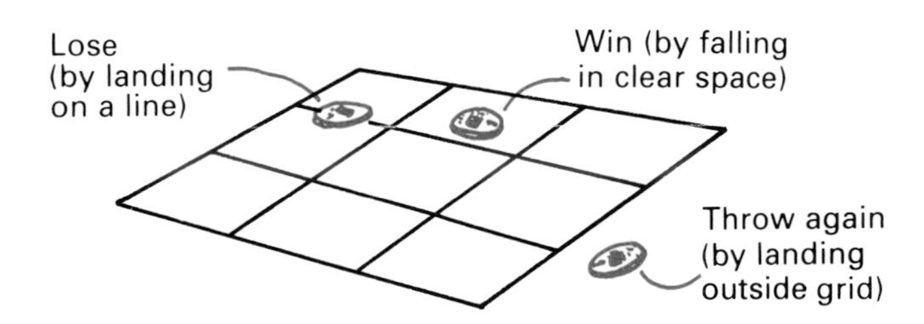

a *Estimate the probability of a win.*
b *Calculate the probability of a win. (Think about the centre of the coin.) Compare your answers.*

INVESTIGATIONS

1 *When two coins are tossed, there is one way of getting no heads, (T, T); two ways of getting one head, (H, T) and (T, H); and one way of getting two heads, (H, H).*
Investigate the number of ways of getting heads with more than two coins.

2 *Can your results in* **1** *help you to solve this problem? There are five sets of traffic lights in Union Street. In how many ways can you be stopped by three of the sets of lights?*

P (A OR B)

Examples

1 When the Wheel of Fortune is spun,
P(winning £10) $= \frac{1}{8}$
P(winning £5) $= \frac{2}{8}$
P(winning £10 or £5) $= \frac{3}{8}$.

Notice that P(winning £10 or £5) $= \frac{3}{8} = \frac{1}{8} + \frac{2}{8} =$ P(winning £10) + P(winning £5).

2

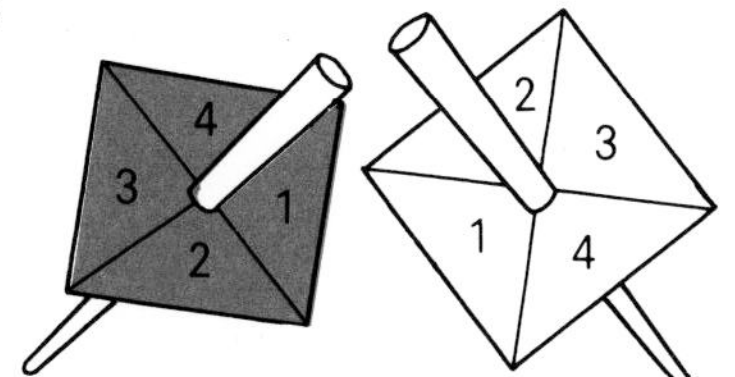

The table shows all possible totals with these two spinners.

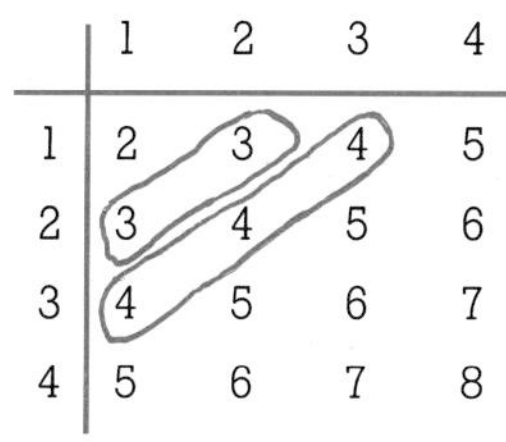

		White spinner			
		1	2	3	4
Red spinner	1	2	3	4	5
	2	3	4	5	6
	3	4	5	6	7
	4	5	6	7	8

P(total of 3) $= \frac{2}{16}$, and P(total of 4) $= \frac{3}{16}$.
P(total of 3 or 4) $= \frac{5}{16}$ (from the table).
Notice that P(total of 3 or 4) $= \frac{5}{16} = \frac{2}{16} + \frac{3}{16} =$ P(total of 3) + P(total of 4).

You can **add** the probabilities when the two events cannot happen at the same time, that is, the events are **mutually exclusive**.

In *Example* **1**, you cannot win £10 and £5 in one spin; in *Example* **2**, you cannot score a total of 3 and a total of 4 in one pair of throws.

For two mutually exclusive events, P(A or B) = P(A) + P(B)

EXERCISE 2A

1 A coin is tossed. Calculate:
a P(H) **b** P(T) **c** P(H or T).
Explain your answer to **c**.

2 A dice is rolled. Calculate:
a P(1) **b** P(2) **c** P(1 or 2).

3 A pack of cards is cut. Calculate:
a P(Ace of Hearts) **b** P(Ace of Spades)
c P(Ace of Hearts or Ace of Spades).

4 A box contains 3 packets of Crinkly Crisps and 5 packets of Crunchy Crisps. A packet is taken out at random. Calculate:
a P(Crinkly) **b** P(Crunchy)
c P(Crinkly or Crunchy).

5 Now, 2 packets of Curly Crisps are put in the box with 3 Crinkly and 5 Crunchy packets. A packet is taken out at random. Calculate:
a P(Curly) **b** P(Crinkly) **c** P(Crunchy)
d P(Curly or Crinkly or Crunchy).

6 Michelle estimates that the probability that her hockey team will win their next game is 0.2, and the probability that they will draw is 0.5.
Calculate:
a P(win or draw) **b** P(lose).

7 The weather forecaster says that the probability of rain tomorrow is 0.3 and of snow is 0.1.

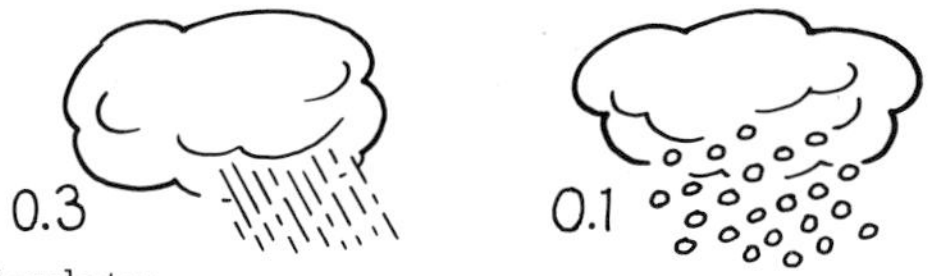
0.3
0.1

Calculate:
a P(rain or snow) **b** P(no rain and no snow).

8 Peter chooses a date at random in April for a party. Calculate the probability that he chooses:
a a Saturday **b** a Sunday
c a Saturday or Sunday

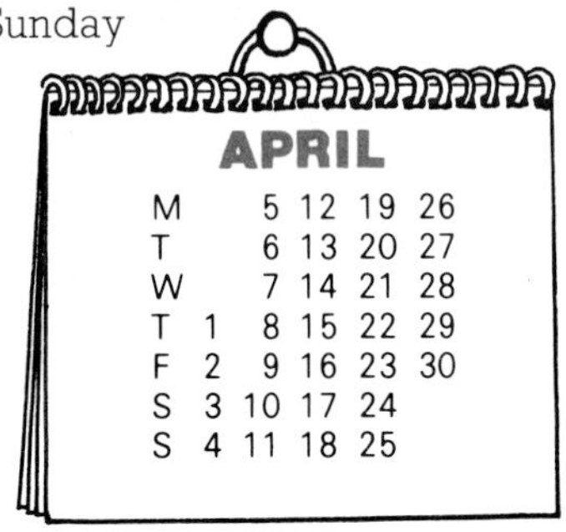
APRIL
M 5 12 19 26
T 6 13 20 27
W 7 14 21 28
T 1 8 15 22 29
F 2 9 16 23 30
S 3 10 17 24
S 4 11 18 25

9 In a 40 period week, Ian has 6 periods of Mathematics, 3 of Games and 1 of Careers. Next week he has to miss 1 period for an eyesight test.
a What is the probability that he will miss:
(i) Mathematics (ii) Mathematics or Games
(iii) Mathematics or Games or Careers?
b What assumptions did you make?

10 A square spinner is given two spins.
a Make a table of all possible outcomes, with headings *First spin* and *Second spin*.
b Calculate the probability of scoring:
(i) the same number each time
(ii) a total of 6 or less
(iii) a total of 5 or 6.

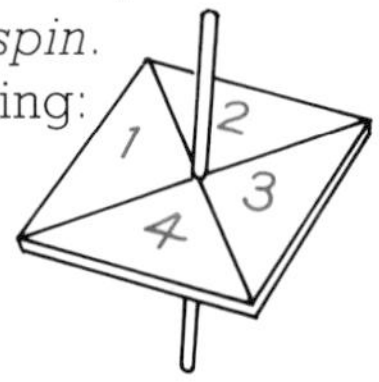
1
2
3
4

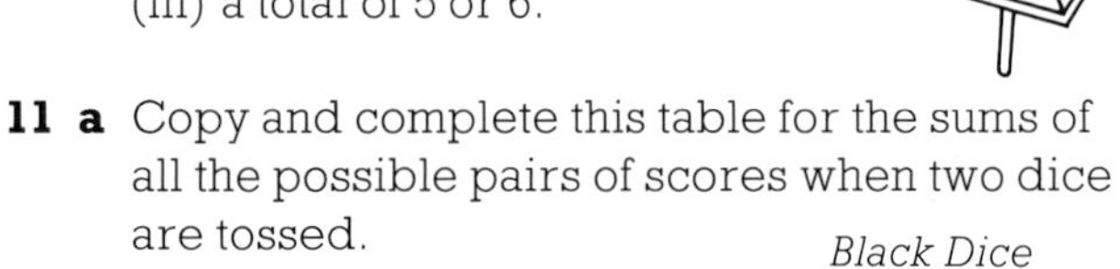

11 a Copy and complete this table for the sums of all the possible pairs of scores when two dice are tossed.
b Which sum is:
(i) most likely
(ii) least likely?
c Calculate: P(2), P(3), P(4), . . ., P(12).
d Calculate also:
(i) P(10 or over) (ii) P(4 or less).

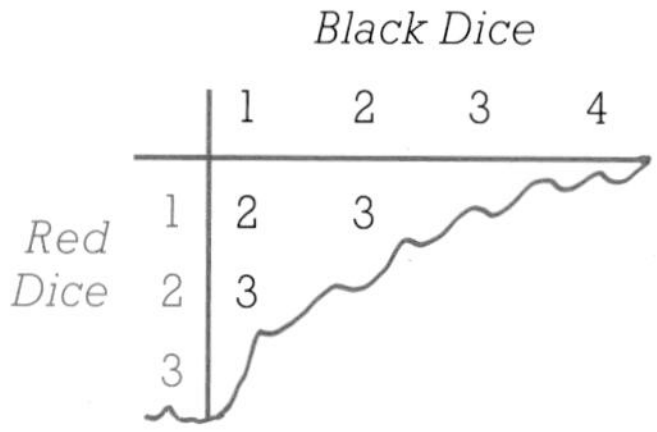
Black Dice
1 2 3 4
Red Dice
1 2 3
2 3
3

EXERCISE 2B

1 For each of these activities say whether or not the two given events are mutually exclusive.
a Tossing a coin: showing a head, showing a tail.
b Playing a rugby match: ending in a win, ending in a draw.
c Choosing a school child at random: getting a girl, getting a child 13 years old.
d Choosing a person at random: getting a child, getting a person 85 years old.
e Throwing a dice: obtaining a multiple of 3, obtaining an even number.
f Throwing a dice: obtaining an odd number, obtaining a 6.
g Throwing a pair of dice: getting a total of 6, getting a total of 7.
h Throwing a pair of dice: obtaining a 1 on one dice, obtaining a 6 on the other.

2 To play Monopoly you need two dice, and you have to pay rent if your total score on the two dice is 7, 9 or 10. Use the table of totals for two dice you made in question **11** of Exercise 2A to calculate:
a P(you have to pay rent)
b P(you avoid paying rent)

3 These three cards are placed face down. One is chosen at random, noted and replaced. The cards are shuffled, and another is chosen at random.

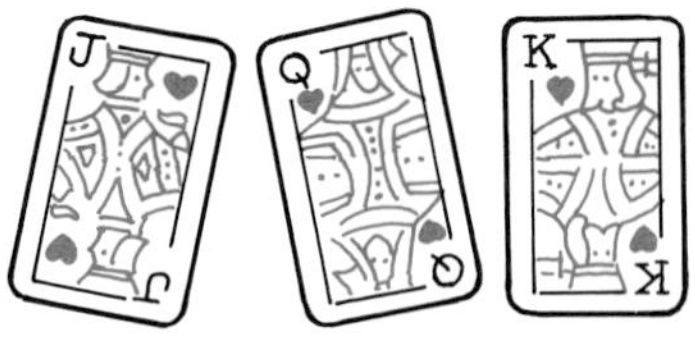
J
Q
K

a Copy and complete the tree diagram.

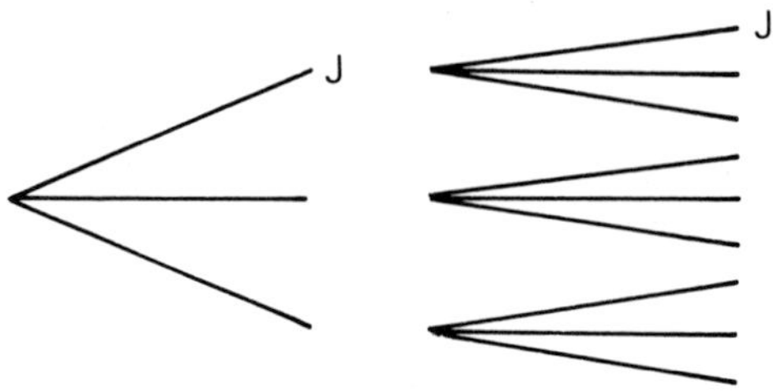
J
J

b Calculate the probability of picking:
(i) a pair of cards which are the same
(ii) a King and a Queen, in either order
(iii) at least one King.

4 The probabilities of a spinner stopping in A, B, red and blue regions are given.

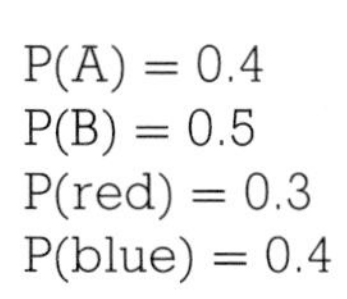
P(A) = 0.4
P(B) = 0.5
P(red) = 0.3
P(blue) = 0.4

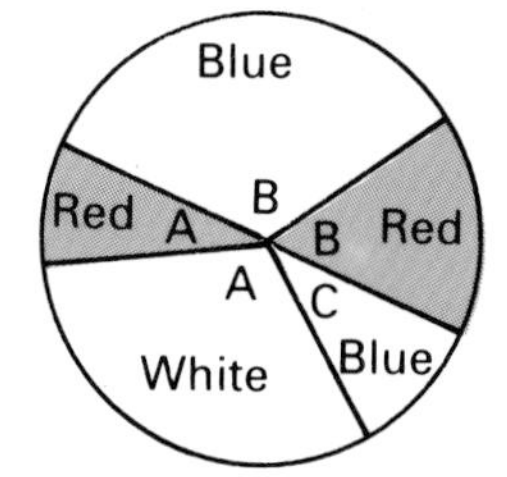

a Calculate:
(i) P(C) (ii) P(white) (iii) P(red or blue)
(iv) P(A or B)

b Explain why P(blue or C) is not 0.5. What is P(blue or C)?

PRACTICAL PROJECTS

1 *Roll two dice at least 120 times, and make a bar graph of the sums of the two scores. How does the result compare with the one you might have expected?*

2 *Repeat **1**, but this time calculate the differences of the two scores on the dice.*

BRAINSTORMER

P(Ace) = $\frac{4}{52}$; *P(3)* = $\frac{4}{52}$
P(Ace or 3) = $\frac{4}{52} + \frac{4}{52}$

P(Ace) = $\frac{4}{52}$; *P(Red card)* = $\frac{26}{52}$
P(Ace or Red) = $\frac{28}{52} \neq \frac{4}{52} + \frac{26}{52}$

What has gone wrong?

RELATIVE FREQUENCY AND PROBABILITY

EXERCISE 3/CLASS DISCUSSION

1 a A problem. *What fraction of the population of Britain is left-handed?* Is it 0.1, or 0.5, or 0.7, or are people equally likely to be left or right-handed? Have a guess.

b (i) How many students are in your class?
(ii) How many are left-handed?
Calculate the **relative frequency** of left-handed students in your class, that is

$$\frac{\text{Number who are left-handed}}{\text{Total number in class}}$$

2 Calculate the relative frequencies of left-handed people in these larger samples:

	a	b	c
Left-handed	11	36	148
Right-handed	89	164	852
Total number			

3 Draw a bar graph of the four results in questions **1** and **2**, using the scales shown.

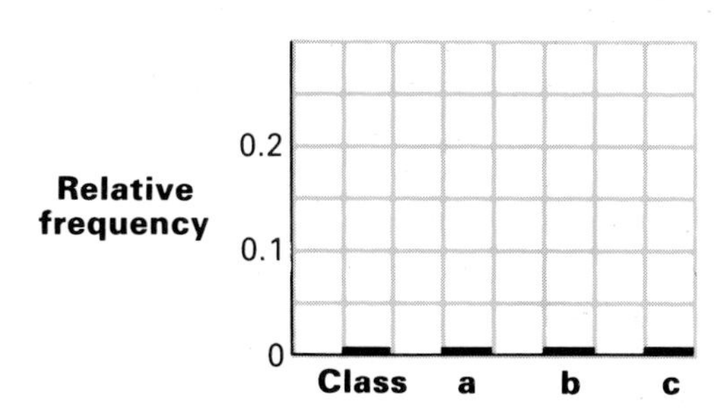

The larger the sample, the more likely the relative frequency will be closer to 0.17. This is the **probability** that a person chosen at random will be left-handed.

Experiment → calculation of relative frequency → probability

4 Here is another problem that can be solved by collecting and analysing data. *What is the probability of success with this new medical treatment?*

	Year 1	Years 1 and 2	Years 1–3	Years 1–4	Years 1–5
Number of successes	7	24	46	46	80
Number of patients	10	30	50	60	100

a Calculate the relative frequency of success in each set of years.
b Estimate the probability of success.

5 Nick records the number of 'pin-ups' in every 40 drops he makes of a drawing pin.

Number of pin-ups so far	18	32	43	51	64
Total number of drops	40	80	120	160	200

a Calculate the relative frequencies of pin-up.
b Estimate P(pin-up) for his drawing pin.
c How many pin-ups should he expect to have in 1000 drops?

6 This table shows the results of a survey of the lunchtime drinks chosen by 50 students.
a Calculate the relative frequency of each choice.
b How many drinks of each type would you expect a canteen to provide for 800 students?

Drink	Number
Tea	6
Cola	22
Milk	10
Orange	12

FORETELLING THE FUTURE: EXPECTATION

Example
A dice is rolled 90 times. How many 6s would you expect?
The number of 6s expected $= P(6) \times$ number of rolls
$= \frac{1}{6} \times 90$
$= 15$

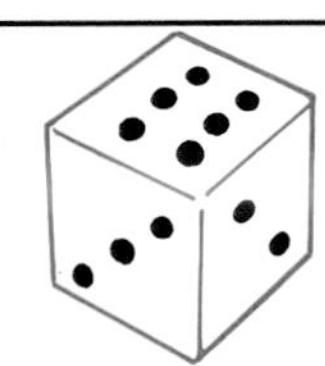

EXERCISE 4A

1

Event	Outcome	Probability	Number of events	Expected number of outcomes
A coin is tossed	Head	a(i)	50	a(ii)
A dice is rolled	2	b(i)	300	b(ii)
A month is chosen at random	Starts with J	c(i)	120	c(ii)

Find the entries for a(i), a(ii), b(i), b(ii) and c(i), c(ii) in the table.

2 a What is the probability that a person chosen at random was born on a Monday?
b How many students in a class of 35 would you expect to have been born on a Monday?

3 In the Modern Music factory a sample of 80 compact discs are selected and tested, and three are found to be faulty.
a Estimate the probability of a faulty disc in their production line.
b 4800 of the discs are distributed to High Street stores. How many faulty ones should Modern Music expect to replace?

4 This tree diagram shows the possible outcomes when two coins are tossed.

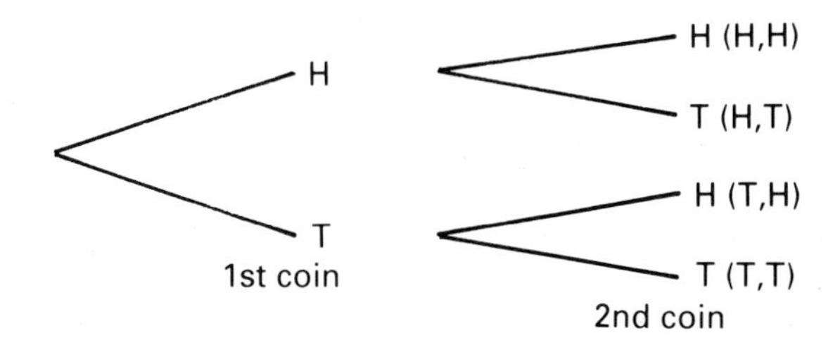

a Calculate the probability of (H, H) or (T, T).
b How many outcomes (H, H) or (T, T) would you expect in 100 tosses of the two coins?

5 Based on last season's results, Rovers' manager predicts that in the new season P(win) = 0.5 and P(draw) = 0.2.
a Write down P(defeat).
b Rovers have 40 games to play. How many wins, how many draws and how many defeats should the manager expect?

EXERCISE 4B

1 A pack of 52 cards is cut 200 times, and shuffled after each cut. How many times would you expect to cut:
a a club **b** an ace **c** a black King?

2 The Supersafe Insurance Company uses these probabilities to predict the cost of car accident claims. In 2000 claims, estimate the number that will be:
a under £1000
b between £1000 and £2000
c £500 or over.

Claim (£)	Probability
<500	0.15
500–999	0.25
1000–1499	0.20
1500–1999	0.15
>2000	0.25

3 Standard Superstore attracts a lot of customers. It has eight exits, and customers are equally likely to use any of these.
a What is the probability of a customer leaving by:
(i) exits D or E
(ii) Kelvin Road exits
(iii) Cart Street or Levern Lane exits?

b 2000 customers pass through in a particular day. Estimate how many leave by:
(i) Clyde Street exits
(ii) Levern Lane or C exits.
c A fast food stall sets up, hoping to catch some of the customers using Standard Superstore. Suggest, with reasons, the best site for the stall.

4 This table gives the probabilities of a car driver being killed or injured during a lifetime of driving.

	Male	Female
Killed or seriously injured	0.04	0.02
Slightly injured	0.13	0.12

a Estimate the number of males, in a sample of 1000, who will be killed or injured while driving their cars.
b Repeat part **a** for female drivers.

5 In a probability/expectation experiment, Joseph coloured five cards red, yellow, green, black and white on one side. He turned them over, selected one, noted the colour, put it back and mixed them up. He then selected and noted another card.
a Make a table of all the possible pairs of colours he could choose.
b Calculate the probability that exactly *one* of the cards will be black or white.
c Estimate the number of times he will choose exactly one black or white card in 100 trials.

PRACTICAL PROJECT

Two dice are rolled 100 times.
a *Estimate the number of times you would get:*
(i) double 4 (ii) both scores the same
(iii) a total of 8 or more.
Explain why you expect these results.
b *Carry out the experiment, and compare the results with your estimates.*
c *Make up an experiment on your own with two dice, and follow the steps in* **a** *and* **b**.

CHECK-UP ON PROBABILITY

1 Jimmie's in a five-a-side football team. The team draw lots to see who will be goalkeeper. What is the probability that Jimmie himself will:
a be in goal **b** not be in goal?

2 Mrs Baker asks a random sample of 40 students if they will come to a fifth-year disco. 28 reply 'Yes'.
a Calculate the relative frequency of students who said 'Yes'.
b How many of the 180 fifth-year students should she expect to attend?

3 Safety First Insurance calculate that the probability of a person being injured on a skiing holiday is 0.02. How many injuries should they expect from 1500 skiing clients?

4 Stephan reckons that the probability of success with his first serve is 0.75. What is the probability that his first serve fails?

5 Anya's kitchen drawer contains a mixture of knives, forks and spoons. By experiment she finds that when she picks one at random P(knife) = $\frac{1}{2}$ and P(fork) = $\frac{1}{3}$. Calculate:
a P(knife or fork) **b** P(spoon).

6 The coin is tossed and the spinner is spun.

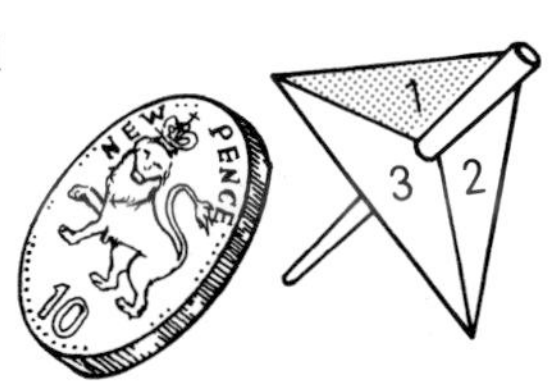

Here is the table of possible outcomes:

Coin \ *Spinner*	1	2	3
H	(H, 1)	(H, 2)	(H, 3)
T	(T, 1)	(T, 2)	(T, 3)

a Copy and complete the tree diagram.

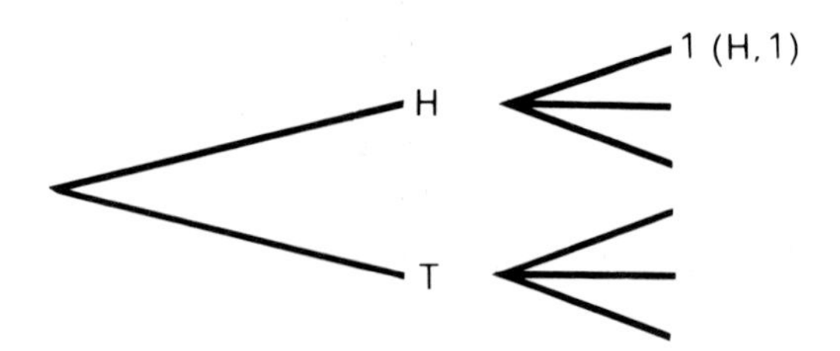

b How many possible outcomes are there?
c Calculate: (i) P(H, 3) (ii) P(T, 2)
(iii) P(either a head or a 3, or both).

7 Bill picks a card at random from these four.

Calculate:
a P(Ace) **b** P(King) **c** P(Ace or King).

8 A lucky dip box contains 20 packets of Crinkly Crisps and 30 packets of Curly Crisps. Amy chooses one at random. Calculate:
a P(Curly) **b** P(Crinkly)
c P(Curly or Crinkly).

9 This table gives the number of boys and girls under 15 years of age, to the nearest 1000, involved in accidents one year, while on foot.

Age (years)	Boys	Girls
0–4	2000	1000
5–9	6000	3000
10–14	5000	4000

Estimate the probability that a young person under 15 who is involved in an accident will be:
a a boy under 5 **b** a boy
c aged from 10 to 14 **d** a girl aged 5 to 14.

10 Two dice are rolled and the scores are added.
a Without making a table, calculate:
(i) P(2) (ii) P(3) (iii) P(4).
b Estimate the number of totals less than 5 in 100 trials.

11 A letter of the alphabet is chosen at random.
a For diagram 1, calculate:
(i) P(letter is in set *A*) (ii) P(letter is in set *B*)
(iii) P(letter is in set *A* or *B*).
b Repeat **a** for the second diagram.
c In which diagram are the sets of letters mutually exclusive?

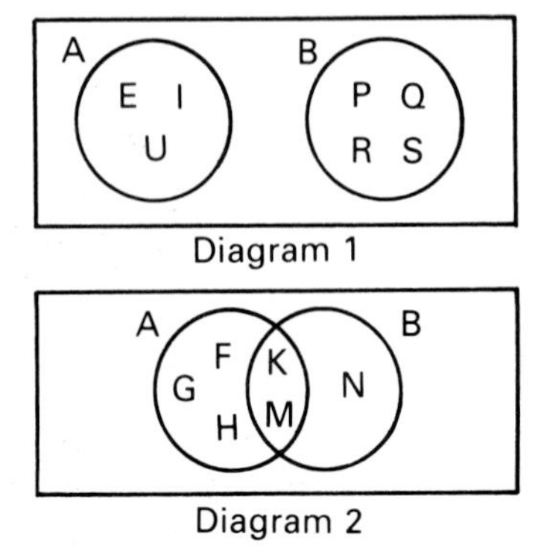

TOPICS TO EXPLORE

1 How to win a magazine competition

Win £1000! All you have to do is choose the eight items you would find most useful, and list them in order, 1 to 8.
How can you be sure of submitting a correct entry?

Think it out! Two items A and B can be listed in two ways, AB and BA.
In how many ways can three items A, B, C be listed, two at a time? Make lists to find out.
There's a quicker way! Think of filling the first place, then the second, and so on. Work out a system.
How many entries would you have to send to the competition to be sure of winning? A magazine costs £1.50. Any snags?

2 Football scores

a List all possible half-time scores in the Rovers v United game. Try to arrange them in a methodical way.
Repeat this for two or three different full-time scores.

b If the final score is 'Rovers m United n' can you make a formula in terms of m and n for the number of possible half-time scores?

3 Leaning towers

a There are many leaning towers in Italy. The most famous one is at Pisa.
Calculate its angle of tilt.

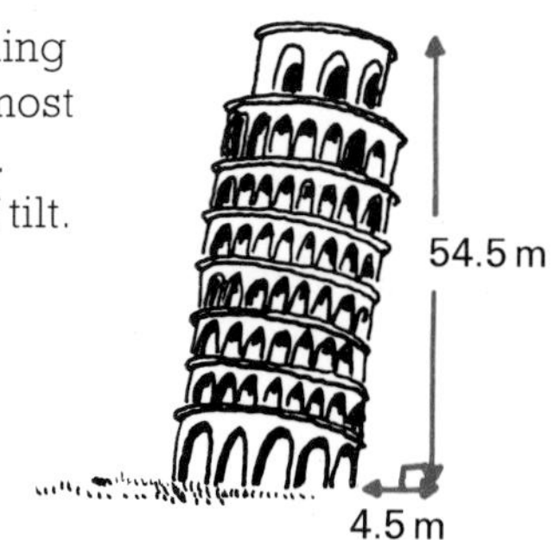

b At some stage this empty box will topple.

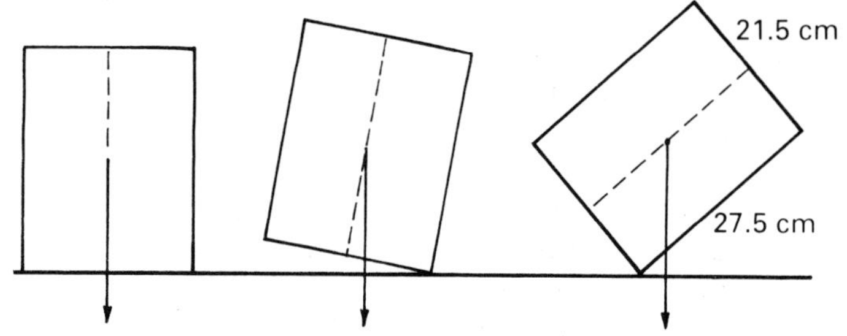

(i) Sketch the box when it is on the point of toppling over.
(ii) Calculate its angle of tilt at this point.
(iii) Check your answer by tilting your mathematics book about one corner, and measuring the angle of tilt with a protractor.

c A crate is $3x$ m by $2x$ m by x m.
Calculate the angles at which it will topple over each edge.

4 A fixture fix

Phil's in a fix. He is secretary of the Bluebell Badminton League. There are six teams, each of which plays the others once. The teams play one game a week, on Wednesdays at 7 pm. What is the least number of weeks needed to complete the fixtures? Make out a fixture list.

5 A magic square

a Copy and complete this magic square. Check that the numbers in each row, column and diagonal add up to the same total when a, b and c have the values shown.

$a - b$	$a + b - c$	$a + c$
$a + b + c$	a	$a - b - c$
$a - c$	$a - b + c$	$a + b$

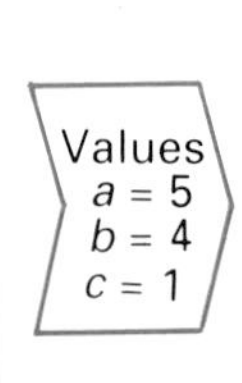

1	8	

b Try it again for $a = 2$, $b = -1$, $c = 3$.

c Explain how it works.

d For all entries to be different, are there any restrictions on a, b, c?

e If you like a *real* challenge, find a 3×3 magic square whose entries are all powers of 2 and whose *products* of the entries in each row, column and diagonal are all equal.

6 Design symmetry

a Copy this pattern on squared paper, or use tracing paper.

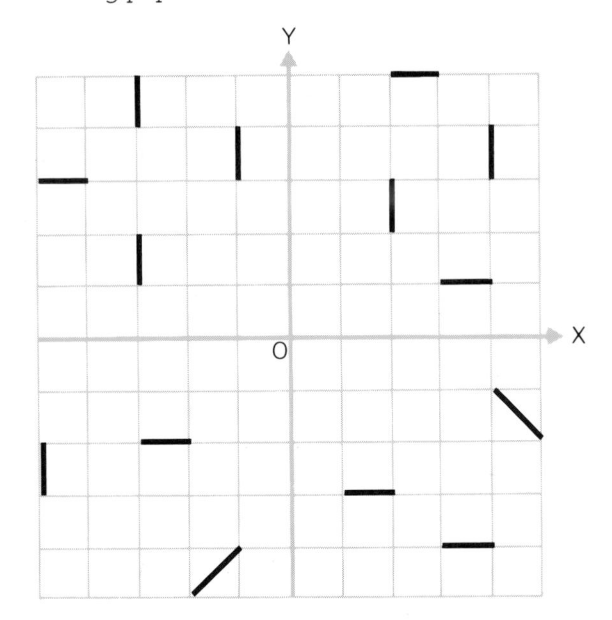

b (i) Choose one of the black lines. Reflect it in the x-axis, then in the y-axis, then in the x-axis, and so on, until you are back to the start.

(ii) Do this for every black line until you have four polygons.

c Design your own shape on a squared grid, and break it up into different parts. It could be a geometric shape, or an animal, or . . . Ask a friend to try to put it together again as in **b**.

7 A house of cards

Mike is building a house of cards. For one storey he uses two cards, and makes one triangle of one unit.

For two storeys he uses five *more* cards, a total of seven. He makes four triangles of one unit and one triangle of four units.

a For three storeys, how many:
(i) more cards
(ii) cards altogether
(iii) 1–unit triangles
(iv) 4–unit triangles
(v) 9–unit triangles?

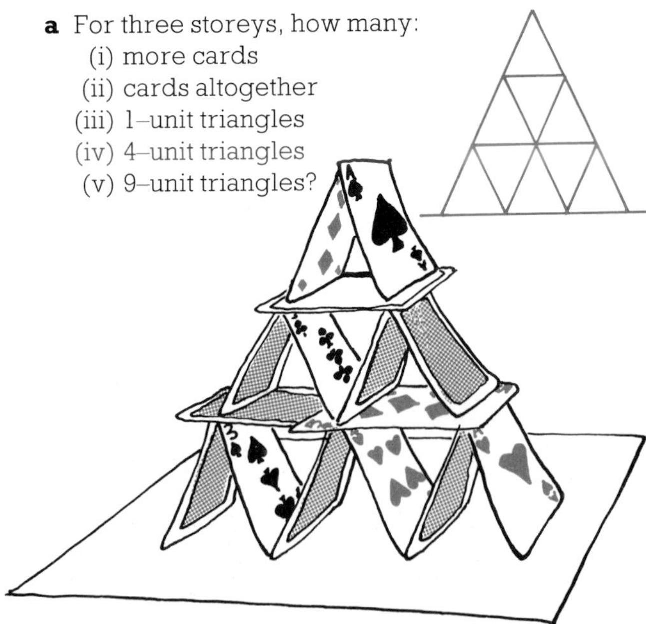

b Arrange all the data in a table, and extend it to four and five storeys.

c Find formulae for:
(i) the number of 1–unit triangles in n storeys
(ii) the number of extra cards needed for the nth storey.

d How many storeys can Mike build with two packs of cards (52 in each)?

CHAPTER REVISION EXERCISES

REVISION EXERCISE ON CHAPTER 1: CALCULATIONS AND CALCULATORS

1 $a = 15.5$, $b = 12.2$ and $c = 9.9$. Calculate, correct to 3 significant figures, the values of:

a $\frac{ab}{c}$ **b** $\frac{a-b}{c}$ **c** $a(b+c)$

d $a^2+b^2+c^2$ **e** $\sqrt{(a+b+c)}$ **f** $\frac{a-b}{b-c}$

2 Express each of the following as a decimal fraction of an hour, correct to 4 decimal places:
a 1 minute **b** 1 second

3 **a** Which of these numbers do you think could equal 68×9.6?
(i) 65.28 (ii) 6526 (iii) 652.8 (iv) 652.6
b Check with your calculator.

4 Write these numbers in standard form, $a \times 10^n$:
a 19.6 **b** 5680 **c** 0.0046 **d** 580 000

5 Write the numbers in these sentences in scientific notation.
a The diameter of the Earth is 12 680 km.
b The half-life of the element Polonium is 0.000 16 s.
c The volume of the Sun is 1330 million million km^3.

6 Write these numbers:
(i) in standard form (ii) in floating point form.

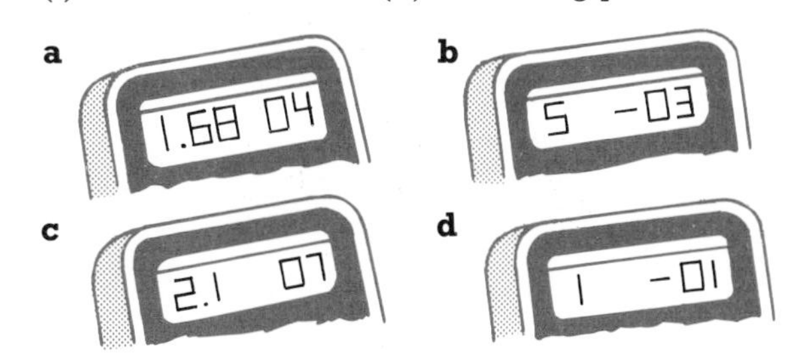

7 Write these as mixed numbers:
a $\frac{5}{2}$ **b** $\frac{10}{7}$ **c** $\frac{8}{3}$ **d** $\frac{12}{5}$ **e** $\frac{22}{3}$

8 Write these as fractions:
a $1\frac{1}{4}$ **b** $2\frac{1}{8}$ **c** $5\frac{2}{3}$ **d** $8\frac{3}{4}$ **e** $1\frac{7}{8}$

9 Calculate in their simplest form:
a $1\frac{3}{10}+2\frac{7}{10}$ **b** $2\frac{3}{5}-1\frac{1}{5}$ **c** $\frac{7}{8}-\frac{3}{4}$ **d** $3\frac{1}{2}+1\frac{2}{3}$

10 In a local election, $\frac{3}{8}$ of the electorate voted Conservative, $\frac{2}{5}$ Labour and $\frac{1}{10}$ Liberal Democrat. The remainder (125 people) did not vote.
a Which candidate was elected?
b How many people were eligible to vote?

11 This picture has to hang the same distance from each edge of the wall. What distance is this?

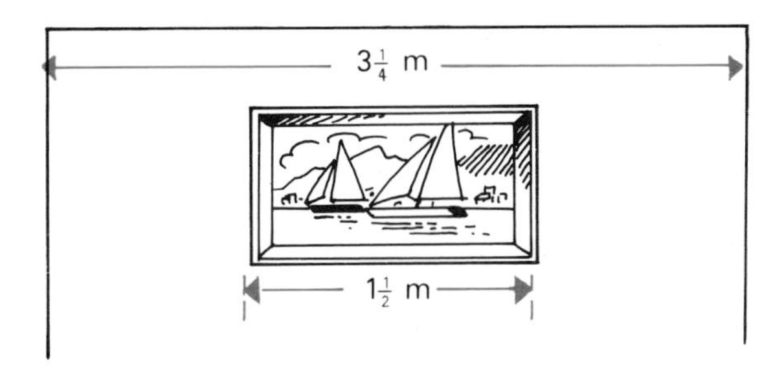

12 Calculate in simplest form:
a $\frac{1}{3}\times\frac{1}{2}$ **b** $\frac{2}{3}$ of $\frac{3}{4}$ **c** $9\times 2\frac{1}{3}$ **d** $1\frac{1}{4}\times 1\frac{3}{5}$

13 $\frac{1}{4}$ of the spectators left after 10 minutes, $\frac{1}{3}$ of the rest left 5 minutes later. 300 spectators remained. How many were there to begin with?

14 Calculate in simplest form:
a $\frac{1}{2}\div\frac{1}{8}$ **b** $3\frac{1}{2}\div 1\frac{3}{4}$ **c** $1\frac{1}{2}\div 4\frac{1}{2}$ **d** $5\div 1\frac{1}{4}$

15 The frequency of a light wave is given by the formula $f = \lambda \div c$, where c is the speed of light, 3×10^{10} cm/s. Calculate the frequency of each light colour in the table. (Remember the constant facility or memory in your calculator.)

Colour	Wavelength (λ)
Red	6.5×10^{-5}
Orange	6.0×10^{-5}
Yellow	5.8×10^{-5}
Green	5.4×10^{-5}
Blue	4.7×10^{-5}
Violet	3.9×10^{-5}

REVISION EXERCISE ON CHAPTER 2: SIMILAR SHAPES

1

Every TV screen is similar to every other TV screen.

a Why is this?

b The height of every screen is $\frac{3}{4}$ of its width. Calculate:

(i) the height of a screen 24 cm wide

(ii) the width of a screen 24 cm high.

2 **a** What is the scale factor of the reduction from the large arrow to the small one?

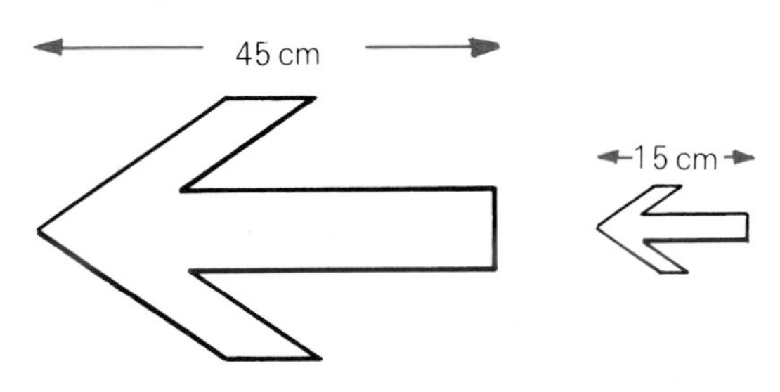

b The large arrow is 24 cm high. Calculate the height of the small one.

c An edge of the small arrow is 7.5 cm long. Calculate the length of the corresponding edge of the large one.

3 The two rectangles are similar. Calculate x.

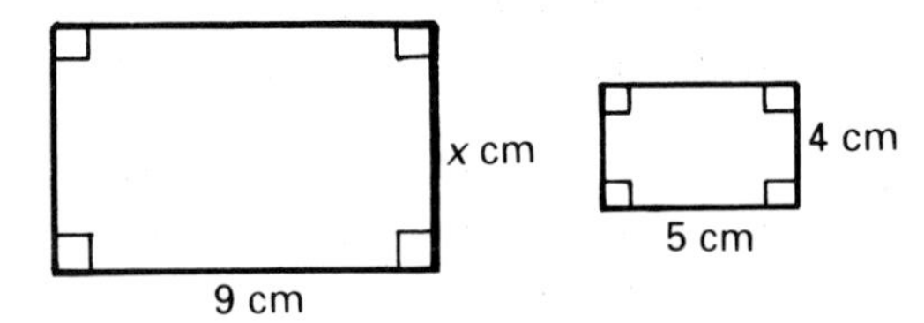

4 The page of Tessa's French book is 24 cm by 18 cm. Find the length and breadth of:

a a rectangle similar to her page, with one side 12 cm long

b a different rectangle (not congruent to the one in **a**), with one side 12 cm long.

5 The three tiers on the cake-stand are similar. Lengths are in millimetres.
Calculate: **a** x **b** y **c** z

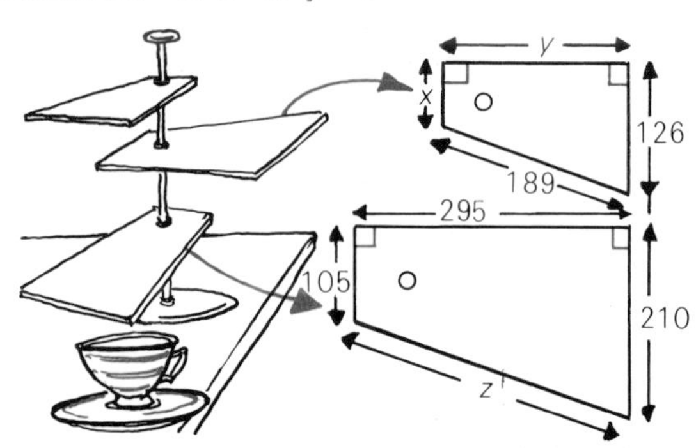

6 This is a view from above of an outdoor clothes drier. There are three pairs of parallel ropes.

a Copy the diagram, and mark pairs of equal angles.

b Which triangle is equiangular to $\triangle$ABC?

c If OB = BE and BC = 2.4 m, calculate EF.

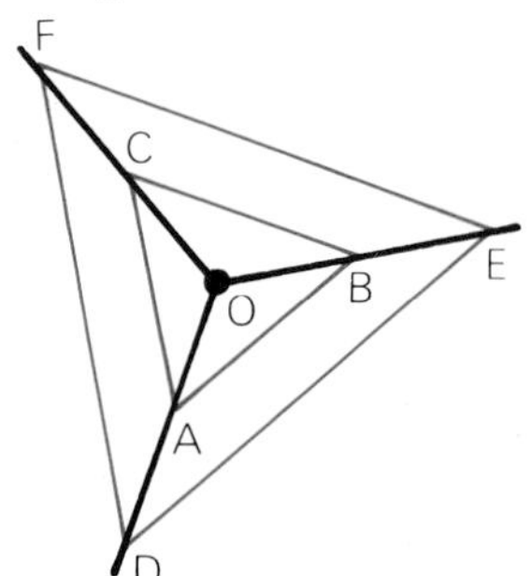

7 **a** Explain why DEFC is a parallelogram.

b Write down the lengths of EF and FC.

c Calculate BC and BF.

8 A ferryboat plies between A and B. If FB = 30 m, calculate x.

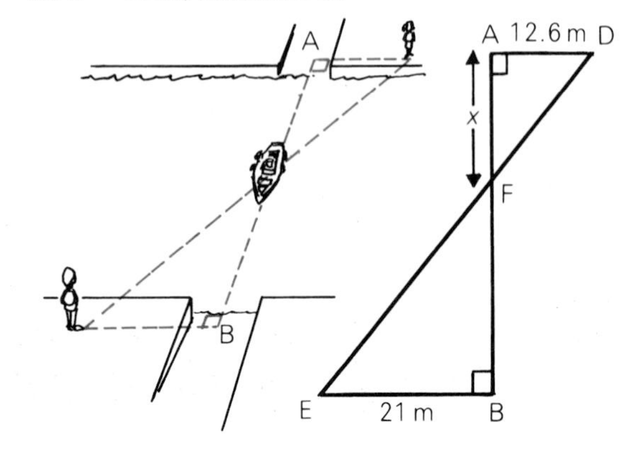

REVISION EXERCISE ON CHAPTER 3: GOING PLACES

1 **a** Calculate the average speed of a journey of 180 km which takes five hours.
b How long does it take to travel 336 km at an average speed of 48 km/h?
c Calculate the distance travelled in two hours six minutes at an average speed of 50 mph.

2 An elephant, a kangaroo and a zebra decide to have a race. The elephant can manage a speed of 20 km/h, the zebra 50 km/h and the kangaroo 60 km/h.

a How far does each animal go in 15 minutes?
b How long does each take to 'run' 10 km?

3 At full speed a tortoise can move at 80 centimetres per minute. How long does it take to cross a road 5 metres wide?

4 John cycles 15 km in 25 minutes, and then walks 3 km in 50 minutes. Calculate:
a the total distance he goes, and the time he takes
b his average speed from start to finish.

5 For the motorist's journey calculate:
a his times of travel from Ayton to Beeton, and from Beeton to Seton
b his overall average speed.

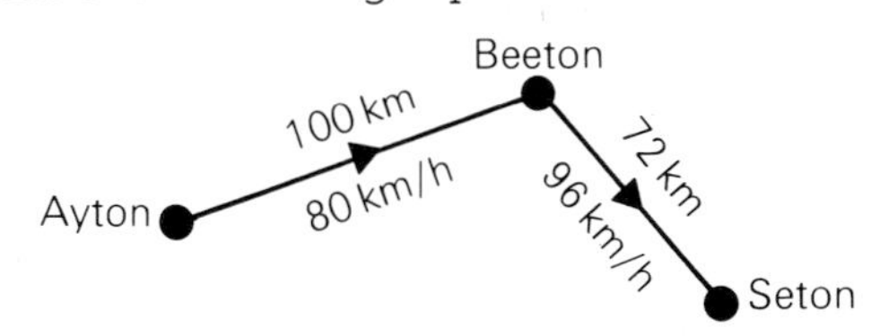

6 Which of the cars is stopped by the police?
Car A: 12 miles in 10 minutes.
Car B: 17 miles in 15 minutes.
Car C: 7.5 miles in 6 minutes.

7

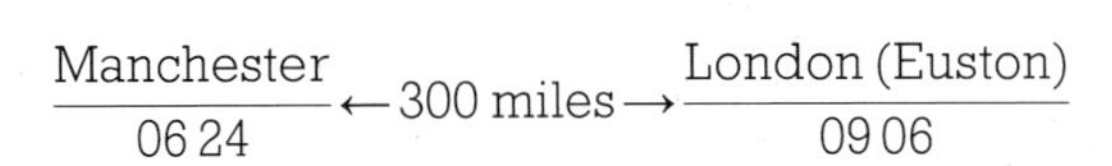

Calculate the average speed of the train, correct to 2 significant figures.

8 A lorry leaves the garage to travel to Ayr.

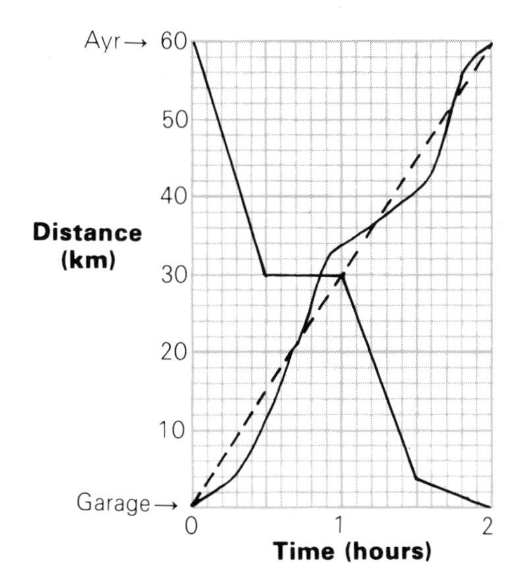

a How far does it travel in two hours?
b Calculate its average speed.
c Why does the graph 'wriggle' around the dotted line?

A bus leaves Ayr at the same time as the lorry leaves the garage.
d For how long does the bus stop on its journey?
e How far from Ayr do the bus and lorry pass each other?
f Calculate the average speeds of the bus on the fastest and slowest parts of its journey.

9 A car travels 60 km from A to B at an average speed of 20 km/h, and returns at 40 km/h. Guess its average speed for the whole journey. Calculate the average speed, to the nearest mph. Are you surprised?

10 A spaceship bound for Mars, 72 million km away, will average a speed of 16 thousand km/h. The launch was at 0000 h on Wednesday 21st June. Find the day and date of its arrival.

REVISION EXERCISE ON CHAPTER 4: MONEY MATTERS—SAVING AND SPENDING

1 Ina had good qualifications, and applied for this job.

> ***JOB VACANCY***
>
> *School leaver required as filing assistant in lawyer's office*
> *Applicants should be keen, able to type and have GCSE/SCE qualifications*
> *£4800 – £6600*
> *Write to T. Law, 55 High Street*

a £4800 a year—how much each:
(i) month (ii) week?

b £6600 a year—how much *more* in:
(i) a year (ii) a month (iii) a week?

2 Bill Andrews is a salesman. He earns £14 000 a year, plus commission of $2\frac{1}{2}\%$ of his sales. How much did he earn in a year when his sales totalled £84 000?

3 Caird Electrical use this flowchart to calculate wages:

START
↓
$X = N \times 8.40$
↓
Any overtime Monday to Friday? — No → (skip to next decision)
↓ Yes
$X = X + W \times 12.60$
↓
Any overtime Saturday or Sunday? — No → (skip to STOP)
↓ Yes
$X = X + S \times 16.80$
↓
STOP

N hours at £8.40 an hour,
W hours at time-and-a-half,
S hours at double time.
Calculate the wages for:

a Jim, 42 hours, no overtime
b Kim, 36 hours, plus 6 hours' overtime on Monday
c Tim, 38 hours, plus 8 hours' overtime on Friday and 5 hours' overtime on Saturday.

4 Stronghold Building Society offers an interest rate of 8% per annum. Calculate the interest on £500 for: **a** 6 months **b** 200 days.

5

This is just the three-piece suite Viv wants. But how will she pay for it? Take out a loan of £950 for a year at 12% p.a. or use HP? Which way would be cheaper, and by how much?

6 a Mr Muir invests £1500 in the North East Bank for three years at 9% p.a. interest. He takes out the interest at the end of each year. How much interest does he get altogether?

b If he leaves the interest in the account he will get compound interest. Calculate the compound interest on £1500 for three years at 9% p.a.

7 Calculate the VAT at $17\frac{1}{2}\%$ on this hotel bill:
5 nights bed & breakfast at £42.50 a night
2 lunches at £8.50 each
4 dinners at £16.75 each.

8 A new car costs 105 500 francs in France, and 2 616 000 pesetas in Spain. Which price is lower in £s, given £1 = 8.57 francs or 222 pesetas? By how much?

9 Calculate the total amount due on this gas bill.

DATE OF READING	METER READING PRESENT	METER READING PREVIOUS	GAS SUPPLIED CUBIC FEET (100s)	GAS SUPPLIED CUBIC METRES	GAS SUPPLIED kWh	CHARGES £
1 MAY	9124	8888	236	668	7422	*
STANDING CHARGE 91 days at 11.05p per day						
				SUBTOTAL		
* 1.62p per kWh				VAT at 17.5%		
				TOTAL AMOUNT DUE	£	

REVISION EXERCISE ON CHAPTER 5: POSITIVE AND NEGATIVE NUMBERS

1 Sort these cards into nine pairs like this:

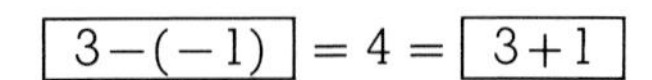

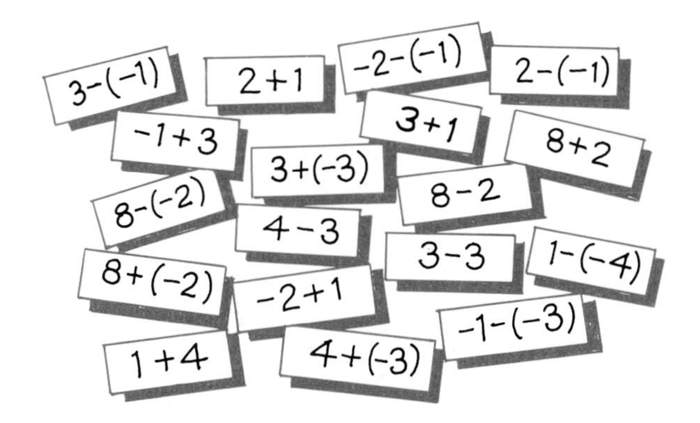

2 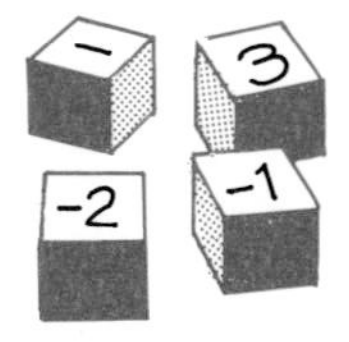

Arrange two of the numbered cubes like this:

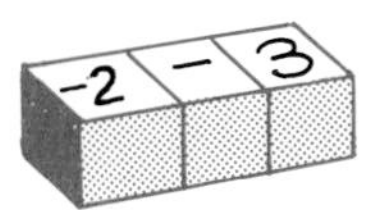

The score for this arrangement is -5.
Make another five arrangements.
Calculate the score for each one.

3 $m = 2$ and $n = -1$. Calculate the values of:

a $m-2n-1$ **b** m^2-n^2 **c** $4mn-\frac{m}{n}$

4 Simplify:
a $8a-2a-(-4a)$ **b** $4-x^2-4-x^2$

5 Find three more step calculations on each flight of stairs, and check them.

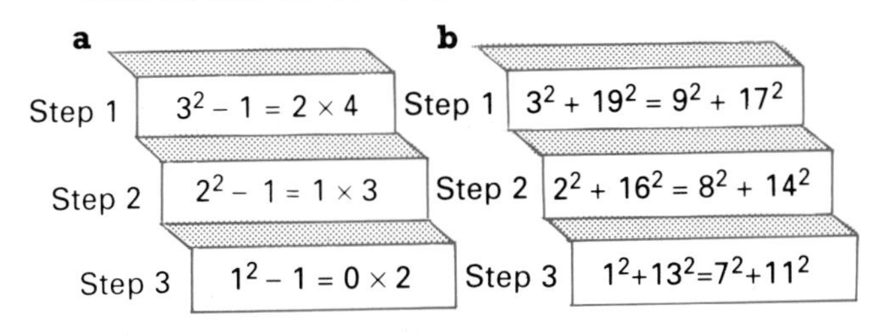

6 Calculate:
a $-4\times(-7)$ **b** $2(-10)^2$ **c** $-108\div 12$

7 Simplify:
a $3k\times(-2k)$ **b** $y^2-(2y)^2$ **c** $-7t^2\times(-6t)$

8 The vertical velocity of a rocket in m/s is $v = 20-10t$, where t seconds is the time from fuel burnout at P.

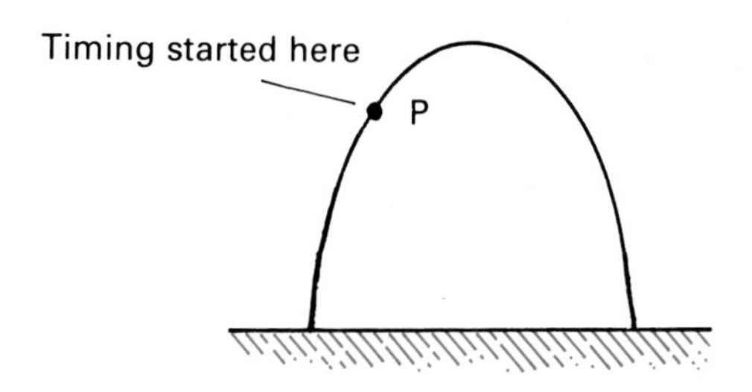

Calculate:
a the speed at
(i) $t=-1$ (ii) $t=0$ (iii) $t=3$
b the time when
(i) $v=0$ (ii) $v=10$ (iii) $v=-20$.

9 Solve these equations.
a $4x+1=-3$ **b** $4-t=-6$ **c** $2-2y=12$
d $6p-2=3p+1$ **e** $1-2k=10+k$

10 The levels of oil in a tank are noted on a scale, as follows:
18 mm, 13 mm, −12 mm, −19 mm, −15 mm, 15 mm. Calculate the mean level.

11 The height h metres of the rocket above P in question **8** t seconds from fuel burnout at P is given by the formula $h = 10+20t-5t^2$.
Calculate its height at: **a** $t=0$ **b** $t=-1$.

12 Simplify:
a $3x+(-2x)+7x$ **b** $2a\times 3a\times(-4a)$
c $t^2-2t^2-3t^2$ **d** $m\times n-2m\times(-n)$

REVISION EXERCISE ON CHAPTER 6: PYTHAGORAS

1 Calculate x in each right-angled triangle.

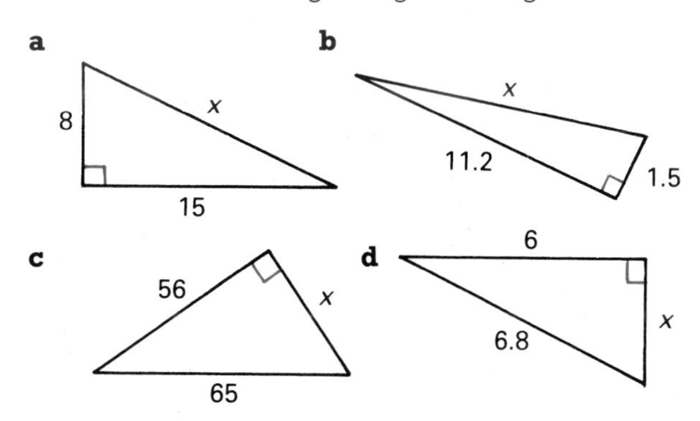

2 Find the length of the third side of each triangle, correct to 1 decimal place.

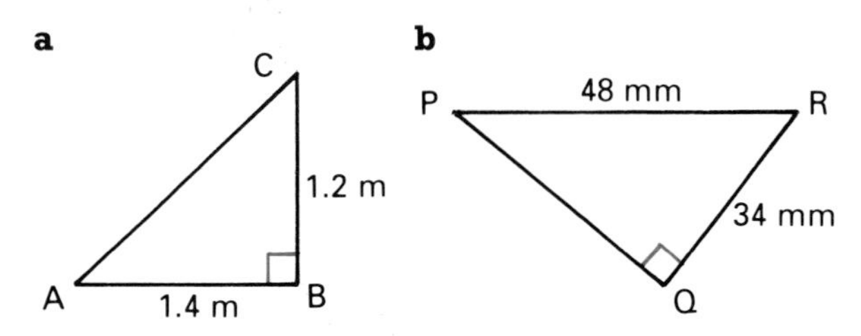

3 Guy and Gavin race each other round the field. They both start at A, Guy in the direction A to B, Gavin in the direction A to C.

a Calculate the length of BC.
b Both boys should run the same distance.
(i) How far is this?
(ii) Where should the finishing line be on BC?

4 Golf Yankee and Victor Echo are stacked above S, waiting to land. At 1200 hours, control at C sees them on radar 9 km and 10 km away. CS = 8 km. Calculate the difference between the aircrafts' heights, to the nearest 10 m.

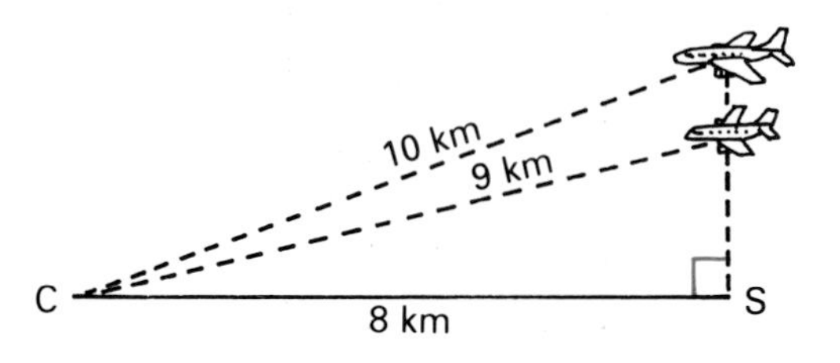

5 Omar is drilling four holes in plywood which has 10 cm squares marked on it.

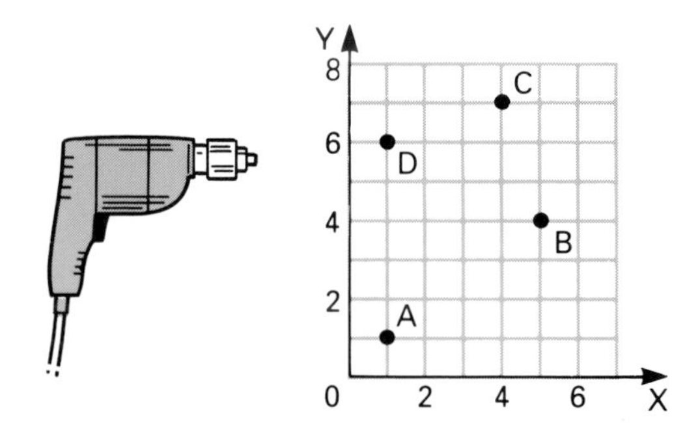

a Calculate AB, BC, CD and DA, to the nearest cm.
b What shape is ABCD?
c Write down the coordinates of the midpoint M of BD.
d Use the converse of Pythagoras' Theorem to check that the diagonals of ABCD cross at right angles.

6 PQRSTUVW is a cuboid.

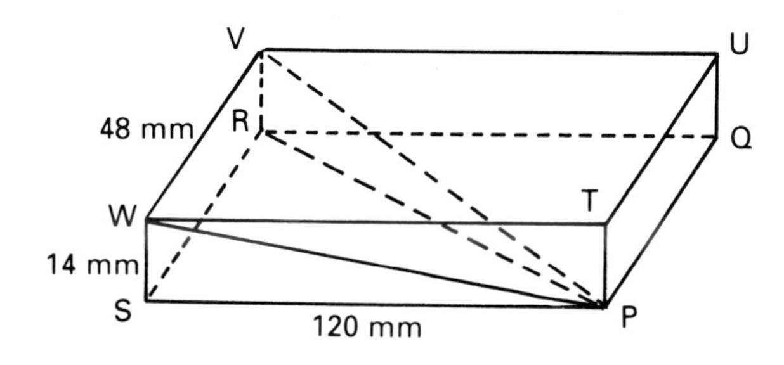

a Use Pythagoras' Theorem to show that PV = 130 mm.
b Use the converse of Pythagoras' Theorem to show that △PVW is right-angled.

7 △ABC is isosceles. Calculate:
a its height, or altitude, AD
b its area
c the lengths of its altitudes from B to AC and C to AB, correct to 1 decimal place. (Your result in **b** might help.)

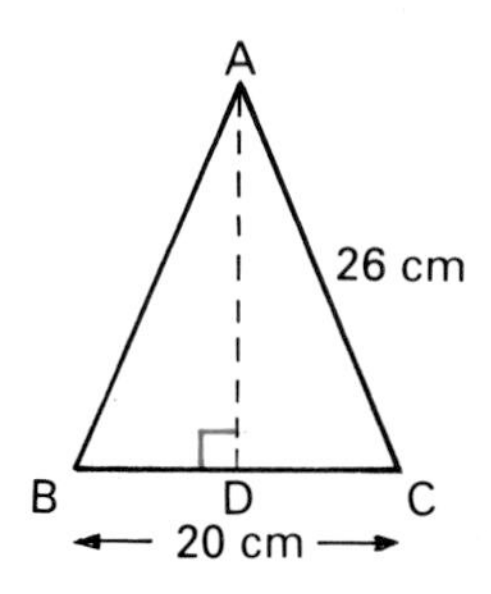

REVISION EXERCISE ON CHAPTER 7: BRACKETS AND EQUATIONS

1 Remove the brackets:
a $8(4-x)$ **b** $2(y-1)$ **c** $3(4t+1)$
d $a(x+y)$ **e** $p(q-r)$ **f** $-2(a-5)$
g $-5(5+c)$ **h** $-(1-x)$ **i** $-x(-2-x)$

2 Multiply out and simplify:
a $7+5(2-x)$ **b** $-3-2(y+1)$ **c** $1-(p-1)$

3 Find the expressions for:
a the total number of coins in collections (i) and (ii)
b the difference between the number of coins in the two collections, (i)–(ii).

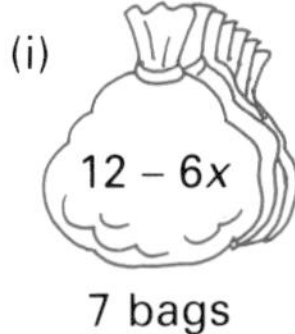

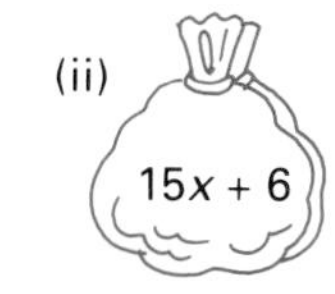

4 The path round the flower bed is 1 m wide. Find an expression for the area of the path in its simplest form.

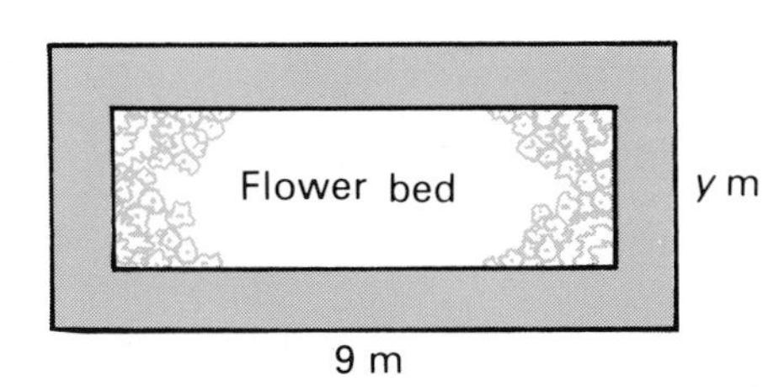

5 Multiply out, then simplify:
a $6x-2(1-x)$ **b** $4(2t-1)-3(1+2t)$

6 Solve these equations:
a $4+2(3t-7)=t$ **b** $3(x-1)-2(1-x)=0$

7 Altogether there are 64 coins in the collection of bags. How many are in each bag?

8 Mary made four phone calls to Pat, each $2x+2$ minutes long. Pat made eight calls to Mary, each $3x-9$ minutes long. The total lengths of time were the same. Make an equation, and find the length of call each girl made.

9 Multiply out:
a $(p+2)(p+8)$ **b** $(t-2)(t-7)$
c $(k+3)(k-3)$ **d** $(a-2)(2a+1)$
e $(4x-3y)(3x-4y)$ **f** $(1-x)(x-1)$
g $(x-6)^2$ **h** $(2a+1)^2$
i $(3c-4d)^2$

10 Solve:
a $y(4y-3)=4(y^2+6)$
b $3p(2p-1)-9=6(p+1)^2$

11 Make an equation, and find the length of each edge of this right-angled metal plate.

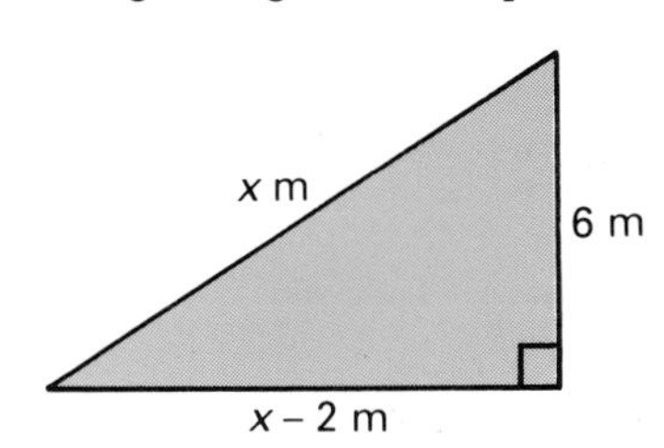

12 A hollow concrete pillar has a square cross-section. Its outer side is 20 cm longer than its inner side, and the cross-sectional area of the concrete is 800 cm². Make an equation, and calculate the lengths of the outer and inner sides.

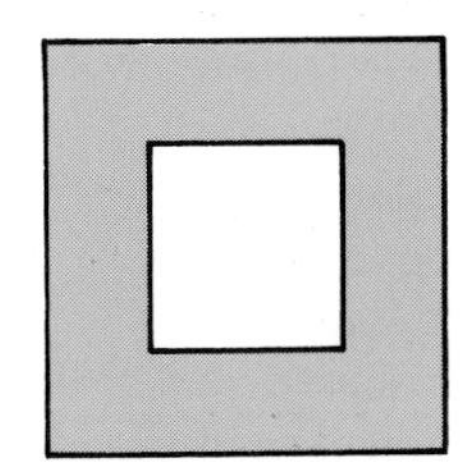

REVISION EXERCISE ON CHAPTER 8: STATISTICS

1 **a** Calculate the mean, median and mode of each set of data below.

b Which of the measures represents the data best?

(i) 1, 1, 1, 1, 2, 2, 3, 3, 40, 59

(ii) 2, 2, 2, 3, 3, 6, 7, 8, 12, 12

2 A sample of 100 people attending a 'Stop smoking' clinic were asked about the number of cigarettes they smoked each day.

a Copy and complete the table.

Number per day	Mid-value	Frequency	Mid-value × frequency
1–5		4	
6–10		8	
11–15		14	
16–20		18	
21–25		20	
26–30		17	
31–35		6	
36–40		13	

b Calculate the mean number of cigarettes smoked per day.

c Write down the median class interval.

d Al Reekie smokes 30 a day, paying £2.40 for each packet of 20. How much of his money goes up in smoke each year?

3 **a** Add a cumulative frequency column to your table in question **2**.

b Draw the cumulative frequency curve.

c Use your graph to estimate:

(i) the lower quartile

(ii) the upper quartile

(iii) the interquartile range.

4

Two surveys are carried out at the 18th hole of the Greenside Golf Course, comparing the number of strokes taken with:

(i) the golfer's height

(ii) the speed of the wind that day.

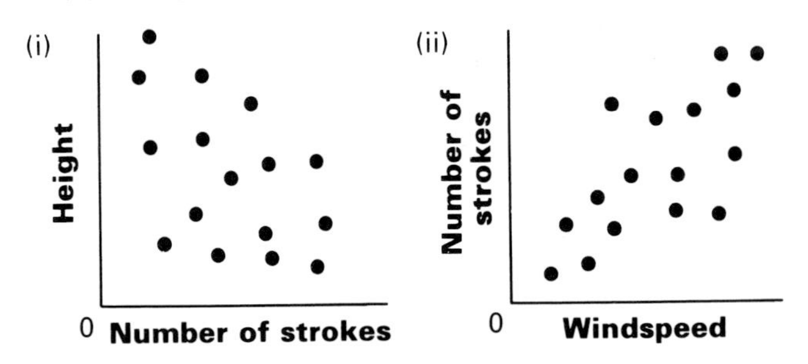

a What do the diagrams indicate?

b What type of correlation is there?

5 The data in the table was compiled over eight days at Sunsea holiday resort.

Day	1	2	3	4	5	6	7	8
Number of bathers	10	40	20	60	1	48	4	10
Hours of sunshine	1	8	4	12	0	10	10	3

a Draw a scatter diagram.

b Calculate the mean number of:

(i) bathers (ii) hours of sunshine per day.

c Draw the best-fitting straight line, and estimate the number of bathers on a day with 5 hours of sunshine.

6 **a** Use class intervals of 1–10, 11–20, . . . to make a cumulative frequency table of these charges, to the nearest £, at a supermarket till between 9 am and 10 am.

31 2 86 17 5 56 35 6 68 33 44 15 77
27 49 21 65 3 29 9 48 22 40 36 8 52
7 45 61 59 38 45 41 55 24

b Use the table to:

(i) calculate the mean bill, to the nearest £

(ii) draw a cumulative frequency curve, and find the interquartile range.

REVISION EXERCISE ON CHAPTER 9: TRIGONOMETRY

1 **a** In right-angled △ABC write down ratios for:
(i) sin A (ii) tan B.

b Calculate the sizes of:
(i) angle A (ii) angle B.

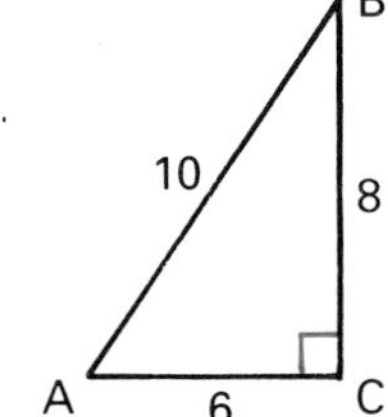

2 Write down the values, correct to 2 decimal places, of:
a sin 78° **b** tan 54.7° **c** 1 − cos 60°.

3 Calculate x, given:
a $\sin x° = 0.46$ **b** $\cos x° = 0.96$
c $\tan x° = \frac{4}{7}$ **d** $\tan x° = 0.50$

4 Calculate x for each triangle.

Reminder **SOH - CAH - TOA**

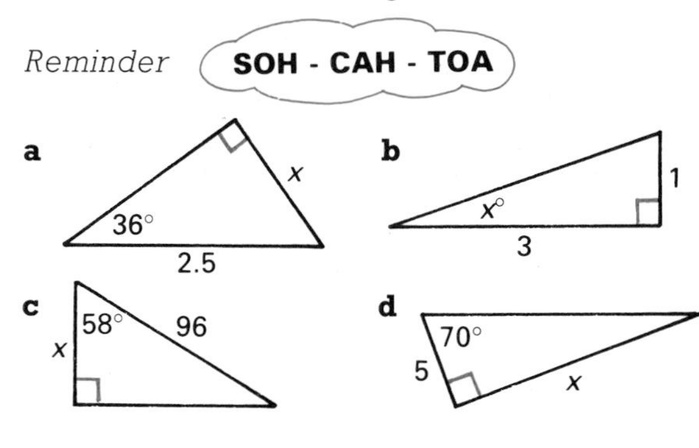

5 Calculate the area and perimeter of this rectangle.

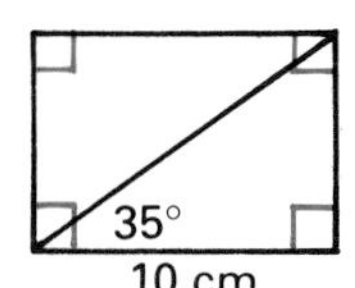

6 The ladder will slip if the angle it makes with the ground is less than 40°. Does it slip?

4 m
6 m
x°

7 Architect Sue is designing a house, and is trying to decide what the slope of the roof should be.

a What is the angle of the slope in the first diagram, to the nearest degree?

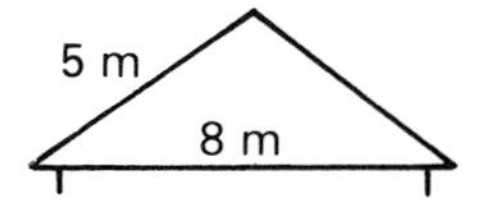

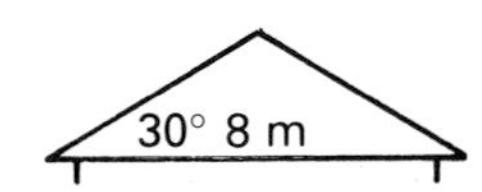

b She decides to reduce the angle to 30°. Calculate the length of the sloping edge now, to the nearest cm.

8 Sue designs a new art gallery in Uptown with an unusual roof.

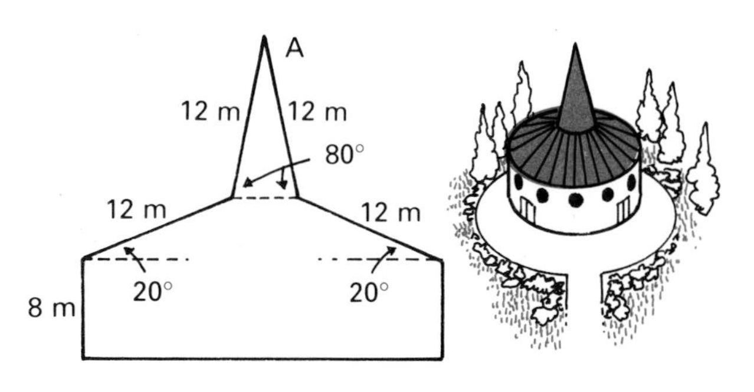

Make a sketch of the roof, and calculate the height of the apex A above the ground.

9 Calculate the width of the art gallery in question **8**.

10 This support can be fixed at the foot between 1 and 2 m from the wall. Find the range of possible angles between the support and the ground, to the nearest degree.

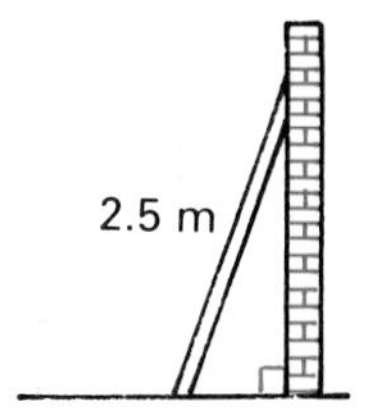

11 From the cliff-top (95 m high) the angles of depression of the two boats are 40° and 20°. Calculate the distance between the boats.

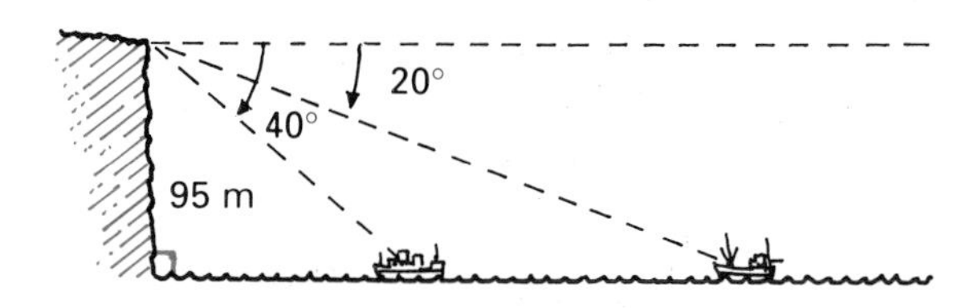

12 A goods container is in the shape of a cuboid. Sketch the right-angled triangles, and calculate:
a AC, in △ABC
b AH, in △ACH
c ∠HAC.

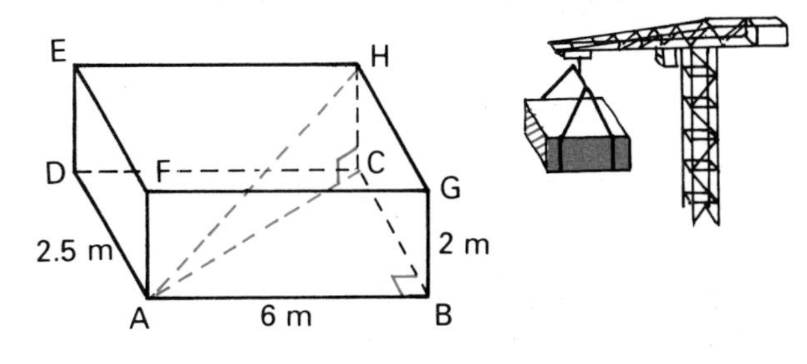

REVISION EXERCISE ON CHAPTER 10: SIMULTANEOUS EQUATIONS

1 a Copy and complete these tables:

(i)

x	0	1	2	3	4
$y = 2x+1$	1				

(ii)

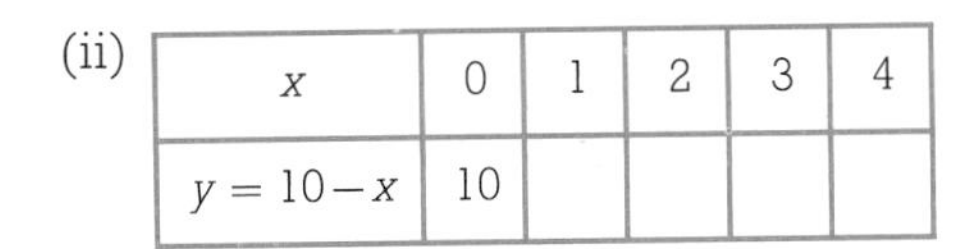

x	0	1	2	3	4
$y = 10-x$	10				

b For what values of x and y is $y = 2x+1$ and $y = 10-x$?

c Illustrate by drawing the graphs of the two equations.

2 Solve these simultaneous equations by drawing their graphs on squared paper:
$x+y = 9$ and $2x+y = 15$.

3 *Sketch* the graphs of these lines on plain paper:
a $y = 2x-8$ **b** $x-3y = 9$ **c** $4x+3y = 24$

4 Solve these pairs of equations by substitution:

a $y = 4x$
$y+5x = 27$

b $y = -x$
$4x+y = 12$

5 Solve these simultaneous equations by first eliminating x or y:

a $2x-3y = 5$
$4x+3y = 1$

b $9x+3y = 3$
$x+3y = 11$

c $2x-5y = -6$
$3x+2y = 10$

d $7x-9y = 2$
$2x+5y = -7$

6 A formula connecting Fahrenheit and Celsius temperatures is $F = aC+b$. Water boils at 212°F and 100°C, and freezes at 32°F and 0°C.

a Write down two equations in a and b, and solve them simultaneously.

b Write down the formula for F in terms of C.

c Draw a graph for values of C from -10 to 110.

d Estimate F when $C = 40$, using:
(i) the graph (ii) the formula.

7 The floor plan of the Simpsons' house is L-shaped. The front is 6 m longer than the side, and the perimeter of the plan is 68 m. Make two equations and solve them to find x and y.

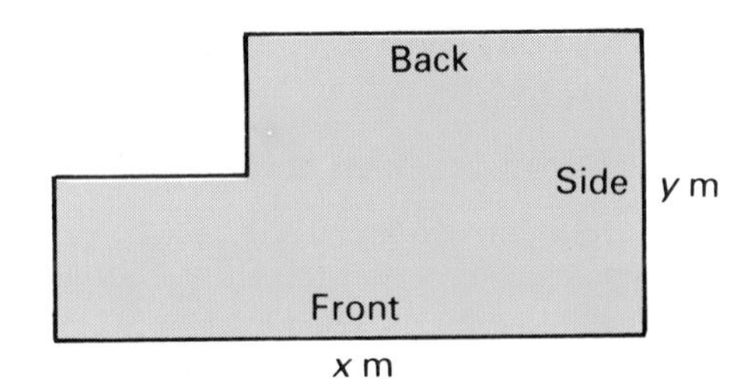

8 In the 'Bonus and Penalty' game there are two types of token:

Alex scored 9 points with these tokens:

Stefanie scored 17 points with these tokens:

a What is each type of token worth?

b Investigate the least number of tokens needed to make scores of 1, 2, 3, 4, . . . 10.

REVISION EXERCISE ON CHAPTER 11: AREAS AND VOLUMES

1 Calculate the areas of these shapes.

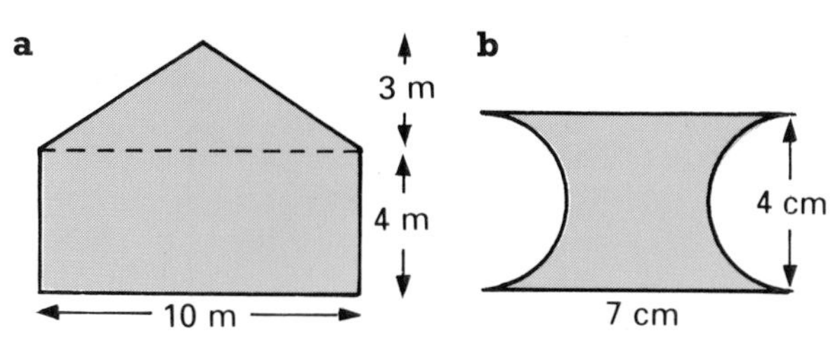

2 Creative Crafts make 'flower' coaster mats. The diagonals of each rhombus are 20 mm and 35 mm long. Find the area of a coaster.

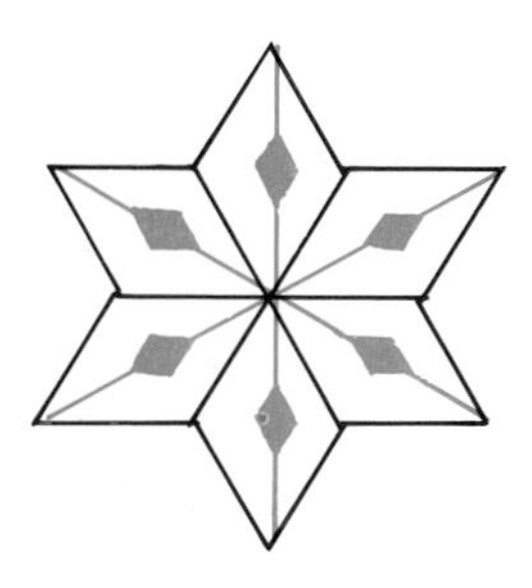

3 Stock cubes of side 2 cm are packed in cubic boxes of side 4.3 cm.

a How many cubes can a box hold?

b What percentage of space is used for packing etc?

4 Calculate:

a the total surface area of this prism

b the volume of the prism.

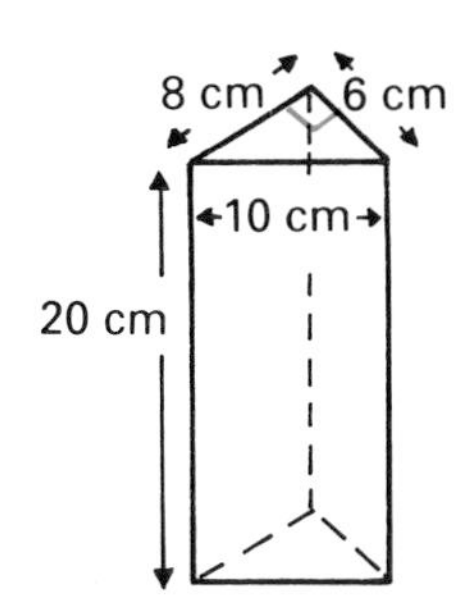

5

The cylindrical tea urn is 90 cm high, and has a base of radius 30 cm. Calculate:

a its capacity in litres

b the number of 200 ml cups it can fill.

6 Calculate:

a the area of each end of the shed

b its volume

c the area of the roof

d the angle of slope of the roof.

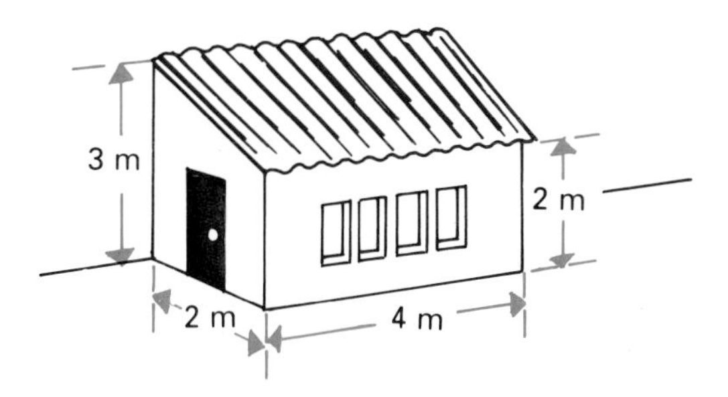

7 The lid of the biscuit tin is a square of side 25 cm, with corners which are quarter circles of radius 1 cm. Calculate the volume of the tin.

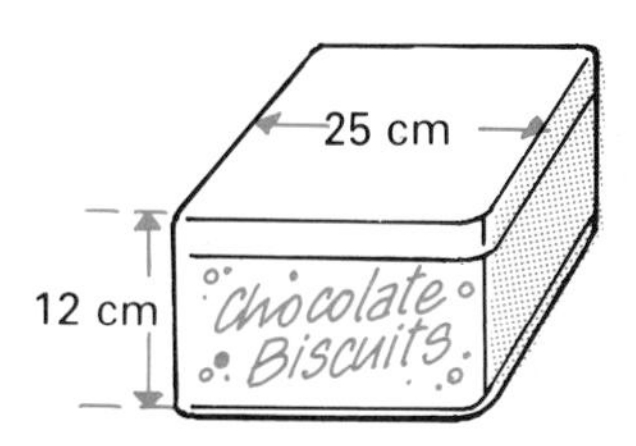

8 Sandra's sketch is for a kennel she plans to make for her collie. Calculate:

a the space inside it in cm^3

b the area of wood she will need, including a floor (and before the entrance is cut out) in m^2.

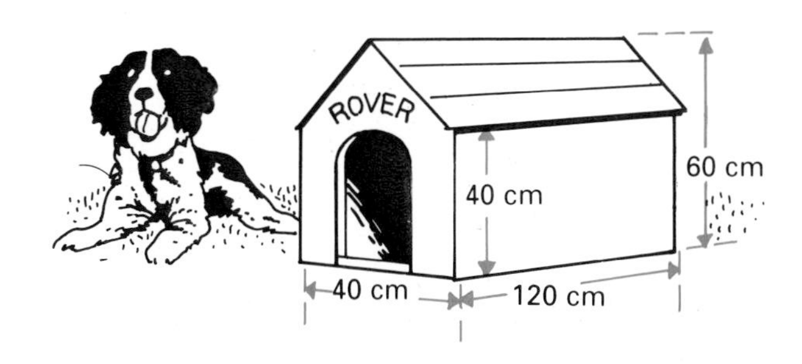

REVISION EXERCISE ON CHAPTER 12: FACTORS

1 Write down all the factors of:
a 40 **b** 50 **c** $9x$ **d** $7x^2$ **e** x^3

2 Write down the highest common factor of:
a 4 and 10 **b** 6 and 15 **c** 6 and 12
d $9x$ and 36 **e** $2x^2$ and x **f** $4t$ and $6t^2$

3 Factorise:
a $6t-18$ **b** $9u+24$
c $11-33t$ **d** $2u^2+u$
e $2y-3y^2$ **f** $6k^2-10k$
g $3a+12b-18c$ **h** $ax+bx-cx$

4 Factorise:
a a^2-b^2 **b** y^2-x^2 **c** n^2-1
d $100-m^2$ **e** $4t^2-1$ **f** $16-9y^2$
g p^2-25q^2 **h** $9r^2-s^2$ **i** $49u^2-36v^2$

5 Use factors, not calculators, for these:
a $7\times 49+7\times 51$ **b** $87\times 49-87\times 39$
c $7.5^2-2.5^2$ **d** 1111^2-111^2
e $0.9\times 8+0.9\times 7-0.9\times 5$

6 Factorise fully:
a $9a^2-9$ **b** $7p^2-28$ **c** $32-8r^2$

7 Factorise:
a $p^2+17p+72$ **b** q^2+2q-3
c $r^2+7r+12$ **d** s^2-s-42
e a^2-6a+9 **f** $1+2b+b^2$
g $27-6x-x^2$ **h** $18+7y-y^2$

8 Factorise:
a $6x^2-5x-4$ **b** $6y^2-11y-10$
c $12t^2+17t+6$ **d** $18u^2+43u-5$

9 Factorise fully:
a $2m^2+10m+8$ **b** $3n^2-18n+24$
c $9-36y^2$ **d** $4x^2y+xy^2$
e x^3-x **f** $12x^2-7x-12$

10 a Prove that $b=\sqrt{(a-c)(a+c)}$.
b Use part **a** to calculate b, when $a=82$ and $c=18$.

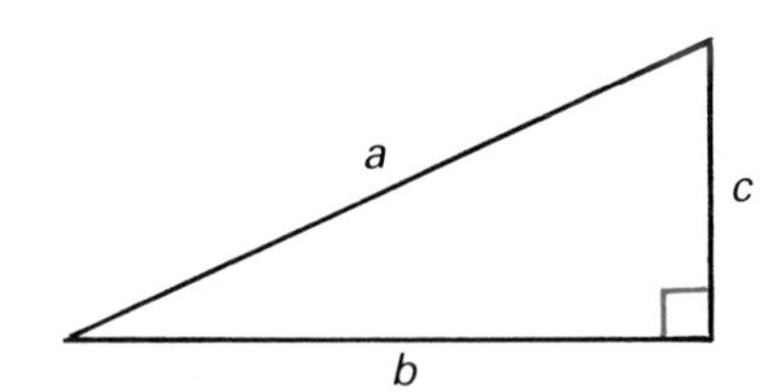

11 Find formulae for the nth terms of these sequences, and use them to prove that the terms are multiples of the given numbers.
a 54, 84, 114, . . . multiples of 6
b 52, 80, 108, . . . multiples of 4.

12 The *sum* of the first n terms of the sequence 1, 5, 9, 13, . . . is $2n^2-n$.
a Check this for $n=1$, 2 *and* 3.
b Calculate the average of the first:
(i) 2 (ii) 3 (iii) 4 (iv) 5 terms of the sequence.
c Prove that the average of the first n terms is always odd.

13 A mixture to factorise:
a $ax+ay$ **b** $3x+6y-9z$
c u^2-v^2 **d** w^2-100
e $x^2-3x-10$ **f** $y^2+14y+49$
g $tu+tv+tw$ **h** $2p-4p^2$
i $a-a^2$ **j** $a-a^3$
k $\pi m^2-\pi n^2$ **l** $\pi y^2+2\pi y$
m $1-x^2$ **n** $1-x^4$
o x^2+6x+9 **p** x^4+6x^2+9
q x^4+2x^2-8 **r** x^4-2x^2-8
s $(u+v)^2-w^2$ **t** $(a-b)^2-c^2$
u $ax+ay+bx+by$ **v** $cx-cy+dx-dy$
w $6-7p+p^2$ **x** $8-7m-m^2$
y $10x^2+7x-12$ **z** $8y^2+6y-9$

REVISION EXERCISE ON CHAPTER 13: MONEY MATTERS—PERSONAL FINANCE

1 Calculate Jack Barr's gross pay, total deductions and net pay for week 36.

Employee 1242	JACK BARR		**Tax Code** 287H	**Week** 36
Basic Pay 173.36	**Overtime** —	**Commission** 26.34	**Bonus** —	**Gross Pay**
Income Tax 38.31	**Superannuation** —	**National Insurance** 14.85	**Other Deductions** 3.86	**Total Deductions**
				Net Pay

2

Liz, a physics teacher in Overtown High School, has an annual salary of £18 200. 6% is deducted for superannuation. How much is left per month, before tax etc?

3 Kevin James earns £248 a week. He pays 6% superannuation and 5% National Insurance.
- **a** Calculate how much he pays for each of these weekly.
- **b** He also pays £47.50 tax and £8.25 union dues each week. What is his weekly take-home pay?

4 At one time, income tax rates were 25% on the first £3000 of taxable income, then 45% on the rest. How much tax was paid by Simon, whose taxable income was £15 375?

5 At 29, Jean Sharp, a non-smoker, took out a 10 year endowment policy worth £4500. Using the insurance tables on page 168, calculate:
- **a** the amount she paid in premiums over 10 years
- **b** the amount she received from the insurance company at the end of 10 years, assuming that the profits paid were £2655.

6 Habib's hobby is hang-gliding. He is 28, and doesn't smoke. Safe and Secure will give him whole life cover of £10 000 provided he pays 40% extra for his premiums.
- **a** Why has he to pay 40% extra?
- **b** Calculate his annual premium. (See page 168.)

7 Robert is saving for a car. He puts £1500 into a special account at 10% p.a. He leaves the interest in so that his money grows with compound interest. How much will he have after:
a one year **b** two years **c** three years?

8 A house valued at £87 000 in 1993 was valued at £91 500 a year later. What was the percentage appreciation in its value?

9 The monthly interest rate Susie has to pay on her credit card is 2.4%.
- **a** Calculate the interest she has to pay on £100 after:
  (i) one month (ii) two months
  (iii) three months (iv) one year.
- **b** What is the APR?

REVISION EXERCISE ON CHAPTER 14: FORMULAE

1 The sum of the numbers in a series like $1+3+5+7+\ldots+15$ is given by the formula $S = \frac{1}{2}n(a+l)$, where a is the first term, l is the last term and n is the number of terms.

a Substitute the values in the series for n, a and l and calculate the sum of the numbers.

b Check your answer by adding the numbers.

2 Make x the subject in each of these:

a $x+b=c$ **b** $ax-d=e$

c $p(x+q)=r$ **d** $\frac{1}{3}(ux-v)=w$

3 A doll's house doorway is in the shape of a square and a semi-circle.

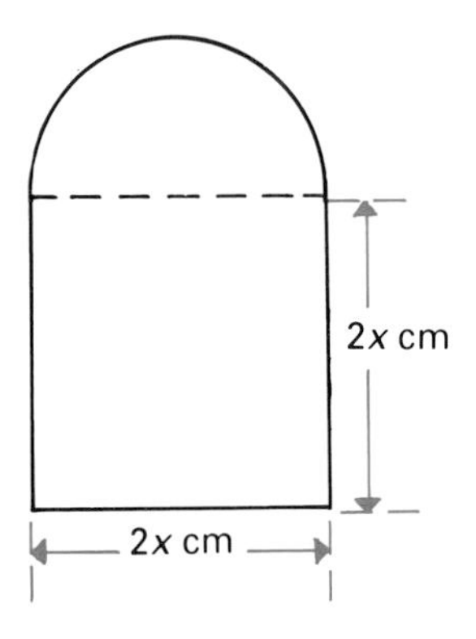

a Write down formulae for:
(i) its perimeter P cm (ii) its area $A\ \text{cm}^2$.

b Calculate P and A, correct to 2 significant figures, when $x=2$.

4 In each formula below, make the letter in brackets the subject of the formula.

a $P=4x+a$ (x) **b** $Q=\dfrac{3t}{s}$ (t)

c $R=\dfrac{4p}{a}$ (a) **d** $\sqrt{(x-a)}=t$ (x)

e $ax+by+c=0$ (y) **f** $\dfrac{x}{a}=\dfrac{b}{c}$ (x)

5 The sum of the numbers in a series like $32+16+8+4+\ldots$ to an 'infinite' number of terms is given by the formula $S=\dfrac{a}{1-r}$, where a is the first term and r is the ratio of any term to the one before it.

a Use the formula to sum the series.

b Add the terms to find the sum of 4, 5, 6, 7 and 8 terms.

c Make r the subject of the formula, and calculate r when $S=121.5$ and $a=81$.

6 Solve for x:

a $x-a=b-x$ **b** $ax+b=cx+d$

c $\dfrac{x}{a}=\dfrac{b}{x}$ **d** $\dfrac{1}{x}=\dfrac{1}{u}-\dfrac{1}{v}$

7 The electrical resistance R ohms of w metres of wire with circular cross-section of diameter d millimetres is $R=\dfrac{2w}{d^2}$.

a Calculate R when $w=4$ and $d=0.4$.

b Make d the subject of the formula, and calculate the diameter of 2.25 m of wire with resistance 2 ohms.

8 A formula for the approximate length of this circular arc is $L=\frac{1}{3}(8h-c)$. Change the subject to:
a c **b** h.

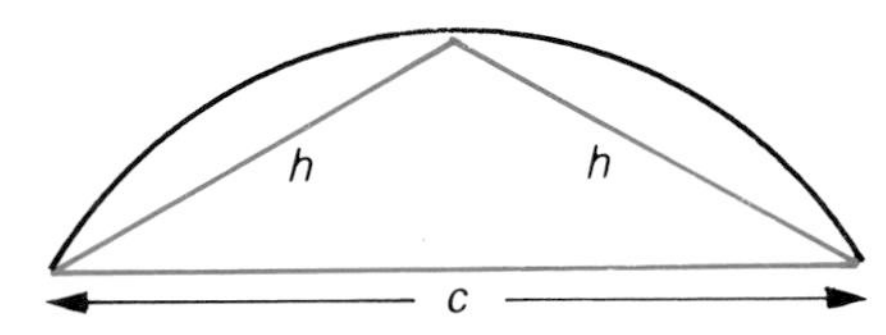

9 A formula for an electric current is $I=\dfrac{nE}{R+nr}$.

Make n the subject of the formula, and calculate n when $I=1.2$, $R=3$, $r=0.5$ and $E=1.5$.

REVISION EXERCISE ON CHAPTER 15: PROBABILITY

1 In an experiment, there are three possible outcomes: success, failure, no conclusion. P(success) = 0.8 and P(failure) = 0.15. Calculate:
a P(no conclusion) **b** P(success or failure).

2 Bill sleeps for eight hours a day. Jack phones him at a random hour from America. Calculate:
a P(Bill is asleep) **b** P(Bill is awake)
c P(Bill is asleep or awake).

3 Mo's torch needs two batteries, placed end to end in the torch with the '+' signs towards the bulb. Without thinking, she drops the batteries into the torch.

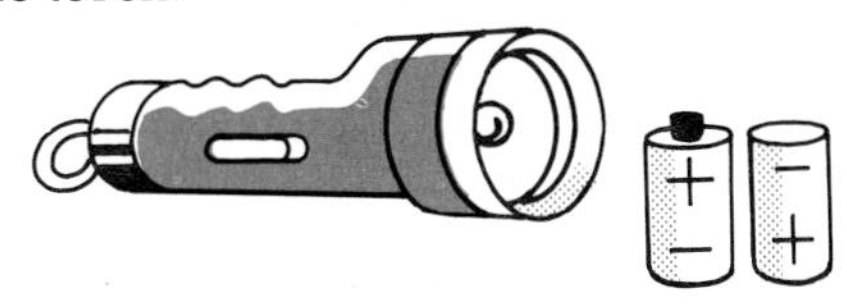

a Sketch all possible positions of the batteries in the torch.
b Calculate the probability that the bulb in the torch lights up first time.

4 Tim, Neil, Nicki and Pippa arrive at the tennis courts to play doubles.
a How many different pairs are possible?
b Calculate the probability that a pair chosen at random are Tim and Nicki.

5 Squeezy toothpaste comes with or without fluoride, and in four colours—red, white, blue and green. Those are supplied in equal numbers.
a Make a table showing the possible combinations of fluoride/not fluoride with the four colours.
b If a tube is chosen at random, calculate:
(i) P(fluoride) (ii) P(fluoride, blue).

6 Two dice, each in the shape of a tetrahedron (regular triangular pyramid) are tossed. The score is the number on the base.

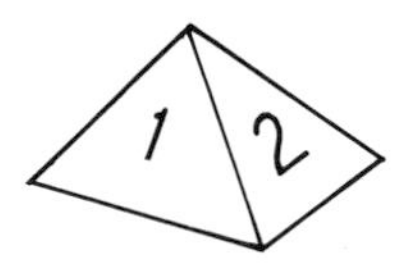

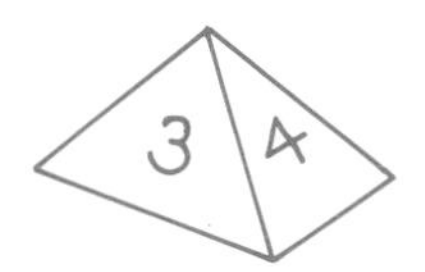

a Show all the outcomes of scores in a tree diagram.
b Calculate:
(i) P(4, 4) (ii) P(odd total score)
(iii) P(score over 5).

7 Bill passes two sets of traffic lights daily. They work independently, and red, amber, green are of equal duration.

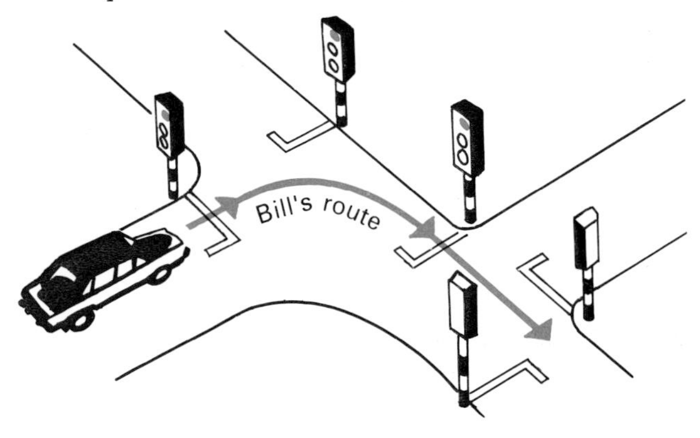

a Copy and complete the table. R and r = red, A and a = amber, G and g = green

		Second lights		
		R	A	G
First lights	r	(r, R)		
	a			
	g			

b How many possible outcomes are there?
c Calculate:
(i) P(both lights show green)
(ii) P(at least one light shows green).
d In a 30 day period, how often can Bill expect to pass through both sets of lights without stopping?

8 E shows the set of factors of eight, and T the set of factors of 12. A whole number is chosen at random from 1 to 12. Calculate:
a P(factor of 8) **b** P(factor of 12)
c P(factor of 8 or 12).

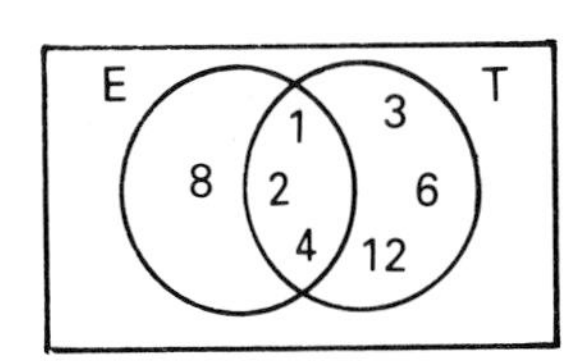

9 Draw a diagram like the one in question **8**, and answer similar questions, for the factors of 6 and 15.

10 In a board game, each player rolls two dice. Sophie is about to throw. Calculate:
a P(12) **b** P(11) **c** P(12 or 11)
d P(12 or 11 or 10).

GENERAL REVISION EXERCISES

GENERAL REVISION EXERCISE 1

1 Calculate:

a $-1+3$ **b** $-4+3$ **c** $2+(-4)$
d $-1-2$ **e** $3-(-1)$ **f** $5\times(-2)$
g 0×9 **h** -4×4 **i** $-3\times(-2)$
j $(-1)^2$ **k** $-2\times(-2)\times(-2)$

2 John Barratt paid £2800 for a second-hand car in 1988. One year later its value had fallen to £2380. Calculate the depreciation as a percentage of the 1988 price.

3 Make an equation, and solve it. How many weights are in each bag?

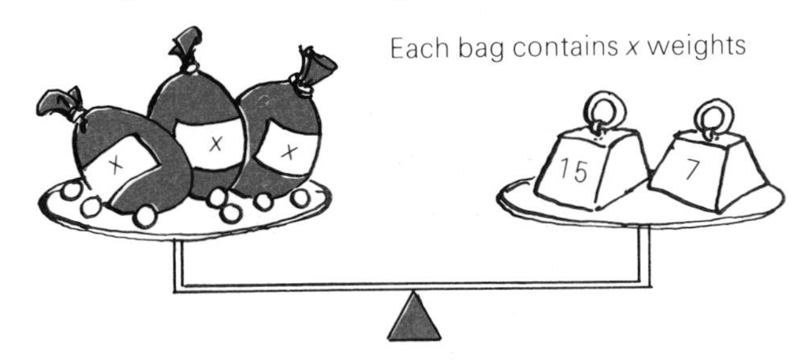

4 The empty spindle of the cassette recorder has diameter 5 mm, and the full reel has diameter 50 mm. Calculate, correct to 3 significant figures:

a the circumference of the empty spindle
b the area of the full reel (including its spindle).

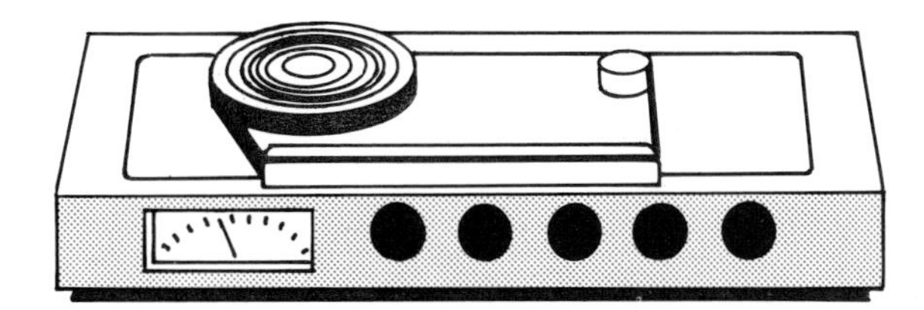

5 Use Pythagoras' Theorem to calculate:

a x **b** y.

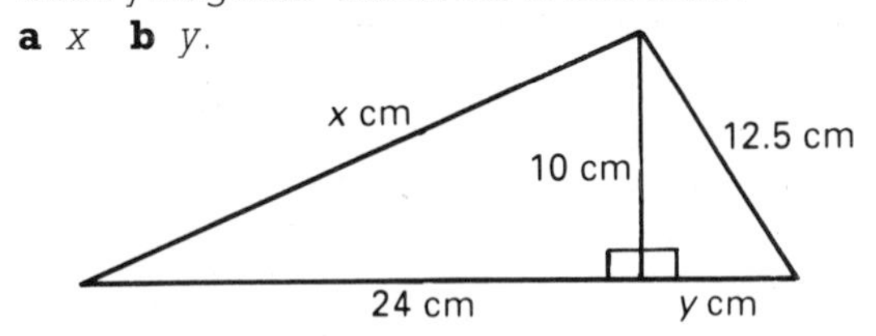

6 Solve these equations:

a $5x = 30$ **b** $x+7 = 12$
c $2x-1 = 9$ **d** $4x-2 = 2x+4$
e $x+5 = 4x-4$

7 Ally goes to America on holiday, and Milly goes to Canada.

Foreign Exchange for £1 Sterling	
US dollars	1.56
Canadian dollars	2.02

a How much foreign currency would each get for £250?
b On return, how much British currency would Ally be given for 20 US dollars?

8 There are four flight paths from Riverside Airport.

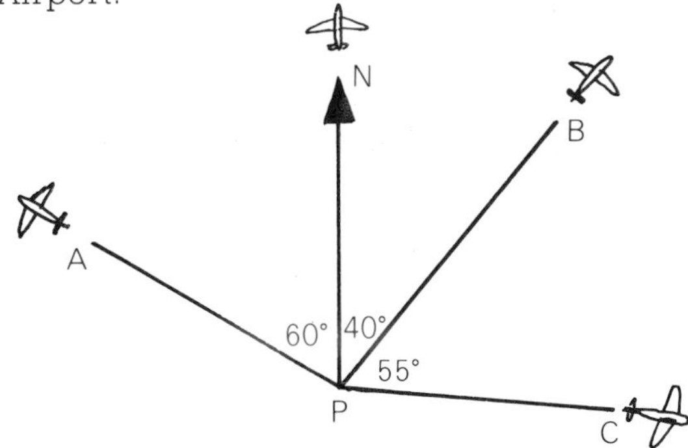

a Name the acute angles which have NP as one arm.
b Calculate the size of the largest obtuse angle.
c PN points north. Write down the three-figure-bearings from P of:
(i) B (ii) C (iii) A.

9 Factorise:

a $2q+2r$ **b** $3p-6$ **c** $8-10y$ **d** $ab-ac$

10 Calculate:

a $\frac{2}{3}+\frac{1}{2}$ **b** $\frac{2}{3}\times\frac{1}{2}$ **c** $1\frac{3}{4}+2\frac{1}{2}$
d $3\frac{3}{5}-2\frac{1}{2}$ **e** $3\frac{1}{2}\times1\frac{1}{2}$ **f** $8\frac{1}{2}\div2\frac{1}{2}$

11 a From the farm to the village is 3 km, and the farm tractor's top speed is 36 km/h. Can the tractor make the journey in less than 5 minutes?
b The tractor covers the first 2 km in 4 minutes, then the next 5 km in 11 minutes. Calculate its average speed for the 7 km.

GENERAL REVISION EXERCISE 2

1 Make an equation for each picture, then solve it to find the number of weights in each bag.

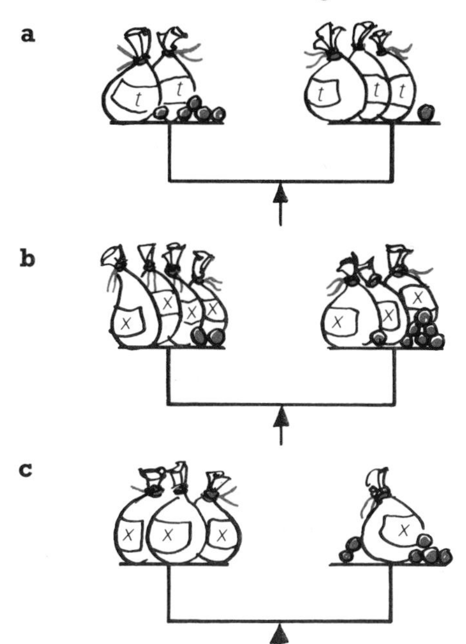

2 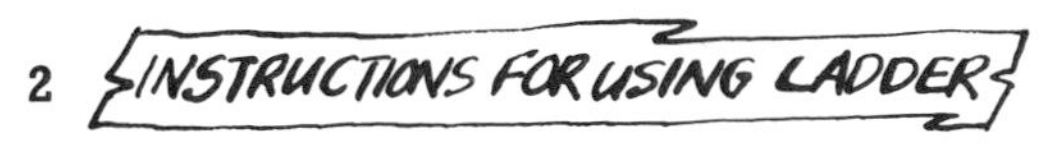

'The bottom should be one quarter as far from the wall as the top is up the wall.'
What angle will the ladder make with the ground, to the nearest degree? (Draw a diagram.)

3 Calculate mentally:
a 60×30 **b** $60 \div 30$ **c** 500×20
d $500 \div 20$ **e** 0.2×10 **f** 0.8×100
g $7 \div 10$ **h** $90 \div 1000$

4 a Sketch each shape, and draw its axes of symmetry. How many axes does each one have?

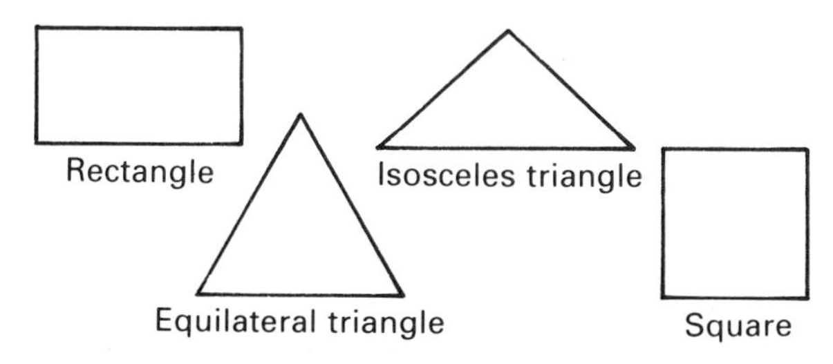

b If each shape is cut out, in how many ways can it be put back in its space?
c Can you see any connection between your answers to **a** and **b**?

5 *Sketch* the straight lines with equations:
a $2x+3y=12$ **b** $y=3x-6$ **c** $4x+y=0$

6 A billion is a thousand million. Write 5 billion:
a in full **b** in standard form.

7 Solve these equations:
a $2x+1=19$ **b** $3x+5=2$ **c** $5x-1=x-9$.

8 L mm of wire are needed to make this skeleton cuboid.

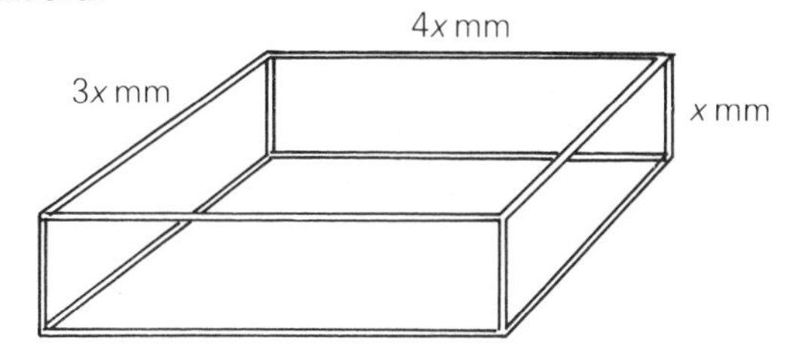

a Make a formula for L.
b Calculate L if $x=9$.

9 Helena's front garden is a 15 m by 9 m rectangle. The lawn is a similar rectangle 6 m broad.

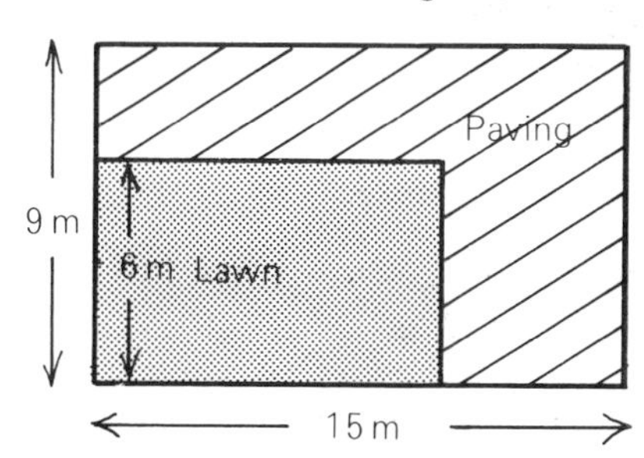

Calculate:
a the scale factor from garden to lawn
b the length of the lawn
c the area of the paved part.

10 Fay won the ice-skating competition with these marks for technical merit:

5.4	5.7	5.3	5.6	5.4	5.3	5.5	5.7	5.3

a Calculate her:
(i) mean
(ii) median
(iii) modal mark.
b What was the range of the marks?

GENERAL REVISION EXERCISE 3

1 Solve each equation for x:
a $2x+3=1$ **b** $4x-1=x+5$ **c** $3(x-2)=6$

2 Which is greater in each pair, and by how much?
a 1.02 or 0.92 **b** 3.14159 or 3.14160 **c** $\frac{3}{4}$ or $\frac{4}{5}$

3 Write down the nth term in each sequence:
a 1, 4, 7, 10, ... **b** −5, −2, 1, 4, ...
c 0.2, 0.3, 0.4, ...

4 Sam's sportscar is 4.5 m long. He wants to make a scale model, with scale factor $\frac{1}{100}$. What length would the model be, in centimetres?

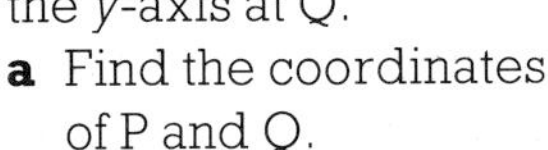

5 The line $5x-12y=60$ cuts the x-axis at P and the y-axis at Q.
a Find the coordinates of P and Q.
b Calculate the length of PQ.

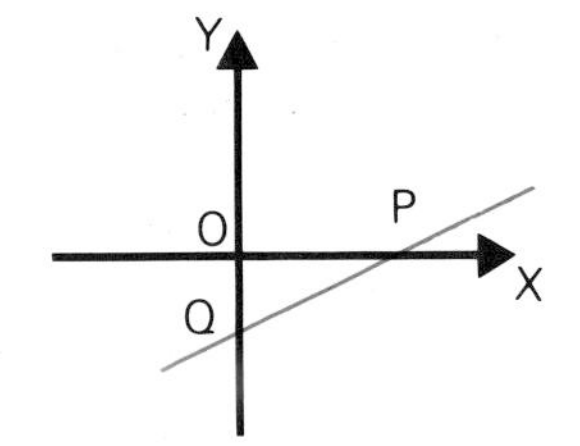

6 An observation tower has been built for the Garden Festival. From a point 26 m from the foot Janie estimates that the angle of elevation of the top is 60°. Calculate (to the nearest metre):
a BC(sin, cos or tan?) **b** AC.

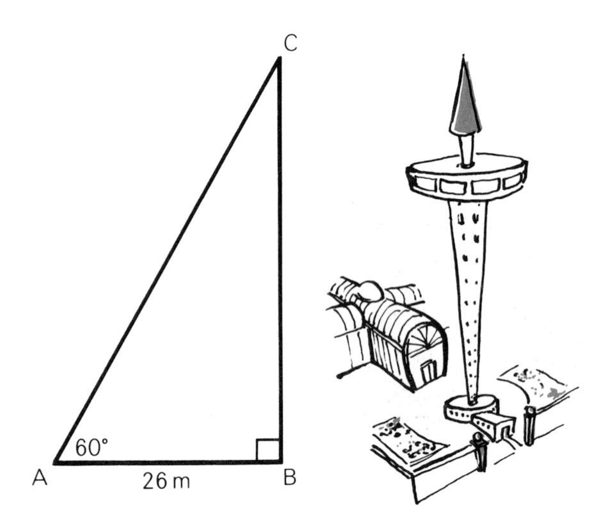

7 Multiply out, and simplify **b** and **e**.
a $u(2u-1)$ **b** $5k-2(k+1)$ **c** $(p-1)^2$
d $(m+4)(m+1)$ **e** $3(1-t)+2(t-3)$

8 a Copy the diagram, and fill in the sizes of all the angles.

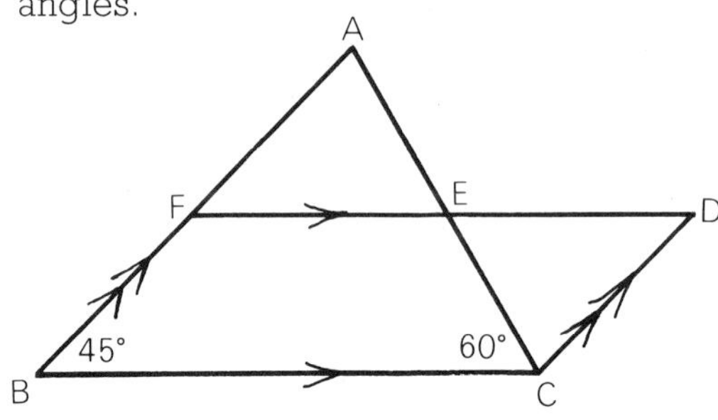

b What two special types of quadrilateral are in the diagram?
c Why are △s AFE and CDE similar?

9 Amy measures the length and breadth of a rectangle to the nearest 0.1 cm. Her measurements are 8.2 cm and 5.7 cm. Calculate the greatest and least perimeters of the rectangle.

10 Solve: **a** $x(x+5)=x^2+10$ **b** $(y-3)^2=y(y-3)$

11 This is a distance/time graph for a train journey from Glasgow to Ayr and back.

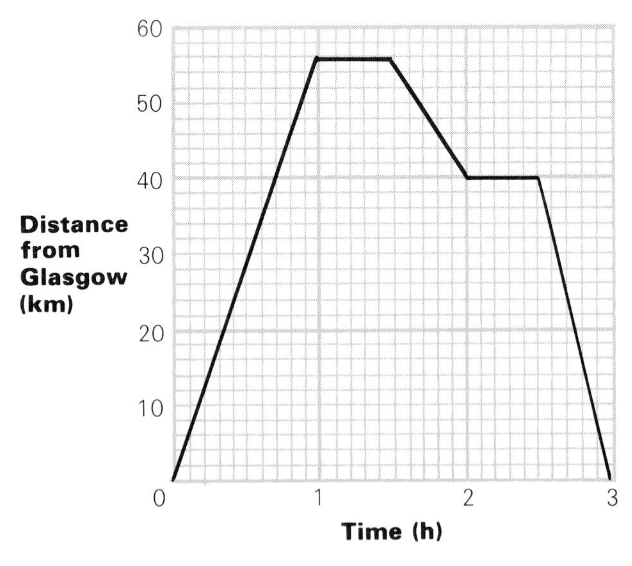

a Find how long the train:
(i) took to reach Ayr
(ii) remained at Ayr
(iii) was held up by a signal failure on the return journey.
b Calculate the average speeds of the journeys to and from Ayr.

GENERAL REVISION EXERCISE 4

1 Solve these pairs of simultaneous equations.

a $y = x$, $y = 6 - 2x$ **b** $x + y = 7$, $x - y = 2$ **c** $2x - 3y = 5$, $3x - y = 4$

2 Calculate the volume and the surface area of a cylindrical oil storage tank with height 20 m and diameter of base 12 m. Give your answers correct to 3 significant figures.

3 Calculate, correct to 3 significant figures:
a sin BAC **b** cos BAC **c** tan BAC **d** AD.

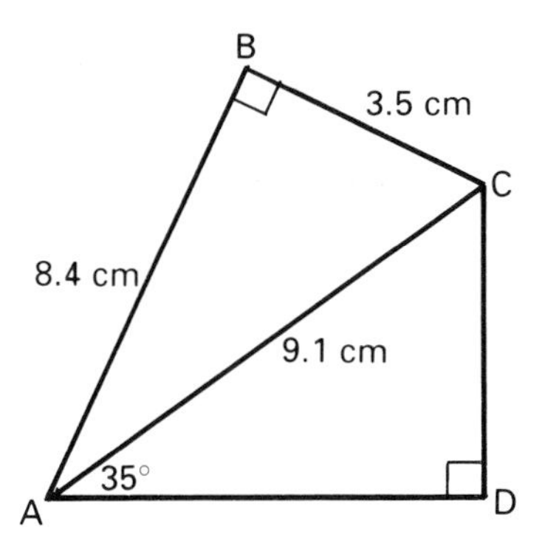

4 Multiply out:

a $(x+2)(x+3)$ **b** $(t+5)(t-1)$
c $(k-2)(k-4)$ **d** $(a+b)^2$
e $(p-q)^2$ **f** $(2n+1)(2n-1)$

5 *Starshine* is 23 km south of the headland H. She is on a course 051°. Calculate how close she comes to the headland, correct to 3 significant figures.

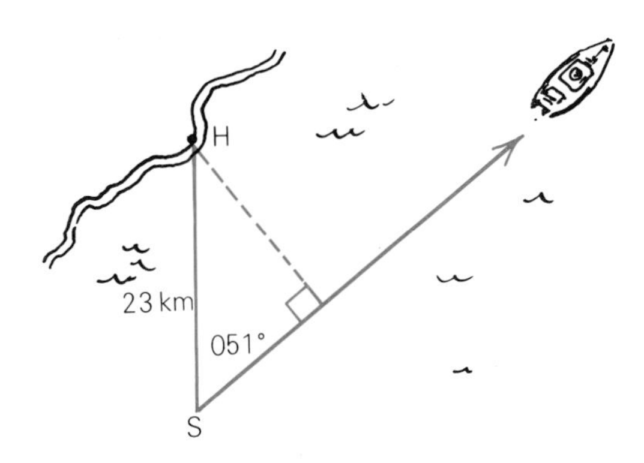

6 a Draw the graphs of the equations $x + 2y = 10$ and $y = 2x$ on squared paper.
b Use the graphs to solve the simultaneous equations, $x + 2y = 10$ and $y = 2x$.
c Check the solution by another method.

7 a Plot the points A(2, 1), B(4, 2), C(4, 4), D(2, 3).
b What shape is ABCD?
c P, Q, R, S are the images of A, B, C, D under reflection in the y-axis. Write down their coordinates. What shape is PQRS?

8 a Make an equation and find x.
b What is the dog's height?

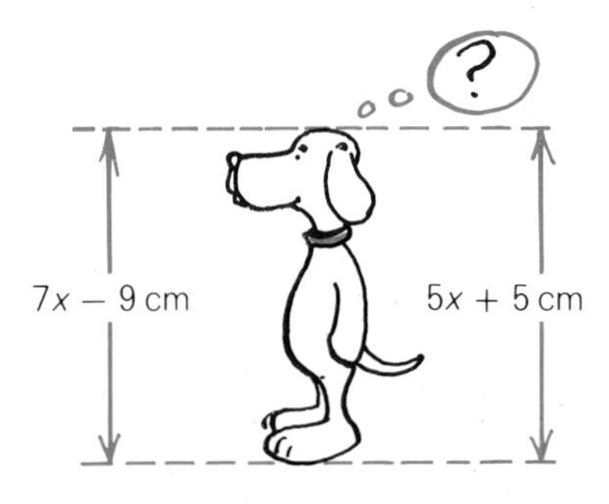

9 An oil tank in the shape of a cuboid is 250 cm long, 180 cm broad and 120 cm high.
a Calculate the volume of the tank in cm³.
b How many litres can it hold?

10 Pete is on a long delivery journey—two hours at an average speed of 50 km/h, then three hours at an average speed of 45 km/h. Calculate:
a the total time he takes
b the length of his journey in kilometres
c his average speed for the whole journey.

11 Draw a distance/time graph of Pete's journey in question **10** on squared paper, using these scales.

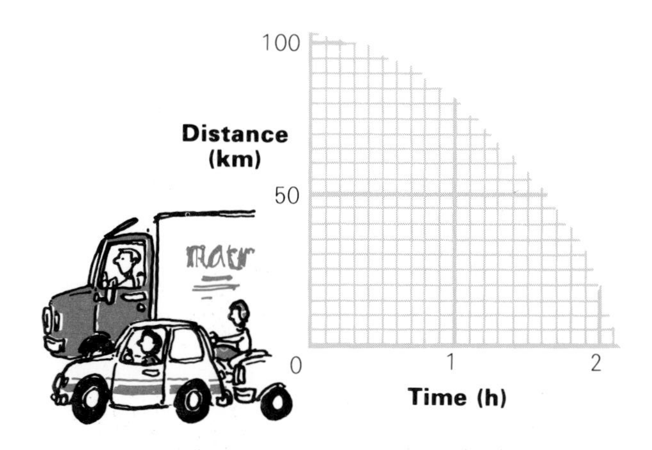

GENERAL REVISION EXERCISE 5

1 All the angles are right angles.

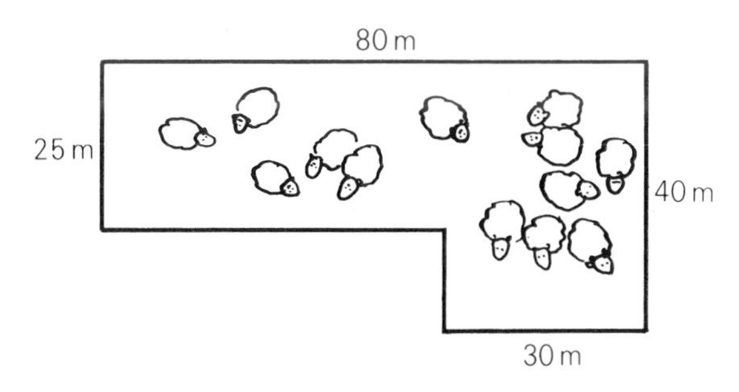

a Calculate the perimeter and area of the field. (Make a sketch.)
b Express the area as a fraction of a hectare (1 hectare = $10\,000\,\text{m}^2$).

2 a *Estimate* the value of $\dfrac{31.05 \times 18.77}{31.05 - 18.77}$.
b Calculate the value, correct to 2 decimal places, using:
(i) the calculator memory (ii) brackets.

3 a Write down a formula for T, the nth term of the sequence 5, 11, 17, 23, . . .
b Make n the subject of the formula, and calculate n when $T = 329$.

4 The frame of a swing is held rigid by horizontal bars BE and CD, and struts BD and CE.

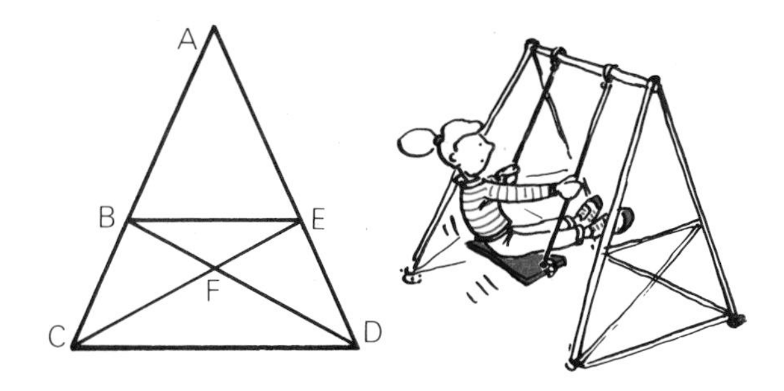

a Name two pairs of:
(i) corresponding angles
(ii) vertically opposite angles
(iii) alternate angles.
b The frame has a vertical axis of symmetry. $\angle$EBD = 30° and $\angle$EBC = 110°. Copy the diagram, and fill in the sizes of as many angles as you can. Write down the size of:
(i) $\angle$ECD (ii) $\angle$BAE.

5 Saturn is 1430 million km from the Sun, and the eccentricity of its orbit is 0.0558. Write both of these numbers in standard form, $a \times 10^n$.

6 a Write down formulae for the perimeters (P) and areas (A) of the square, circle and rectangle.
b Calculate the area of the equilateral triangle, correct to 1 decimal place.

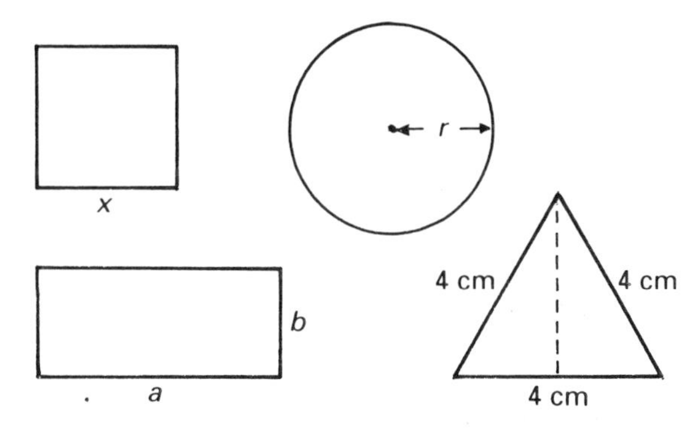

7 a Calculate the height of this chimney, to the nearest 0.1 m.

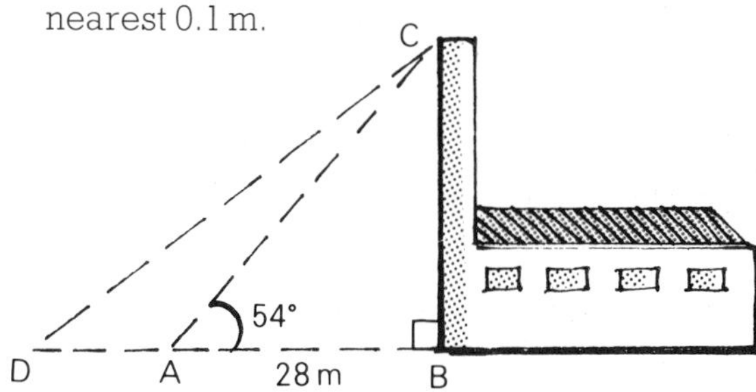

b Calculate, to the nearest degree, the angle of elevation of the top from D, where AD = 9 m, and B, A, D are in a straight line.

8 Triangle ABC is right-angled at B.

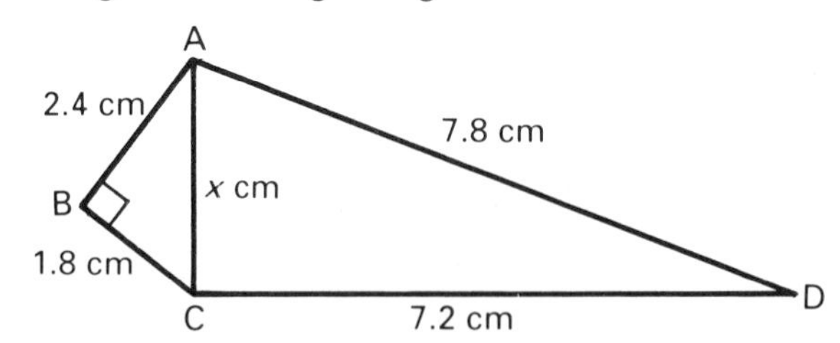

a Calculate x (Pythagoras' Theorem).
b Prove that $\angle$ACD = 90° (converse of Pythagoras' Theorem).

9 A clock is set in a wooden stand. Its clockface diameter is 12 cm, and its base is 36 cm long. Calculate:
a (i) AC (ii) CB
b the angle AB makes with the base (to the nearest degree).

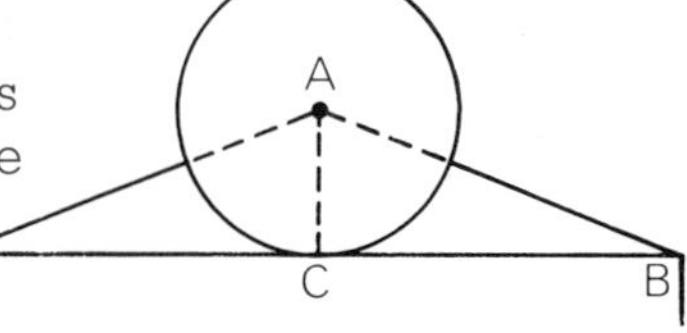

GENERAL REVISION EXERCISE 6

1 Copy the diagram, and fill in the sizes of all the angles.

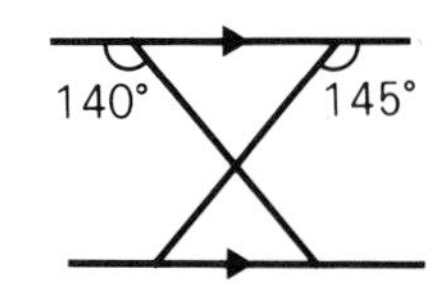

2 Safety First Insurance Co calculate that the probability of a driver under 25 years of age having an accident in any year is 0.03. How many of their 1575 clients under 25 would they expect to submit claims this year?

3

Roof Renewals charge for repairs using the formula $P = C + NR$, where
P = price charged
C = call-out charge
N = number of hours
R = rate per hour.

a Calculate the price of a job which takes three hours at £16.50 an hour, with a call-out charge of £12.50.
b Make N the subject of the formula, and calculate N when $P = 110$, $C = 20$ and $R = 15$.

4 Copy and complete this pattern on squared paper so that it is symmetrical about the broken line.

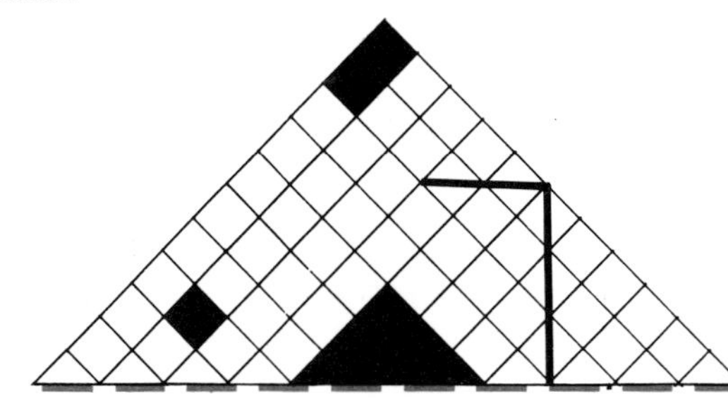

5 Factorise:
a $x^2 - x$ **b** $a^2 - 4$ **c** $x^2 + 6x + 9$
d $t^3 + t^2$ **e** $y^2 + 2y - 8$ **f** $2n^2 - 5n + 3$

6 a Simplify:
(i) $-7 + 2$ (ii) $-3 - 4$ (iii) $4 \times (-3)$
b $x = 1$, $y = -2$, $t = 0$. Calculate the value of:
(i) $x + y$ (ii) $x - y$ (iii) xy (iv) xyt
(v) $x^2 + y^2 + t^2$

7 Calculate, correct to 1 decimal place:
a the length of the wire supporting the telephone pole
b the angle between the wire and the ground.

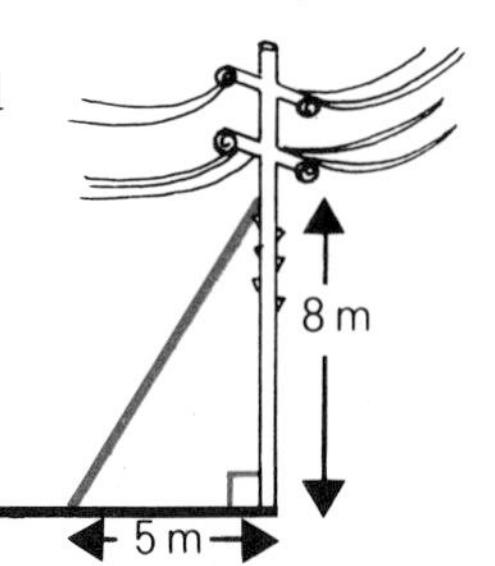

8 a Solve the equations:
(i) $3(x - 2) = 12$ (ii) $3(2x - 1) + 2 = 47$
(iii) $x(x + 4) = x^2 + 12$
b Factorise:
(i) $2y + 8y^2$ (ii) $m^2 + 3m + 2$ (iii) $t^2 - 5t - 6$

9 a Make an equation, and find the lengths of the sides of the triangle.
b Calculate the area of the triangle.

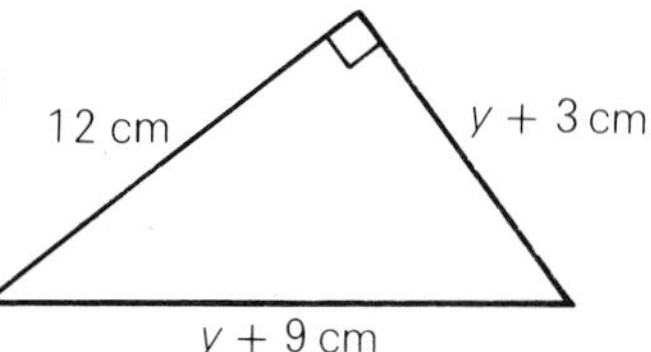

10 Colin is making a circular coffee table with a diameter of 90 cm.
a Calculate the area of the top of the table, to the nearest cm^2.
b He puts a strip of veneer round the edge. What length of strip will he need to buy?

11 This pie chart tells you about the nationality of the students at one Scottish University.

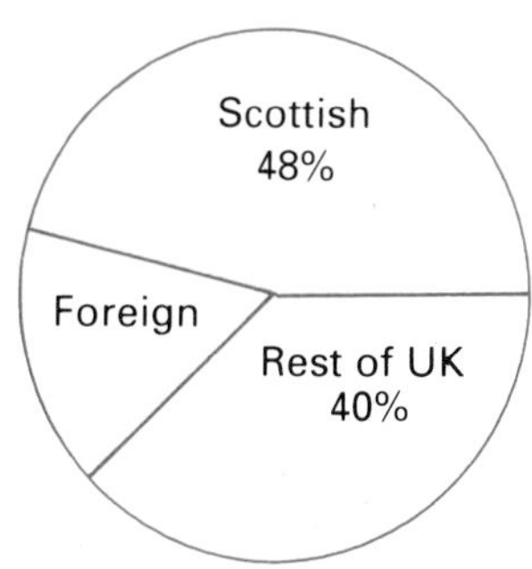

a Calculate:
(i) the percentage of foreign students
(ii) the angle at the centre for the 'rest of UK' slice.
b There are 4500 students altogether. How many are Scottish?

GENERAL REVISION EXERCISE 7

1 Claire puts £600 into her Building Society Account at 8% p.a. rate of interest. She leaves the interest in her account, so that it earns compound interest.
Calculate the total amount in her account after:
a 1 year **b** 2 years **c** 5 years.

2 Calculate all possible:
a sums **b** products
c differences **d** quotients
of pairs of numbers on the cards.

3 If square BADC is rotated clockwise about B, and is then enlarged, it can be made to coincide with square BDEF. Find:
a the size of the angle of rotation
b the scale factor of the enlargement.

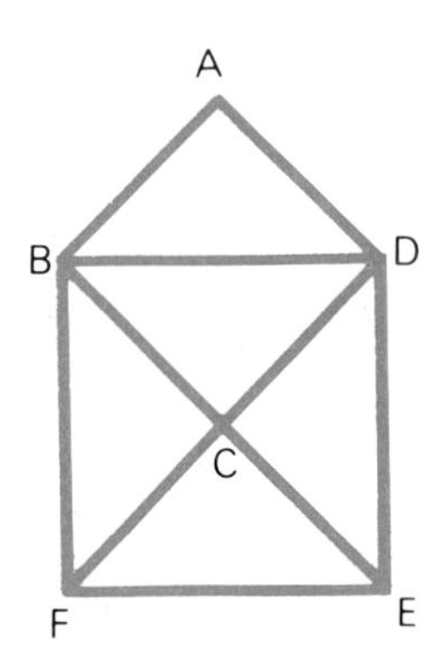

4 The work dockets of 24 employees of Doonvale District Council showed that they worked the following overtime (hours) in the week ending 3rd April:

0 3 8 10 7 6 5 6 8 4 0 6
2 4 5 7 8 10 4 6 2 7 4 6

a Construct a frequency table of the number of hours overtime they worked.
b Draw a frequency diagram.
c Find the mean (to 1 decimal place), median and modal number of hours overtime they worked.

5 The spinner is equally likely to land in any sector. Calculate:
a P(W) in one spin
b (i) P(W, W) (ii) P(L, L)
(iii) P(a Win and a Lose), in two spins.

6 On a hiking holiday, Teresa walks nine miles a day for three days and x miles a day for four days. She walks 79 miles in the 7 days.
a Make an equation and solve it for x.
b Solve:
(i) $6a+3 = -3$ (ii) $5-3t = 16-2t$
(iii) $3(x-1)-2(x+1) = 1$ (iv) $y^2-(y-1)^2 = 2$

7 A hot water pipe has a diameter of 2 cm.

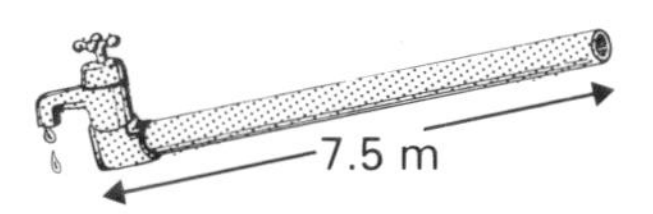

a Calculate the cross-sectional area of the pipe, to 3 significant figures.
b When the tap is turned fully on, water flows through a 7.5 m length of pipe every minute. Calculate the volume of water that comes out of the tap every minute, in:
(i) cm^3
(ii) litres, correct to 3 significant figures.

8 The lengths in the parallelogram are in metres. Calculate the area of the parallelogram.

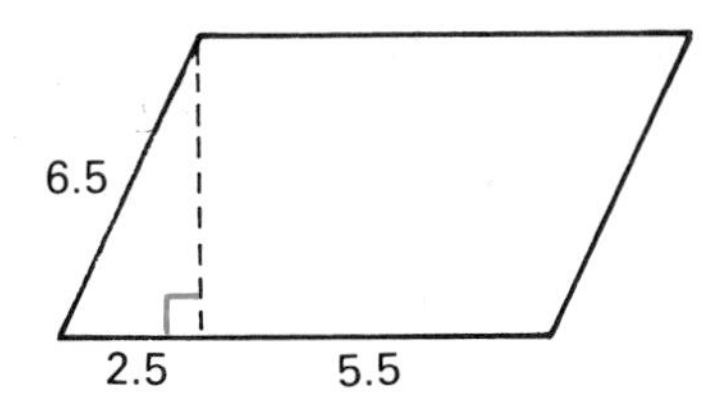

9 a (i) Name two similar triangles in this figure.

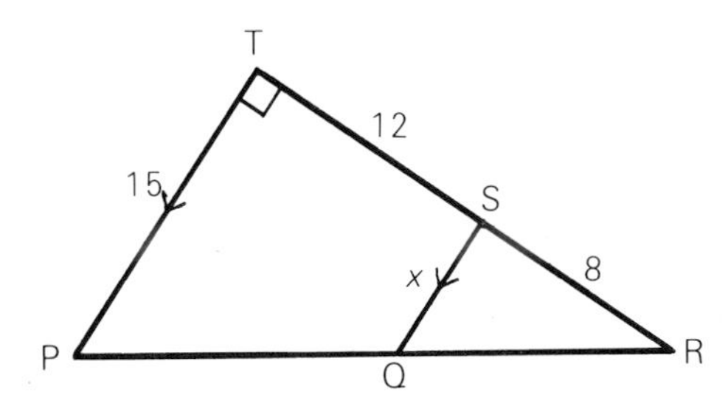

(ii) Write down the value of $\frac{RS}{RT}$.
(iii) Calculate x. Lengths are in cm.
b Use Pythagoras' Theorem to calculate PR.

10 A sports arena consists of a rectangle with semi-circles on two opposite ends.

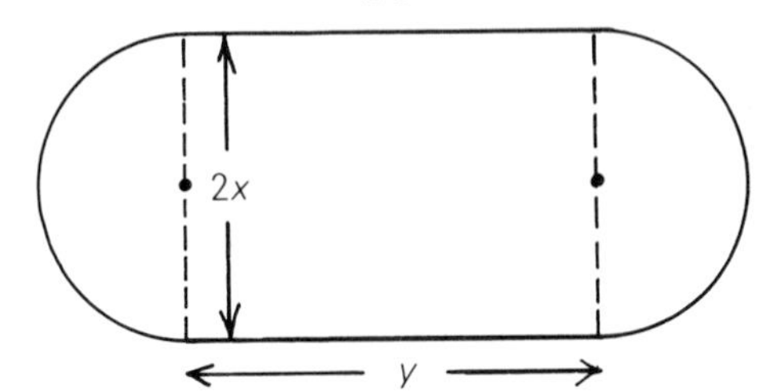

a Find the formulae for:
(i) the perimeter (P)
(ii) the area (A), of the arena.
b Change the subject of each formula to y.

GENERAL REVISION EXERCISE 8

1 **a** Write the numbers below in full:
(i) The population of England is about 5×10^7.
(ii) Mount Everest is 8.85×10^3 m high.
(iii) This page has an approximate thickness of 1.12×10^{-2} cm

b Write the numbers below in standard form:
(i) The Mississippi river is nearly 6050 km long.
(ii) The population of China increased from 838 000 000 to 1 000 000 000 over a ten year period.
(iii) The Sun is about 0.000 016 of a light year from the Earth.

c Calculate $9.5 \times 10^8 \div 1.9 \times 10^{-2}$.

2 Two expeditions in the desert leave base at the same time. One travels north to P at 45 km/h, and the other travels east to Q at 55 km/h.
a Make a sketch which shows their positions after 1 hour.
b How far apart are they then, to the nearest km?
c Calculate the bearing of Q from P, to the nearest degree.

3 One of the sails on *Sea Bobbin* is triangular. Its angles are in the ratio 2:3:5. Calculate the sizes of the angles.

4 Factorise:
a $5x+10y$ **b** $6k-6$ **c** x^2+2x^3
d $2x^2-18$ **e** x^2-x-12 **f** $2x^2+4x-30$

5 The mean (m) of the scores p, q and r can be found by using the formula $m = \frac{1}{3}(p+q+r)$.
a Use the formula to calculate the mean of 69, 74 and 76.
b Make r the subject of the formula.
c Use this new formula to calculate r when $m = 12$, $p = 4$ and $q = 5$.

6 Change the subject to the letter in brackets.
a $2u+v=3$ (u) **b** $ax+y=z$ (x)
c $mx^2=n$ (x) **d** $\frac{m}{n}=p$ (m)
e $\frac{s+t}{u}=v$ (t) **f** $\frac{x}{a}+\frac{y}{b}=c$ (x)

7 Calculate:
a the area of the kite
b the size of $\angle$BAC, to the nearest degree.

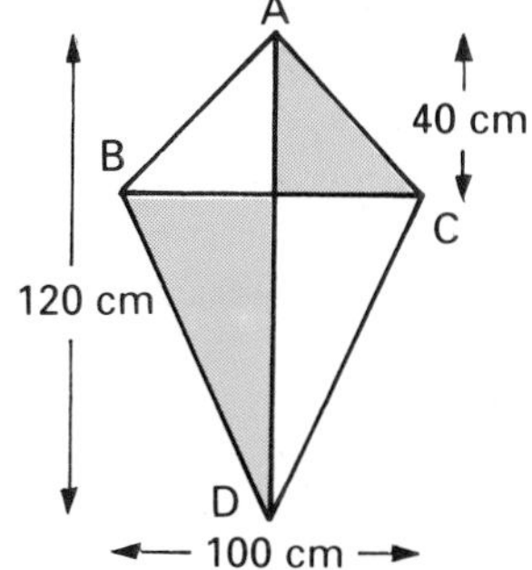

8 A tin of soup has height 10 cm and diameter 8 cm.
a How many tins can be packed into the carton?

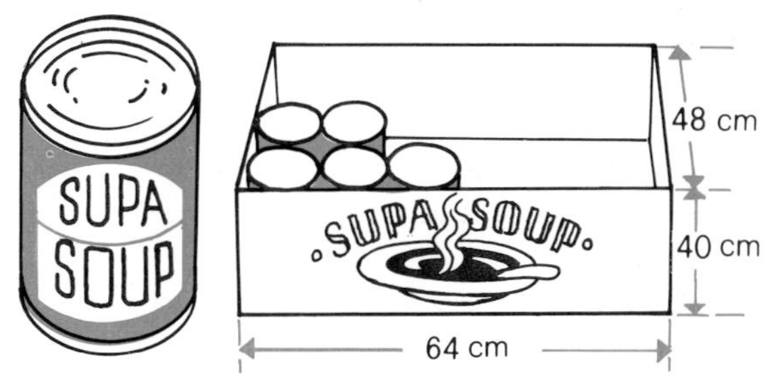

b Calculate the volume of:
(i) the carton (ii) all the tins
(iii) the wasted space.

9 **a** For the yacht *Robin*, $\triangle$ABC and $\triangle$DBE are similar. Can you explain why?
b Calculate the length of BE.

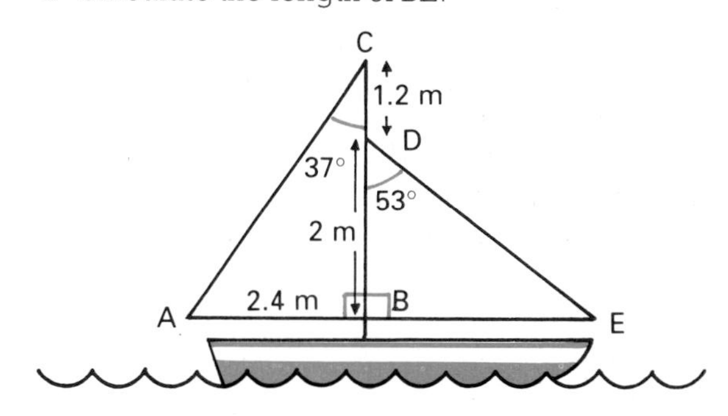

10 The sports bag is roughly cylindrical. Estimate its capacity to the nearest cm³.

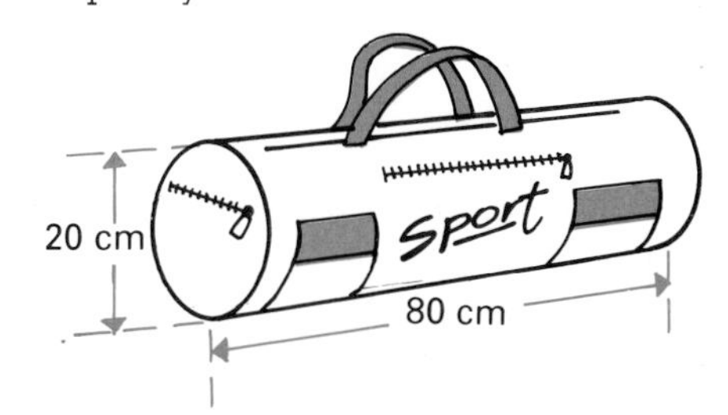

ANSWERS

1 CALCULATIONS AND CALCULATORS

Page 1 Looking Back

1a 57 218 000 **b** 234; 230; 200 **2a** \$1.497; 1.50; 1.5 **b** (i) \$14.97 (ii) 149.72 (iii) 1497.20 **3** 423 cm^2 **4a** 160 cm; 1600 cm^2 **b** 170 cm, 1800 cm^2 **5a** 100 **b** 1000 **c** 250 **d** 5 **e** 3.6 **f** 0.005 **6a** 3 **b** 2 **c** 6 **7** $\frac{2}{3}, \frac{6}{9}; \frac{4}{10}, \frac{2}{5}; \frac{9}{12}, \frac{3}{4}$ **8a** $\frac{1}{3}$ **b** $\frac{4}{5}$ **c** $\frac{3}{4}$ **d** $\frac{1}{2}$ **e** $\frac{4}{5}$ **9a** 20 **b** 18 **c** 24 **10a** 70; 82.36 **b** 50; 53.65 **c** 2; 2.25

Page 2 Exercise 1

1 (ii) 0.483 (iii) 0.571 **2** (ii) 1.93 (iii) 123 **3** (ii) 0.782 (iii) 2.40 **4** (ii) 0.738 (iii) 185 **5** (ii) 0.873 (iii) 119 **6** (ii) 11.8 (iii) 42 500 **7** (ii) 0.0878 (iii) 0.00411 **8** (ii) 0.167 (iii) 0.345 **9** (ii) 18.5 (iii) 111 000 **10** (ii) 3.22 (iii) 244

Page 3 Exercise 2

2a 546.25 **b** 5376.25 **c** 12.17 **d** 34.78 **3** 14.7, 25, 18.6, 24.6, 30.8—all cm **4** 5.9, 6.6, 7.1, 7.8, 8.4—all minutes **5** £82.25, £35.84, £141, £916.50, £1180.88 **6a** \$64.35, \$852.64, \$1462 500 **b** £6.84, £157.26, £6736.41 **7a** 243 **b** 20 736 **c** 3 796 416 **d** 0.000 976 562 **e** 306.019 6848

Page 4 Exercise 3

1a 9000 **b** 60 **c** 10 **d** 70 **e** 30 **f** 0.8 **2a** 8 **b** 100 **c** 100 **d** 500 **e** 200 **f** 3000 **3a** 1600 **b** 1458 **4a** 800 **b** 851 **5a** 2700 cm^2 **b** 2800 cm^2; 2, same as data **6a** 1200, 1080 **b** 210, 203 **c** £8, £8.25 **d** 25 kg, 23.5 kg **7a** 45, 44.1 **b** 250, 220

Page 5 Exercise 4

1a 4×10^1 **b** 4×10^2 **c** 4×10^3 **d** 2.3×10^2 **e** 1.23×10^3 **f** 2.04×10^2 **g** 6.7×10^4 **h** 1.1×10^5 **2a** 4×10^{-1} **b** 4×10^{-2} **c** 4×10^{-3} **d** 6.5×10^{-1} **e** 1.03×10^{-1} **f** 5×10^{-4} **g** 1×10^{-1} **h** 2×10^{-6} **3a** 9.7×10^1 **b** 4.2×10^{-1} **c** 1.08×10^3 **d** 4×10^{-2} **e** 6.3×10^4 **f** 6×10^{-3} **g** 1.23×10^6 **h** 1.1×10^{-5} **4a** 3×10^8 **b** 1.268×10^4 **c** 1.13×10^{-1} **d** 2.15×10^9 **e** 5×10^{-8} **f** 1.49125×10^5 **g** 8×10^{-7} **h** 6.6×10^{21} **i** 1×10^9 **j** 1×10^{-2} **5** 5×10^9 **6a** 400 **b** 6000 **c** 0.07 **d** 0.009 **e** 0.98 **f** 5 **7a** 240 000 000 000 **b** 8 600 000 **c** 0.0025 **d** 165 000 **e** 2450 **f** 0.000 000 000 000 000 000 000 000 001 675 **8a** 600 **b** 52

Page 6 Exercise 5A

1d 10 000 000 000, or 100 000 000 **2b** 1.2×10^{10} or 1.2×10^8; 12 000 000 000 or 120 000 000 **3a** 21 300 **b** 110 **c** 901 000 **d** 58 080 000 000 **e** 6 070 000 **f** 8300 **4b** 5.3×10^{-8} or 5.3×10^{-10}; 0.000 000 053 or 0.000 000 000 53 **5a** 0.000 6 **b** 0.9 **c** 0.032 **d** 0.004 98 **e** 0.000 000 082 6 **f** 7.5 **6a** 7000 **b** 150 **c** 780 000 **d** 2 310 000 **e** 0.314 **f** 0.000 21 **g** 0.0861 **h** 0.000 053 **7** 1.2×10^{20}; 1.2×10^{-8} **8** (i) 1 116 000 000 (ii) 1.116×10^9 **9a** (i) 7 980 000 000 (ii) 7.98×10^9 **b** (i) 2 912 700 000 000 (ii) 2.9127×10^{12} **10** (i) 4 340 000 (ii) 4.34×10^6

Page 7 Exercise 5B

1a 1.69×10^9 **b** 1.3×10^0 **c** 4×10^{-3} **d** 1.3×10^9 **2** 9.6135×10^2 **3a** 5×10^{12} **b** 3.75×10^{13} **4** 1.419×10^{16} km **6a** 5.9×10^{24} kg **b** 82:1 **7a** 3.0×10^{18} km **b** 1.8×10^{18} km

Page 8 Exercise 6

1a $1\frac{1}{2}$ **b** $2\frac{1}{4}$ **c** $1\frac{1}{3}$ **d** $3\frac{1}{2}$ **e** $2\frac{2}{5}$ **2a** $\frac{7}{2}$ **b** $\frac{7}{4}$ **c** $\frac{16}{5}$ **d** $\frac{16}{3}$ **e** $\frac{11}{10}$ **3Ca** $1\frac{3}{8}$ **b** $5\frac{1}{3}$ **c** $4\frac{4}{5}$ **d** $3\frac{3}{10}$ **e** $11\frac{1}{9}$ **4Ca** $\frac{14}{3}$ **b** $\frac{43}{8}$ **c** $\frac{27}{10}$ **d** $\frac{9}{8}$ **e** $\frac{79}{8}$ **5C** $\frac{100}{3}$ **6C** Rows: $\frac{11}{8}, \frac{17}{4}, \frac{35}{16}, \frac{31}{8}, \frac{23}{4}$; 11, 17, 35, 31, 23 **7Ca** $\frac{3}{2}$ **b** $\frac{4}{3}$ **c** $\frac{7}{5}$ **d** $\frac{13}{9}$ **8Ca** 1.5, 1.25 **b** 1.33. . . , 1.2 **c** 1.3, 1.4 **d** 1.44 . . . , 1.36 . . . Using decimals. **9C** 24

Page 9 Exercise 7A

1a $\frac{3}{5}$ **b** $\frac{1}{2}$ **c** $\frac{1}{2}$ **d** $\frac{3}{4}$ **e** $\frac{5}{6}$ **f** $\frac{11}{12}$ **g** $\frac{1}{10}$ **h** $\frac{1}{3}$ **2a** $2\frac{4}{5}$ **b** $3\frac{1}{2}$ **c** $1\frac{1}{2}$ **d** $1\frac{1}{3}$ **3Ca** $1\frac{7}{20}$ **b** $1\frac{1}{12}$ **c** $\frac{5}{12}$ **d** $\frac{1}{20}$ **e** $\frac{11}{15}$ **f** $1\frac{1}{6}$ **g** $\frac{3}{14}$ **h** $\frac{1}{8}$ **4Ca** $3\frac{3}{4}$ **b** $2\frac{1}{8}$ **c** $7\frac{1}{4}$ **d** $\frac{5}{6}$ **5C** $11\frac{1}{2}$ litres **6Ca** 15 km **b** $15\frac{3}{4}$ km **c** $\frac{3}{4}$ km **7Ca** $2\frac{1}{4}$ hours **b** $1\frac{1}{2}$ hours **8Ca** $\frac{7}{8}$ **b** $\frac{1}{8}$ **9Ca** $\frac{1}{20}$ **b** Janice 50, Lana 25, Sean 20, Andrew 5

Page 9 Exercise 7B

1C $\frac{1}{2}$ hour **2C** $1\frac{2}{5}$ cm, $\frac{4}{5}$ cm **3C** $8\frac{1}{4}$%, $3\frac{3}{5}$%, $8\frac{1}{3}$%, $2\frac{1}{2}$%, $5\frac{1}{2}$% **4Ca** 24 km **b** $18\frac{3}{4}$ km **c** $\frac{3}{4}$ km **d** $4\frac{1}{2}$ km **5Ca** $\frac{9}{16}$ inch **b** $\frac{3}{8}$ inch **6Ca** (i) $59\frac{1}{2}$ cm (ii) $4\frac{3}{4}$ cm **b** (i) $82\frac{1}{4}$ cm (ii) $9\frac{5}{8}$ cm

Page 10 Exercise 8A

1a 4 **b** 4 **c** $1\frac{2}{3}$ **d** $2\frac{1}{2}$ **2a** $\frac{1}{8}$ **b** $\frac{3}{8}$ **c** $\frac{1}{2}$ **d** $\frac{1}{10}$ **e** $\frac{1}{5}$ **f** $\frac{1}{8}$ **g** $\frac{1}{3}$ **h** $\frac{1}{4}$ **3a** $3\frac{3}{4}$ **b** $1\frac{7}{8}$ **c** $2\frac{1}{2}$ **d** 4 **4Ca** $\frac{1}{2}$ **b** 1 **c** $\frac{1}{9}$ **d** $\frac{1}{12}$ **5Ca** 6 **b** $7\frac{1}{2}$ **c** $\frac{15}{16}$ **d** 3 **6Ca** (i) 3 litres (ii) $2\frac{1}{4}$ litres (iii) 9 litres **b** (i) $\frac{1}{4}$ litre (ii) $\frac{1}{8}$ litre (iii) $1\frac{7}{8}$ litres **7Ca** 100 **b** $166\frac{2}{3}$ **8Ca** 60, 30, 15 **b** each is half the one before it **9C** 13, $32\frac{1}{2}$, 117 miles **10Ca** $6\frac{1}{4}$ m^2 **b** $27\frac{3}{16}$ m^2

Page 11 Exercise 8B

1C $15\frac{3}{4}$ **2Ca** $22\frac{1}{2}$ m^2 **b** $7\frac{1}{2}$ m^2 **3C** Dress £43.20, skirt £10.80 **4Ca** 36 cm **b** 15 cm **c** $13\frac{1}{4}$ cm **5C** $38\frac{1}{2}$ cm^2 **6Ca** $2\frac{1}{2}$ **b** $-\frac{5}{12}$

Page 12 Exercise 9B

1Ca 2 **b** 4 **c** $1\frac{1}{2}$ **d** 1 **e** 17 **f** 8 **g** 4 **h** 7 **i** 3 **j** 6 **k** 40 **l** $1\frac{1}{2}$ **2C** 48 **3C** 192 **4Ca** 4 **b** 2 **c** 5 **5Ca** 6 **b** $\frac{5}{8}$ tonne **6Ca** $6\frac{1}{4}$ cm **b** $4\frac{1}{2}$ cm **7C** 20 **8C** 2, 8, 16, 4 **9Ca** $11\frac{1}{4}$ **b** $5\frac{1}{3}$ **c** $11\frac{1}{4}$ **d** $12\frac{1}{2}$ **e** 40 **f** $4\frac{4}{5}$ **10C** $3\frac{1}{3}$ cm

Page 13 Check-up on Calculations and Calculators

1Ca 4.04 **b** 128 **c** 4410 **d** 103 **e** 54 900 **f** 0.003 75 **2Ca** (i) 32 cm^2 (ii) 33.5 cm^2 **b** (i) 280 000 cm^3 (ii) 305 000 cm^3 **3Ca** 9750 **4Ca** 2.7×10^7 **b** 1×10^{-6} **5Ca** 6×10^9 **b** 1.5×10^{11} **6C** 2.2×10^{-6} **7C** 3.22×10^{23}, 4.83×10^{24}, 6.38×10^{23}, 1.90×10^{27}, 5.68×10^{26}

8C Last column: 10, 16, 21, 27, 32, 38 **9Ca** $\frac{7}{8}$ **b** $4\frac{7}{15}$ **c** $\frac{1}{4}$ **d** $2\frac{11}{20}$ **e** $\frac{2}{15}$ **f** $1\frac{1}{5}$ **g** $4\frac{4}{5}$ **h** $7\frac{1}{2}$ **10C** $\frac{13}{16}$ inch **11C** 40 m^2 **12C** $\frac{2}{5}$ **13C** $2\frac{1}{2}$ m

2 SIMILAR SHAPES

Page 14 Looking Back

1a 10 cm **b** 36 mm **c** 2 m **2a** 24 **b** 80 **c** 180 **d** 12 **e** 84 **f** 192 **3a** $\frac{2}{3}$ **b** $\frac{5}{6}$ **5**(i) **a**, **g**; **d**, **i**; **e**, **k** (ii) **b**, **h**; **c**, **j**; **f**, **l** **6** 4.5 m **7** 2.5 mm **8** **a**; it is most like him

Page 16 Exercise 1A

1a 3 **b** $\frac{1}{3}$ **2a** $\frac{3}{2}, \frac{4}{3}, 2$ **b** $\frac{2}{3}, \frac{3}{4}, \frac{1}{2}$ **3a** 12 cm **b** 2 cm **c** 9 cm **d** 3 cm **e** 6 cm **4a** 6 cm, 3 cm **b** 2 cm, 1 cm **5a** 140 mm by 100 mm **b** 14 cm by 10 cm **8a** $\frac{1}{10}$ **b** 2.5 m **9a** 10 m **b** 6 m **10a** 40 cm **b** 27 cm **11a** 5000 cm **b** 50 m

Page 17 Exercise 1B

2 6.4 mm, 4.8 mm **3a** $\frac{3}{2}$ **b** 60 cm, 102 cm **4a** Enlarge **b** 2 **c** 4 **5a** Reduce **b** $\frac{1}{4}$ **c** 30 **6a** $\frac{3}{2}$ **b** 54 **7a** $\frac{3}{4}$ **b** $\frac{4}{5}$ or 0.8

Page 19 Exercise 2

1 (iv) **2a** (iii) **b** (ii) **3a** (i) and (iii), (ii) and (iv) **b** △ a warning, ○ an order **4** (i), (iv), (v) **5a** (ii)–(vi) **b** all of them **6** 24 cm, 18 cm **7a** Yes—all have the same shape **b** no—they don't all have the same size

Page 20 Exercise 3A

1a All right angles **b** (i) 3 (ii) 3; yes **2a** All right angles **b** (i) 2 (ii) 2; yes **3** Yes—all angles are 90°, ratios of lengths and breadths are $\frac{9}{10}$ **4** **a**, **b**, **d** are similar **5** Yes—angles are 90°, ratios of lengths and heights are $\frac{4}{5}$ **6** No. Ratios of pairs of sides are different **7a** 12 cm **b** (i) 18° (ii) 18° **8a** $\frac{3}{5}$ **b** 7.2

Page 22 Exercise 3B

1a $\frac{4}{3}$ **b** 12 **2a** $\frac{4}{5}$ **b** 12 **3a** $\frac{3}{4}$ **b** 4.5 **4a** $\frac{5}{6}$ **b** 27.5 **5a** $\frac{5}{8}$ **b** 20 **c** 15 **6a** (i) 200 (ii) 180 **b** 40 mm, 60 mm **7a** $\frac{2}{1}$ **b** yes **8a** $\frac{2}{1}$ **b** no; corresponding angles are not equal **9a** They are not all in corresponding positions

Page 25 Exercise 4A

1 All of them **2a** They are equiangular **b** 52 cm, 30 cm **3a** 3 **b** 15 **4a** $\frac{2}{3}$ **b** 8 **5** 30 cm **6a** They are equiangular **b** 75 **7** 35 m **8a** They are equiangular **b** BC **c** $3\frac{1}{3}$

Page 26 Exercise 4B

1a (i) and (ii) **b** (i) $\frac{u}{h}=\frac{v}{i}=\frac{w}{g}$ (ii) $\frac{a}{f}=\frac{b}{d}=\frac{c}{e}$ **2** (i), (iii) **3a** yes **b** 19.3 **4a** yes **b** 3.6 m **5a** 4 **b** 9 **6a** 3.68 m **b** 6.9 m

Page 27 Exercise 5

1a (i) $\frac{1}{2}$ (ii) 3 **b** (i) $\frac{3}{2}$ (ii) 3 **c** (i) $\frac{3}{5}$ (ii) 6 **d** (i) $\frac{8}{5}$ (ii) 9.6 **2a** (i) $\frac{13}{20}$ (ii) 20.8 m **b** (i) 83 m (ii) 29 m **3a** 225 cm **b** 135 cm **4a** 6 m **b** 3.75 m

5a

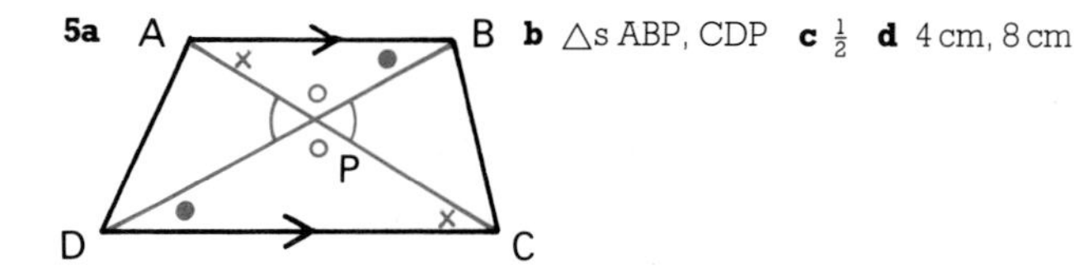

b △s ABP, CDP **c** $\frac{1}{2}$ **d** 4 cm, 8 cm

Page 28 Check-up on Similar Shapes

2a 1.5, 1.25 **b** 0.6, 0.5 **3a** 40 cm **b** 8 m **4a** $\frac{2}{3}$ **b** 3 **c** 2.7 **5a** 250 mm **b** $\frac{5}{3}$ **c** 350 mm **d** 70 mm **6** **a** and **b** **7a** 7.2 m **b** $7\frac{1}{3}$

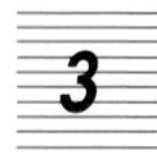

3 GOING PLACES

Page 29 Looking Back

1a 525 km **b** 391 km **2a** 1 h 15 min **b** 1 h 20 min **c** 1 h 10 min **d** 1 h 20 min **3a** 0.5 **b** 0.75 **c** 0.4 **d** 0.1 **4a** (i) 20 m (ii) 70 m **b** (i) 1 s (ii) 4 s (iii) 9 s **5a** 9 h 24 min **b** 5 h 40 min **6a** 20 mph **b** 25 mph **7a** 96 (i) 92 (ii) 92.32 (iii) 92 **b** 6 or 7 (i) 7 (ii) 6.89 (iii) 6.9

Page 30 Exercise 1A

1a 10 km/h **b** 55 km/h **c** 60 km/h **d** 47 km/h **2a** 6.67 m/s **b** 5.56 m/s **c** 4.55 m/s **3a** 1840 km/h **b** 83 km/h **4a** 10 mph **b** 30 mph **c** 50 km/h **d** 800 km/h **e** 5 m/s **f** 0.5 m/s **5a** 94 miles per day **b** 4 mph **6a** 55 mph **b** 45 mph **c** 50 mph **7a** 1.58 **b** 2.24

Page 31 Exercise 1B

1a 0.05 h **b** 0.67 h **c** 0.2 h **d** 0.38 h **e** 6.58 h **f** 2.1 h **2** 67 km/h **3** 95 km/h; 57 km/h; 76 km/h **4a** 3600 km, $2\frac{1}{4}$ h **b** 1600 km/h **5a** 26 miles; 205 min **b** 7.6 mph **6a** (i) 2 h 20 min (ii) 1 h 10 min (iii) 1 h 30 min **b** (i) 209 miles (ii) 90 miles (iii) 102 miles **c** (i) 90 mph (ii) 77 mph (iii) 68 mph **7a** 50 km/h **b** 12.8 km/h **c** 27 km/h **d** 0.004 km/h **e** 36 km/h **f** 60 km/h; kangaroo, zebra, skier, cyclist, runner, snail

Page 33 Exercise 2

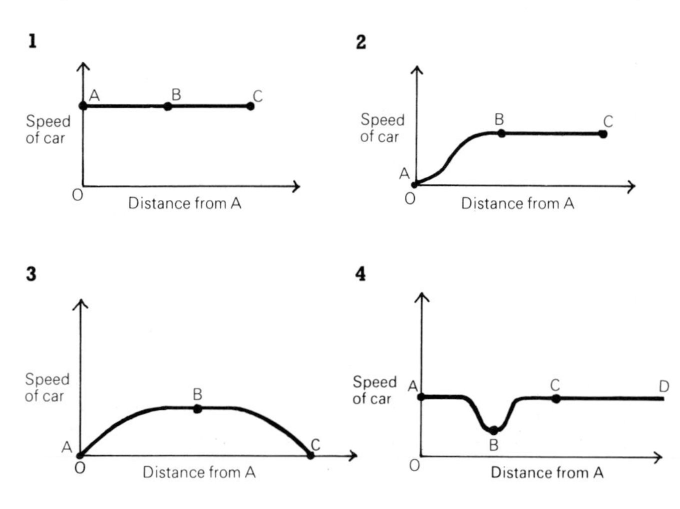

5

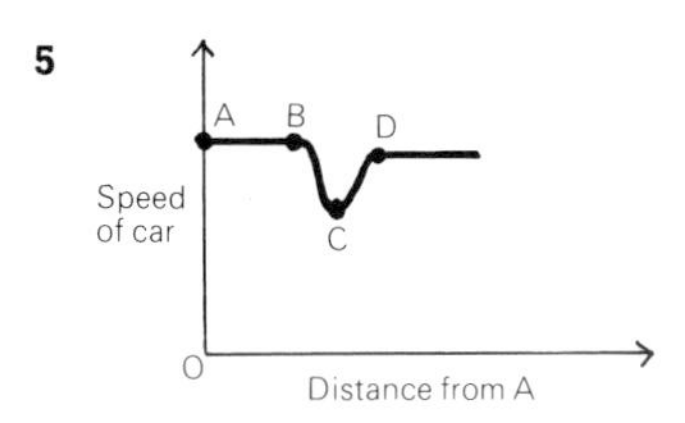

6 There are traffic lights at B, or car stops at the crossroads
7a The graph shows the route via B to D.
b

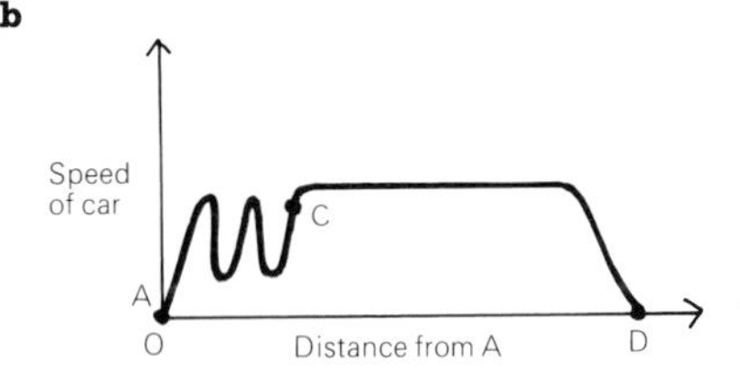

Page 34 Exercise 3

1a 6 h **b** 5 h **c** $3\frac{1}{2}$ h **d** $12\frac{1}{2}$ s **2** $2\frac{1}{4}$ h **3a** 1 h 30 min **b** 30 min **c** 2 h 20 min **4a** 3 h **b** $5\frac{1}{2}$ h **c** 50 s **5** $7\frac{1}{2}$ min **6a** 57 s **b** 4 s **7a** 225 **b** Friday 28th **8** 10 min **9** 6.22 pm **10** Yes **11** 1.2 s **12** 500 s, or 8 min 20 s

Page 35 Exercise 4

1a 420 miles **b** 50 km **c** 162 m **d** 3000 km **2** 135 km **3a** 675 km **b** 352 m **c** 0.198 m **4** 222 km **5** 900 miles **6** 7475 km **7** 6800 km **8** 62.5 cm **9** 13.3 km **10** 6.0 km **11** 583 miles **12a** 171 m **b** 190 cm/s

Page 36 Exercise 5A

1a 300 miles **b** 60 mph **c** after driving for about 2 hours **d** a steady speed throughout the journey **2b** (ii) the plane **c** coach 50 mph, train 80 mph, plane 400 mph
3a Distances (km): 20, 40, 0, 30;
Average speeds (km/h): 20, 40, 0, 30 **b** the car stopped **d** 3.30 pm **4a** 12 min **b** 48 min **c** going to Leeds **d** 20 km/h, 10 km/h
5

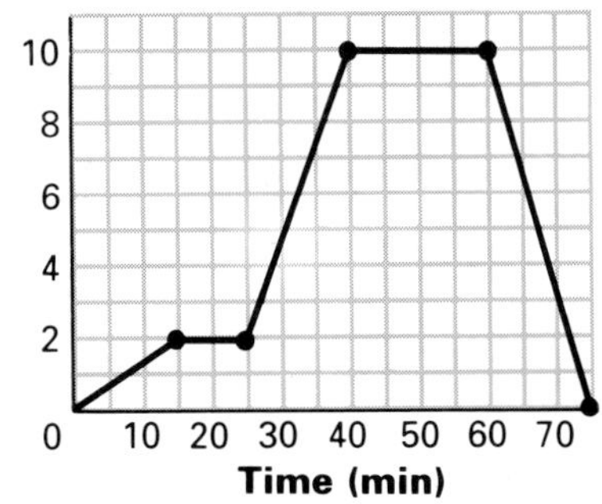

Page 38 Exercise 5B

1a OA running, BC walking, CD bus **b** 30 min **c** 12 km/h, 2 km/h, 28 km/h **2a** 6 h **b** 260 km **c** Swindon $\frac{1}{2}$ h, Reading $\frac{3}{4}$ h **d** 52 km/h, 65 km/h, 52 km/h **3a** 90 km **b** 1 h 48 min, 2 h 36 min **c** 90 km, 130 km **d** 30 min **4a** 11.30 am, 40 km **b** Steve, last part, 40 km/h
5 They start at the same time. Ian runs steadily, gradually getting faster all the time. He overtakes Alan after 8 seconds at the 50 m mark. He finishes in 11 s, 1.5 s ahead of Alan, who started off quickly but slowed down in the middle and at the end of the race **6b** 15–16 km **c** 10.09–10.10 am **d** Joan 13 km/h, Mandy 26 km/h

Page 40 Exercise 6A

1a 70 km/h **b** 12 h **c** 3.9×10^4 m **d** 10 400 km/h **2** 12.5 m/s **3** A 2 h, B 147 km, C $4\frac{1}{2}$ h **4a** 1.7 s **b** 0.9 s **c** 0.5 s **5a** 54 miles **b** 26 mph **6a** 198 km **b** 33 km **c** 55 km **7a** 15 05 **b** 31 mph

Page 41 Exercise 6B

1a 36 min **b** 947 km/h
2 Men's 11.90 m/s; women's 11.54 m/s **3** 1 h 32 min
4b 3.97×10^{13} km, 8.14×10^{13} km, 2.46×10^{17} km **5** 36.4 km/h

Page 42 Check-up on Going Places

1a 150 km **b** 5 m/s **2a** 56 km/h **b** 4.5 h **c** 165 km **3a** 13 mph **b** balloon **4** 890 miles **5a** 36 km/h **b** 60 km/h **c** 45 km/h **6a** 56.8 km/h **b** 56 km/h **c** (i) 8 pm (ii) 60 km/h **7a** About 12 22 or 12 23 **b** 120 km **c** 60 km/h **d** 45 min **e** 76 km/h

4 MONEY MATTERS - SAVING AND SPENDING

Page 43 Looking Back

1a £0.35, or 35p **b** £1.60 **c** £9.02 **d** £0.01, or 1p **2a** £345.45 **b** £125 **c** £12.12 **3a** 0.25 **b** 0.5 **c** 0.175 **d** 1.175 **4a** 50% **b** 75% **c** 25% **d** 20% **5** 2159 **6a** £6 **b** £9.50 **7a** £21.60 **b** £26 **c** £39.96 **8a** £4 **b** £54 **9a** £48 **b** £19.50 **c** £33 **10a** £12 **b** 30% **11** £465

Page 44 Exercise 1

1 £750 **2a** £8840 **b** £170 **3a** £138.75 **b** £832.50 **4a** £4.50 **b** £702 **5a** £2250 **b** £30 600, £2550 **6a** £1084 **b** to encourage staff to sell more goods or services
7 Sleepsound, by £25 **8a** £500 **b** £1000 **c** £1300
9a £332.50 **b** £57 **c** it is usually at night, or weekends
10 £203.40

Page 46 Exercise 2

2a (i) £50 paid in (ii) £25.10 paid out to Top Ten; Balance £24.90 (iii) £475 paid in by Regional Council; Balance £487.40 **b** (i) £24.90 (ii) £487.40 (iii) £387.40 **3a** £14 **b** £49 **c** £87.50 **d** £2.45 **4a** £76.80 **b** £19.20 **c** £7.38 **5** 9 **6a** 6 times **b** (i) £105 (ii) £40.50 (iii) £108.75 **7a** 7.3%, 8%, 8.25%, 8.45% **b** higher rates of interest are paid for larger deposits **8a** (i) £145 (ii) £11.60 **c** £182.10 **d** it has a higher rate of interest
9a (i) £520 (ii) £542.90 (iii) £661.25 **b** interest rate increases each year **10a** £100 **b** £172.50

Page 48 Exercise 3

1 Scotthol £6.38 dearer **2a** 39p, £2.43 **b** £2.59, £16.29
3 £20.25, £71.28, £91.53, £16.02, £107.55
4 Cash and Carry by 36p **5** Alroyds £1392, Thomsons £1204
6a £9 **b** $17\frac{1}{2}\% - 15\% = 2\frac{1}{2}\%$, so increase is $2\frac{1}{2}\%$ of £360 = £9
7a £5.24, £11.46 **b** £22.32, £48.82

Page 49 Exercise 4

1a 6 pm–8 am Mon–Fri and all day Sat and Sun
b 8 am–6 pm Mon–Fri **2a** 15p **b** 15p **3** 50p
4 A £69.82, B £12.22, C £82.04 **5** £57.09, £9.99, £67.08

Page 50 Exercise 5

1 A 1122, B £91.21, C £15.96, D £107.17 **2a** 793 **b** £59.87 **c** £66.37, £11.61, £77.98 **3a** A 117, B £9.20, C 814, D £24.83, E £41.18, F £7.21, G £48.39 **b** electicity supplied at times when the demand is low, for example at night **4** A £272.24, B £47.64, C £319.88 **5** £279.10 **6** A £240.94, B £10.24, C £251.18, D £43.96, E £295.14
7a (i) C, B, A (ii) C, A, B (iii) A, C, B **b** (i) 250 (ii) 550 **c** (i) £30 (ii) 4p **d** C, for the first 50 units, and after the first 100 units **8a** £351.07 **b** £29.26

Page 52 Exercise 6

1a You usually pay some of the money, the deposit, then pay the rest by regular instalments **b** You don't have to pay it all at once, but it usually costs more **2a** £161.40 **b** £13.56 **3a** £175 **b** £27 **4** £7.61 **5** £8.21 **6a** £325.32 **b** £60.36 **7a** £1013.74, £1118, £1222 **b** this depends on how much they can pay weekly, and overall **8** £46.55, £71.95, £80.70 **9** £19.69

Page 53 Exercise 7A

1 656.80, 697.85, 533.65, 779.95 **2** 64 102.50 **3a** (i) 3200 (ii) 1600 **b** 6400 **4a** £6 **b** £12.50 **c** £3 **d** £5 **e** £11 **f** £7 **5a** £0 = $0, £100 = $150 **b** Rows: 10, 40, 70, 60, 27, 47, 84, 75; 15, 60, 105, 90, 40, 70, 126, 112

Page 54 Exercise 7B

1a 1854 **b** £18.87 **2** £75.75 **3a,b** Karl (Germany) £131.05, Pierre (France) £127.27, Andrew (UK) £125, Manuel (Spain) £117.89, Wilbur (USA) £116.13 **4** £22.55 **5a** Canada $3, Australia $2.90, New Zealand $3.26 **b** rates of exchange vary and prices change

Page 55 Check-up on Money Matters—Saving and Spending

1a £625 **b** £144.23 **2a** £254.60 **b** £304.85 **3** £81 **4** Dawn BS, £3.75 **5** $4\frac{1}{3}$% **6** £52.29 **7** £89.46 **8** £34.71 **9** £54.90 **10** £81.38

5 POSITIVE AND NEGATIVE NUMBERS

Page 56 Looking Back

1a −6 −5 −4 −3 −2 −1 0 1

b −20 −15 −10 −5 0 5 10

2a −12°C, −4°C, −3°C, 0°C, 1°C, 5°C, 10°C **b** 22°C **3a** 1, 0, −1 **b** −5, −10, −15 **c** 0, 2, 4 **4a** (i) (3, −2), (2, −1), (1, 0) (ii) (−3, 2), (−2, 1), (−1, 0) **b** (0, −1), (−1, −2), (−2, −3), (−6, −7) **5a** −2 **b** −7 **c** 2 **d** −1 **e** 5 **f** 3 **g** −1 **h** 1 **7** 1, −1, −4 **8a** $x = -1$ **b** $x = -5$ **c** $x = 2$ **d** $x = 7$ **e** $x = -2$ **f** $x = -4$ **9a** −4 **b** 2 **c** −8 **d** 2 **10a** T **b** F **c** T **d** T **11a** −1 **b** 6 **12a** $-2x$ **b** $3k$ **c** $-5t$ **d** $3x$ **e** 0 **f** $8p$

Page 57 Exercise 1

1a 9 **b** 2 **c** −2 **d** −6 **e** 2 **f** 3 **g** 0 **h** −5 **i** −5 **j** −4 **k** −1 **l** −6 **m** −6 **n** 2 **o** −5 **p** 6 **q** 7 **r** −2 **3a** −5 **b** −2 **c** 10 **d** −7 **e** 5 **f** −2 **g** 10 **h** −14 **i** −1 **j** 0 **k** 6 **l** −3 **m** −8 **n** 0 **o** 28 **p** −14 **q** 16 **r** 0 **4** 1, 2, −3 **5** 9, −9, 6, −6, 3, −3 **6** 25.00, 10.00, 17.00, −2.00, +13.00, 1.00 **7a** Aug, Apr **b** £5000 **c** losses **8a** −1, −3, −5 **b** −2, 0, 2 **c** −6, −5, −4 **d** 0, 5, 11
9a Moira's scores: 0, +3, −1, 0, −1, +2; total +3
b Hassan's scores: −1, +3, +2, 0, +1, +1, +2, 0, −1, +1, +4, +1; total +13
Moira's scores: −1, +2, +3, −2, +2, 0, +4, +1, 0, −1, +2, 0; total +10 **c** Moira

Page 58 Exercise 2

1a −2 **b** 8 **c** 19 **d** −6 **e** 2 **f** 13 **g** −5 **h** 21 **i** −7 **j** 5 **k** −3 **l** 2 **m** −2 **n** −7 **o** −13 **p** 1 **q** −3 **r** 2 **s** 1 **2a** (i) 10 m (ii) 0 m (iii) −15 m **b** (i) 10 m above cliff edge (ii) level with edge (iii) 15 m below edge **3a** (i) −£7000 (ii) −£2000 (iii) £3000 (iv) £8000 **b** 120 **4a** h: 0, 3, 4, 3, 0, −5, −12, −21 **c** 6 seconds **5a** $3x, -2y, 6z, 5a, a, -b, -5b, 5c, -c$ **b** $-9x, -5x, -2p, -3q, 5r, 0, 3x^2, -2y^2, 2z^2$ **6a** $5a$ **b** 0 **c** 0 **d** $6d$ **e** $-3e$ **f** 0 **g** $-2g$ **h** $-2h$ **i** $4x^2$ **j** $3x^2$ **k** $2y^2$ **l** $3z^2$ **m** $3x-y$ **n** $2p-2q$ **o** $-m$ **p** $-s-2t$ **q** $-4u+v$ **r** $-b^2$ **s** $-2x^2+2y^2$ **t** $2x^2-4x$ **u** $2ab+2a^2$ **v** $-2y^2$

Page 60 Exercise 3A

1b

9	6	3	0	−3	−6	−9
6	4	2	0	−2	−4	−6
3	2	1	0	−1	−2	−3
0	0	0	0	0	0	0
−3	−2	−1	0	1	2	3
−6	−4	−2	0	2	4	6
−9	−6	−3	0	3	6	9

2a −6 **b** −2 **c** −3 **d** −1 **e** −4 **f** −9 **4a** 1 **b** 2 **c** 3 **d** 2 **e** 6 **f** 9 **6a** 35, −35, −35, 35 **b** 24, −24, −24, 24 **c** 18, 18, −18, −18 **d** −49, 49, −49, 49 **e** 50, −50, −50, 50 **7a** 5 **b** −10 **c** −15 **d** 20 **e** 0 **f** −21 **g** 1 **h** −36 **i** 42 **j** −36 **k** 18 **l** −24 **m** −24 **n** 0 **o** 25 **p** 4 **q** 9 **r** 16 **s** 25 **t** 1 **u** 63 **v** −63 **w** 63 **x** −63 **8a** −6, −8, 12 **b** 2, 5, 10 **c** 6, −14, −21 **d** 20, 24, 30 **e** 7, −4, −28 **f** 15, −3, −5
9 Blue Co: −10, −12, −15, −8, −1, 0, 5, 16, 36; £11 000 profit
Red Co: −15, −4, −18, −2, −2, 0, 6, 18, 30; £13 000 profit

Page 61 Exercise 3B

1a 45 **b** −45 **c** −45 **d** 45 **e** 0 **f** 4 **g** −24 **h** −24 **i** 0 **j** −1 **k** 54 **l** 54 **m** −90 **n** −49 **o** 56 **p** −21 **q** −49 **r** 48 **s** 1 **t** 1000 **u** 0
2a Multiply by −2; 16, −32, 64 **b** multiply by −1; −5, 5, −5 **c** multiply by $-\frac{1}{2}$; 1, $-\frac{1}{2}$, $\frac{1}{4}$

3a 1 **b** 9 **c** −8 **d** 6 **e** 6 **f** −24 **g** 0 **h** −60 **i** −12
4a 5 **b** 6 **c** 1 **d** −3 **e** 13 **f** −6 **g** 9 **h** 13 **i** 27 **j** −9
5a 4 **b** 4 **c** −9 **d** 6 **e** 7 **f** −5 **g** 12 **h** 18

Page 62 Exercise 4A

1a −2 **b** −10 **c** 8 **d** 2 **e** 2 **f** 0
2a 20 **b** 32 **c** −10 **d** 4 **e** −3 **f** 12
3a 24 **b** −5 **c** 0 **d** 4 **e** −6 **f** 10 **g** 54 **h** 22 **i** −24
4a −6 **b** −10 **c** 14 **d** −10 **e** −48 **f** −14
5a (i) 10 m (ii) −30 m (iii) 40 m (iv) 90 m; the joint is 30 m below sea-level **b** (i) 30 m (ii) 30 m (iii) 110 m
6a (i) 20 m/s (ii) 10 m/s **b** 6 s
c (i) 3 s before passing Ron (ii) 45 m/s **d** 65 m/s
7a 95°F, yes **b** 14°F, yes **c** −28°F, yes **d** −28°F, yes
e −109°F, no **f** −126°F, no

Page 63 Exercise 4B

1a 0 **b** −2 **c** 10 **d** −3 **e** 4 **f** 20 **g** −26 **h** −8 **i** −8
2 Right side: **a** $3x, 2x, x, 0, -x, -2x$ **b** $3x, 2x, x, 0, -x, -2x$
3a uv **b** $-uv$ **c** $-uv$ **d** uv **e** a^2 **f** $4b^2$ **g** $9c^2$ **h** d^3
i k^2 **j** $-k^3$ **k** $12pq$ **l** $12pq$ **m** a^2b^2 **n** a^2b^2 **o** $-a^2b^2$
p $-a^2b^2$ **4a** $6a$ **b** 0 **c** $24c^2$ **d** $6t^3$ **e** m^2+n^2 **f** $2ab$ **g** 0
h $2c^2$ **i** $-a^2b^2$ **j** $-24x^3$ **5a** $ab+ac$ **b** $pq-pr$ **c** $2my-ny$
d $2ax+3bx$ **e** a^2-ab **f** n^3-mn
6 (i) **a** (−3, −6), (−2, −4), (−1, −2), (0, 0), (1, 2), (2, 4), (3, 6)
b (−3, 6), (−2, 5), (−1, 4), (0, 3), (1, 2), (2, 1), (3, 0)
(ii)

a

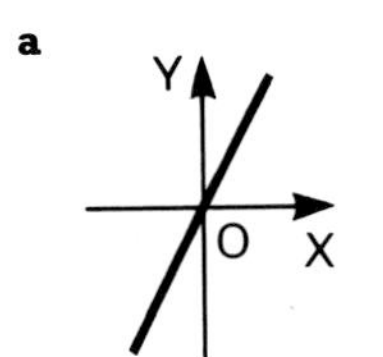

b

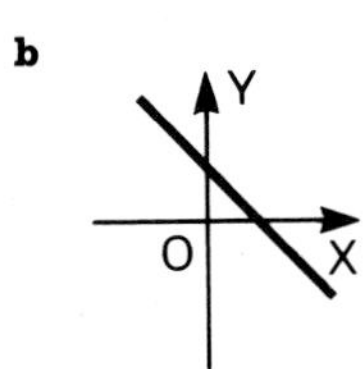

Page 64 Exercise 5

1a Neg. **b** pos. **c** neg. **d** pos. **e** neg. **f** pos. **g** neg.
h pos. **2a** 4 **b** −4 **c** −4 **d** 4 **e** −2 **f** −3 **g** 5
h −9 **i** 9 **3a** −4 **b** −4 **c** 1 **d** −15 **e** 3 **f** −10
g 4 **h** −13 **i** 9 **j** 3 **4** 1 **5** −2.5°C **6a** (i) 5°C
(ii) −5°C (iii) −15°C **b** (i) 14°F (ii) 5°F (iii) −238°F
7a −2 **b** −4 **c** 2 **d** 16 **e** 1 **f** 4 **g** −6 **h** −1
8a (1, 1) **b** (−3, 3) **c** (−5, 4) **d** (−3, −2)
9a (i) 5 pm (ii) 8 pm (iii) 3 am next day
(iv) 10 am same day **b** 8 hours

Page 65 Exercise 6A

1a 13 **b** −8 **c** −3 **d** −4 **e** −1 **f** −5 **g** 6 **h** 0
i −9 **j** 5 **2a** 5 **b** −3 **c** −2 **d** 4 **e** −11 **f** −4
g −7 **h** 0 **i** −1 **j** −5 **k** −8 **l** 6 **3a** 11 **b** 3 **c** −10
d −2 **e** 19 **f** −8 **g** 5 **h** −5 **i** −1 **j** 5 **k** 4 **l** 5
m 0 **n** −1 **o** −2 **4a** 2 **b** −2 **c** −3 **d** 3 **e** 9 **f** −2
g 0 **h** −5 **i** −1 **j** −1 **k** 2 **l** −2

Page 66 Exercise 6B

1a −2 **b** 2 **c** −2 **d** −3 **e** 3 **f** −2 **g** −5 **h** 3 **i** −2
j −4 **k** 1 **l** −3 **m** −5 **n** −4 **o** 1 **p** 1
2 Rows: Card 7, 12, 1, 14; 2, 13, 8, 11; 16, 3, 10, 5; 9, 6, 15, 4

Page 67 Check-up on Positive and Negative Numbers

1a 1 **b** −9 **c** 1 **d** $-4x$ **e** $6a$ **f** 0 **2a** 2 **b** 3
3a 0, −5, −10 **b** 27, −81, 243 **c** −2, −4, −5
4 5°, 3°, 1°, 4°, 10°, 2° **5a** −12, −9, −15
b 7, −7, 4, −4, 3, −3, 10, −10 **6a** −2, −4, −5
b 13, 15, 16 **c** −9, −11, −12 **d** −11, −9, −8 **7a** −8
b 8 **c** −8 **d** $-6t$ **e** k **f** $4a^2$ **8a** 40 **b** 5 **c** $-3a^2$
d $2x^2$ **9a** Columns: 1 × (−2), 6 × (−1), 5 × 0, 5 × 1, 1 × 2; −2, −6, 0, 5, 2; −1 **b** +1 **c** Lauren, by 2 strokes
10a −4 **b** 1 **c** 2 **11a** −3 **b** −9 **c** −5 **d** −1 **e** −2
12a $-4a$ **b** $6b^3$ **c** $-2c^2$ **d** $2x^3$

6 PYTHAGORAS

Page 68 Looking Back

1a 25 **b** 49 **2a** 64 **b** 100 **3a** 6 **b** 8 **c** 9 **4a** 4 **b** 12
c 2 **5a** 9 **b** 16 **c** 81 **d** 2.25 **e** 0.04 **6a** 10 **b** 11
c 3.5 **d** 36 **7a** 17.64 cm² **b** 12.25 cm² **c** 3.61 cm²
8a 81 **b** 144 **c** 225 **d** 15 **9a** △s PSR, QSR, PQR
b △s ABC, ADC, ACG **10a** 49 cm², 25 cm² **b** 5 cm

Page 69 Exercise 1/Class Discussion

1a 4.6 m **b** 4.3 m **c** 2.2 m **2a** 16, 16, 32
b 32 squares in each **3c** Each is 25 cm²
4b (i) 7.2 cm (ii) 7.1 cm **c** (i) 36 cm², 16 cm², 52 cm²
(ii) 25 cm², 25 cm², 50 cm²

Page 71 Exercise 2A

1a $p^2 = q^2+r^2$ **b** $x^2 = y^2+z^2$ **c** $x^2 = 3^2+7^2$ **d** $a^2 = 2^2+5^2$
2a (i) $13^2 = 5^2+12^2$ **b** (i) $15^2 = 12^2+9^2$
c (i) $29^2 = 21^2+20^2$ **d** (i) $5^2 = 4^2+3^2$ **3a** 17 **b** 50 **c** 74
d 20.5 m **4a** 2.2 **b** 11.2 **c** 12.2 **d** 17.0 m **5a** 12 **b** 30
c 15 **6a** 8.7 **b** 33.2 **c** 8.5 **7a** 9.6 **b** 5 **c** 2.6 **8a** 5
b 12.5 **c** 260 **d** 8 **9a** 17 **b** 36 **c** 20 **10** 7.5 m
11a 30 m

Page 72 Exercise 2B

1a 7 m **b** 7.5 m **2a** 6.5 m **b** 30 cm **3a** 68 m **b** 1224 m
c 127 m **4a** 4 m **b** 1 m **c** (i) 346 cm (ii) 387 cm
(iii) 312 cm **5** 600 m, 1600 m **6a** A rhombus **b** 25 cm
7a A kite **b** 92 cm

Page 74 Exercise 3

1 10.4 cm **2a** 11.0 cm **b** 5.8 cm **3a** 85 m **b** 87 m
4a 75 m **b** 65 m **c** 32 m **5a** ∠ABC, 25 cm
b ∠ABF, 30 cm **c** ∠AFG, 30.8 cm **6** Yes **7** 173 cm

Page 75 Exercise 4A

1a 5.4 **b** 8.5 **c** 9.2 **d** 6.4 **2a** 5 **b** 10 **3** 447 cm
4a 10 **b** 10 **c** 13 **d** 5 **5** A 806 cm, B 806 cm, C 781 cm, D 860 cm, E 762 cm; D
6 AB = 5, BC = 5.4, CD = 7.3, DE = 4.5, EA = 5
7a 8.06 km **b** 9.49 + 4.00 + 10.05 km = 23.54 km **c** 15.48 km

Page 77 Exercise 4B

1a (6, 5) **b** 5 **c** on **2a** OA² = OB² = 26, so OA = OB
b (3, 3) **c** 12 sq. units **3b** △s ABC, ADC, ADE, CDE **c** BD
4a No **b** yes, for the plant at (8, 14) **5a** 801 m **b** 785 m

Page 79 Exercise 5A

1 b, c, e, f, h **2 a, c, e, f** **3a** (i) 40 cm (ii) 30 cm

(iii) 50 cm **b** yes **4a** (i) 45 m (ii) 60 m
b $PQ^2+QR^2 = 5625$, $PR^2 = 5625$, so $PR^2 = PQ^2+QR^2$, and $\triangle PQR$ is right-angled by converse of Pythagoras' Theorem
5a PS = 58 m, SR = 60.9 m **b** $PS^2+SR^2 = 7072.81$, $PR^2 = 7072.81$, so $PR^2 = PS^2+SR^2$, etc **6a** (i) 5.6 cm
(ii) 3.3 cm **b** $AC^2+BC^2 = 42.25$, $AB^2 = 42.25$, so $\angle ACB = 90°$

Page 80 Exercise 5B

1 Areas of squares on sides of large triangles are:
a 225, 400, 625 **b** 20, 80, 100 **c** 14.0625, 25, 39.0625
d 56.25, 100, 156.25; etc.
2 $EF^2+FC^2 = 355.3225$, $EC^2 = 355.3225$, etc.
3a $AB^2+AC^2 = 65 = BC^2$; $\angle A$
b $PQ^2+PR^2 = 50 = QR^2$, $\angle P$, etc.
4 Prove angles are right angles, or opposite sides equal and one angle right, or diagonals equal and bisecting
5 $HA^2+AJ^2 = 196 = HJ^2$, etc.

Page 81 Check-up on Pythagoras

1a $r^2 = p^2+q^2$ **b** $p^2 = r^2-q^2$ **c** $q^2 = r^2-p^2$ **2a** 50 **b** 6
3 40 cm **4a** 24 m **b** 252 m² **5** 58 cm **6a** 17 **b** 15
7a 5 cm **b** 5.4 cm
8a If in $\triangle ABC$ $a^2 = b^2+c^2$, then $\angle A = 90°$ **b** (iv)
9a 30 cm **b** no; $BC^2 = 900$, $AB^2+AC^2 = 578$, so $BC^2 \neq AB^2+AC^2$ **10** $x^2 = a^2+h^2$, $y^2 = b^2+h^2$, and subtract

7 BRACKETS AND EQUATIONS

Page 82 Looking Back

1a 6 **b** −2 **c** 2 **d** −6 **e** 8 **f** −8 **g** −8 **h** 8 **i** $\frac{1}{2}$
j $-\frac{1}{2}$ **k** $-\frac{1}{2}$ **l** $\frac{1}{2}$ **3a** $2x$ **b** $-4y$ **c** $2t$ **d** a^2 **e** $2b^2$ **f** 49
g 25 **h** $3x+3$ **i** $4y$ **4a** $8x+16$ **b** $3-3x$ **c** $6x+2$
d $15-10x$ **e** x^2-x **f** y^2+xy **5a** $x=1$ **b** $x=-4$
c $x=-2$ **d** $x=2$ **e** $y=-3$ **f** $y=2$ **g** $k=4$ **h** $x=4$
6a $x=4$ **b** $y=6$ **7a** 17 cm **b** 3
8a $5(y-2)$ cm² **b** $y=5$
9a (i) $x-2$ (ii) $x-4$ **b** (i) $4x-4$ (ii) x (all cm)

Page 83 Exercise 1A

1a $2x+10$ **b** $3x+6$ **c** $4y+4$ **d** $5y+15$ **e** $6b+12$
f $3x-6$ **g** $2x-10$ **h** $4y-4$ **i** $6m-12$ **j** $8n-8$ **k** $6+2x$
l $3+3y$ **m** $20-4a$ **n** $14-7b$ **o** $6-6c$ **p** $5a+5b$
q $3x-3y$ **r** $45-9x$ **s** $10t+10w$ **t** $11y-11$ **u** $1-x$
2a $5(x+2), 5x+10$ **b** $4(x-1), 4x-4$ **c** $6(y+3), 6y+18$
d $3(t-3), 3t-9$ **e** $7(5-x), 35-7x$ **f** $2(10-w), 20-2w$
g $8(x-7), 8x-56$ **h** $9(k+6), 9k+54$ **i** $4(t-11), 4t-44$
3 $5(y+3)$ g, $5y+15$ g **4** $6(x+1)$ cm², $6x+6$ cm²
5 $3(12-j)$ ml, $36-3j$ ml **6** $4(t-11)$ m, $4t-44$ m

Page 83 Exercise 1B

1a $6t+8$ **b** $10x-6$ **c** $3+6x$ **d** $8-8y$ **e** $12p-6$
f $ax+3a$ **g** $by+2b$ **h** $ck-c$ **i** $dm-5d$ **j** $en+2e$
k $ab+ac$ **l** $ax+ay$ **m** $bx-by$ **n** x^2+xy **o** a^2-ab
p r^2-4r **q** $x-x^2$ **r** y^2-4y **s** x^2+2xy **t** $3y^2-4y$
u $3x^2-5x$ **v** $8xy-7y^2$ **w** $2t^2-7t$ **x** $15w-3w^2$
2a $2x+2y+10$ **b** $5x+5y+15$ **c** $3x-3y-12$
d $4a+8b+4c$ **e** $12a-6b+6c$ **f** $21x-28y+14$ **g** x^3-x
h y^3+y **i** a^3-5a **j** b^3+b^2 **k** c^3-c^2 **l** t^4-3t^3

Page 84 Exercise 2A

1a $-2x-6$ **b** $-3x+12$ **c** $-5x-5$ **d** $-4x+8$
e $-6-6y$ **f** $-6+2m$ **g** $-28-7n$ **h** $-10+10k$
i $-2p-2q$ **j** $-5p+5q$ **k** $-m-n$ **l** $-m-n$ **2a** $7+2x$
b $7-3y$ **c** $6-4x-4 = 2-4x$ **d** $4+5w+15 = 5w+19$
e $2x+6$ **f** $4x+2$ **g** $3x$ **h** $x+7$ **i** $-2y+4$ **j** $-3y-1$
k $-4y+3$ **l** $-y+6$ **m** $n+4$ **n** $-2m+12$ **o** $5k-4$
p $-7x-4$ **q** $2a-1$ **r** $-3b$ **s** $6-c$ **t** $8w-4$
3 $10y-8y+16$ m² $= 2y+16$ m² **4a** $10x, 8(x-2), 2x+16$
b $7t, 5(t-2), 2t+10$ **c** $5x, 3(x-2), 2x+6$
d $12w, 10(w-2), 2w+20$ (all m²) **5** $2x+12$ m²

Page 85 Exercise 2B

1a $x+14$ **b** $-x+18$ **c** $2x-2$ **d** $7x-17$ (all cm²)
2a $x+18$ **b** $3x+12$ **c** 18 **d** $-20y+55$ **3a** $x-8$
b $2y+21$ **c** $5t-13$ **d** $w+3$ **e** $x-7$ **f** $6y-17$
g $-4x+2$ **h** $-3x$

Page 85 Exercise 3A

1a $x=3$ **b** $x=9$ **c** $x=4$ **d** $x=9$ **e** $y=4$ **f** $y=6$
g $y=6$ **h** $y=0$ **i** $x=2$ **j** $x=7$ **k** $y=1$ **l** $y=5$
2a $x=3$ **b** $x=2$ **c** $y=1$ **d** $y=1$ **e** $y=1$ **f** $u=2$
g $v=4$ **h** $w=6$ **3** 3, 4 kg **4a** 2, 4 kg **b** 4, 5 kg
c 5, 15 kg **d** 3, 7 kg **e** 1, 3 kg **f** 3, 5 kg **g** 0, 3 kg
h 2, 2 kg

Page 86 Exercise 3B

1a $x=4$ **b** $y=1$ **c** $k=6$ **d** $p=2$ **2a** $x=7$; 6, 9 coins
b $x=5$; 12, 6 coins **c** $x=4$; 3, 5 coins **d** $x=3$; 4, 14 coins
3 $x=5$ **4** 42 km/h, 40 km/h **5** 85 km/h, 80 km/h
6a $x=1$ **b** $x=-10$ **c** $x=0$ **d** $p=5$ **e** $m=-3$ **f** $x=0$
g $y=8$ **h** $k=2$ **i** $x=5$ **j** $x=2$ **k** $y=3$ **l** $x=-1$

Page 87 Exercise 4A

1a x^2+4x+3 **b** x^2+3x+2 **c** $x^2+9x+20$ **d** $a^2+8a+12$
e $b^2+8b+16$ **f** $y^2+7y+12$ **g** m^2+2m+1 **h** t^2+5t+6

2a

	x	3
x	x^2	$3x$
1	x	3

b

	x	2
x	x^2	$2x$
1	x	2

c

	x	4
x	x^2	$4x$
5	$5x$	20

d

	a	2
a	a^2	$2a$
6	$6a$	12

e

	b	4
b	b^2	$4b$
4	$4b$	16

f

	y	3
y	y^2	$3y$
4	$4y$	12

g

	m	1
m	m^2	m
1	m	1

h

	t	2
t	t^2	$2t$
3	$3t$	6

3a y^2-3y+2 **b** w^2-5w+6 **c** $u^2-11u+30$ **d** b^2-6b+5
e c^2-6c+9 **f** x^2-5x+6 **g** n^2-4n+4 **h** $s^2-7s+12$
4a $x^2+3x-10$ **b** p^2+3p-4 **c** t^2-2t-3 **d** x^2-x-2
e y^2-y-6 **f** z^2+z-20 **g** m^2-36 **h** n^2-1
5a x^2+7x+6 **b** m^2-2m-8 **c** $r^2-4r-21$ **d** $t^2-7t+10$
e $y^2+18y+81$ **f** $x^2+4x-60$ **g** $y^2+5y-14$ **h** n^2+4n+3
i $q^2+4q-12$ **j** t^2-5t+4 **k** $x^2+20x+100$ **l** $y^2+4y-77$
m $n^2+24n+144$ **n** a^2-81 **o** $b^2-5b-24$ **p** $c^2-10c+25$

6 $(x+1)(x-2) = x^2-x-2$, $(x+1)(x+3) = x^2+4x+3$, $(x+1)(x-4) = x^2-3x-4$, $(x+1)(x+5) = x^2+6x+5$; $(x-2)(x+3) = x^2+x-6$, $(x-2)(x-4) = x^2-6x+8$, $(x-2)(x+5) = x^2+3x-10$; $(x+3)(x-4) = x^2-x-12$, $(x+3)(x+5) = x^2+8x+15$; $(x-4)(x+5) = x^2+x-20$

Page 88 Exercise 4B

1a $4a^2-1$ **b** $9x^2-1$ **c** $4t^2-9$ **d** $4b^2-10b+6$ **e** $6c^2-16c+8$ **f** $2s^2+3s-2$ **g** $4t^2-7t-2$ **h** $3w^2+w-4$ **i** $4m^2-9$ **j** $6z^2+16z+8$ **k** $6n^2-19n+15$ **l** $20x^2+2x-6$ **m** $2-3x+x^2$ **n** $15+7y-2y^2$ **o** $1-9t+20t^2$ **2** x^2+x-6, $2x^2-5x+2$, $3x^2-2x-8$, $2x^2+5x-3$, $3x^2+13x+12$, $6x^2+5x-4$ **3** x^3+x^2-3x+9
4a x^3-x^2-5x+2 **b** x^3-2x^2-6x+9 **c** x^3-3x^2-2x+4
5a (i) $xy\,\text{mm}^2$ (ii) $(x+a)(y+b)\,\text{mm}^2$ **b** $bx+ay+ab\,\text{mm}^2$
6 The square, by $a^2\,\text{cm}^2$

Page 89 Exercise 5A

1a $x^2+2xy+y^2$ **b** $m^2+2mn+n^2$ **c** $u^2+2uv+v^2$ **d** $s^2+2st+t^2$ **e** $a^2+2ab+b^2$ **f** $w^2+2wz+z^2$ **g** $x^2-2xy+y^2$ **h** $u^2-2uv+v^2$ **i** $p^2-2pq+q^2$ **j** $c^2-2cd+d^2$ **k** $e^2-2ef+f^2$ **l** $r^2-2rs+s^2$ **m** $x^2+10x+25$ **n** y^2+6y+9 **o** a^2+2a+1 **p** $b^2+20b+100$ **q** w^2+4w+4 **r** $y^2+8y+16$ **s** $c^2-8c+16$ **t** d^2-2d+1 **u** $e^2-12e+36$ **v** $f^2-16f+64$ **w** $x^2-10x+25$ **x** $y^2-18y+81$
2a $x^2+2x+1\,\text{cm}^2$ **b** $y^2-6y+9\,\text{cm}^2$ **c** $w^2+4w+4\,\text{cm}^2$
3a $x^2+8x+16$, $x^2+10x+25$, $x^2+12x+36$, $x^2+14x+49$, $x^2+16x+64$ **c** $(x+n)^2 = x^2+2nx+n^2$ **4b** 19.7 **5b** 15

Page 89 Exercise 5B

1a $4x^2+12x+9$ **b** $9y^2+6y+1$ **c** $9a^2-24a+16$ **d** $25b^2+20b+4$ **e** $4c^2-12c+9$ **f** $9a^2-30a+25$ **g** $4x^2+4xy+y^2$ **h** $x^2-6xy+9y^2$ **i** $4a^2+12ab+9b^2$ **j** $16c^2-16cd+4d^2$ **k** $25p^2+20pq+4q^2$ **l** $16u^2-40uv+25v^2$
2a $x^2+2+\frac{1}{x^2}$ **b** $y^2-2+\frac{1}{y^2}$ **c** $4m^2+2+\frac{1}{4m^2}$
3a $9x^2-6x+1-x^2+2x-1 = 8x^2-4x$ **b** (i) $2x^2+8x+10$ (ii) $4x+8$ (iii) $5x^2-8x+5$ (iv) $10x-3$
4a $2x^2+12x+20$; $4x+12$ **b** $8x^2+16x+26$; $24x+24$
5a $x^2+2x\,\text{m}^2$ **6a** $\text{OM} = r-h$ **b** $h^2-2rh+a^2 = 0$ **d** 25

Page 90 Exercise 6A

1a $x = 2$ **b** $x = -2$ **c** $x = -2$ **d** $x = 4$ **e** $x = -1$ **f** $y = -2$ **g** $y = 1$ **h** $y = -2$ **2a** $(x+3)^2 = x^2+6^2$ **b** 7.5 m
3a 15, 20, 25 m **b** 10, 24, 26 m **4a** 99 m **b** 101 m

Page 91 Exercise 6B

1a 20 by 30 cm, 24 by 25 cm **b** 20 by 20 cm, 25 by 16 cm **c** 50 by 60 cm, 40 by 75 cm **d** 25 by 24 cm, 30 by 20 cm
2a $x = 2$ **b** $x = -2$ **c** $x = 2$ **d** $t = -2$ **e** $t = -3$ **f** $t = 3$ **g** $p = 2\frac{1}{2}$ **h** $p = 5$ **i** $p = 2$ **j** $y = 1$ **k** $y = 5$ **l** $y = -4\frac{1}{2}$

Page 92 Check-up on Brackets and Equations

1a $5x+10$ **b** $3y-12$ **c** $8k+4$ **d** $7-7n$ **e** $ab+ac$ **f** x^2-xy **g** $-2t-2$ **h** $-3t+3$ **i** $-5x-5y$ **2a** $2x+1$ **b** $-2x-1$ **c** $2x-1$ **d** $12-3w$ **3a** $x = 7$ **b** $y = 3$
4a 4, 9 **b** 3, 5 **5** 5 cm **6a** $3(x-1) = 2(x+2)$ **b** (i) £6 (ii) £9 **7a** x^2+4x+3 **b** y^2-3y+2 **c** $n^2-3n-10$ **d** $6m^2+13m+6$ **e** $20n^2-19n+3$ **f** $u^2+10u+25$ **g** $a^2-8a+16$ **h** $4w^2+12w+9$

8a $x = 2$ **b** $y = -2$ **c** $m = -1$ **d** $t = 5$
9 5 cm, 12 cm, 13 cm **10** 36 by 25 cm, 30 by 30 cm
11a 1, 4, 9, 16, . . .
b square numbers; the expression simplifies to n^2
12a 0, 15, 40, 75, . . . **b** all are multiples of 5; the expression is $5n^2-5$, of the form $5(n^2-1)$

8 STATISTICS

Page 93 Looking Back

1a 90 km/h from 2 to 2.30 pm **b** 4 hours **c** (i) 11–11.30 am (ii) 3.30–4 pm **2a** Frequencies 1, 3, 7, 12, 7
b

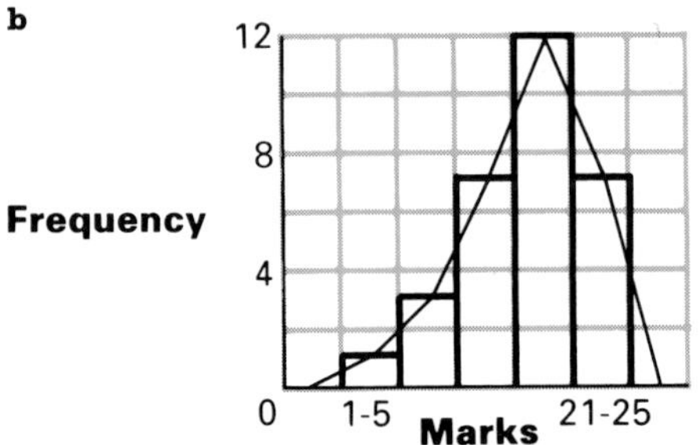

c 16–20
3a (i) 2 years (ii) 8 years **b** (i) 3 years (ii) 3.1 years
4a Fairly strong negative correlation **b** no correlation **c** fairly strong positive correlation

Page 94 Exercise 1

1a (i) Radio 1 (ii) Radio 3 **b** people might listen to some programmes, like the music on Radio 3, for much longer than 5 minutes **2a** Reduction in solid fuel, increase in gas and petroleum **b** (i) 60–65% (ii) 10%
3a Kenmere 10, Greenover 5 **b** Greenover, 25 mm
4a Round 1: 73, 76, 77.2; Round 2: 77, 75, 74.3 **b** Round 2 **c** the player with scores of 72 and 69 **d** mean
5a (i) 10.8 h (ii) 9.7 h **c** 9 h
b

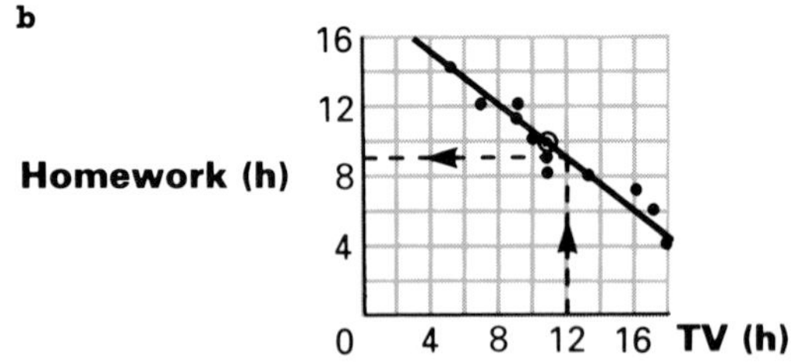

6a Positive **b** 7.6, 167 cm **d** $10\frac{1}{2}$
c

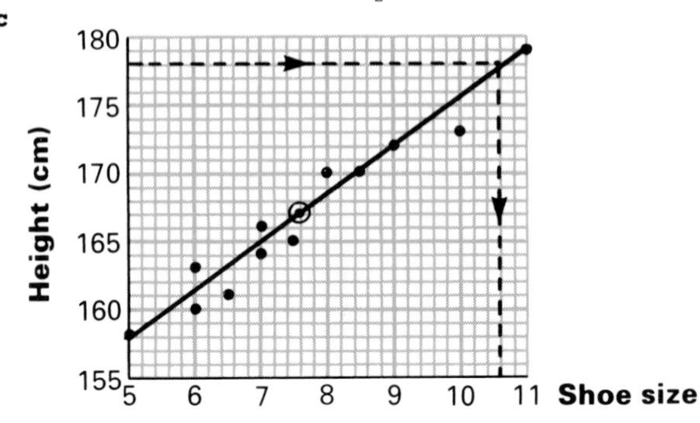

Page 96 Exercise 2A

1a 2, 5, 8, 11, 14; 140; 114, 230, 224, 66, 42; 676 **b** 1–3 letters **c** 4–6 letters **d** 4.8 letters **2a** (i) 26–30 (ii) 26–30 **b** 27.3 **3a** 60.6 g **b** 3 **4** 43.9 years

Page 97 Exercise 2B

1b 24.5 **2a** Girls: 3, 8, 9, 16, 12, 2;
Boys: 1, 2, 7, 10, 14, 14, 5, 5, 2 **b** 166 cm, 174 cm
3a Rows: 42.5, 2, 85; 47.5, 5, 237.5; 52.5, 4, 210; 57.5, 5, 287.5; 62.5, 9, 562.5; 67.5, 19, 1282.5; 72.5, 14, 1015; 77.5, 7, 542.5; totals 65, 4222.5 **b** 65–70 **c** 65–70 **d** 65.0 m

Page 98 Exercise 3A

1a 6, 15, 23, 28, 30, 31 **c** 9–10 mm
2a 2, 7, 19, 35, 43, 49, 52 **c** 315–320 **d** 32.7%
3a 1, 2, 3, 8, 7, 3, 3, 3; 1, 3, 6, 14, 21, 24, 27, 30
4a History 3, 19, 41, 80, 123, 147, 160;
Geography 0, 14, 37, 62, 97, 131, 152, 160
b History 75–77%, Geography 77–79%
c (i) 23.1% (ii) 39.4%

Page 100 Exercise 3B

1a 57.1 **b** 61–70 **c** 7, 17, 35, 59, 87, 116, 138, 156, 160
d (i) 72 − 43 = 29 (ii) 58 **e** Suggestions: give a mean mark over 50% to encourage the student; spread the marks to distinguish attainments; allow high scores to get close to 100%. This test satisfies these points, although 7 students below 20% is a disappointing result
2a Girls 3, 11, 20, 36, 48, 50; Boys 1, 3, 10, 20, 34, 48, 53, 58, 60
b 171 − 161 = 10, 179 − 168 = 11; similar spread of heights, though middle 50% of girls are smaller
3a Helen 6, 10, 18, 28, 40; Sally 1, 4, 24, 33, 40
c Helen 51 − 24 = 27, Sally 43 − 29 = 14 **d** 35.4 cm, 35.4 cm
e Helen is more likely to hit the bull's-eye, but is more erratic

Page 103 Check-up on Statistics

1a About 15 ohms **b** About 50 ohms
2a 7.56 m, 0.75 m **b** 10.8, 10.85, 10.9 s **c** (ii) About 7.8 m
3a 71–75 **b** 71–75 **c** 74.2
4a 5, 15, 29, 56, 92, 126, 152, 166, 169, 170 **c** (i) 74 (ii) 81 (iii) 68 (iv) 12

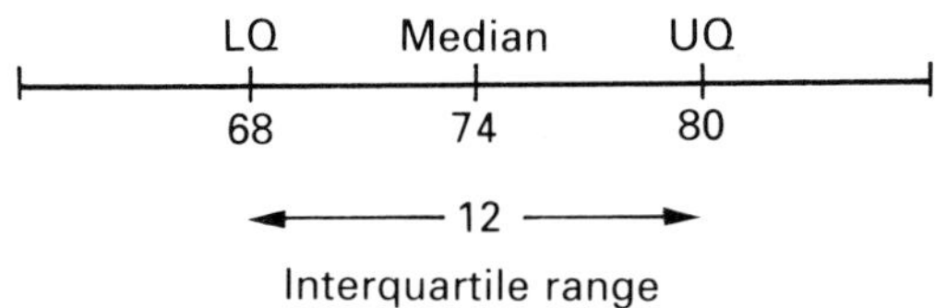

9 TRIGONOMETRY

Page 104 Looking Back

1a 1.85 **b** 15.52 **2a** 65 **b** 32 **3a** $\frac{1}{2}$ **b** $\frac{1}{2}$ **c** $\frac{1}{2}$
4a PR **b** 8.5 m **5a** (i) ∠BAC (ii) ∠PQR **b** 38°
6a 12 **b** 42 **c** 12 **7** (i) 20 (ii) same as in **6**
8a 14 m **b** takes time, may not be accurate enough

Page 105 Exercise 1/Class Discussion

1a 0.5 **b** ratio has same value for each point
2a 75 m **b** 125 m **c** 175 m **d** 500 m **3a** 2, 1, 0.25 **b** 1 **c** (i) 50 m (ii) 150 m (iii) 3000 m **4** 0.2, 0.4, 0.6

Page 106 Exercise 2

1 (i) **a** **b**

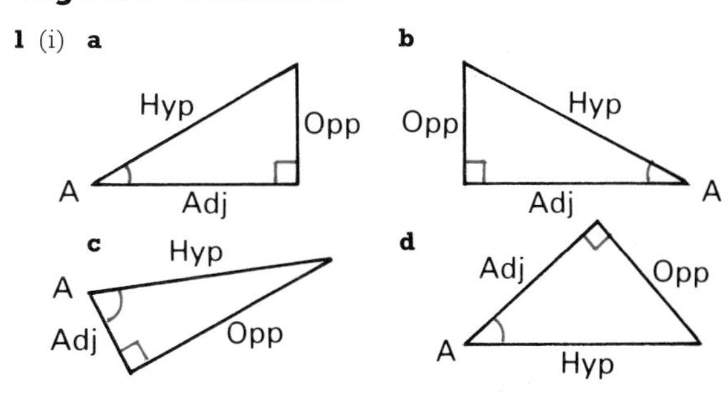

(ii) **a** $\frac{3}{4}$ **b** $\frac{5}{12}$ **c** $\frac{24}{7}$ **d** $\frac{21}{20}$ **2a** $\frac{7}{9}$ **b** $\frac{6}{11}$ **c** $\frac{13}{5}$ **d** $\frac{6}{8}$
3a 0.56 **b** 1.17 **c** 3.43 **d** 1.33
4 0.18, 0.36, 0.58, 0.84, 1.19, 1.73, 2.75, 5.67
5a 0.67 **b** 1.07 **c** 1.84 **d** 4.59 **e** 0.05 **6a** 0.51 **b** 5.91 **c** 0.03 **7a** 35.0° **b** 24.7° **c** 45° **d** 60.9° **e** 71.3° **f** 89.9°
8a (i) $\frac{9}{5}$ (ii) 60.9° **b** (i) $\frac{6}{7}$ (ii) 40.6° **c** (i) $\frac{7}{24}$ (ii) 16.3°
d (i) $\frac{3}{4}$ (ii) 36.9°

Page 108 Exercise 3A

1 $h = 8 \times tan\,58° = 12.8, 13\,m$ **2a** 4.8 **b** 12.6 **c** 13.7 **d** 204.0
3a 26.0 **b** 12.2 **c** 7.7 **d** 26.7 **4** 156 m **5** 35 m
6a 137 m **b** 149 m **c** 300 m **d** 413 m **7** 84 m **8** 44.2 m
9 32 m **10** 25 cm

Page 110 Exercise 3B

1a 10.2 **b** 8.8 **c** 6.4 **d** 7.9 **2a** 1.6 **b** 6.9 **c** 11.2 **d** 1.7
3 308 m **4a** 5.1 m **b** 71.4 m² **5** 12 km

Page 111 Exercise 4

1a 26.6° **b** 35.2° **c** 42.8° **d** 19.4°
2a 59° **b** 24° **c** 50° **d** 20° **3** 26.6° **4a** 57.4° **b** 32.6°
5 100.4°, 36.8°, 111.4°, 111.4°

Page 112 Exercise 5

1 0.17, 0.34, 0.50, 0.64, 0.77, 0.87, 0.94, 0.98, 1
2 30°, 64.2°, 5.7°, 14.5°, 50.4°, 28.0°, 72.9° **3** No
4 (i) **a** **b**

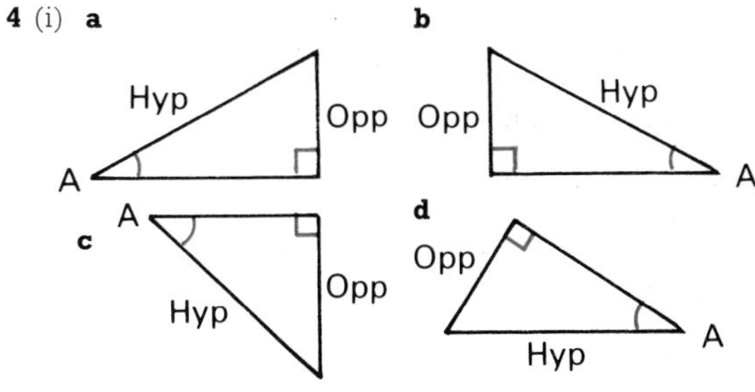

(ii) **a** $\frac{2}{3}$ **b** $\frac{1}{2}$ **c** $\frac{4}{5}$ **d** $\frac{5}{13}$
5a (i) $\frac{4}{9}$ (ii) 26.4° **b** (i) $\frac{6}{10}$ (ii) 36.9° **c** (i) $\frac{15}{17}$ (ii) 61.9°
d (i) $\frac{12}{15}$ (ii) 53.1° **6a** 30 **b** 16.9 **c** 44.9 **d** 17.0

Page 113 Exercise 6A

1a 8.6 **b** 5.8 **c** 16.9 **d** 38.4 **2a** 5.7 **b** 17.5 **c** 2.3 **d** 13.2
3a 38.6 m **b** 25.7 m **4** 6.4° **5a** 19.3 m **b** 30°
6a 90° **b** 4.9 km **7a** 120° **b** 60° **c** 30° **d** 4 cm **e** 9.8 cm

Page 114 Exercise 6B

1a 12.4 **b** 12.9 **c** 59.1 **d** 9.2
2a 39.6 **b** 2.1 **c** 153.8 mm **d** 6.0 m **3a** 6.22 m **b** 8.77 m
4a 61.3 km **b** 80 km **c** 102.8 km

Page 115 Exercise 7A

1 0.98, 0.94, 0.87, 0.77, 0.64, 0.5, 0.34, 0.17, 0
2 60°, 25.8°, 84.3°, 75.5°, 39.6°, 62.0°, 17.1° **3a** (i) $\frac{12}{13}$ (ii) 22.6°
b (i) $\frac{15}{25}$ (ii) 53.1° **c** (i) $\frac{15}{17}$ (ii) 28.1° **d** (i) $\frac{24}{25}$ (ii) 16.3°
4a 11.8 **b** 4.1 **c** 118.9 **d** 36.7 **5** 11.7° **6** 33.6° **7** 36.9°
8a 10.4 km **b** 6 km

Page 116 Exercise 7B

1 24.6 m **2** 7.9 m **3a** 78.4 mm **b** 98.7 mm **4** 9.2, 14.3

Page 117 Exercise 8A

1a 11.0 **b** 7.9 **c** 87.5 **d** 5.7
2a 39.8° **b** 38.5° **c** 53.1° **d** 36.9° **3** 381 m **4** 11 m 41 cm
5 *t*, by 7.6 mm **6** 59.0° **7** Angle at A is 36.9°, at B 31.9°.
So easier for football? **8** 404 m **9a** 14.1 km **b** 5.1 km
10a 36.9 **b, c** 2 **d** 3 m² **11a** 66.4° **b** (i), (ii) 4.6 m
12a 45 cm **b** 30 cm **c** 33.7° **d** 54.1 cm

Page 119 Exercise 8B

1 124.7 m **2b** 3.4 m **3b** (i) 10 cm (ii) 24 cm (iii) 240 cm²
(iv) 67.4°, 67.4°, 45.2° **4a** 38.7°, 5.13° **b** 77.4°, 102.6°
5a 4.6 **b** 6.6 **c** 30.4 m² **6a** 3.9 m **b** 3.5 m
7a 0.4, 21.8° **b** (i) 0.2, 11.3° (ii) 1.5, 56.3° **8a** 63.4° **b** 5.6 m
9a $\tan x° = \frac{w}{d}$, so $w = d\tan x°$; $H = h + w = h + d\tan x°$ **b** 41.9
10a $\sin C = \frac{AD}{b}$, so $AD = b\sin C$
b Area $= \frac{1}{2}CD \times AD = \frac{1}{2}ab\sin C$ **c** 18.7 cm² **11a** 16 **b** 58

Page 121 Check-up on Trigonometry

1a $\frac{3}{5}, \frac{4}{5}$ **b** $\frac{24}{25}, \frac{24}{7}$ **c** $\frac{12}{13}, \frac{5}{12}$ **d** $\frac{8}{17}, \frac{8}{17}$
2a 51.1 **b** 56.3 **c** 6.1 **d** 7.0 **3a** 2.1 cm **b** 23.6 cm
c 24.8 cm² **4a** 8.7 cm **b** 104 cm² **5a** 21.9 cm **b** 41.3 cm
6a (i) 1.8 m (ii) 3.6 m **b** (i) 10.2 m² (ii) 40.7 m²
7 ∠A = 60°, ∠C = 30°; perimeter is 9.5 m; area is 3.5 m²
8a 22.6 **b** 2.5 m **9** 1051 m

10 SIMULTANEOUS EQUATIONS

Page 122 Looking Back

1 Rows: −4, −2, 0, 2, 4; −1, 0, 1, 2, 3; 6, 5, 4, 3, 2
2a 11*x* **b** 0 **c** −4*x* **d** −*x* **3a** 3*y* **b** 11*y* **c** −3*y* **d** −*y*
4 A(3, 0), B(0, 4), C(−3, 0), D(0, −2) **5a** 7 **b** −1 **c** 0
6a 1 **b** −5 **c** 4 **7a** 4 **b** −3 **c** 4 **d** 5 **e** 3 **f** 3
8a $y = -4$ **b** $x = 2$ **9a** A, C **b** B **c** D, F **d** C, D
e A, F **f** E **10a** £36 **b** £3*x* **c** £(5*x*+3*y*). £1, £9; £4, £4

Page 123 Class Discussion/Investigation

1 49 **2** At (3, 3) **3a** (2, 2) **b** (3, 5) **c** (2, 4)

Page 124 Exercise 1

1d (2, 2) **2d** (2, 4) **3d** (2, 3) **4d** (0, 2) **5d** (−2, 2)
6d (1, −2) **7d** (5, 1) **8d** (4, −2)

Page 124 Exercise 2

1 (0, −4), (4, 0) **2** (0, −6), (2, 0) **3** (0, 3), (−3, 0)
4 (0, 5), (5, 0) **5** (0, −8), (8, 0) **6** (0, 4), (−4, 0)
7 (0, 2), (3, 0) **8** (0, −5), (2, 0) **9** (0, 3), (−4, 0)
10 (0, 0), (1, 8) **11** (0, 0), (1, −2)

Page 125 Exercise 3A

1 (4, 2) **2** (3, 1) **3** (6, 2) **4** (5, 1) **5** (4, 1) **6** (−3, 0)
7 (6, −1) **8** (−2, 0) **9** (4, 5)

Page 125 Exercise 3B

1 (−1, −2) **2** (0, 0) **3** (4, 4) **4** (4, 1) **5** (0, 6) **6** (−3, 2)
7 (1.5, 0.5) **8** (−1.5, −3.5) **9** (2.4, 3.6)

Page 126 Exercise 4

1 (10, 20) **2** (5, 15) **3** (9, 10) **4** (0, 12) **5** (−3, 7) **6** (1, 4)
7 ($\frac{1}{2}$, $1\frac{1}{2}$) **8** (−1, 2) **9** (−1, 1) **10** (1, 2) **11** (2, 1) **12** (3, 2)

Page 127 Exercise 5

1 £3, £1 **2** £2, £2 **3** £5, £3 **4** £3, £1 **5** £8, £5 **6** £7, £4
7 £7, £2 **8** £4.50, £2.50

Page 128 Exercise 6

1 (9, 3) **2** (10, 5) **3** ($6\frac{1}{2}$, $3\frac{1}{2}$) **4** (6, −1) **5** (2, 5) **6** (6, −1)
7 (2, −2) **8** ($-\frac{1}{2}$, −1) **9** (−3, $2\frac{1}{2}$) **10** (8, 1) **11** (1, −1)
12 (0, 1) **13** (6, 1) **14** (3, 1) **15** (1, −1) **16** (3, 2)
17 (−2, −1) **18** (3, −1)

Page 129 Exercise 7A

1 (2, 3) **2** (2, 0) **3** (2, 1) **4** (4, 1) **5** (1, 0) **6** (1, −1)
7 (2, −1) **8** (1, −1) **9** (0, 1) **10** (2, −2) **11** (1, −1)
12 (2, 1) **13** (1, −1) **14** (−1, 0)

Page 129 Exercise 7B

1 (5, 0) **2** (2, −1) **3** (−1, −1) **4** (3, 4) **5** (3, −2)
6 (2, −3) **7** (4, 2) **8** (−2, 5) **9** (1, 4) **10** (2, −1)

Page 130 Exercise 8

1 (3, 2) **2** (4, 2) **3** (3, 7) **4** (1, −1) **5** (1, 2)
6 (−2, −8) **7** (−1, 1) **8** ($5\frac{1}{2}$, $-1\frac{1}{2}$) **9** ($1\frac{1}{2}$, $1\frac{1}{2}$)
10 ($5\frac{1}{2}$, $4\frac{1}{2}$) **11** (−3, −4) **12** (1, −2)

Page 131 Exercise 9

1 $y = 2x+4$ **2a** $y = x+3$ **b** $y = 2x-4$ **3** $y = x-2$
4 $y = 3x+2$ **5a** $P = \frac{1}{2}x+15$ **c** $P = 55$
d atmospheric pressure = 15 units
6b $C = 400+10n$ **c** £29

Page 132 Exercise 10A

1a Book £5 **b** pen £1 **2a** Foil £2 **b** glue £1
3a 8 m **b** 3 m **4a** 4 h **b** 5 h **5** £7 **6a** 55p **b** 35p

Page 133 Exercise 10B

1 £6, £3 **2** £12, £8 **3** £4, £2 **4** 9 m, 3 m **5** 75 m, 25 m
6a 6, 1 **b** 1 goal below and 1 above, or 7 above

Page 134 Check-up on Simultaneous Equations

1d (2, 3) **2** Lines cut axes at: **a** (0, −5), (5, 0) **b** (0, 4), (4, 0)
c (0, −3), (5, 0) **3a** (2, 3) **b** (2, 4)
4a (4, 4) **b** (i) $x = 4$ (ii) $y = x$ (iii) $x + y = 8$

5a $(-3, -15)$ **b** $(-2, -6)$ **6a** $(5, -2)$ **b** $(1\frac{1}{2}, 1\frac{1}{2})$
7 $y = x+1$ **8a** $(4, 3)$ **b** $(\frac{1}{2}, 5)$ **c** $(-\frac{1}{2}, 2\frac{1}{2})$ **d** $(1\frac{1}{2}, -2\frac{1}{2})$
9 £35, £45

11 AREAS AND VOLUMES

Page 135 Looking Back

1a (i) 47 cm (ii) 35 cm **b** (i) 120 cm² (ii) 126 cm² (iii) 314 cm² **c** (i) 15 625 cm³ (ii) 17 500 cm³
2a 3 **b** 3.1 **c** 3.14 **3a** 22.5 cm² **b** 24 cm² **4a** 15 **b** 5
5a 3.3 **b** 20.6 **6a** 17 cm **b** 57°

Page 136 Exercise 1A

1a 48 mm, 135 mm² **b** 36 cm, 56 cm² **c** 44.0 mm, 154 mm²
d 8.93 m, 4.91 m² **2a** 20 m² **b** 40.3 m² **c** 20 m² **d** 30.9 m²
3a 4.57 m **b** 1.39 m² **4a** 28 **b** 90 cm edge along 360 cm edge
5a 5 cm edge along 10 cm edge **b** 3 **c** 36

Page 137 Exercise 1B

1a 16.6 m **b** 12.4 m² **2a** 68 m² **b** (i) 5 m (ii) £7200
3a 78.0 cm **b** 4780 cm² **4a** (i) 78.5 cm², 50 cm²
b 78.5%, 63.7% **5a** 14.1 cm² **b** 5150 mm²

Page 138 Exercise 2A

1a 7 sq. **b** 4 sq. **c** 5 sq. **d** 12 sq. **e** 16 sq.
2a 120 cm² **b** 80 cm² **c** 48 m² **d** 300 mm²
3a 4500 cm² **b** 3600 cm² **c** 3300 cm²
d 4000 cm²; order **a**, **d**, **b**, **c** **4a** 8 sq. **b** 10 sq. **c** 12 sq.
d 8 sq. **e** 6 sq. **5a** 54 cm² **b** 21 cm² **c** 21.6 cm²
6a 8 cm, 120 cm² **b** 12 cm, 168 cm² **7** 2250 cm²
8a 30 cm², 20 cm² **b** 50 cm² **9a** 345 mm² **b** 26 cm²
c 8 km² **d** 270 cm² **10a** 6.25 m² **b** 325 cm²

Page 140 Exercise 2B

1a 10 500 cm² (1.05 m²) **b** 60 cm² **c** 20 cm² **d** 32 km²
2a 12 cm, 120 cm² **b** 24 cm, 480 cm²
3a 58.9 m² **b** 75.5 cm² **4a** 4 m² **b** 37.5
5a $A = bh$ **b** $A = \frac{1}{2}d_1d_2$ **c** $A = \frac{1}{2}(a+b)h$
6a 50 cm² **b** 24 cm² **c** 36 cm² **d** 42 cm²
7a To make you slow down
b 54 m² (combine all the parallelograms)

Page 141 Exercise 3A

1a Cuboid **b** cuboid **c** cylinder **2a** 4, 8, 12 cm³
b 5, 10, 20 cm³ **c** 1, 1.5 cm³ **d** 3.5, 7, 10.5 cm³ **3 a**, **c**, **d**, **h**
4a 360 cm³ **b** 96 cm³ **c** 800 cm³ **d** 2500 cm³ **5a** (i) 1 cm² (ii) 5 cm³ **b** (i) 3 cm² (ii) 15 cm³ **c** (i) 2 cm² (ii) 10 cm³
6a 216 cm³ **b** 192 cm³ **c** 160 cm³ **d** 144 cm³
7a (i) 12 cm² (ii) 240 cm³ **b** (i) 50 cm² (ii) 1250 cm³
8 24 m³

Page 143 Exercise 3B

1a 40 m³ **b** 120 m³ **c** 220 m³ **d** 9 m³
2a Triangular prism **b** (i) 216 cm² (ii) 144 cm³
3a 15 m³ **b** 2.5 m **c** 31 m² **4** 12 m³ **5** 45 litres
6a Three right-angled △s, sides about the right angle 24 cm; two rectangles 24 cm by 50 cm **b** (i) 3264 cm² (ii) 14 400 cm³ (iii) 34 cm (iv) 1700 cm²
7a 2.92 cm **b** 29.5 m³

Page 145 Exercise 4A

1a 151 cm³ **b** 283 cm³ **c** 15 700 cm³ **2 a** by 16 cm³
3 4420 litres **4a** 44.1 m² **b** 441 m³ **5** 58 **6a** 109 m³ **b** 11

Page 146 Exercise 4B

1a 150 **b** 21% **2a** 3.27 cm² **b** 6540 (6530) cm³ **c** 76 kg
3a (i) 8π m³ (ii) 16π m³ (iii) 32π m³ **b** (i) $\frac{1}{2}$ (ii) $\frac{1}{4}$
4a 5.64 cm **b** 1.59 cm
5a Make height 31.3 cm **b** make radius 5.6 cm

Page 147 Exercise 5

1a 120 cm² **b** 72 cm² **2a** 251 cm³ **b** 302 cm³ **3** 905 cm²
4a (i) 427 cm² (ii) 653 cm³ **b** 933 cm², 3050 cm³
5a $A = 2\pi rh$ **b** $A = 2\pi r^2$ **c** $A = 2\pi rh + 2\pi r^2$
6a Tin **b** umbrella stand **c** napkin ring
7a 12.7 **b** 15.7 m² **c** 0.196 m³

Page 148 Exercise 6

1a $P = 12x$, $A = 6x^2$, $V = x^3$
b $P = 4u+4v+4w$, $A = 2uv+2vw+2uw$, $V = uvw$
2 Area formulae have r^2 or d^2 in them, i.e. **a**, **d**
3 Perimeters **a**, **c**; area **d**; volume **b**
4 Volume **a**, area **c**, circumference **b**
5 Volume **b**, area **a**, distance **c**
6 Perimeters: K, M, N, areas: G, L, R, volumes: H, J, P
7a $P = 2p+2q+2r+3s$; $A = pq+ps+qs+rs$; $V = \frac{1}{2}pqs$

Page 149 Check-up on Areas and Volumes

1a 96 cm² **b** 60 cm² **c** 135 mm² **d** 41 m² **2a** 288 cm³
b 192 cm³ **3a** 18 cm **b** 351 cm² **4a** 8.82 m **b** 485 m²
5 72 cm³ **6b** 687 cm², 866 cm³ **7** 3.93 cm³, 4.73 cm³
8a 180 cm³, 202 cm² **b** 396 cm³, 320 cm²
c 7500 cm³, 3300 cm² **9b** 2700 tonnes

12 FACTORS

Page 150 Looking Back

1a 8×1, 4×2 **b** 9×1, 3×3, **c** 10×1, 5×2 **d** 11×1,
e 12×1, 6×2, 4×3 **2a** 2 **b** 5
3a $3m \times 1, 3 \times m$ **b** $x^2 \times 1, x \times x$ **4a** 4 **b** y **c** 2 **d** $2x$
5a 6, 8, 10, 12 **b** 0, 3, 6, 9 **6a** (i) 20 m² (ii) 10 m² (iii) 30 m²
b (i) $6x$ m² (ii) 30 m² (iii) $6x+30$ m², or $6(x+5)$ m²
7a $3x+3$ **b** $4y-8$ **c** $10n+5$ **d** $6-6m$ **e** $ab+ac$ **f** x^2-xy
8a $5x-1$ **b** $3y$ **9a** $8(x+y), 8x+8y$ **b** $u(v+w), uv+uw$
10a x^2+5x+6 **b** y^2-5y+4 **c** $6z^2-z-1$ **d** $25u^2-4$
e $v^2+10v+25$ **f** $4w^2-4w+1$
11a (i) $2n-1$ (ii) $3n+1$ **b** (i) 39 (ii) 61

Page 151 Exercise 1

1a 1, 2, 3, 6 **b** 1, 2, 4, 8, 16 **c** $1, x, y, xy$ **d** $1, b, b^2$
e $1, 3, a, 3a$ **2a** 1, 2, 4 **b** 1, 2, 4, 8 **c** 1, 13 **d** 1, 2, 7, 14
e 1, 2, 4, 5, 10, 20 **3a** 9 **b** 1 **c** 2 **d** $2x$ **e** 6 **f** x
4 $7, 4x, 2y, 4, 5, 3a, 4b, y^2$ **5a** 3 **b** 4 **c** 4 **d** 10 **e** 5 **f** 4
g 5 **h** 6 **i** n **j** x **k** $2x$ **l** $7y$ **6a** $2, x, 2x, 2x$ **b** $y, 2, 2y$
c x, xy **d** $2x$ **7a** 3 **b** 5 **c** x **d** $2x$ **e** 3 **f** a

Page 152 Exercise 2A

1a $4, 4(x+3)$ **b** $3, 3(3x+5)$ **c** $8, 8(3a+1)$ **d** $7, 7(3y-2)$
e $5x, 5x(1-2x)$ **f** $x^2, x^2(1+x)$

2a $3(2x+3)$ **b** $2(2y+3)$ **c** $5(2y+3)$ **d** $5(a-2)$
e $7(m-2)$ **f** $6(b+2)$ **g** $3(1+3p)$ **h** $2(2-3q)$ **i** $9(2-r)$
j $10(2s+1)$ **k** $2(9t-5)$ **l** $4(2u-3)$ **m** $9(4v-5)$
n $4(5w+3)$ **o** $2(12-19t)$
3a $x(x+1)$ **b** $y(y-1)$ **c** $n(n+3)$ **d** $m(m-2)$ **e** $2k(k+2)$
f $3t(3t-1)$ **g** $2a(2a-3)$ **h** $b(5b-3)$ **i** $d(4+9d)$
j $6c(2c+5)$ **k** $2d(3d-4)$ **l** $10e(1+5e)$ **m** $8f(3-5f)$
n $9g(2g-5)$ **o** $6w(7+3w)$ **4a** $2(x+2y+3)$
b $3(m+3n+2)$ **c** $3(a-4b+3)$ **d** $5(x^2+2x+4)$
e $4(y^2-y+1)$ **f** $4(2a^2+2a+5)$ **g** $2(p^2-4p+5)$
h $x(x^2+x+1)$ **5a** $6(3+7)=60$ **b** $4\times 100=400$
c $9\times 1000=9000$ **d** $7\times 10=70$ **e** $8\times 100=800$
f $7.5\times 20=150$ **g** $\frac{1}{4}\times 100=25$ **h** $897\times 1=897$
6 $38\times 50\text{p}=£19$ **7** $12\times £1=£12$ **8** $36\times £3=£108$

Page 153 Exercise 2B

1 $\frac{1}{2}ah, \frac{1}{2}bh$; area $=\frac{1}{2}ah+\frac{1}{2}bh=\frac{1}{2}h(a+b)$
2a $A=4r^2+\pi r^2=r^2(4+\pi)$ **b** $A=2xy+\frac{1}{2}h\times 2x=x(2y+h)$
3a Area $=\pi R^2-\pi r^2=\pi(R^2-r^2)$ **b** $377\ mm^2$
4a (i) $4(2+a), 2(4+b), 2(4+3ab)$ (ii) $2(2a+b), 2a(2+3b)$
(iii) $2b(1+3a)$ **b** (i) $p(q-2p), q(p-4q), pq(1-5q)$
(ii) $2(p^2-2q^2), p(2p-5q^2)$ (iii) $q^2(4-5p)$

Page 154 Exercise 3A

1a $(a-b)(a+b)$ **b** $(c-d)(c+d)$ **c** $(p-q)(p+q)$
d $(x-2)(x+2)$ **e** $(y-3)(y+3)$ **f** $(z-4)(z+4)$
g $(5-u)(5+u)$ **h** $(7-v)(7+v)$ **i** $(1-w)(1+w)$
j $(t-3)(t+3)$ **k** $(k-2)(k+2)$ **l** $(m-6)(m+6)$
m $(d-10)(d+10)$ **n** $(10-d)(10+d)$ **o** $(m-n)(m+n)$
p $(a-1)(a+1)$ **q** $(8-n)(8+n)$ **r** $(y-9)(y+9)$
2a $(2x-3)(2x+3)$ **b** $(2y-7)(2y+7)$ **c** $(3a-1)(3a+1)$
d $(2a-5)(2a+5)$ **e** $(3b-2)(3b+2)$ **f** $(4c-3)(4c+3)$
g $(2x-y)(2x+y)$ **h** $(p-3r)(p+3r)$ **i** $(10a-b)(10a+b)$
j $(8c-d)(8c+d)$ **k** $(x-4y)(x+4y)$ **l** $(t-6u)(t+6u)$
m $(2a-3b)(2a+3b)$ **n** $(3c-4d)(3c+4d)$
o $(4e-5f)(4e+5f)$ **p** $(1-10k)(1+10k)$ **q** $(9n-1)(9n+1)$
r $(7x-9y)(7x+9y)$ **3a** $(99-1)(99+1)=9800$
b $(67-33)(67+33)=3400$ **c** $(18.5-8.5)(18.5+8.5)=270$
d $(111-110)(111+110)=221$ **e** $(\frac{3}{4}-\frac{1}{4})(\frac{3}{4}+\frac{1}{4})=\frac{1}{2}$
f $(9.95-0.05)(9.95+0.05)=99$
4a $(a-2b)(a+2b), (a-3c)(a+3c), (a-5d)(a+5d)$;
$(2b-3c)(2b+3c), (2b-5d)(2b+5d)$; $(3c-5d)(3c+5d)$
b $(1-x)(1+x), (1-4y)(1+4y), (1-t)(1+t)$;
$(x-4y)(x+4y), (x-t)(x+t)$; $(4y-t)(4y+t)$

Page 154 Exercise 3B

1a $2(x-2)(x+2)$ **b** $3(y-1)(y+1)$ **c** $5(a-3)(a+3)$
d $4(b-2)(b+2)$ **e** $2(a-b)(a+b)$ **f** $9(c-2d)(c+2d)$
g $10(e-3f)(e+3f)$ **h** $11(g-h)(g+h)$ **i** $a(x-y)(x+y)$
j $\pi(R-r)(R+r)$ **k** $k(a-2b)(a+2b)$ **l** $n(p-3q)(p+3q)$
m $x(x-2)(x+2)$ **n** $b(5-b)(5+b)$ **o** $2y(y-4)(y+4)$
p $3a(3a-4)(3a+4)$ **2a** 2 **b** 0.6 **c** 9 m **d** 7 m
3a Area $=d^2-16=(d-4)(d+4)\text{m}^2$ **b** 14.25m^2
4a $A=\pi R^2-\pi r^2=\pi(R^2-r^2)=\pi(R-r)(R+r)$
b 3140 sq. units
5 $(a-b)(a+b)(a^2+b^2), a^2(a-1)(a+1), (a^2-2)(a^2+2)$;
$(b^2-a)(b^2+a), (b^2-2)(b^2+2)$; $(a-2)(a+2)$

Page 155 Exercise 4A

1a $(x+2)(x+1)$ **b** $(x+5)(x+1)$ **c** $(x+1)(x+1)$
d $(x+5)(x+2)$ **e** $(x+10)(x+1)$ **f** $(x+3)(x+2)$
g $(y+5)(y+3)$ **h** $(t+4)(t+2)$ **i** $(t+8)(t+1)$
2a $(n-1)(n-1)$ **b** $(k-3)(k-3)$ **c** $(p-2)(p-1)$
d $(q-2)(q-2)$ **e** $(r-7)(r-3)$ **f** $(s-6)(s-1)$
g $(u-7)(u-5)$ **h** $(v-8)(v-2)$ **i** $(w-5)(w-1)$
3a $(a+5)(a+5)$ **b** $(a-5)(a-5)$ **c** $(t+4)(t+4)$
d $(t-4)(t-4)$ **e** $(t+8)(t+2)$ **f** $(t-8)(t-2)$
g $(t+16)(t+1)$ **h** $(t-16)(t-1)$ **i** $(p-10)(p-10)$
4a $(x+3)^2$ **b** $(x+1)(x-7)$ **c** $(x+2)^2$ **d** $(y+5)(y+4)$
e $(y-2)(y+3)$ **f** $(y+1)(y-2)$ **g** $(z+2)^2$ **h** $(z-2)(z-8)$
i $(z+2)(z-5)$
5a $(m+1)(m-3)$ **b** $(n-1)(n+5)$ **c** $(p-4)(p+5)$
d $(q+6)^2$ **e** $(r+3)(r-7)$ **f** $(s+2)(s+4)$ **g** $(t+7)(t-8)$
h $(u-8)(u+9)$ **i** $(v+8)^2$
6a $(x-3)(x+2)$ **b** $(x-5)(x+4)$ **c** $(y+2)(y-1)$
d $(m+6)(m-5)$ **e** $(p+5)(p-2)$ **f** $(q+5)(q-3)$
g $(t-5)(t+4)$ **h** $(u-6)(u+4)$ **i** $(v-5)(v+3)$
7a $(x+4)(x+3)$ **b** $(x-4)(x-3)$ **c** $(y+4)(y-3)$
d $(y-4)(y+3)$ **e** $(z+6)(z-2)$ **f** $(z-6)(z+2)$
g $(a-6)(a-2)$ **h** $(b+12)(b-1)$ **i** $(c-12)(c-1)$
8a $(v+10)(v-2)$ **b** $(w-9)(w+4)$ **c** $(x-11)(x+1)$
d $(y+6)(y+3)$ **e** $(z-9)(z-2)$ **f** $(x+7)(x+7)$
g $(m-6)(m-6)$ **h** $(n-15)(n+2)$ **i** $(p+24)(p-3)$

Page 156 Exercise 4B

1a $y-4$ m **b** less than or equal to 4 m **2a** $y-11$ cm
b less than or equal to 11 cm **3** $k+1$ cm, $k-4$ cm
4a $x-1$ m **b** $x+5$ m, $4x+8$ m **c** $x+2$ m, $x^2+4x+4\text{ m}^2$

Page 156 Exercise 5

1a $(2x+1)(x+1)$ **b** $(3y+1)(y+1)$ **c** $(2x+1)(x+2)$
d $(2x+7)(x+1)$ **e** $(5y+1)(y+2)$ **f** $(3a+2)(a+1)$
g $(3b+1)(b+2)$ **h** $(4c+1)(c+1)$ **i** $(2d+1)(2d+1)$
j $(3t+1)(2t+1)$ **k** $(5x+2)(x+1)$ **l** $(11y+2)(y+1)$
m $(6a+1)(a+1)$ **n** $(4m+1)(3m+1)$
2a $(2x-1)(x-1)$ **b** $(3x-1)(x-1)$ **c** $(5a-1)(a-1)$
d $(3y-1)(2y-1)$ **e** $(6b-1)(2b-1)$ **f** $(8x-1)(3x-1)$
g $(2m-1)(m-2)$ **h** $(3t-2)(t-1)$ **i** $(2y-1)(y-3)$
j $(4k-3)(k-1)$ **k** $(3e-1)(3e-1)$ **l** $(9f-1)(f-1)$
m $(5g-1)(g-3)$ **n** $(2x-1)(2x-3)$
3a $(2n-1)(n+1)$ **b** $(3a+1)(a-1)$ **c** $(5b-1)(b+1)$
d $(3x-2)(x+1)$ **e** $(3x+2)(x-1)$ **f** $(5y-2)(y+1)$
g $(6d-1)(4d+1)$ **h** $(8m-1)(3m+1)$ **i** $(4a-1)(2a+3)$
j $(4b-3)(2b+1)$ **k** $(6c-1)(c+3)$ **l** $(3d-1)(2d+3)$
m $(3x+1)(3x-2)$ **n** $(4y-1)(3y+2)$
4a $(5x-1)(5x-1)$ **b** $(1+3x)(1+x)$ **c** $(1-2x)(1+x)$
d $(9a-1)(a-2)$ **e** $(4y-3)(y-2)$ **f** $(3y+2)(2y-3)$
g $(3-2t)(5+t)$ **h** $(1-6u)(1+3u)$ **i** $(5-4v)(1+3v)$
j $(3+2d)(2-3d)$ **k** $(1-4x)(1-4x)$ **l** $(5y+4)(2y-1)$
m $(9b-4)(2b+1)$ **n** $(4y+3)(3y-4)$

Page 157 Exercise 6A

1 $3(x+2)$ **2** $2(4-x)$ **3** $x(x+1)$ **4** $y(1-y)$ **5** $2a(2-a)$
6 $(c-d)(c+d)$ **7** $(p-2)(p+2)$ **8** $(3-y)(3+y)$
9 $3x(3x+2)$ **10** $x(x-y)$ **11** $(6-k)(6+k)$ **12** $b(a+c)$
13 $3(2a^2-b^2)$ **14** $2a(2a+b)$ **15** $2(a-b)(a+b)$
16 $(p+3)(p+3)$ **17** $(t-2)(t-1)$ **18** $(u-3)(u+2)$
19 $(v+3)(v-1)$ **20** $(x+5)(x-3)$ **21** $(a-10)(a+10)$
22 $2(y-3)(y+3)$ **23** $x(x-1)(x+1)$ **24** $x^2(x-2)$
25 $x(x-2)(x+2)$ **26** $3y(x+2a)$ **27** $2n(2m-3p)$
28 $5(p-3)(p+3)$ **29** $2(n-9)(n+8)$ **30** $2m(1+m+m^2)$
31 $3x(x-4)$ **32** $2x(x-3)(x+3)$ **33** $3p(p-1)$
34 $2(2y+5)(y-1)$ **35** $3t(t-3)(t+3)$ **36** $2mn(3m-4n)$

Page 157 Exercise 6B

1 $(6-y)(6+y)$ **2** $2ab(c-4d)$ **3** $(1-x)^2$ **4** $2y(3+8y^2)$
5 $2(7x+3)(x+1)$ **6** $2(x+3)(x-2)$ **7** $(2x-3y)(2x+3y)$
8 $2(1-5d)(1+5d)$ **9** $2(2s+1)(s-1)$ **10** $(4k-3)(k-2)$
11 $(m-3n)(m+3n)$ **12** $xyz(z+x)$ **13** $b^2(a-2)(a+2)$
14 $2(3x+2)(x-2)$ **15** $(3x-4)(2x-3)$ **16** $y(1-y)(1+y)$

17 $4(1+3a)(1-a)$ **18** $x^2y^2(y-x)(y+x)$
19 $2(10a-1)(10a+1)$ **20** $2(1-b)^2$ **21** $(x+y)^2$
22 $(2a-b)(a+2b)$ **23** $u^2(1-u)(1+u)(1+u^2)$
24 $k^2(1-k+k^2)$ **25** $(3p-5)(2p+3)$ **26** $(x+3y)(x-2y)$
27 $(a+b-c)(a+b+c)$ **28** $(p-q-1)(p-q+1)$
29 $(x+y)(a+b)$ **30** $3(p+q)+a(p+q)=(p+q)(3+a)$
31 $(\pi R+2h)(\pi R+h)$ **32** $(6x-5y)(6x+5y)$
33 $(a-c)(a+2b+c)$ **34** $(9m-8)(4m-3)$ **35** $(x^2+1)^2$
36 $(x^2-3)(x^2+2)$

Page 158 Exercise 7

1a $6n+2=2(3n+1)$ **b** $6n+3=3(2n+1)$
c $12n-4=4(3n-1)$ **d** $18n-12=6(3n-2)$
2a (i) 0, 3, 8, 15 (ii) $(n-1)(n+1)$; $0\times1, 1\times3, 2\times4, 3\times5$
b (i) 2, 6, 12, 20 (ii) $n(n+1)$; $1\times2, 2\times3, 3\times4, 4\times5$
c (i) 6, 12, 20, 30 (ii) $(n+1)(n+2)$; $2\times3, 3\times4, 4\times5, 5\times6$
d (i) 0, 1, 4, 9 (ii) $(n-1)^2$; $0, 1^2, 2^2, 3^2$ **e** (i) 1, 3, 6, 10
(ii) $\frac{1}{2}n(n+1)$; $\frac{1}{2}\times1\times2, \frac{1}{2}\times2\times3, \frac{1}{2}\times3\times4, \frac{1}{2}\times4\times5$
3 Sequence 1: $(n+1)(4n+1)$, 40 501; Seq. 2: $(n+1)(4n+3)$, 40 703;
Seq. 3: $(2n+1)(2n+3)$, 40 803; Seq. 4: $(n+1)(4n+5)$, 40 905;
Seq. 5: $n(4n+3)$, 40 300; Seq. 6: $(2n+1)^2$, 40 401;
Seq. 7: $2(n+1)(2n+1)$, 40 602; Seq. 8: $4n^2$, 40 000

Page 159 Class Discussion/Investigation

1 The sum of four consecutive numbers is even.
$n+(n+1)+(n+2)+(n+3)=4n+6=2(2n+3)$
2 The sum of five consecutive numbers is a mulitple of five.
$n+(n+1)+(n+2)+(n+3)+(n+4)=5n+10=5(n+2)$
3 $(n-2)+(n-1)+n+(n+1)+(n+2)=5n$
4 Number added: 3 4 5 6 7 8 9 10 11 12
Multiples of . . . **3** *2* **5** *3* **7** *4* **9** *5* **11** *6*
When consecutive numbers are added, the sum is a multiple of n when n is odd, and of $\frac{1}{2}n$ when n is even.
5a $(2n+1)+(2m+1)=2(n+m+1)$, even
b $2n+2m=2(n+m)$, even
c $2n+(2m+1)=2(n+m)+1$, odd
d $(2n+1)(2m+1)=2(2nm+n+m)+1$, odd
e $2n(2m+1)=2(2nm+n)$, even
f $(2n)(2m)=4(nm)$, multiple of 4
g $(2n+1)+(2n+2)+(2n+3)=6(n+1)$, multiple of 6
h $2n+(2n+1)+(2n+2)+(2n+3)+(2n+4)=10(n+1)$, multiple of 10
i $(2n+1)^2-1=4n(n+1)$, multiple of 4
j $(2n+1)^2-(2m+1)^2=(2n+1-2m-1)(2n+1+2m+1)=4(n-m)(n+m+1)$, multiple of 4

Page 160 Check-up on Factors

1a 1, 2, 3, 4, 6, 8, 12, 24 **b** 1, 2, 5, 10, x, $2x$, $5x$, $10x$
c 1, 2, x, $2x$, x^2, $2x^2$ **2a** 6 **b** $4t$ **3a** 4 **b** 6 **c** $4y$
4a $6(p-4)$ **b** $5(1-q)$ **c** $x(x+2)$ **d** $3(a-2b+3c)$
e $2\pi(R-r)$ **f** not possible
5a $(m-n)(m+n)$ **b** $(k-1)(k+1)$ **c** $(p-4q)(p+4q)$
d $2(a-3)(a+3)$ **e** $\pi(R-r)(R+r)$ **f** $a(x-2y)(x+2y)$
6a 7200 **b** 0.752 **7a** $(x+3)^2$ **b** $(y-5)^2$ **c** $(z-6)(z+5)$
d $(u+7)(u-4)$ **e** $(x-1)^2$ **f** $2(y+1)^2$
8 $t-6$ m; less than or equal to 6 m **9a** $(2a+3)(a+1)$
b $(2b-3)(b-1)$ **c** $(2c-1)(c+3)$ **d** $(2d+1)(d-3)$
e $y(y-1)$ **f** $(y-2)(y+2)$ **g** $(y-2)^2$ **h** $2t(1-2t)(1+2t)$
i $(6a-5)(a+2)$ **10a** $5n+5=5(n+1)$ **b** $9n-3=3(3n-1)$
11a (i) 3, 15, 35 (ii) 12, 20, 30
b (i) $(2n-1)(2n+1)$, $1\times3, 3\times5, 5\times7$
(ii) $(n+2)(n+3)$, $3\times4, 4\times5, 5\times6$
12a Area $=\frac{1}{2}y^2-\frac{1}{2}x^2=\frac{1}{2}(y-x)(y+x)\,\text{m}^2$
b 6 m, 6 m, 8.5 m; 8 m, 8 m, 11.3 m
13a $2n+1+2n+3+2n+5=6n+9=3(2n+3)$
b $2n+2n+2+2n+4=6n+6=6(n+1)$

13 MONEY MATTERS - PERSONAL FINANCE

Page 161 Looking Back

1a 644 **b** 147 **2a** 85% **b** (i) 0.85 (ii) $\frac{17}{20}$
3a £255.30 **b** £319.70 **4a** £240 **b** £825
5 £88.50, £125.12, £213.62, £37.38, £251
6 £10 750, £10 719.14 **7a** £75 **b** £17.30

Page 162 Exercise 1

2a £217.86 **b** £52.34 **c** £165.52
3a £1487.50 **b** £372.89 **c** £1114.61
4 Basic pay £656.85, income tax £93.36, total deductions £178.50
5 Gross pay £959.12, total deductions £254.21, net pay £704.91

Page 163 Exercise 2

1b £832 **2a** £39.60 **b** £475.20 **3** 5.5%
4a £1996.80 **b** £59 904

Page 164 Exercise 3

2a £0 **b** £21.46 **c** £13.18
3a £1025.20 **b** £1502.56 **c** £1761.76
4 Before, he paid no NI, after the rise he pays £1.21, so receives 1p less in his pay **5** 7.3% **6a** £21.75 **b** £1.33 less

Page 166 Exercise 4A

1 £0 **2** £15 080, £3445, £11 635, £500, £2283.75, £2783.75 per year, £53.53 per week
3a £4525 **b** (i) £1006.25 (ii) £19.35
4a £18 200 **b** £14 755 **c** (i) £3563.75 (ii) £68.54
d £281.46 **5a** £3360 **b** £5758.75 **c** £2694.05 **d** £0
6 Gross pay £1496, income tax £256.09, total deductions £373.75, net pay £1122.25

Page 167 Exercise 4B

1 Total allowances £7445, taxable income £30 555, income tax due £8542 **2** £32 942 **3a** £39 536 **b** £87 004
4a (i) £10 000 (ii) £18 000 (iii) £23 000 (iv) £28 000
c (i) 100% (ii) 90% (iii) 77% (to nearest whole number) (iv) 70%

Page 169 Exercise 5

1a A greater risk of client dying **b** females live longer, so will be able to pay more premiums **c** shorter lives, risk of ill-health and difficulty in paying premiums **2a** (i) £1.90 (ii) £2.95 **b** (i) £11.29 (ii) £4.62 **3a** £3 **b** £7.80
4 £56.55 **5** £72.38 **6a** £19 **b** £27.48 **c** £56.45 **d** £52.50
7a £446.40 **b** £244.80 **8** £10 032
9 Whole life £42 100, 10 year endowment £8300, 20 year endowment £17 500. Less, if a smoker
10a £126, £1050 **b** £132.30, £1102.50 **c** £138.92, £1157.63

Page 170 Exercise 6

1 42, 44.1, 926.1, £926.10
2a £605 **b** £441 **c** £1124.86 **d** £138.67 **4** £3666.92 **5** 8 years
6a £919.36 **b** annual amount £1.36 less **7** £314.82

Page 171 Exercise 7

1 £87 480 **2a** £3000 **b** £2400 **c** £2040 **3a** 2% **b** 6%

4a 22.86% **b** 42.86% **5** £830 304
6 −22.22%, +14.29%, −50% **7a** £156 800 **b** −10.4%
8 £6400, £2592, £0 **9** £452.63

Page 172 Exercise 8B

2 42.58% **3** 214% **4a** 12.68% **b** 0%
5 £999.17 **6** 3.44%

Page 173 Check-up on Money Matters—Personal Finance

1 £323, £227.28 **2a** £14.88, £18.40 **b** £172.01 **3a** £2652.50
b £203.44 **4a** £28 900 **b** £12 000 **5a** £49.44 **b** £123.52
6 £2 282 663 **7** £1 022 208 **8** 29.84%

14 FORMULAE

Page 174 Looking Back

1a 36 **b** −4 **c** 5 **d** −4 **e** 16 **f** 2 **g** 16 **h** 144 **i** 2
j 5 **k** 9 **2a** 5 **b** 6 **c** 1 **d** 3 **e** 16 **f** −36
3a $6y-12$ **b** $ax-bx$ **c** $-2x-2$ **d** $2x-3$
4a $P=6x$ **b** $A=\pi y^2$ **c** $V=pqr$ **d** $P=2r+\pi r$
5a $3n-2$ **b** 148 **6a** $5n+1$ **b** 251
7a (i) 60 (ii) 84 **b** (i) 5 (ii) 35
8a $T=m+n-3$ **b** $T=3n$

Page 175 Exercise 1A

1a 15 feet **b** 315 feet **2a** (i) $A=lb$ (ii) 30 cm^2
b (i) $A=\pi R^2$ (ii) 28 cm^2 **c** (i) $C=\pi D$ (ii) 19 cm
d (i) $D=ST$ (ii) 120 miles **e** (i) $V=lbh$ (ii) 160 mm^3
3a 48 **b** 108 **4a** 28 **b** 40.5
5 Both the same **6a** Lens 1 **b** 1 unit
7 Raspberry 15%, strawberry 16%; raspberry

Page 176 Exercise 1B

1a (i) 296 (ii) 444 **b** (i) D (ii) A
2a 0.7 **b** 62 m/s **c** 228 yards

Page 177 Exercise 2

1a $p=s-c$ **b** 4 **2a** $S=9500+250n$ **b** 11 000
3a $A=\dfrac{x+y+z}{3}$ **b** $65\frac{1}{3}$ kg **4a** $S=D+np$ **b** 29
5a $T=60+40x$ **b** 160 **6a** $L=2x+y$ **b** 47 000
7a $C=\dfrac{(L-x)p}{100}$ **b** 17.92 **8a** $T=3n-1$ **b** 149
9a $T=5n+7$ **b** 257
10a $a=\dfrac{360}{n}$ **b** $b=90-\dfrac{180}{n}$ **c** $180-\dfrac{360}{n}$ **d** 45°, $67\frac{1}{2}$°, 135°

Page 178 Exercise 3A

1a $x=b-a$ **b** $x=b-3$ **c** $x=n-m$ **d** $x=p+q$
2a $x=ab$ **b** $x=3b$ **c** $x=36$ **d** $x=st$ **e** $x=uv$
3a $x=\dfrac{b}{a}$ **b** $x=\dfrac{b}{5}$ **c** $x=\dfrac{d}{c}$ **d** $x=\dfrac{n}{m}$ **e** $x=\dfrac{t}{r}$
4a $x=\dfrac{b-3}{2}$ **b** $x=\dfrac{a-b}{2}$ **c** $x=\dfrac{y+1}{3}$ **d** $x=\dfrac{m+n}{3}$
e $x=\dfrac{q}{p}$ **f** $x=\dfrac{q+r}{p}$ **5** $S=\dfrac{D}{T}$ **6** $A=\dfrac{V}{h}$ **7** $A=\dfrac{3V}{h}$
8 $x=\dfrac{A}{y}$, $x=\dfrac{P-2y}{2}$ **9** $r=\dfrac{C}{2\pi}$, $r=\sqrt{\dfrac{A}{\pi}}$ **10** $M=\dfrac{5K}{8}$
11 $V=\dfrac{M}{D}$ **12** $R=\dfrac{100I}{PT}$ **13a** $u=v-ft$ **b** $f=\dfrac{v-u}{t}$
14a $h=\dfrac{V}{x^2}$ **b** $x=\sqrt{\dfrac{V}{h}}$ **15a** (i) $S=\dfrac{D}{T}$ (ii) $T=\dfrac{D}{S}$
b (i) $B=\dfrac{A}{L}$ (ii) $L=\dfrac{A}{B}$ **c** (i) $y=mx$ (ii) $m=\dfrac{y}{x}$
d (i) $C=\pi D$ (ii) $\pi=\dfrac{C}{D}$ **e** (i) $r^2=\dfrac{A}{\pi}$ (ii) $\pi=\dfrac{A}{r^2}$
f (i) $x=\dfrac{y}{2}$ **g** (i) $m=\dfrac{F}{a}$ (ii) $a=\dfrac{F}{m}$ **h** (i) $s=Md$
(ii) $d=\dfrac{s}{M}$ **i** (i) $i=\dfrac{e}{r}$ (ii) $r=\dfrac{e}{i}$

Page 179 Exercise 3B

1a $x=\dfrac{P}{4}$ **b** 55 cm **2a** $x=\sqrt{A}$ **b** 40 cm **3** $H=\dfrac{M}{60}$, 5
4a $D=7W$ **b** $W=\dfrac{D}{7}$, 13 **5a** £110 **b** $n=\dfrac{C-50}{5}$, 8
6a $A=6a^2$ **b** $a=\sqrt{\dfrac{A}{6}}$ **c** 2.4 cm **7a** $r=\sqrt{\dfrac{S}{4\pi}}$ **b** 2.82 cm
8a 110 **b** $b=l-h$, 110 **9a** $c=4F-148$ **b** (i) 52
(ii) 37°F **10a** (i) $T=2n+10$ (ii) $T=n^2+1$
b $n=\dfrac{T-10}{2}$, $n=\sqrt{(T-1)}$ **c** 56, 11

Page 181 Exercise 4A

1a $b=\dfrac{a}{c}$ **b** $b=a-c$ **c** $b=ac$ **d** $b=a+c$
2a $r=\sqrt{A}$ **b** $r=\sqrt{\dfrac{A}{\pi}}$ **c** $r=\sqrt{\dfrac{A}{4\pi}}$ **d** $r=\sqrt{\dfrac{V}{\pi h}}$
3a $h=\dfrac{V}{lb}$ **b** $h=\dfrac{V}{A}$ **c** $h=\dfrac{V}{\pi r^2}$ **d** $h=\dfrac{2A}{b}$
4a $x=\dfrac{P+2}{2}$ **b** $x=\dfrac{Q-2y}{2}$ **c** $x=\dfrac{R-3a}{3}$ **d** $x=\dfrac{an-S}{a}$
5a $R=\dfrac{V}{I}$ **b** $b=\dfrac{1}{a}$ **c** $d=\sqrt{\dfrac{C}{I}}$ **d** $A=\pi r^2$ **e** $c=\dfrac{b}{3a}$
f $b=\dfrac{Q}{al}$ **g** $d=\sqrt{\dfrac{3}{4M}}$ **h** $c=\sqrt{(a^2-b^2)}$
6a $x=a-b$ **b** $x=u-v$ **c** $x=\dfrac{s-t}{2}$ **d** $x=\dfrac{y+1}{2}$
e $x=\dfrac{y-3}{3}$ **f** $x=2y+3$ **g** $x=\dfrac{4y+3}{3}$
7a $V=\dfrac{KT}{P}$ **b** $T=\dfrac{PV}{K}$ **8** $r=\dfrac{v^2}{g}$ **9** $t=\sqrt{\dfrac{2s}{g}}$
10a $d=\sqrt{\dfrac{85\,000a}{n}}$ **b** 11.98 cm

Page 182 Exercise 4B

1 $d=\sqrt{\dfrac{GMm}{F}}$ **2** $r=\sqrt[3]{\dfrac{3V}{4\pi}}$ **3** $l=\dfrac{gT^2}{4\pi^2}$
4a $P=2x+2x+2\pi x=4x+2\pi x=2x(2+\pi)$ **b** $x=\dfrac{P}{2(2+\pi)}$

5 $h = \dfrac{2A}{a+b}$ **6** $u = \dfrac{4L+108}{5}$ **7** $h = \dfrac{A-2\pi r^2}{2\pi r}$

8a $x = \dfrac{c}{a+b}$ **b** $x = \dfrac{r}{p-q}$ **c** $x = \dfrac{2y-n}{m}$ **d** $x = \dfrac{a}{1+y}$

e $x = \dfrac{a-ay}{1+y}$ **9** $v = \dfrac{uf}{2u-f}$ **10a** $f_1 = \dfrac{f_2}{pf_2-1}$ **b** 0.6

Page 183 Exercise 5

1a Increases **b** decreases **c** doubled **d** halved
2a Decreases **b** increases **c** halved **d** doubled
3a Increases **b** decreases
c A is 4 times as large, C is doubled
4a Decreases **b** increases **c** halved
5a Doubles **b** decreases
6a Decreases **b** 4 times as large

Page 184 Check-up on Formulae

1a $A = x^2$, 225 mm² **b** $a^2 = b^2+c^2$, 12.5 cm
c $P = 2x+2y$, 426 m **d** $A = \frac{1}{2}bh$, 72 cm²

2a $A = 2ab+2ac+2bc$ **b** 208 **3a** $t = 7n-2$ **b** $n = \dfrac{t+2}{7}$, 13

4a $A = a^2+4\times\frac{1}{2}a\times a = 3a^2$ **b** (i) 432 (ii) 5 **c** $a = \sqrt{\dfrac{A}{3}}$

5 $r = \sqrt[3]{\dfrac{3V}{\pi}}$ **6a** $T = 6+5x$ **b** $x = \dfrac{T-6}{5}$ **c** 3 points

7 $R = \dfrac{3V+\pi h^3}{3\pi h^2}$ **8a** $x = -y-1$ **b** $x = \dfrac{y+2}{5}$ **c** $x = \dfrac{y+6}{3}$

d $x = \dfrac{4y-3}{3}$ **e** $x = \sqrt{(a^2+b^2)}$

f $x = \dfrac{1}{p^2}$ **g** $x = \dfrac{1}{y-2}$ **h** $x = \dfrac{y+1}{2(y-1)}$

9a (i) $A = r^2(4+\pi)$ (ii) 11 427 **b** (i) $r = \sqrt{\dfrac{A}{4+\pi}}$ (ii) 50

10a Increases **b** decreases **c** it is halved

d it is divided by 4 **11** $R_1 = \dfrac{RR_2}{R_2-R}$

15 PROBABILITY

Page 185 Looking Back

1 P('R' in the month) = 0.67
2a (i) 0.4 (ii) 0.2 (iii) 0.7 (iv) 0.5 (v) 0 (vi) 1

b

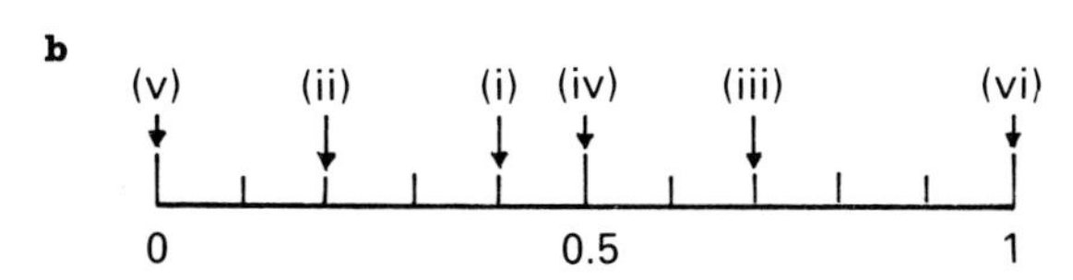

3a P(Red) = $\frac{3}{5}$, P(White) = $\frac{2}{5}$ **b** P(Red) = $\frac{1}{3}$, P(White) = $\frac{2}{3}$
4a $\frac{1}{2}$ **b** $\frac{1}{6}$ **c** $\frac{2}{3}$ **5a** Past data **b** survey **c** counting equally likely outcomes **d** experiment
6a 0.2 **b** (i) 100 (ii) 60 (iii) 40
7 About 1, 5, 6 or 7, 4, 3 or 4

Page 186 Exercise 1A

1a

	S	R
S	(S, S)	(S, R)
R	(R, S)	(R, R)

b (i) $\frac{1}{4}$ (ii) $\frac{1}{4}$ (iii) $\frac{1}{2}$
c Diagrams as in introductory box. Outcomes (S, S), (S, R); (R, S), (R, R)

2a

	£5	£10
W	(W, 5)	(W, 10)
L	(L, 5)	(L, 10)

b (i) $\frac{1}{4}$ (ii) $\frac{1}{4}$ (iii) $\frac{3}{4}$
c Outcomes (W, 5), (W, 10); (L, 5), (L, 10)

3a

	W	D	L
H	(H, W)	(H, D)	(H, L)
A	(A, W)	(A, D)	(A, L)

b (i) $\frac{1}{6}$ (ii) $\frac{1}{3}$ **c** $\frac{2}{3}$
d Outcomes (H, W), (H, D), (H, L); (A, W), (A, D), (A, L)

4a

	B	T	J
Sh	(Sh, B)	(Sh, T)	(Sh, J)
Sk	(Sk, B)	(Sk, T)	(Sk, J)
Sl	(Sl, B)	(Sl, T)	(Sl, J)

b (i) $\frac{2}{3}$ (ii) $\frac{2}{9}$ (iii) $\frac{4}{9}$

5a Outcomes (1, 1), (1, 2), (1, 3); (2, 1), (2, 2), (2, 3); (3, 1), (3, 2), (3, 3) **b** (i) $\frac{1}{9}$ (ii) $\frac{2}{9}$ (iii) $\frac{1}{3}$ (iv) $\frac{1}{3}$

Page 187 Exercise 1B

1a

	K	F	S
K	(K, K)	(K, F)	(K, S)
F	(F, K)	(F, F)	(F, S)
S	(S, K)	(S, F)	(S, S)

b (i) $\frac{1}{9}$ (ii) $\frac{1}{3}$ (iii) $\frac{2}{9}$
c Diagram as in question **5** of text.
Outcomes (K, K), (K, F), (K, S); (F, K), (F, F), (F, S); (S, K), (S, F), (S, S)

2a Rows: (1, 1), . . . , (1, 6); (2, 1), . . . , (2, 6); (3, 1), . . . , (3, 6); (4, 1), . . . , (4, 6); (5, 1), . . . , (5, 6); (6, 1), . . . , (6, 6) **b** 36 **c** (i) $\frac{1}{36}$ (ii) $\frac{11}{36}$ (iii) $\frac{1}{6}$ (iv) $\frac{5}{12}$
3a Rows: (H, 1), (H, 2), . . . , (H, 6); (M, 1), (M, 2), . . . , (M, 6); (L, 1), (L, 2), . . . , (L, 6) **b** (i) $\frac{1}{3}$ (ii) $\frac{1}{18}$ (iii) $\frac{2}{9}$
4a Outcomes (H, H, H), (H, H, T); (H, T, H), (H, T, T); (T, H, H), (T, H, T); (T, T, H), (T, T, T) **b** 8 **c** (i) $\frac{1}{8}$ (ii) $\frac{1}{8}$ (iii) $\frac{3}{8}$ (iv) $\frac{7}{8}$ **5** Rows: (M, M), . . . ,(M, Su); (Tu, M), . . . ,(Tu, Su); (W, M), . . . ,(W, Su); (Th, M), . . . ,(Th, Su); (F, M), . . . ,(F, Su); (Sa, M), . . . ,(Sa, Su); (Su, M), . . . ,(Su, Su) **b** (i) $\frac{1}{49}$ (ii) $\frac{1}{7}$ (iii) $\frac{24}{49}$ (iv) $\frac{25}{49}$

Page 189 Exercise 2A

1a $\frac{1}{2}$ **b** $\frac{1}{2}$ **c** 1; result must be H or T **2a** $\frac{1}{6}$ **b** $\frac{1}{6}$ **c** $\frac{1}{3}$
3a $\frac{1}{52}$ **b** $\frac{1}{52}$ **c** $\frac{1}{26}$ **4a** $\frac{3}{8}$ **b** $\frac{5}{8}$ **c** 1 **5a** $\frac{1}{5}$ **b** $\frac{3}{10}$ **c** $\frac{1}{2}$ **d** 1
6a 0.7 **b** 0.3 **7a** 0.4 **b** 0.6 **8a** $\frac{2}{15}$ **b** $\frac{2}{15}$ **c** $\frac{4}{15}$
9a (i) $\frac{3}{20}$ (ii) $\frac{9}{40}$ (iii) $\frac{1}{4}$ **b** all are equally likely
10a Rows :(1, 1), (1, 2), (1, 3), (1, 4); (2, 1), (2, 2), (2, 3), (2, 4); (3, 1), (3, 2), (3, 3), (3, 4); (4, 1), (4, 2), (4, 3), (4, 4) **b** (i) $\frac{1}{4}$ (ii) $\frac{13}{16}$ (iii) $\frac{7}{16}$ **11a** Rows: 2, 3, 4, 5, 6, 7; 3, 4, 5, 6, 7, 8; 4, 5, 6, 7, 8, 9; 5, 6, 7, 8, 9,10; 6, 7, 8, 9, 10, 11; 7, 8, 9, 10, 11, 12
b (i) 7 (ii) 2 and 12 **c** $\frac{1}{36}, \frac{1}{18}, \frac{1}{12}, \frac{1}{9}, \frac{5}{36}, \frac{1}{6}, \frac{5}{36}, \frac{1}{9}, \frac{1}{12}, \frac{1}{18}, \frac{1}{36}$ **d** (i) $\frac{1}{6}$ (ii) $\frac{1}{6}$

Page 190 Exercise 2B

1 Yes: **a**, **b**, **d**, **f**, **g** **2a** $\frac{13}{36}$ **b** $\frac{23}{36}$
3a Outcomes: (J, J), (J, Q), (J, K); (Q, J), (Q, Q), (Q, K); (K, J), (K, Q), (K, K) **b** (i) $\frac{1}{3}$ (ii) $\frac{2}{9}$ (iii) $\frac{5}{9}$
4a (i) 0.1 (ii) 0.3 (iii) 0.7 (iv) 0.9 **b** P(blue) and P(C) are not mutually exclusive; 0.4

Page 191 Exercise 3/Class Discussion

2a 0.11 **b** 0.18 **c** 0.148 **4a** 0.7, 0.8, 0.92, 0.77, 0.8 **b** 0.8
5a 0.45, 0.4, 0.36, 0.32, 0.32 **b** 0.32 **c** 320
6a 0.12, 0.44, 0.2, 0.24 **b** About 100, 350, 160, 200 respectively

Page 192 Exercise 4A

1a (i) $\frac{1}{2}$ (ii) 25 **b** (i) $\frac{1}{6}$ (ii) 50 **c** (i) $\frac{1}{4}$ (ii) 30 **2a** $\frac{1}{7}$ **b** 5
3a 0.0375 **b** 180 **4a** $\frac{1}{2}$ **b** 50 **5a** 0.3 **b** About 20, 8 and 12

Page 193 Exercise 4B

1a 50 **b** 15 or 16 **c** 7 or 8 **2a** 800 **b** 700 **c** 1700 **3a** (i) $\frac{1}{4}$ (ii) $\frac{1}{4}$ (iii) $\frac{3}{8}$ **b** (i) 750 (ii) 500 **c** at the corner of Clyde Street and Kelvin Road. Most customers will pass this way
4a 170 **b** 140 **5a** Rows: (R, R), . . . ,(R, W); (Y, R), . . . ,(Y, W); (G, R), . . . ,(G, W); (B, R), . . . ,(B, W); (W, R), . . . ,(W, W)
b $\frac{12}{25}$ **c** 48

Page 194 Check-up on Probability

1a $\frac{1}{5}$ **b** $\frac{4}{5}$ **2a** 0.7 **b** 126 **3** 30 **4** 0.25 **5a** $\frac{5}{6}$ **b** $\frac{1}{6}$
6a Outcomes (H, 1), (H, 2), (H, 3); (T, 1), (T, 2), (T, 3) **b** 6
c (i) $\frac{1}{6}$ (ii) $\frac{1}{6}$ (iii) $\frac{2}{3}$ **7a** $\frac{1}{4}$ **b** $\frac{1}{2}$ **c** $\frac{3}{4}$
8a 0.6 **b** 0.4 **c** 1 **9a** 0.095 **b** 0.62 **c** 0.43 **d** 0.33
10a (i) $\frac{1}{36}$ (ii) $\frac{1}{18}$ (iii) $\frac{1}{12}$ **b** about 17
11a (i) $\frac{3}{26}$ (ii) $\frac{4}{26}$ (iii) $\frac{7}{26}$ **b** (i) $\frac{5}{26}$ (ii) $\frac{3}{26}$ (iii) $\frac{6}{26}$ **c** diagram 1

CHAPTER REVISION EXERCISES

Page 197 Revision Exercise on Chapter 1: Calculations and Calculators

1a 19.1 **b** 0.333 **c** 343 **d** 487 **e** 6.13 **f** 1.43
2a 0.0167 **b** 0.0003 **3a** (iii) 652.8 **4a** 1.96×10
b 5.68×10^3 **c** 4.6×10^{-3} **d** 5.8×10^5 **5a** 1.268×10^4
b 1.6×10^{-4} **c** 1.33×10^{15} **6a** (i) 1.68×10^4 (ii) 16 800
b (i) 5×10^{-3} (ii) 0.005 **c** (i) 2.1×10^7 (ii) 21 000 000
d (i) 1×10^{-1} (ii) 0.1 **7a** $2\frac{1}{2}$ **b** $1\frac{3}{7}$ **c** $2\frac{2}{3}$ **d** $2\frac{2}{5}$ **e** $7\frac{1}{3}$
8a $\frac{5}{4}$ **b** $\frac{17}{8}$ **c** $\frac{17}{3}$ **d** $\frac{35}{4}$ **e** $\frac{15}{8}$ **9a** 4 **b** $1\frac{2}{5}$ **c** $\frac{1}{8}$ **d** $5\frac{1}{6}$
10a Labour **b** 1000 **11** $\frac{7}{8}$m **12a** $\frac{1}{6}$ **b** $\frac{1}{2}$ **c** 21 **d** 2
13 600 **14a** 4 **b** 2 **c** $\frac{1}{3}$ **d** 4 **15** 2.2×10^{-15}, 2×10^{-15}, 1.9×10^{-15}, 1.8×10^{-15}, 1.6×10^{-15}, 1.3×10^{-15}

Page 198 Revision Exercise on Chapter 2: Similar Shapes

1a To show the size of picture transmitted **b** (i) 18 cm (ii) 32 cm **2a** $\frac{1}{3}$ **b** 8 cm **c** 22.5 cm **3** 7.2
4a, b 16 cm by 12 cm or 9 cm by 12 cm
5a 63 **b** 177 **c** 315 **6b** △DEF **c** 4.8 m
7a Its opposite sides are parallel **b** 4 cm, 3 cm
c 9 cm, 6 cm **8** 18 m

Page 199 Revision Exercise on Chapter 3: Going Places

1a 36 km/h **b** 7 h **c** 105 miles
2a 5 km, $12\frac{1}{2}$ km, 15 km **b** 30 min, 12 min, 10 min **3** $6\frac{1}{4}$ min
4a 18 km, 75 min **b** 14.4 km/h **5a** $1\frac{1}{4}$ h, $\frac{3}{4}$ h **b** 86 km/h
6 A and C **7** 110 mph **8a** 60 km **b** 30 km/h
c the speed varies **d** 30 min **e** 30 km **f** 60 km/h, 8 km/h
9 27 km/h **10** Monday 25 December

Page 200 Revision Exercise on Chapter 4: Money Matters—Saving and Spending

1a (i) £400 (ii) £92.31 **b** (i) £1800 (ii) £150 (iii) £34.62
2 £16 100 **3a** £352.80 **b** £378 **c** £504 **4a** £20 **b** £21.92
5 Loan, by £39 **6a** £405 **b** £442.54 **7** £51.89
8 In Spain, by £527 **9** £153.10

Page 201 Revision Exercise on Chapter 5: Positive and Negative Numbers

1 $3-(-1) = 3+1$, $-1-(-3) = -1+3$, $-2+1 = -2-(-1)$, $2+1 = 2-(-1)$, $1+4 = 1-(-4)$, $4+(-3) = 4-3$, $3-3 = 3+(-3)$, $8-(-2) = 8+2$, $8+(-2) = 8-2$
2 $-1-3 = -4$, $-2-(-1) = -1$, $-1-(-2) = 1$, $3-(-1) = 4$, $3-(-2) = 5$ **3a** 3 **b** 3 **c** −6
4a $10a$ **b** $-2x^2$ **5a** $0^2-1 = -1 \times 1$, $(-1)^2-1 = -2 \times 0$, $(-2)^2-1 = -3 \times (-1)$ **b** $0^2+10^2 = 6^2+8^2$, $(-1)^2+7^2 = 5^2+5^2$, $(-2)^2+4^2 = 4^2+2^2$
6a 28 **b** 200 **c** −9 **7a** $-6k^2$ **b** $-3y^2$ **c** $42t^3$
8a (i) 30 (ii) 20 (iii) −10 **b** (i) 2 (ii) 1 (iii) 4
9a −1 **b** 10 **c** −5 **d** 1 **e** −3
10 0 mm **11a** (i) 10 m (ii) 15 m below P
12a $8x$ **b** $-24a^3$ **c** $-4t^2$ **d** $3mn$

Page 202 Revision Exercise on Chapter 6: Pythagoras

1a 17 **b** 11.3 **c** 33 **d** 3.2 **2a** 1.8 m **b** 33.9 mm
3a 50 m **b** (i) 56 m (ii) 8 m from C **4** 1.88 km or 1880 m
5a AB = 50 cm, BC = 32 cm, CD = 32 cm, DA = 50 cm
b a kite **c** (3, 5) **d** for example, $AM^2+MD^2 = 25 = AD^2$, etc
6a $PV^2 = 120^2+48^2+14^2$, so PV = 130 cm
b $PW^2+WV^2 = 16\,900 = PV^2$, etc
7a 24 cm **b** 240 cm² **c** 18.5 cm

Page 203 Revision Exercise on Chapter 7: Brackets and Equations

1a $32-8x$ **b** $2y-2$ **c** $12t+3$ **d** $ax+ay$ **e** $pq-pr$
f $-2a+10$ **g** $-25-5c$ **h** $-1+x$ **i** $2x+x^2$
2a $17-5x$ **b** $-2y-5$ **c** $2-p$ **3a** $96-12x$ **b** $72-72x$
4 $2y+14m^2$ **5a** $8x-2$ **b** $2t-7$ **6a** $t = 2$ **b** $x = 1$ **7** 6, 2
8 12, 6 minutes **9a** $p^2+10p+16$ **b** $t^2-9t+14$ **c** k^2-9
d $2a^2-3a-2$ **e** $12x^2-25xy+12y^2$ **f** $2x-1-x^2$
g $x^2-12x+36$ **h** $4a^2+4a+1$ **i** $9c^2-24cd+16d^2$
10a $y = -8$ **b** $p = -1$ **11** 6 m, 8 m, 10 m **12** 30 cm, 10 cm

Page 204 Revision Exercise on Chapter 8: Statistics

1a (i) 11.3, 2, 1 (ii) 5.7, 4.5, 2 **b** (i) median (ii) mean
2a 3, 8, 13, 18, 23, 28, 33, 38; 12, 64, 182, 324, 460, 476, 198, 494 **b** 22.1 **c** 21–25 **d** £1314
3a 4, 12, 26, 44, 64, 81, 87, 100 **c** (i) 14 (ii) 26 (iii) 12
4a (i) The taller the golfer, the fewer the strokes (ii) the stronger the wind, the higher the scores **b** (i) negative (ii) positive **5b** (i) 24 (ii) 6 **c** 25 **6a** 7, 2, 5, 6, 6, 4, 3, 1, 1; 7, 9, 14, 20, 26, 30, 33, 34, 35 **b** (i) £36 (ii) 32

Page 205 Revision Exercise on Chapter 9: Trigonometry

1a (i) $\frac{8}{10}$ (ii) $\frac{6}{8}$ **b** (i) 53.1° (ii) 36.9°
2a 0.98 **b** 1.41 **c** 0.5 **3a** 27.4 **b** 16.3 **c** 29.7 **d** 26.6
4a 1.5 **b** 18.4 **c** 50.9 **d** 13.7 **5** 70 cm², 34 cm **6** No
7a 37° **b** 4 m 62 cm **8** 23.9 m **9** 26.7 m **10** 37°–66°
11 148 m **12a** 6.5 m **b** 6.8 m **c** 17.1°

Page 206 Revision Exercise on Chapter 10: Simultaneous Equations

1a (i) 1, 3, 5, 7, 9 (ii) 10, 9, 8, 7, 6 **b** $x = 3, y = 7$ **2** (6, 3)
3 Lines cut axes at: **a** (0, −8), (4, 0) **b** (0, −3), (9, 0)
c (0, 8), (6, 0) **4a** (3, 12) **b** (4, −4)
5a (1, −1) **b** (−1, 4) **c** (2, 2) **d** (−1, −1)
6a $212 = 100a+b$, $32 = 0a+b$; $b = 32$, $a = 1.8$
b $F = 1.8C+32$ **d** (i) and (ii) 104 **7** $x = 20, y = 14$
8a Bonus 5, penalty 3 **b** 2, 3; 1, 1; 3, 4; 2, 2; 1, 0; 3, 3; 2, 1; 4, 4; 3, 2; 2, 0 . . .

Page 207 Revision Exercise on Chapter 11: Areas and Volumes

1a 55 m² **b** 15.4 cm² **2** 2100 mm² **3a** 8 **b** 19.5%
4a 528 cm² **b** 480 cm³ **5a** 254 litres **b** 1272
6a 5 m² **b** 20 m³ **c** 8.94 m² **d** 26.6° **7** 7490 cm³
8a 240 000 cm³ **b** 2.52 m²

Page 208 Revision Exercise on Chapter 12: Factors

1a 1, 2, 4, 5, 8, 10, 20, 40 **b** 1, 2, 5, 10, 25, 50
c $1, 3, 9, x, 3x, 9x$ **d** $1, 7, x, 7x, x^2, 7x^2$ **e** $1, x, x^2, x^3$
2a 2 **b** 3 **c** 6 **d** 9 **e** x **f** $2t$
3a $6(t-3)$ **b** $3(3u+8)$ **c** $11(1-3t)$ **d** $u(2u+1)$
e $y(2-3y)$ **f** $2k(3k-5)$ **g** $3(a+4b-6c)$ **h** $x(a+b-c)$
4a $(a-b)(a+b)$ **b** $(y-x)(y+x)$ **c** $(n-1)(n+1)$
d $(10-m)(10+m)$ **e** $(2t-1)(2t+1)$ **f** $(4-3y)(4+3y)$
g $(p-5q)(p+5q)$ **h** $(3r-s)(3r+s)$ **i** $(7u-6v)(7u+6v)$
5a 700 **b** 870 **c** 50 **d** 1 222 000 **e** 9
6a $9(a-1)(a+1)$ **b** $7(p-2)(p+2)$ **c** $8(2-r)(2+r)$
7a $(p+9)(p+8)$ **b** $(q+3)(q-1)$ **c** $(r+4)(r+3)$
d $(s-7)(s+6)$ **e** $(a-3)(a-3)$ **f** $(1+b)(1+b)$
g $(9+x)(3-x)$ **h** $(9-y)(2+y)$ **8a** $(3x-4)(2x+1)$
b $(3y+2)(2y-5)$ **c** $(4t+3)(3t+2)$ **d** $(9u-1)(2u+5)$
9a $2(m+4)(m+1)$ **b** $3(n-4)(n-2)$ **c** $9(1-2y)(1+2y)$
d $xy(4x+y)$ **e** $x(x-1)(x+1)$ **f** $(3x-4)(4x+3)$
10a By Pythagoras' Theorem, $a^2 = b^2+c^2$, so $b^2 = a^2-c^2 = (a-c)(a+c)$ **b** 80 **11a** $30n+24 = 6(5n+4)$
b $28n+24 = 4(7n+6)$ **12b** (i) 3 (ii) 5 (iii) 7 (iv) 9
c $(2n^2-n)\div n = n(2n-1)\div n = 2n-1$, which is always odd
13a $a(x+y)$ **b** $3(x+2y-3z)$ **c** $(u-v)(u+v)$
d $(w-10)(w+10)$ **e** $(x+2)(x-5)$ **f** $(y+7)^2$
g $t(u+v+w)$ **h** $2p(1-2p)$ **i** $a(1-a)$ **j** $a(1-a)(1+a)$
k $\pi(m-n)(m+n)$ **l** $\pi y(y+2)$ **m** $(1-x)(1+x)$
n $(1-x)(1+x)(1+x^2)$ **o** $(x+3)^2$ **p** $(x^2+3)^2$
q $(x^2+4)(x^2-2)$ **r** $(x-2)(x+2)(x^2+2)$
s $(u+v-w)(u+v+w)$ **t** $(a-b-c)(a-b+c)$
u $(a+b)(x+y)$ **v** $(c+d)(x-y)$ **w** $(6-p)(1-p)$
x $(8+m)(1-m)$ **y** $(5x-4)(2x+3)$ **z** $(4y-3)(2y+3)$

Page 209 Revision Exercise on Chapter 13: Money Matters—Personal Finance

1 Gross pay £199.70, deductions £57.02, net pay £142.68
2 £1425.67 **3a** Superannuation £14.88, NI £12.40
b £164.97 **4** £6318.75 **5a** £5211 **b** £7155
6a High risk sport **b** £327.60
7a £1650 **b** £1815 **c** £1996.50 **8** 5.17%
9a (i) £2.40 (ii) £4.86 (iii) £7.37 (iv) £32.92 **b** 32.92%

Page 210 Revision Exercise on Chapter 14: Formulae

1a, b 64 **2a** $x = c-b$ **b** $x = \dfrac{d+e}{a}$ **c** $x = \dfrac{r-pq}{p}$
d $x = \dfrac{v+3w}{u}$ **3a** (i) $P = 6x+\pi x$ (ii) $A = 4x^2+\frac{1}{2}\pi x^2$
b $P = 18, A = 22$ **4a** $x = \dfrac{P-a}{4}$ **b** $t = \dfrac{Qs}{3}$ **c** $a = \dfrac{4p}{R}$
d $x = a+t^2$ **e** $y = \dfrac{-ax-c}{b}$ **f** $x = \dfrac{ab}{c}$
5a 64 **b** 60, 62, 63, $63\frac{1}{2}$, $63\frac{3}{4}$ **c** $r = \dfrac{S-a}{S}$, $\frac{1}{3}$
6a $x = \dfrac{a+b}{2}$ **b** $x = \dfrac{d-b}{a-c}$ **c** $x = \sqrt{(ab)}$ **d** $x = \dfrac{uv}{v-u}$
7a 50 **b** $d = \sqrt{\dfrac{2w}{R}}$, 1.5 mm **8a** $c = 8h-3L$ **b** $h = \dfrac{3L+c}{8}$
9 $n = \dfrac{IR}{E-Ir}$, 4

Page 211 Revision Exercise on Chapter 15: Probability

1a 0.05 **b** 0.95 **2a** $\frac{1}{3}$ **b** $\frac{2}{3}$ **c** 1
3a

− + | − +

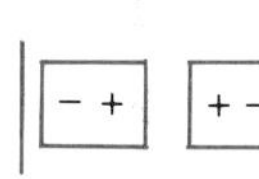

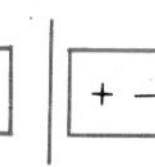

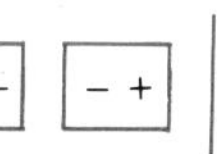

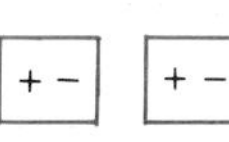

b $\frac{1}{4}$
4a 6 **b** $\frac{1}{6}$ **5a** Rows: (F, R), (F, W), (F, B), (F, G); (N, R), (N, W), (N, B), (N, G) **b** (i) $\frac{1}{2}$ (ii) $\frac{1}{8}$
6a Outcomes (1, 1), (1, 2), (1, 3), (1, 4); (2, 1), (2, 2), (2, 3), (2, 4); (3, 1), (3, 2), (3, 3), (3, 4); (4, 1), (4, 2), (4, 3), (4, 4) **b** (i) $\frac{1}{16}$ (ii) $\frac{1}{2}$ (iii) $\frac{3}{8}$ **7a** Rows: (r, R), (r, A), (r, G); (a, R), (a, A), (a, G); (g, R), (g, A), (g, G) **b** 9 **c** (i) $\frac{1}{9}$ (ii) $\frac{5}{9}$ **d** 3 or 4
8a $\frac{1}{3}$ **b** $\frac{1}{2}$ **c** $\frac{7}{12}$ **9a** P(factor of 6) = $\frac{4}{15}$ **b** P(factor of 15) = $\frac{4}{15}$
c P(factor of 6 or 15) = $\frac{2}{15}$ **10a** $\frac{1}{36}$ **b** $\frac{1}{18}$ **c** $\frac{1}{12}$ **d** $\frac{1}{6}$

GENERAL REVISION EXERCISES

Page 212 General Revision Exercise 1

1a 2 **b** −1 **c** −2 **d** −3 **e** 4 **f** −10 **g** 0 **h** −16
i 6 **j** 1 **k** −8 **2** 15% **3** $3x+7 = 22$, $x = 5$; 5
4a 15.7 mm **b** 1960 mm **5a** 26 **b** 7.5
6a 6 **b** 5 **c** 5 **d** 3 **e** 3
7a 390 US dollars; 505 Canadian dollars **b** £12.82
8a ∠s NPB, NPA **b** 155° **c** (i) 040° (ii) 095° (iii) 300°
9a $2(q+r)$ **b** $3(p-2)$ **c** $2(4-5y)$ **d** $a(b-c)$
10a $1\frac{1}{6}$ **b** $\frac{1}{3}$ **c** $4\frac{1}{4}$ **d** $1\frac{1}{10}$ **e** $5\frac{1}{4}$ **f** $3\frac{2}{5}$ **11a** No **b** 28 km/h

Page 213 General Revision Exercise 2

1a 4 **b** 5 **c** 3 **2** 76° **3a** 1800 **b** 2 **c** 10 000 **d** 25
e 2 **f** 80 **g** 0.7 **h** 0.09
4a

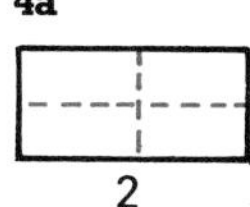

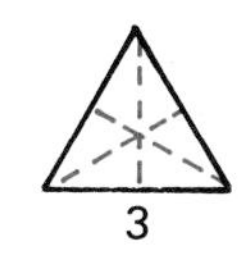

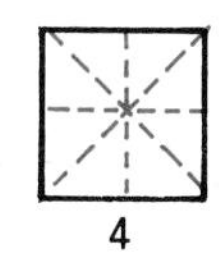

b 4, 2, 6, 8 **c** **b**'s answers are twice **a**'s

5a

Y, 4, 0, 6, X

b

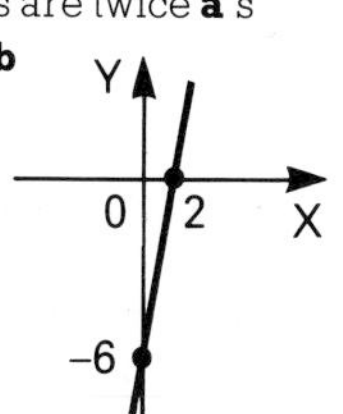

c

Y, 0, X, (1,−4)

6a 5 000 000 000 **b** 5×10^9 **7a** 9 **b** -1 **c** -2
8a $L = 32x$ **b** $L = 288$ **9a** $\frac{2}{3}$ **b** 10 m **c** 75 m^2
10a (i) 5.5 (ii) 5.4 (iii) 5.3 (iv) 0.4

Page 214 General Revision Exercise 3

1a -1 **b** 2 **c** 4 **2a** 1.02, by 0.1 **b** 3.141 60, by 0.000 01
c $\frac{4}{5}$ by $\frac{1}{20}$ **3a** $3n-2$ **b** $3n-8$ **c** $0.1n+0.1$ **4** 4.5 cm
5a P(12, 0), Q(0, -5) **b** 13 units **6a** 45 m **b** 52 m
7a $2u^2-u$ **b** $3k-2$ **c** p^2-2p+1 **d** m^2+5m+4 **e** $-t-3$
8a

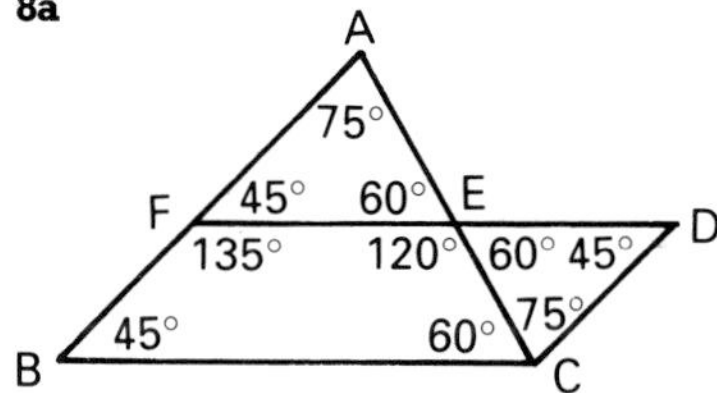

b Parallelogram and trapezium **c** they are equiangular
9 28 cm, 27.6 cm **10a** 2 **b** 3
11a (i) 1 hour (ii) 30 minutes (iii) 30 minutes
b To Ayr 56 km/h; from Ayr $37\frac{1}{3}$ km/h

Page 215 General Revision Exercise 4

1a (2, 2) **b** $(4\frac{1}{2}, 2\frac{1}{2})$ **c** (1, -1) **2** 2260 m^3, 980 m^2
3a 0.385 **b** 0.923 **c** 0.417 **d** 7.45
4a x^2+5x+6 **b** t^2+4t-5 **c** k^2-6k+8 **d** $a^2+2ab+b^2$
e $p^2-2pq+q^2$ **f** $4n^2-1$ **5** 17.9 km
6b, c (2, 4) **7b** A parallelogram
c (-2, 1), (-4, 2), (-4, 4), (-2, 3); a parallelogram
8a $7x-9 = 5x+5$, $x = 7$ **b** 40 cm
9a 5 400 000 cm^3 **b** 5400 litres
10a 5 hours **b** 235 km **c** 47 km/h

Page 216 General Revision Exercise 5

1a 240 m, 2450 m^2 **b** 0.245 **2a** 60 **b** 47.46

3a $T = 6n-1$ **b** $n = \dfrac{T+1}{6}$, 55

4a (i) ∠s ABE, ACD; ∠s AEB, ADC (ii) ∠s BFE, CFD; ∠s BFC, EFD (iii) ∠s BEC, ECD; ∠s EBD, BDC
b (i) ∠ECD = 30° (ii) ∠BAE = 40°
5 1.43×10^9, 5.58×10^{-2} **6a** $P = 4x$, $A = x^2$; $P = 2\pi r$, $A = \pi r^2$; $P = 2(a+b)$, $A = ab$ **b** 6.9 cm^2 **7a** 38.5 m **b** 46°
8a 3 **b** $AD^2 = AC^2+CD^2$ **9a** (i) 6 cm (ii) 18 cm **b** 18°

Page 217 General Revision Exercise 6

1

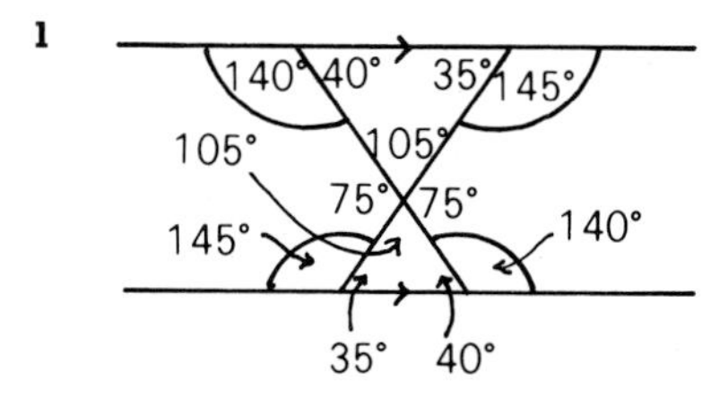

2 47 **3a** £62 **b** $N = \dfrac{P-C}{R}$, 6

4

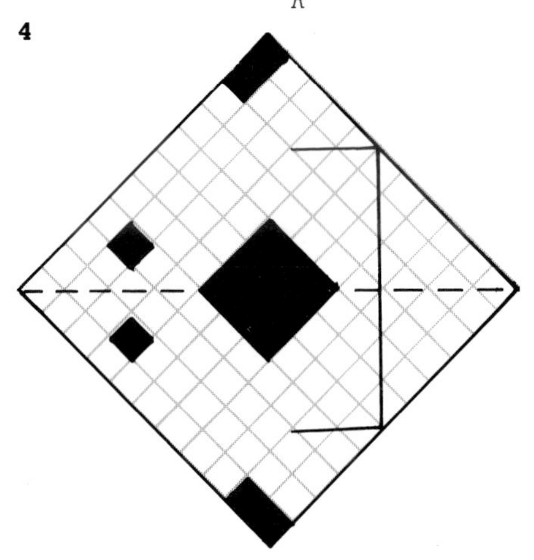

5a $x(x-1)$ **b** $(a-2)(a+2)$ **c** $(x+3)^2$ **d** $t^2(t+1)$
e $(y+4)(y-2)$ **f** $(2n-3)(n-1)$
6a (i) -5 (ii) -7 (iii) -12 **b** (i) -1 (ii) 3 (iii) -2 (iv) 0 (v) 5 **7a** 9.4 m **b** 58.0°
8a (i) 6 (ii) 8 (iii) 3 **b** (i) $2y(1+4y)$ (ii) $(m+1)(m+2)$ (iii) $(t-6)(t+1)$ **9a** 9 cm, 12 cm, 15 cm **b** 54 cm^2
10a 6362 cm^2 **b** 283 cm **11a** (i) 12% (ii) 144° **b** 2160

Page 218 General Revision Exercise 7

1a £648 **b** £699.84 **c** £881.60 **2a** 0, 2, 6 **b** -8, -4, 8
c -6, -4, -2, 2, 4, 6 **d** -2, -1, $-\frac{1}{2}$, $\frac{1}{2}$, 2
3a 45° **b** $\sqrt{2}$, or 1.4 correct to 1 decimal place
4a Frequency total = 24, hours × frequency total = 128
c 5.3 hours, 6 hours, 6 hours **5a** $\frac{1}{4}$ **b** (i) $\frac{1}{16}$ (ii) $\frac{9}{16}$ (iii) $\frac{3}{8}$
6a $27+4x = 79$, -13 **b** (i) -1 (ii) -11 (iii) 6 (iv) $1\frac{1}{2}$
7a 3.14 cm^2 **b** (i) 2360 cm^3 (ii) 2.36 litres **8** 48 m^2
9a (i) △s RSQ, RTP (ii) $\frac{2}{5}$ (iii) 6 **b** 25
10a (i) $P = 2y+2\pi x$ (ii) $A = 2xy+\pi x^2$

b $y = \dfrac{P-2\pi x}{2}$, $y = \dfrac{A-\pi x^2}{2x}$

Page 219 General Revision Exercise 8

1a (i) 50 000 000 (ii) 8850 m (iii) 0.0112 cm
b (i) 6.05×10^3 km (ii) 8.38×10^8, 1×10^9 (iii) 1.6×10^{-5}
c 5×10^{10} **2b** 71 km **c** 129° **3** 36°, 54°, 90°
4a $5(x+2y)$ **b** $6(k-1)$ **c** $x^2(1+2x)$ **d** $2(x-3)(x+3)$
e $(x+3)(x-4)$ **f** $2(x+5)(x-3)$
5a 73 **b** $r = 3m-p-q$ **c** 27

6a $u = \dfrac{3-v}{2}$ **b** $x = \dfrac{z-y}{a}$ **c** $x = \sqrt{\dfrac{n}{m}}$ **d** $m = np$

e $t = uv-s$ **f** $x = \dfrac{abc-ay}{b}$ **7a** 6000 cm^2 **b** 103°

8a 192 **b** (i) 122 880 cm^3 (ii) 96 510 cm^3 (iii) 26 370 cm^3
9a They are equiangular **b** 2.7 m **10** 25 133 cm^3